F. H. Prager and H. Rosteck
Polyurethane and Fire

Related Titles

Krause, U.
Fires in Silos
Hazards, Prevention, and Fire Fighting

approx. 250 pages with approx. 115 figures
Hardcover
ISBN 3-527-31467-9

Quintiere, J.
Fundamentals of Fire Phenomena

448 pages
Hardcover
ISBN 0-470-09113-4

Fitzgerald, R. W.
Building Fire Performance Analysis

534 pages
Hardcover
ISBN 0-470-86326-9

Franz H. Prager and Helmut Rosteck

Polyurethane and Fire

Fire Performance Testing under Real Conditions

WILEY-VCH Verlag GmbH & Co. KGaA

The Autors

Dr.-Ing. Franz H. Prager
Fürstenbergstr. 1b
51379 Leverkusen
Germany

Dipl.-Ing. Helmut Rosteck
Hufelandstr. 106
51061 Köln
Germany

■ All books published by Wiley-VCH are carefully produced. Nevertheless, authors, editor, and publisher do not warrant the information contained in these books, including this book, to be free of errors. Readers are advised to keep in mind that statements, data, illustrations, procedural details or other items may inadvertently be inaccurate.

Library of Congress Card No.: Applied for

British Library Cataloging-in-Publication Data:
A catalogue record for this book is available from the British Library

**Bibliographic information published by
Die Deutsche Bibliothek**
Die Deutsche Bibliothek lists this publication in the Deutsche Nationalbibliografie; detailed bibliographic data is available in the
Internet at <http://dnb.ddb.de>.

© 2006 WILEY-VCH Verlag GmbH & Co. KGaA, Weinheim

All rights reserved (including those of translation into other languages). No part of this book may be reproduced in any form – nor transmitted or translated into machine language without written permission from the publishers. Registered names, trademarks, etc. used in this book, even when not specifically marked as such, are not to be considered unprotected by law.

Printed in the Federal Republic of Germany

Printed on acid-free paper

Composition   Mitterweger & Partner Kommunikationsgesellschaft mbH, Plankstadt
Printing   betz-Druck GmbH, Darmstadt
Bookbinding   Litges & Dopf Buchbinderei GmbH, Heppenheim

ISBN-13   978-3-527-30805-7
ISBN-10   3-527-30805-9

Table of Contents

1	**Introduction** *1*	
2	**Fire-Protection Problems** *3*	
2.1	Definition of the Fire-Performance Criteria *5*	
2.1.1	Formation of Flames *6*	
2.1.2	Course of Fire *7*	
2.1.3	Characteristics of Fire Behavior *9*	
2.2	Essential Fire Scenarios *21*	
2.2.1	Pyrolysis *21*	
2.2.2	Oxidative Decomposition *22*	
2.2.3	Propagation of a Self-sustained Smoldering Fire *24*	
2.2.4	Fully Developed Fire with a Flashover Situation *28*	
3	**Research of Causes of Fires** *35*	
3.1	General Experience of Fire Statistics *35*	
3.2	Knowledge of Fire Risks *37*	
3.3	Experience with Fire Spread – Fire-Detection Units and Sprinklers *46*	
3.4	The Time of the Fire Initiation as a Classification Criterion *55*	
4	**Preventive Fire Protection – National Requirement and Classification Systems** *57*	
4.1	Building Section *60*	
4.1.1	United States of America *62*	
4.1.2	European Requirement and Classification Profiles *73*	
4.1.3	Eastern Europe and Far East *104*	
4.1.4	International Standardisation – European Harmonisation *112*	
4.2	Furniture and Furnishing *121*	
4.2.1	American Testing and Evaluation Criteria *121*	
4.2.2	European Test and Evaluation Criteria *125*	
4.2.3	Far East, e.g. Australia *132*	
4.3	Transportation *133*	
4.3.1	Road Traffic *133*	
4.3.2	Aviation Sector *134*	
4.3.3	Shipbuilding Sector *137*	

Polyurethane and Fire. Franz H. Prager and Helmut Rosteck
Copyright © 2006 WILEY-VCH Verlag GmbH & Co. KGaA, Weinheim
ISBN: 3-527-30805-9

4.3.4	Railway Sector *140*	
4.4	Electrical Engineering *146*	
4.4.2	Proof of the Resistance against Overheated (Hot and Glowing) Wiring *148*	
4.4.3	Thermal Stress with Flames *152*	
4.4.4	Side Effects of a Fire *154*	
5	**Material-specific Fire-Performance Characteristics of PUR** *157*	
5.1	Polyurethane Production *157*	
5.2	Risk of Ignition in the Production and Storage Area *161*	
5.2.1	PUR Raw Materials – Basic Characteristics *161*	
5.2.2	Laboratory Research Work with the Tewarson Apparatus *163*	
5.2.3	Drum Tests and Supplementary Pool-Fire Experiments *165*	
5.3	Polyurethanes *176*	
5.3.1	Material-related Igniting Risks of Polyurethanes *177*	
5.3.2	Material-specific Burning Behavior of Polyurethanes *193*	
6	**Use and Interpretation of PUR-Test Results Determined under Enduse Conditions** *267*	
6.1	Relevance of Combustion Systems *267*	
6.1.1	Relevance of the Procedures for the Risk of Ignition *268*	
6.1.2	Relevance of the Procedures for the Assessment of the Side Effects of a Fire *306*	
6.2	Relevance of the Evaluation Criteria *366*	
6.2.1	Risk of Ignition *368*	
6.2.2	Burning Dripping *371*	
6.2.3	Relevance of the Criteria for the Evaluation of the Heat Release *371*	
6.2.4	Relevance of the Evaluation Criteria of the Smoke Potential *378*	
6.3	Relevance of the Requirement Profiles *417*	
6.3.1	Relevance of Full-scale Field Trials *422*	
6.3.2	Relevance of the Investigation Results Concerning the Cause of Fires *441*	
6.3.3	Orders Concerning the Risk of Ignition *452*	
6.3.4	Demands Concerning the Heat and Fire-Gas Liberation *458*

Summary *473*

References *477*

Abbreviations list *491*

Norm list *495*

Index *501*

1
Introduction

This book gives advice of the relevance of the most important characteristics of fire reaction testing in the various national and international laboratory, bench scale and large scale fire reaction test methods. A detailed analysis is given concerning the fire hazard and the characteristics ignitability, flame spread and side effects like burning dripping, heat and smoke release rate. The various facets of fire reaction testing and evaluation/classification systems are handled on the basis of meaningful fire statistic records on the example of flexible and rigid polyurethane foams and raw material as well. It is fact, that most lives being lost through fires occurring in the home, especially by selfsustained smoldering fires initiated by smokers material. Because ignition is the start of the fire, it is necessary to know how fires start and to be able to prevent starting a fire, to get the knowledge to reduce the chance of starting a fire. Advice is given, that a risk-relevant estimation needs the linkage to meaningful decomposition models and risk-relevant assessment criteria. The prediction of fire hazard by mathematical modelling has been advanced to a great extent during the last decades, very helpful for fire science but still not valid for regulatory purposes. The limitation of pure material data and the need for classification criteria related to end use conditions is demonstrated.

The temperature history and the smoldering propagation rate are very important parameters for the concentration of smoke, which is the amount of smoke produced in a fire mixed with air and distributed in the endangered volume.

The number of publications concerning fire testing and fire research has considerably increased during recent years. In general, only limited aspects of fire are considered, because of the complexity of the fire phenomena. The numerous results and the extensive research show the importance of these topics for the protection of persons and properties. The basic requirements of fire prevention, which guarantee priority to the fire protection of people, can be optimized if the findings of the fire statistics and the outcome of natural fires are properly considered. Because of fire disasters, fire research is orientated as a rule towards the two key questions:

Have existing requirements been violated? and

Have the legal requirements been updated and adopted?

As a rule no fundamental examination of the relevance of testing and evaluating criteria for the estimation of danger takes place.

The key question of every preventive fire-safety action should be whether laboratory test procedures are able to simulate the main/most important fire scenarios and

Polyurethane and Fire. Franz H. Prager and Helmut Rosteck
Copyright © 2006 WILEY-VCH Verlag GmbH & Co. KGaA, Weinheim
ISBN: 3-527-30805-9

to determine the relevance of the characteristic fire scenarios. The importance of this question increased all the more as large-scale investigations are evidence that laboratory tests do not deliver meaningful results. Based on the outcome of relevant fire-statistics data as well as on the findings of fire disasters, the most used fire-reaction tests including classification systems/criteria are discussed in detail. The significance of the requirements concerning the reaction to fire parameters will not be fully discussed due to major inputs like shape and arrangement, ventilation and possibly thermal feedback effects. The latter, preheating effects, will be more dominant than material-related data. The possible reduction of visibility by thermal decomposition products will also be a key aspect as will the danger of the toxic and corrosive acting effluents.

While the pure material-related classification is given in all sections of the design construction part of an effective quality control and meaningful supply arrangement, the safety requirements are dominated by realistic tests under enduse conditions. The adherence of preventive fire requirements to pure material test procedures neglects any safety aspects that have been recommended, e.g. by the commissions final report [1]. The proposed harmonized European fire-reaction test procedures and classification criteria will, in the future, apply to the majority and also to products classifications and material-related classifications. Neither the regulation nor the assurance fire-reaction test data will be dominated by realistic risk evaluation.

2
Fire-Protection Problems

Due to extraordinary fire disasters, as demonstrated by Fig. 1, meaningful, effective preventive and defensive life prevention measures have to be cogently recorded. Burning theatres, cinemas, dance halls and products as well as aeroplanes and ships, etc. and especially spectacular fires of high-rise buildings have proven that safe rescue ways are the key problem [2–8]. The basic requirements of the building codes according to Table 1 have been accepted worldwide for all fields of human life.

Table 1 Basic fire protection requirements

- The occurrence of fire must be prevented
- The spread of fire and smoke must be prevented
- Emergency services and fire – extinguishing measurement must be possible

Figure 1 Skyscraper Fire Sao Paulo

Polyurethane and Fire. Franz H. Prager and Helmut Rosteck
Copyright © 2006 WILEY-VCH Verlag GmbH & Co. KGaA, Weinheim
ISBN: 3-527-30805-9

They demonstrate how far preventive measures and fire precautions can guarantee sufficient safety for all fields of human life in the case of a fire. These basic requirements will deal sufficiently with all aspects, if the measurements of the preventive fire safety in the developing stage of the fire guarantee a high resistance against possible ignition sources and the danger of fire spread is minimized. Safety requirements often depend on the negative experiences of the community. Therefore they will be associated with actual fire disasters. This is why the basis of regulatory requirements is not a common framework but a tailor-made solution depending on the actual fire situation.

Weak points within this safety concept are the intentions to approach a solution by preselective testing. The complexity of the whole fire-safety problem will be generally neglected. Since the actual fire case requires ad hoc solutions, the results of time-consuming basic investigations can not be awaited. With the presentation of the ad hoc test results the political pressure for basic investigations will be avoided, insofar as appropriate solutions can not be awaited due to the complexity. The variety of the fire-safety requirements can be sufficiently met if the test and safety concepts can be adjusted to the findings of actual fire statistics. Within harmonized research programs, the potential risks have to be investigated and the resultant test requirements derived. Rare expected disasters have to be excluded and hazards within the high-risk areas only taken into account for the worst case. The problem of smoke toxicity has become the focus of the public discussion within recent years. The second essential requirement of the planned, harmonized furniture regulation [9] is one of these reasons. The discussion referring to this is going on with quite varied arguments. One side is arguing that the fire effluents of modern plastic materials are poisonous and the opposition is arguing that decomposition products are totally nontoxic. Both sides are not in line with reality, because incomplete combustion always guarantees toxic compounds. The range of possible risks will be a function of the developing gas concentration, a fact that has been already stated by Paracelsus with: "The dose is responsible for the poison". Generalisation without any specific scientific proof does not have scientific value and is therefore not valid for the basis of regulatory requirements. Consolidated final conclusions have to be predicted on the basis of a detailed risk analysis and hazard. The terms risk and hazard give evidence to the gap between safety and danger. The term risk will be used to indicate the combination of the likelihood of a fire probability and the fire hazards. Requirements concerning possible danger by smoke production will normally be defined on the basis of nonrelated facts. The interplay of the production of effluents with the important fire-reaction criteria is as a rule neglected. The relevance of the chosen decomposition models with regard to the specific fire scenarios is also neglected. An extensive evaluation of the relevant findings gives evidence that the danger to humans by smoke inhalation will be of particular importance/significance in the case of a fire. If these effluents are being produced over a longer period of time, e.g. under smoldering conditions or by a flashover situation, e.g. by fiercely burning with forced decomposition producing large amounts of effluents. Under both conditions, the times to lethal concentrations are reached will be significantly influenced by the ventilation conditions. The "exposure time" has no importance with regard to the reduction of visibility. The dangerous situation will happen just after an inadmissible reduction of visibility occurs. It is not the duration of smoke exposure that is of great importance but the knowledge of escape routes. The conditions relating to space and the input of the enduse conditions will have to be taken into account. The time range being available for rescue and extinguishing actions will be as important.

2.1 Definition of the Fire-Performance Criteria

The phenomenon of fire has engaged mankind from the beginning. The "Duden" [10] defines fire by combustion with simultaneous heat and light release and oxidation of the fuel. A simple representation of the phenomenon of fire is given by the fire triangle according to Fig. 2, similar to Emmons [11, 12], Thomas [13] and others [14, 15]. The fire triangle demonstrates that the interaction of the three components heat, fuel and oxygen is needed. If one of these components is withdrawn, the fire will not arise or it will extinguish. While heat and oxygen have to be stated as quantitative variables of the fire, the fuel might be available in gaseous, liquid or solid form.

The fuel had to be vaporized or decomposed by heat supply because under evaporated conditions only oxidation will occur. There is a range of temperatures where different materials give off enough decomposition products to form an ignitable mixture with the atmospheric oxygen. Solids need to be decomposed and liquids evaporated by the heat supplied. If under pyrolytic conditions, the tested fuel produces more heat than is lost by the cooling process, e.g. the heat release of the fuel is larger than the loss energy, then the burning process will be stabilized and strengthened by feedback effects. If the heat release of the fuel is smaller than the loss of energy, the fire will be extinguished. The ignitability of a material is the possibility to ignite the decomposition products under defined conditions. Ignition is the starting point of the burning process.

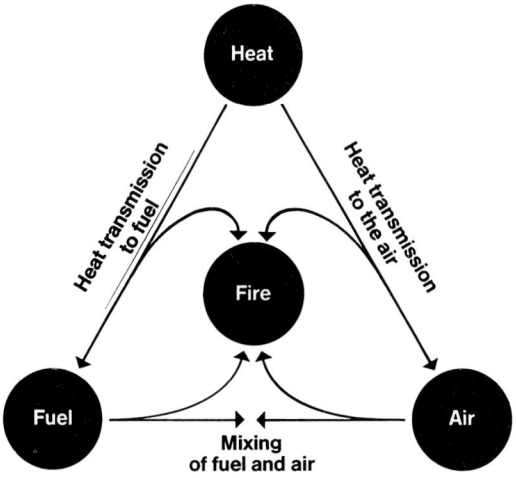

Figure 2 Fire triangle

2.1.1
Formation of Flames

The nature of the flames is determined by the decomposition products [16]. According to [17] the mixture of decomposition products that burn above the decomposing surface generates high temperatures. The temperature that initiates continuous flaming of the fuel is the fire point, according to Rasbash [18]. The exposed surface is the sample surface under thermal stress. There is a differentiation between diffusion flames of natural fires and premixed flames of many burners. Natural fires are diffusion flames in which air is getting into the flame from the surrounding atmosphere. The fuel–air mixture of premixed flames will be produced within the burner. The resultant fuel mixture leads to a stabilized burning process. Paul et al. [19] discuss possible differences of diffusion and premixed flames to a certain extent. Premixed flames were used for model investigations to obtain the possibility of an extensive variation of the most important parameters. Ventilation conditions and room configuration will significantly influence the temperature profile of diffusion flames. The candle flame according to Fig. 3 demonstrates the different temperature fields inside the flame zone. The pyrolysis zone has a luminous core of more than 1000 °C. Above the core is the diffusion flame with a temperature of max. 1400 °C. Within the outer layer of the flame temperature is decreasing below 1000 °C. The required oxygen will diffuse from the surrounding air in the stream of decomposition products. The radiating heat from the burning candle is too small to decompose enough fuel (wax) and the additional input of the convection and conduction phase are needed. Reference [20] states that the diffusion flame will extinguish if the flame temperature within the reactive zone decreases below 600 K. The temperature spectrum has been investigated by different authors [17, 21–24]. Numerous investigations with wooden cribs demonstrate the relation of diffusion flames and natural fires. The estimation of the flame length was part of various studies. According to [25–27] the deflecting of flames, e.g. by the ceiling, leads to significant lengthening of the flames. Thomas [28] discusses the extensiveness of natural flames. The theoretical explanations are based on experimental data of crib fires. Key parameters that have to be taken into account are the burning area, the porosity of the crib and the ventilation effects. Kordina et al. [29] studied the size of flames produced by the hot gases of room fires emerging from the windows with large-scale tests in Lehrte.

The importance of different parameters like oxygen content and concentration of the decomposition products to the flame configuration has been part of some major investigations [30, 31]. The appearance of the flame can be significantly changed by chemical as well as physical inputs, as proven by some studies [17, 32–34]. Seeger [35] discusses the dependence of the size of flames on viscous, inertia and buoyancy forces. The possible input of feedback effects to the temperature levels of the flame have been part of various studies. Detailed information concerning temperature zones, e.g. cool flames with slow oxidation and hot flames with temperature levels above 800 °C are given in [14]. Questions concerning soot formation are discussed. Similar questions about the development of flames have been part of the basic work of Emmons [12]. The findings are that the flame formation in the case of a fire will

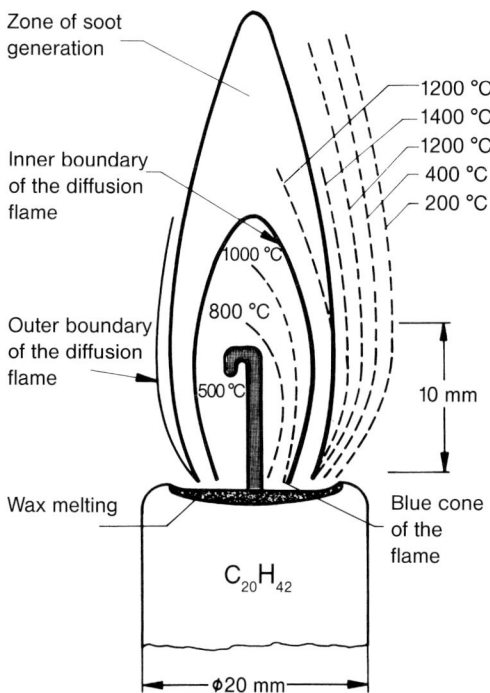

Figure 3 Scheme of a diffusion flame

be a chemical reaction based on the diffuse mechanism. Warren [37] discusses a complex reciprocal effect of formation of radicals, feedback and aerodynamical diffusion. The geometry of burner flames has been the object of various investigations, e.g. by Zubrowski et al. [38]. The size of the burner and the composition of the fuel gas are key parameters. The aim of the investigation was to assess the shape of the flames, the temperature profile and the layer formation of the gas stream.

2.1.2
Course of Fire

The development of the fire is characterized by different phases, which are contained within by the two end positions: ignition and extinguishment. Figure 4a demonstrates the typical course of a fire. After the ignition, flame spread occurs, according to [14] a chain of ignition as a result of the flame formation and a significant increase of the temperature level. The surface spread of flame will be dominated by the speed of temperature increase of the burning material. The size of the flames is crucial for the feedback effect. In large flames [36] the radiant portion is dominant. The convection part decreases with an increase in size of the flames unless the convection part becomes dominant due to ventilation effects. After a specific temperature has been reached, flashover occurs if enough oxygen is available. Any combustibles within the compartment will be set on fire. In Fig. 4b, after the flashover situation, the developing stage of the fire, the growth phase accelerates to the fully developed stage of the fire The fully developed stage of the fire produces temperature levels of 800 °C to 1000 °C. After the fire load has been consumed, the fully developed fire passes over to the decay phase.

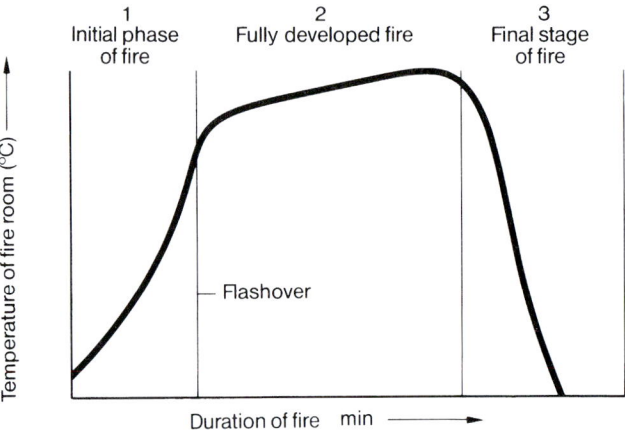

Figure 4a Stages of a fire

Figure 4b Flashover situation – large-scale test Wuppertal

According to Friedmann [39, 40] as a rule, the fully developed fire will progress by a power law or exponential function. Two thirds of the heat release will assist the fire spread. Preheating can significantly increase the spread, 12 kW/m^2 can lead to a 40-fold acceleration. Ventilation effects have been investigated by Gross and Loftus [41] in detail. Flame spread within rooms will be strongly influenced by the radiation of the smoke layer. The dominant influence of coverings to the surface spread of flame has been documented by the studies of Karwaller [42]. The fire development within channels has been investigated extensively by Ris [43]. He underlines the dimension of the channel. This has also been documented by Roberts [44]. Thomas et al. explain in [45] important divergencies of the fire performance of pool and crib fires. Feedback effects will be discussed as well as the input of ventilation. Due to the increasing oxygen availability the fuel-controlled fire passes over to a ventilation controlled one. The flashover situation, as characterized by Fig. 4b occurs at a temperature level of approx. 520°C. According to [46], the ventilation is the most important parameter. The spread of the fire to neighboring rooms is especially dependent on the effectiveness of the fire resistance of doors, windows, etc.

2.1.3
Characteristics of Fire Behavior

The fire behavior is not a material property [15, 40, 47]. Reaction to fire is characterized by

- the time of ignition
- the contribution to flame spread and
- the contribution to other factors like dripping, heat-release rate and smoke generation

that will be influenced by both material and environmental factors [14, 48, 49]. The most important factors are:

- distribution of the material in the room, e.g. height and density, shape, design and construction of products
- physical and technological properties like heat transfer, orientation or preheating

- surface roughness, textile construction, porosity, melting behavior
- size and arrangement of the ignition source as well as
- ventilation and oxygen availability

Ignitability
How far the ignitability of a material can be influenced by the shape and arrangement as well as the preheating process can be demonstrated by the example of paper. A single sheet will be ignited by a burning match and will be consumed rapidly. This is not valid for a book or a pile of paper. According to Fig. 5a ignition will be effected by the spacing between the sheets of paper and the size and duration of the ignition source.

Van Kreveler [50] discusses the possible influences of the chemical nature of the materials like content of aromatics and flame retarding additives. The influence of the material compound was studied extensively [51, 52]. Long-lasting preheating can produce, smoldering as well as flaming ignition of a material. Glowing cigarettes may initiate a self-sustained smoldering fire in connection with upholstery furniture. Smoldering outer coverings based on cellulosics favor the development of smoldering fires. The use of noncombustible interlayers, e.g. glass fabric does not prevent the risk of a self-sustained smoldering fire. Figure 5b illustrates that the nature of the ignited material will significantly influence the risk of ignition.

The self-ignition of hydrocarbons takes place in the gas phase, e.g. at relatively low temperature levels. The problem of self-ignition as a result of long-lasting preheating has been the focus of many investigations. A very critical parameter is the heat conduction. The possible self-ignition process of charcoal- or oil-impregnated cleaning rags are important. The extent of spread will be determined by both the heat capacity of the ignition source and the impingement time, as well as heat released by the burning material [14, 53]. In the actual fire situation, the extent of the spreading fire will be determined by the interaction of the relative portions of convection, conduction and radiation heat in relation to the environmental conditions [54, 55]. The complete burning behavior of upholstery furniture can be significantly influenced by the surface spread of flame of the outer covering or by local burning situations on the splitting of the protective covers [56]. The burning behavior of decorations and furnishings can be changed, e.g. by the design [57].

Figure 5a Ignition risk of paper

Figure 5b Influence of the chemical nature on the fire risk

Figure 5c Forest fire – wind effect

Contribution to the flame spread

Design-related factors like textile construction (height of pile, density of pile and texture) as well as the backing of the textile can become dominant during the course of the fire. The surface spread of flames of curtains and draperies will be significantly influenced by the texture and the mass per unit area. Reference [58] discusses the significance of the size of the ignition source on the surface spread of flame. Wilde [59] emphasizes the possible influence of ventilation. The rate of heating of the surface of the material is obviously of major importance. This effect is especially known to affect forest fires and is illustrated by Starrett who describes in [60] how strongly the contribution to the surface spread of flame will be influenced by chemical and thermophysical material properties as well as by environmental conditions. He points out that, as a rule, geometrical factors will sometimes exert a much stronger influence over the spread than material-related properties. Admissible floor coverings complying with the fire-safety requirements, are unacceptable to wall coverings. This was the situation in the Star Dust Fire [61] and the MGM Hilton Fire in Las Vegas [62] that prove these statements. Williamson discusses in [63] all the important aspects like heat transfer, smoldering and flaming fire, ventilation and the input of the ignition source, which can all influence the development of a fire. The effectiveness of specific parameters like texture, chemical nature of the fiber or the design of clothes is discussed in [64] in detail. Reference [65] shows how the contribution to the surface spread of flame can be changed, e.g. by the coating of incombustible mineral fiberboards. The paper covering was ignited during hot work and due to heat build up flame spreading and strong smoke release occurred. Evacuation of this part of the hospital was necessary. The risk of a self-sustained smoldering fire with mineral-fiber products is the subject of reference [66]. The inclination of the flames is crucial in determining if the radiation or the convection part of the heat-release rate becomes dominant. The schematic drawing of Fig. 6 illustrates this aspect, which can influence the degree of the fire spread.

The flame can reach, according to Fig. 7a, a larger portion of the material. How far fire spread is influenced by the material compound has been proven by numerous enduse-related investigations. The nature of the substrate as well as the method of

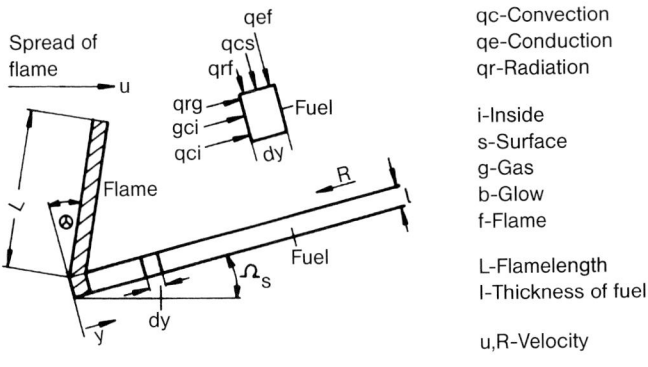

Figure 6 Illustration – Preheating by the radiation of flames

Figure 7a Fuse effect

Figure 7b Burning and hot dripping

fixings, the surface coating and melting processes can change the spreading of the fire decisively. The influence of facings on the surface spread of flames is pointed out, e.g. in [60, 64, 67].

Flame spread within the gas–air mixture below the ceiling in room–corridor experiments achieved velocities up to 2 m/s according to Quintiere [68]. Theoretical considerations of the surface spread of flame are described by Pagni et al. [69]. The influence of the radiation level was discussed by Kashiwagi [70] for flooring materials and was related to the relevant scenario. Increased radiation levels cause due to preheating effects an accelerated fire spread. Similar conclusions are given by Starrett [60]. According to Christian [71] the content of the building is up to 90% responsible for the course of the fire. The protection of the construction mode of action was summarized in [69].

Side effects of a fire
The side effects of a fire include, as a rule, the dripping behavior as well as the smoke production. These material-related potency data will be changed considerably due to the course of the fire. Because the heat-release rate behaves simultaneously the heat-release rate is likewise part of the side effects. A general classification of these material data can be done only if these data have been gained with regard to the various parameters of the different phases of a fire.

Burning and hot dripping
The side effects of burning and hot dripping are materials characteristics, which change at a specific temperature level the melting point [72]. The melting heat is the amount of heat that is required to initiate the melting process from 1 g of a material. Material and environmental factors will play a major role. The ignition source and impingement time can significantly change the dripping behavior. The extent to which the arrangement in the room will be important for the classification can be postulated by the comparison of the risk situation due to ceiling and flooring linings.

Contribution of heat-release rate
The heat of combustion of a combustible material is one of the most important thermodynamic properties. The heat of combustion, the so-called calorific value, is the heat output of complete combustion of a mass or volume unit. There is a difference between the lower and the upper calorific values. The upper value takes into consideration the heat of vaporisation of the water content. These secondary effects of a fire are multipoint data and might be changed by the temperature range and the oxygen availability. With increasing test temperatures, the heat-release rate and the smoke-production rate approaches the maximum value, which is fixed by the calorific value. As a rule, the material-related potency value is not the dominant parameter of the heat-release rate, which is effected by the fire scenarios, see Fig. 8a. The graphical representations of Fig. 8b demonstrates that volume-based material-related net calorific values lead to a quite different ranking to mass-based data. The net calorific values of latex, cotton and PUR are relatively comparable, but wood compares unfavorably on a volume basis.

The chemical nature of the material decides if the thermal attack leads to decomposition or to a melting process. The decomposition temperature is the temperature level where the decomposition products form an ignitable mixture in combination with the atmospheric oxygen. This specific material-related data is effected by the test

Figure 8a Heat-release rate

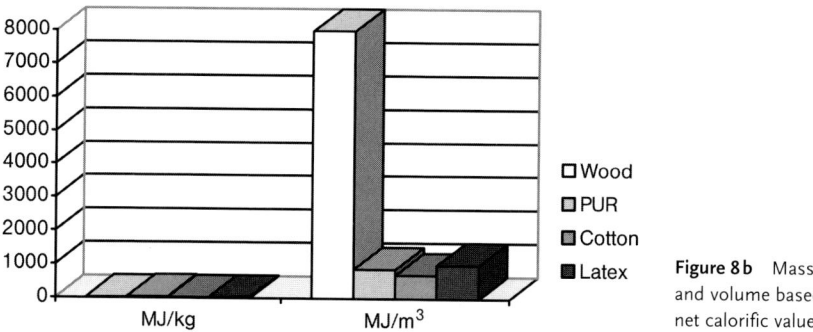

Figure 8b Mass- and volume based net calorific values

method, especially by the mass of the test piece, the air supply and the ignition-source size. The possible influence of material- and environmental-related factors to these criteria have been established in various studies. The chemical nature of the decomposition products can influence the efficiency of the course of the burning process considerably. The extent to which the burning processes and the heat-release rate will be influenced by the material compound has been proven by risk-related large-scale testing with upholstery furniture [56]. Fundamental research investigations have been carried out by Tewarson [73]. In the test apparatus, see Fig. 9, the sample was horizontally arranged and the combustion initiated by the impact of variable, external radiation. The test sample had dimensions of 100 mm diameter and

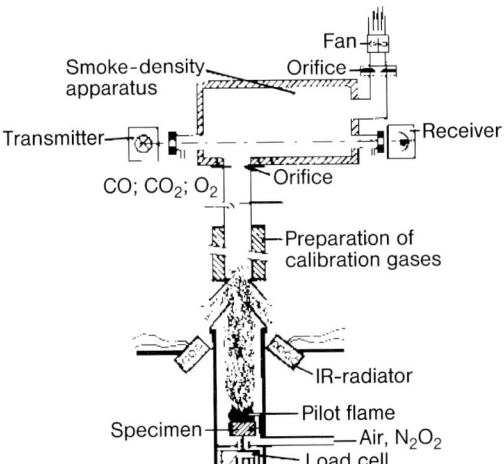

Figure 9 Tewarson apparatus

20 mm thickness. The flaming will be initiated by a small pilot flame. The heat-release rate is estimated by the continuous determination and registration of the temperature of the stream of combustion gases.

Tewarson et al. demonstrated in [74] that the sequence of liquid, foam and solid produces a minor increase in heat-release rate. Fang demonstrates in [75] that the size and the arrangement of the ignition source, the distribution of the fire load and the ventilation can significantly change the burning rate. The chemical nature of the material to be classified and whether it leads to a pool-fire situation or to the formation of a protecting carbonaceous layer, will decisively alter the burning rate. Similar conclusions concerning carbonisation have been reached by Martin at SRI [76]. Martin also showed that the propensity to stress crack and delamination effects will enlarge the attacked surface area and thereby significantly change the relevant material-related data. Quintiere demonstrates by his work that the very strong influence of preheating effects documented the external radiation as the dominant parameter [77]. The burn rate of furniture can be significantly increased by heat produced from already burning items. These studies were carried out with a 2-chair arrangement (see Fig. 10).

Temperature measurements of the combustion gases are made by thermocouples in the fire room as well as in the neighboring corridor. The variation of outer coverings and fillings and the use of interlayers caused significant changes to the temperature levels. The maximum temperature of the combustion gases within the fire room could be reduced from 800 °C to below 100 °C by careful selection of upholstery furniture. The heat-release rate seems to be the dominant parameter.

The test results gained by meaningful large-scale tests demonstrate that the area-burning rate is responsible for the heat release per unit time. The protecting effects of the upholstery combination are important. Reference [78] illustrates a project where the use of interlayers alone increase the resistance against ignition sources of different intensity (cigarette, match, paper and burning item) and decrease signifi-

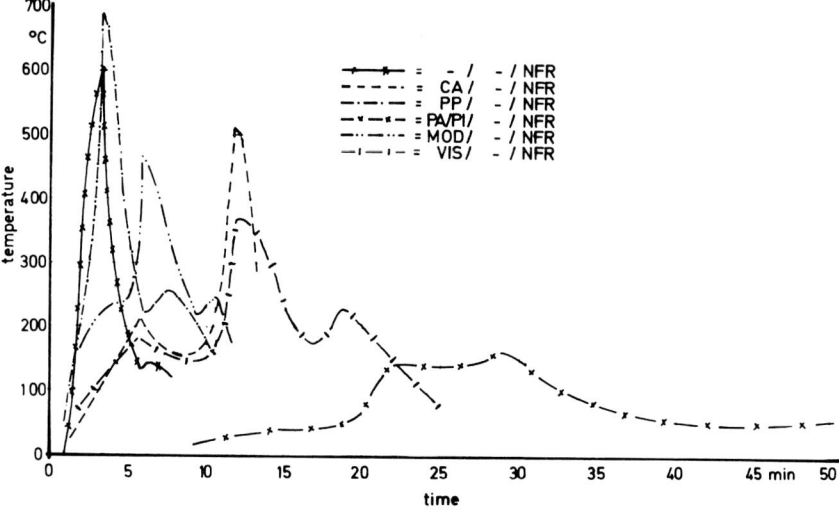

Figure 10 Smoke-layer temperature course

cantly the heat-release rate of upholstery items. Thomas et al. [45] discussed the input of ventilation for pool and crib fires. With developed fires feedback effects will play a dominant role with respect to the room geometry. Heskestad [46] demonstrated that porosity and ventilation effects will be the dominating parameters and will determine the heat-release rate. Figure 4b illustrates that the heat-release rate as well as the fire load, the ventilation and preheating effects will dominate the thermal preparation of the environment.

Smoke density
Smoke density is defined as the reduction of visibility by the influence of gaseous, liquid and solid decomposition and combustion products. The degree of smoke production is linked with the fire scenario. Figures 11a and 11b are illustrating that the extent of smoke production is dominated by the fire scenario and the burning material. By a careful selection of the fuel as well as the fire parameters a nearly smokeless fire can be obtained.

The principle of smoke-density measurement is primarily the optical reduction of the light beam between a giver (lamp) and the receiver (photometer, diode, selenium cell) by smoke particles. The dependence of the smoke-density figures from parameters like test-temperature level or particle size and distribution has been investigated in many studies. Test parameters like thermal attack of sample, exposed surface area of the sample and ventilation conditions, can change the results considerably [79]. Smoke-density measurements, based on specific laboratory test methods to simulate the reduction of visibility of the human eye, did not always correlate with the results of risk-related large-scale tests in the transportation field [80, 81]. Where limited reduction of visibility by gaseous, liquid or solid decomposition of combustion products occur in the areas adjacent to the fire room or in the escape routes, escape is still

Figure 11a Smoke formation/tire fire

Figure 11b Smoke formation/charcoal fire

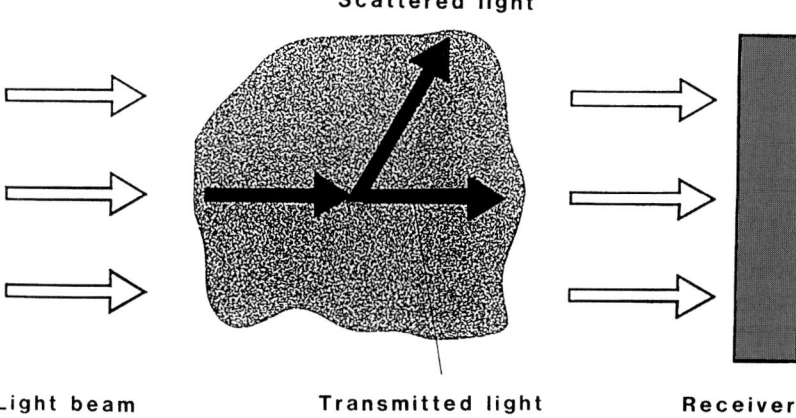

Figure 11c Scheme – smoke-density measurement

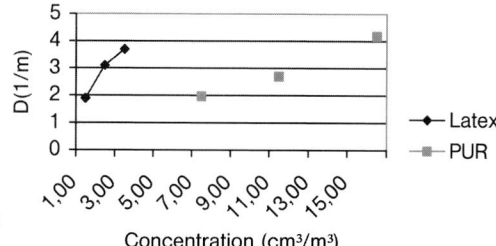

Figure 11d Influence of the concentration on the smoke density

possible, providing these areas are not subject to reduced oxygen or high thermal load. Grimberger et al. [82] demonstrated for air pollution incineration tests that the combustion temperature can significantly change the degree of soot and tar production. Figure 11c demonstrates the principle of smoke-density measurement. The room shape must be used as the basis for calculating reality-related smoke results, the possible changes of fire load can be assessed.

The graphs of Fig. 11d use as examples flexible polyurethane foam and latex foam to illustrate the influence of the concentration of the effluents produced at 400 °C on the smoke-density data. The possible smoke danger in natural fires is related to the sample volume taking part in the fire and to the material-related test results. Rasbash et al. explained in [83] that smoke-density data have to be expressed as mass-, volume- or area-based data. Rasbash et al. discuss in [84] that the smoke distribution and the smoke-layer formation in real-fire situations can significantly change the degree of visibility reduction. The interpretation of test results to size and nature of the particles, dilution effects, reflection and light conditions have to be considered. Berman gave advice in [85] that the possible danger of smoke and the required rescue provisions will be influenced by the smoke production and the environmental conditions. Rasbash [86] has suggested that, as far as possible, smoke problems should be categorized as risk-based assessment. 10% of the 0.5 kg fuel are sufficient to produce dangerous visibility conditions within a room of 400 m³ volume.

Acute toxicity of effluents

The acute toxicity of effluents can be interpreted as the inhalation toxicity of the fire-gas mixture. Every time after fire disasters, there is an attempt to assess how much the decomposition and combustion products of modern materials caused more severe health problems than natural organic products. The schematic of Fig. 12 illustrates that in relation to the fire situation, smoke might become dangerous in areas that are far away from the room of origin. This statement is backed by various experiences of natural fire situations.

A classification of the acute toxicity of effluents can be done, according to toxicologists, only on the basis of bioassay testing [87–92]. This seems to be the only way to cover possible additive, synergistic or antagonistic effects of single components. Toxicologists are of the opinion that neither the identification of the chemical composition of the combustion products nor the analytically measured concentrations of the main components are sufficient to determine acute toxic potency. Experience of studies has shown that analytically gained concentration data might offer the chance to classify the toxic potency if a basis correlation with the bioassay data determined with the same piece of equipment, e.g. the DIN 53436 tube furnace [93]. The LC_{50} value, one of the most important criteria, can be determined, by mathematical modelling.

Realistic solutions for similar problems are:

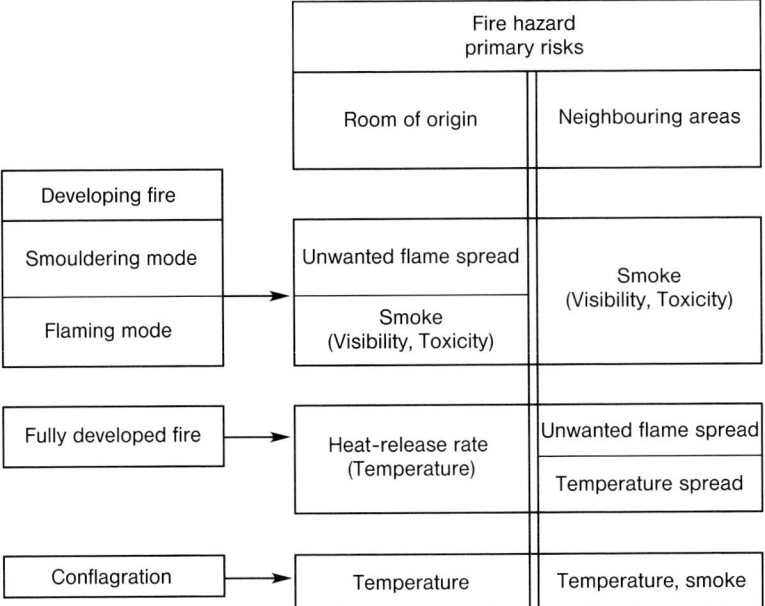

Figure 12 Danger of smoke in the case of a fire

- preheating of the insulating material to make sure that the possible decomposition products are removed
- the use of conditions with sufficient ventilation
- the use of breathing devices

Corrosive effects of effluents

How much the corrosive effects of fire gases are dependant on material- and environmental-relating factors have been assessed after some spectacular fires like the power plant station of Pleinting [94]. The distribution of the effluents by the ventilation ducts was one reason for the large extent of the damage. The contribution to the smoke production and especially to the distribution of the fire gases can be limited by fire-resistant barriers in the ventilation system. Smoke detectors play a very important part. In the case of a fire the ventilation system can be stopped and fire shutters initiated.

2.2 Essential Fire Scenarios

During the research work of TC 92 SC3 Toxic Hazard in Fire the guidelines for the five essential fire scenarios have been defined as illustrated by Fig. 13.

2.2.1 Pyrolysis

Pyrolysis is defined as thermal decomposition without oxygen availability. As for the chemical degradation fragments will be formed. The interrelation of the compounds/fragments is characteristic and depends on the decomposition conditions and

	Oxygen (%)	CO_2/CO	Temperature (°C)

1. Decomposition
  a) Smouldering
     (self-sustained)
  b) Non-flaming (oxidative)
  c) Non-flaming (pyrolytic)
	
	21
	5 to 21
	< 5
		
		N/A
		N/A
		N/A
			
			< 100
			< 500
			< 1000

2. Developing fire (flaming)	10 to 15	100 to 200	400 to 600

3. Fully developed (flaming)
  a) Relatively low ventilation
  b) Relatively high ventilation
	
	1 to 5
	5 to 10
		
		< 10
		< 100
			
			600 to 900
			600 to 1200

N/A = not applicable

Figure 13 Characterization of effluents

the material. The course of natural fires may be similar if an external heat source is producing thermal decomposition products behind metal facing of sandwich panels of cold-store elements. The time-dependent temperature will be a function of the heat source and the time of impingement. A very practice-related example for the pyrolysis scenario is the localized decomposition in processing machinery. The materials will be decomposed by thermal stress under oxygen exclusion, e.g. by injection molding or the production of pressed pieces. These decomposition products will be dominated by the manufacturing temperature. Requirements of accident prevention will guarantee that no dangerous concentrations will arise. Figure 14a illustrates with respect to Japanese research work that increasing temperatures will lead to a significant increase of the smoke component HCN. In relation to the nitrogen content there is a continual increase of the HCN component of smoke from nitrogen-containing materials.

2.2.2
Oxidative Decomposition

Oxidative decomposition is the thermal degradation in an air stream and takes place, e.g. under the influence of external radiation. Radiation from naked flames of burning items generates degradation of neighboring areas. The extent of the damage will be dominated by the intensity and the duration of the radiation. The oxidative degradation will, according to Fig. 14b increase until ignition occurs and the maximum

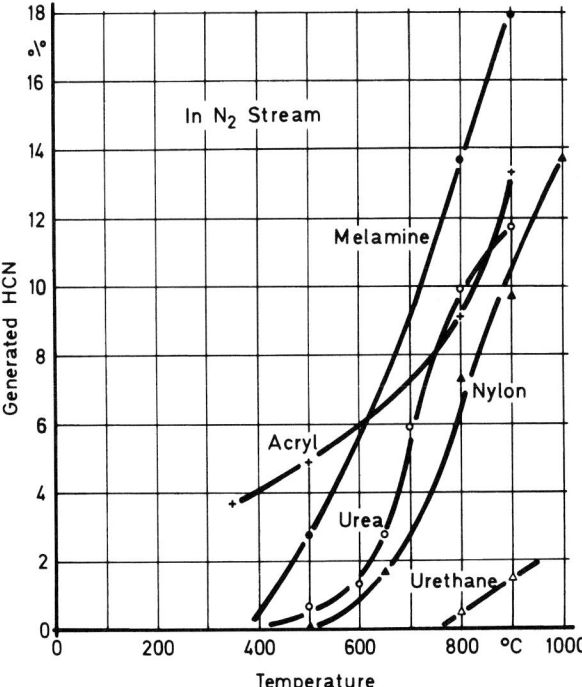

Figure 14a Pyrolysis products

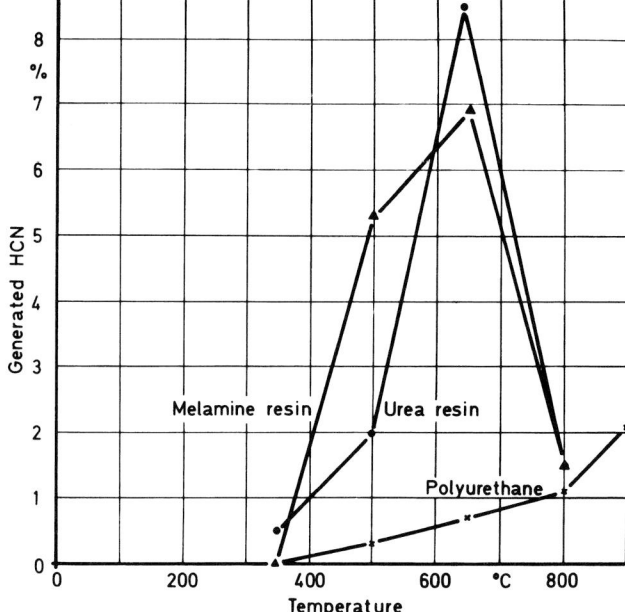

Figure 14 b Oxidative decomposition

value is reached when it will significantly decrease. In the case of complete combustion the decomposition products will be partly combusted. The present results show that flexible polyurethane foam has to be favorably judged in relation to nitrogen-

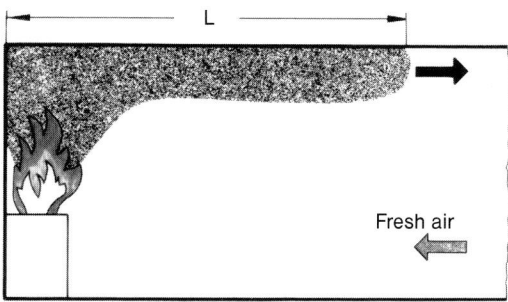

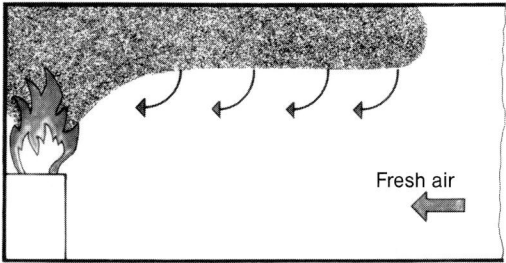

Figure 15 Scheme – oxidative decomposition by a smoke layer

containing reference products, because in the critical temperature range only, minor HCN concentrations will be formed. Figure 15 illustrates the degradation due to the heating from the smoke layer.

2.2.3
Propagation of a Self-sustained Smoldering Fire

After a self-sustained smoldering fire has been initiated by an ignition source, it will propagate without any further external heat supply. This kind of smoldering fire will be initiated in upholstered furniture by glowing tobacco. The possibility of air circulation is assessed for such self-sustained smoldering fires. Figure 16a illustrates the self-sustained smoldering process in the upholstery section. This kind of fire is typical for permeable smoldering materials.

Smoldering tests may give poorly reproducible results but it is not the reproducibility of the test method that is crucial but the question of whether the test method will be able to simulate the risk situation adequately. Figure 16b shows very clearly that the stream of effluents is characterized by a low level of the CO_2/CO ratio.

Studies described in [95–97] show the relevant risk of self-sustained smoldering fires. In the latter, the risk of a smoldering fire of chipboard is discussed.

The graphics of Figs. 17a and b show the time dependence of the smoke temperature level of two different upholstery combinations. The smoldering temperature of the PUR composite measured by thermocouples reached 400 °C. The smoldering fire of a latex composite initiated by a glowing cigarette, is characterized according to

Figure 16a Smoldering fire in the upholstery furniture field

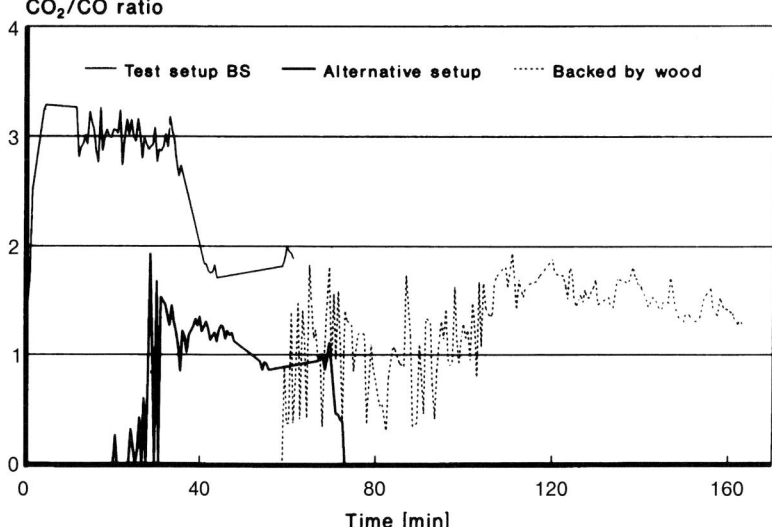

Figure 16b Cigarette test – CO_2/CO – relationship of effluents

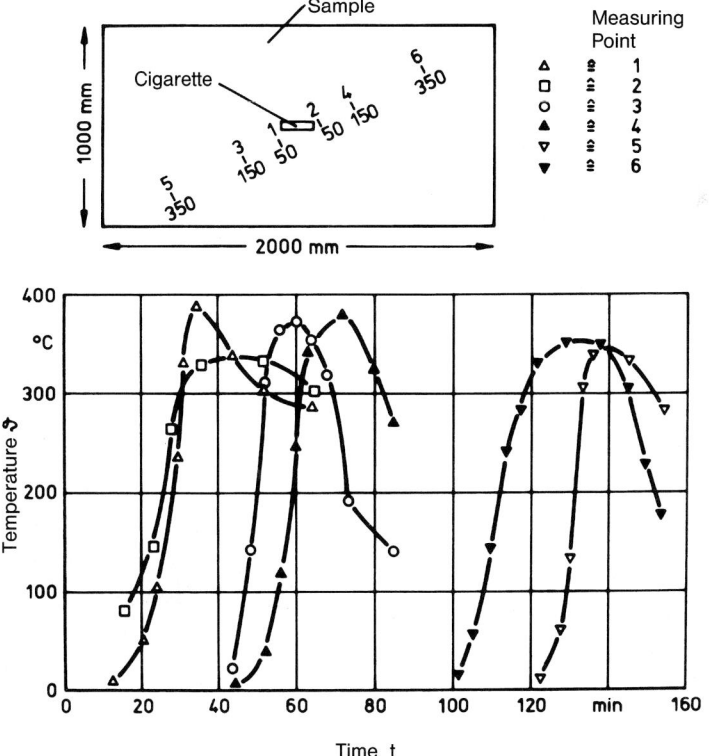

Figure 17a Time–temperature course of PUR upholstery

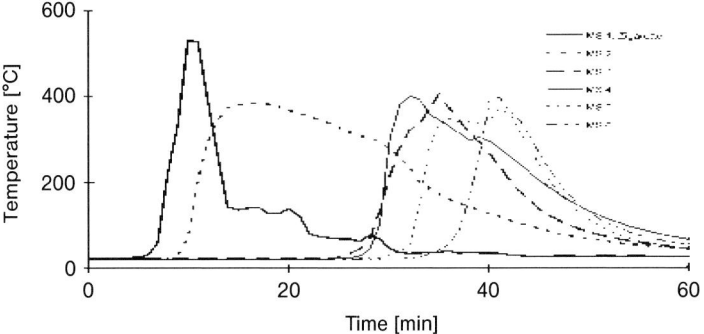

Figure 17 b Time – temperature course of latex foam

Fig. 17b by a significantly higher temperature level. Figures 18 a, b and c demonstrate the time–temperature curve of a self-sustained smoldering process indicated by exothermic reactions in the mineral fiber [66].

The probability of such kinds of fire will be dominated by the permeability and the nature and concentration of the binder used.

The concentration of effluents given off during a self-sustained smoldering fire will be influenced by the various parameters that dominate the course of the fire. Beneath the material-related influence such as, e.g. density, porosity of the compound, especially preheating effects, ventilation and environmental space conditions play an important role. The possible smoke layer can be estimated as the fuel contribution to the fire. The drawings of Fig. 19 demonstrate how the spread of the initiated fire takes place. Smoldering velocity, space and ventilation conditions will be the basis for mathematical modelling of the degree of danger and the estimation of the build up of the effluent concentration.

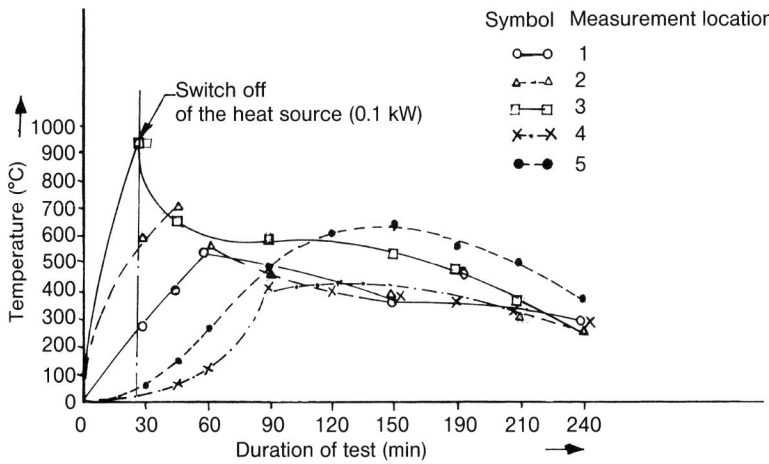

Figure 18 a Time–temperature course of mineral fiber

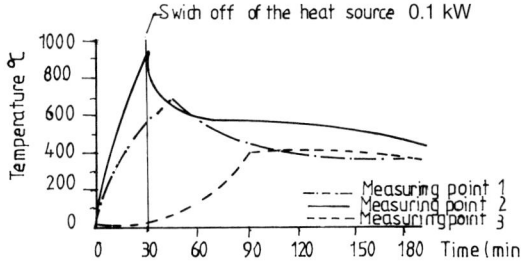

Figure 18b Time–temperature course of mineral fiber

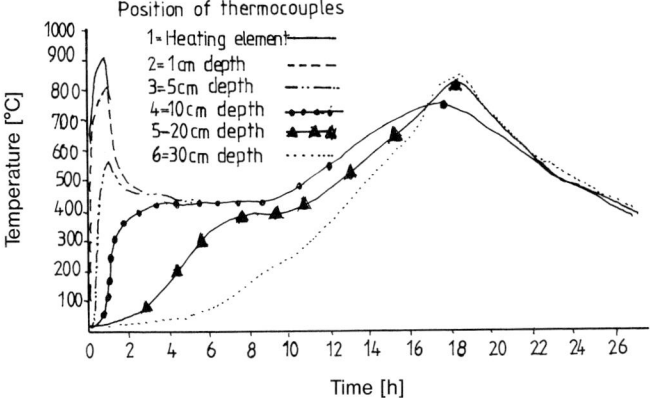

Figure 18c Time–temperature course of mineral fiber

The results of the investigations of natural fires, given by Treitmann [98] in Salt Lake City 1974, explain (Fig. 19b) the material-related differences of the smoldering velocity. The data of Fig. 19c demonstrates that the material-related differences in comparison to the ventilation effects are of minor importance. Ventilation effects can change the smoldering velocity significantly. The differences from the above-mentioned results are because of the changes in the upholstery combination. The outer coverings can become the dominant parameter.

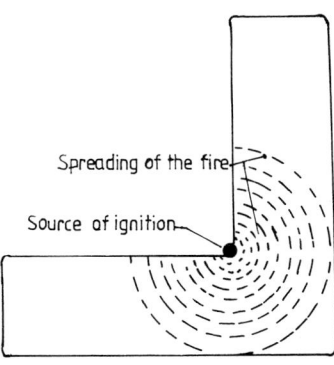

Figure 19a Scheme – Spreading of a smoldering fire

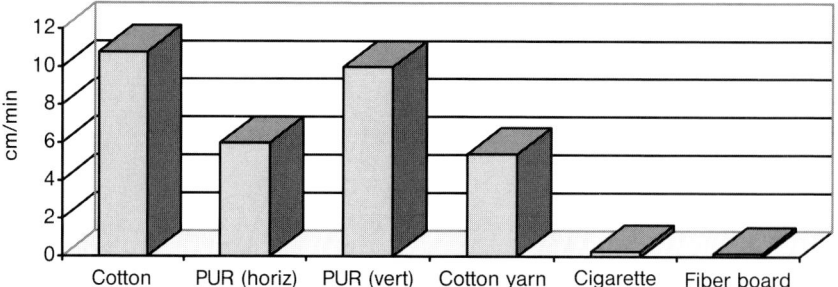

Figure 19b Smoldering velocity

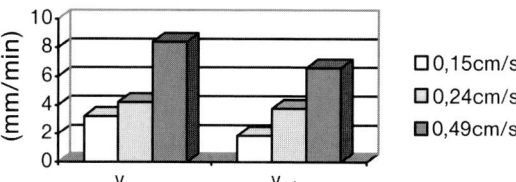

Figure 19c Smoldering velocity

The self-smoldering process, which is spreading without any external heat supply, is increasing the thermal preheating of the upholstery item. As a rule, the smoldering process will change to a flaming process after a long-lasting phase.

2.2.4
Fully Developed Fire with a Flashover Situation

Burgess et al. [98] reported details of a study of dwelling fires that had been investigated in cooperation of Harvard University and the fire department of Boston. The analytical measurements of the stream of fire effluents documented, as shown in Fig. 20a, that the CO_2/CO ratio of the effluents is characterized by a level of below 50.

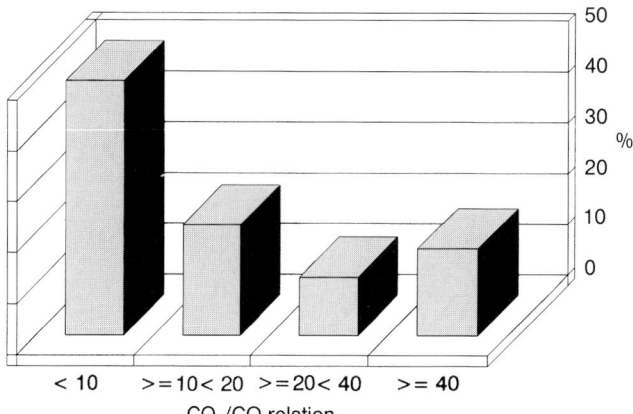

Figure 20a CO_2/CO ratio of effluents from residential fires.

This interpretation has been confirmed by the results of risk-related large-scale tests [99]. The studies were carried out by the order of NASA with furniture of different fire-performance classifications and demonstrated, see Fig. 20b, that apart from the developing stage of a fire CO_2/CO values below 30 were achieved. The aldehyde concentration and especially acrolein have to be specially considered, even the concentrations of CO are much higher.

Large-scale tests with fully furnished rooms, conducted also in closed rooms as in ventilated ones (see Fig. 20c) can deliver low CO_2/CO ratios. Considering the basic equation

$$X\, C_nH_n + Y\, O_2 = Z CO_2 + H_2O + N\, kcal$$

a higher portion of CO will be released in the fully developed fire. Realistic large-scale tests that have been conducted with upholstered furniture at the Rapra testing station in Shawbury and fully furnished room burns in Moreton in the Marsh have been initiated by the III (International Isocyanate Institute). Some of these investigations have been conducted as part of the FTC 5-million dollar program concerning the fire performance of polyurethane foams ordered by the PRC (Product Research Committee) [14, 100–104]. Figures 20d and e illustrate that the CO_2/CO ratio of effluents from the developing stage of a fire of upholstered furniture reached values of less then 50 within 2 min. The results gained in large-scale tests with fully furnished room burns in Moreton in the Marsh confirmed the American findings. Independent of the chosen ventilation, except for the initial stages of a fire, the CO_2/CO level was below 30.

Possible qualitative statements concerning material-related data, which have been based on a laboratory test, e.g. the cone-calorimeter method (see Fig. 21) may not be

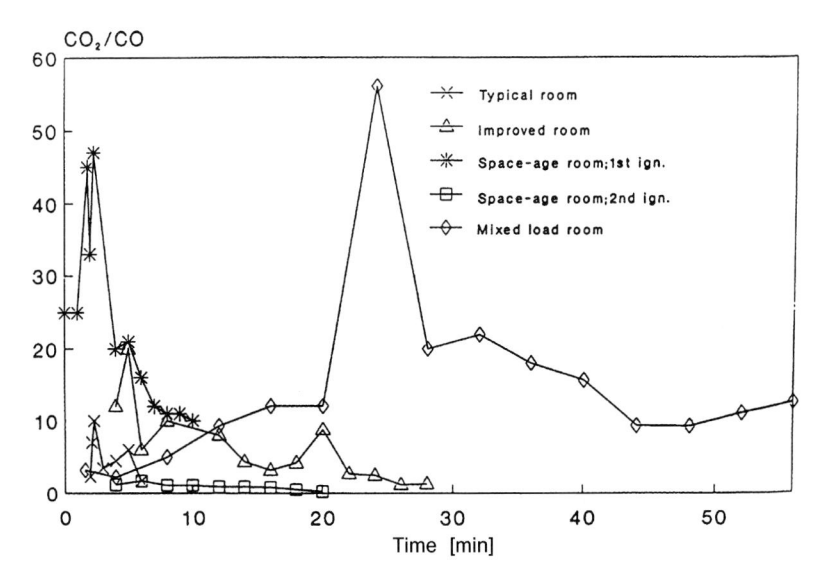

Figure 20b Large-scale tests with fully furnished rooms

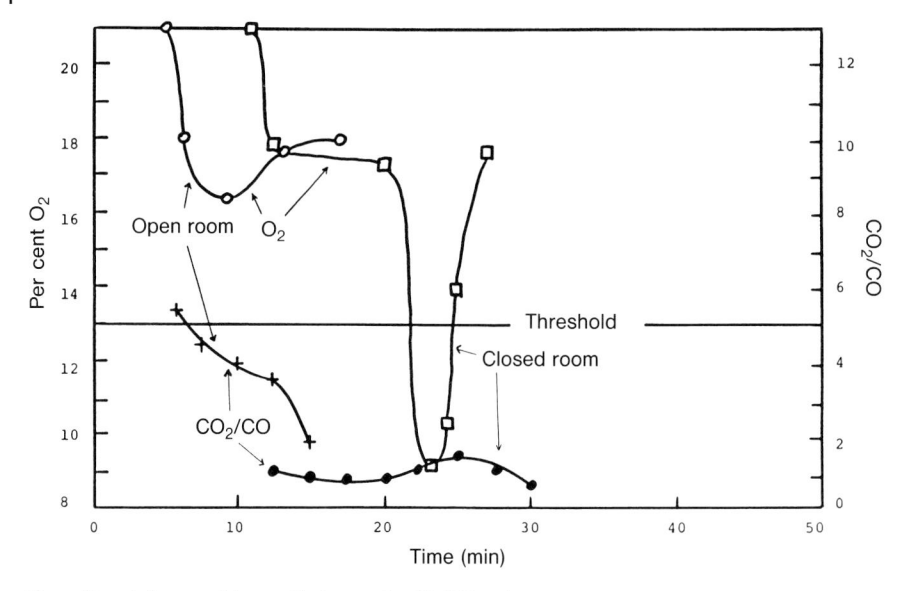

Figure 20c Influence of the ventilation on the CO_2/CO ratio

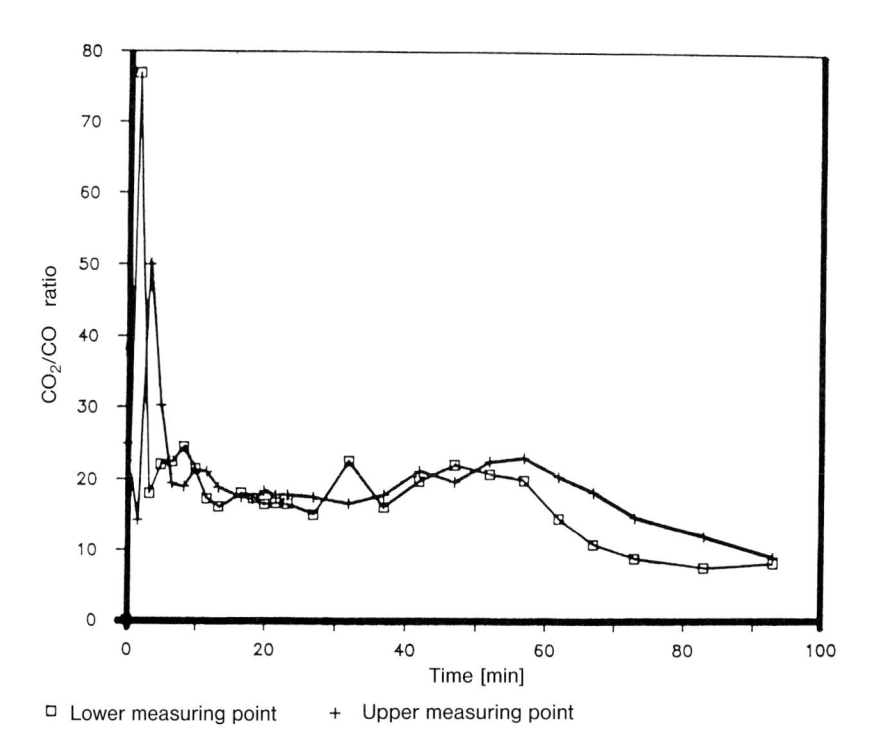

Figure 20d Moreton 1:1 tests – CO_2/CO ratio in the flue-gas stream

□ Lower measuring point   + Upper measuring point

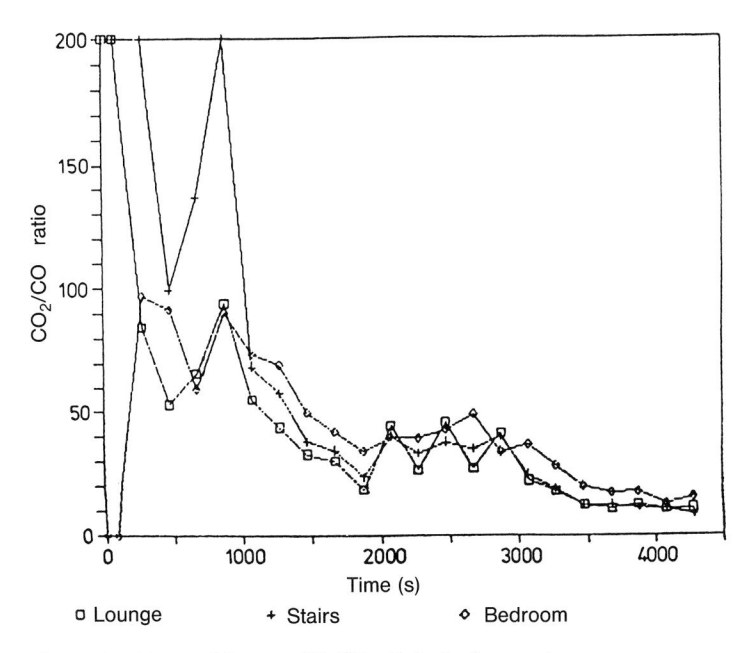

Figure 20 e Moreton 1:1 tests – CO_2/CO ratio in the flue-gas stream

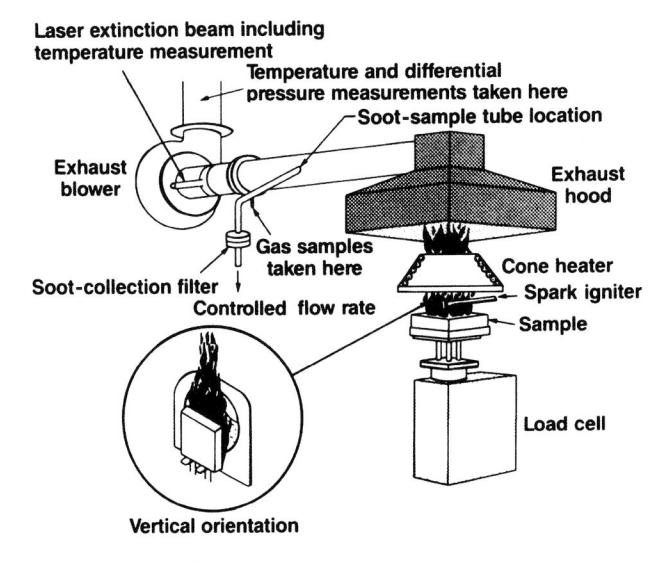

Figure 21 Cone calorimeter

valid. In these tests, CO_2/CO values were extremely high and therefor can not be obtained by natural dwelling fires. This statement will be underlined also by the findings of the NBS investigations, with upholstery furniture as shown by Fig. 22a [105].

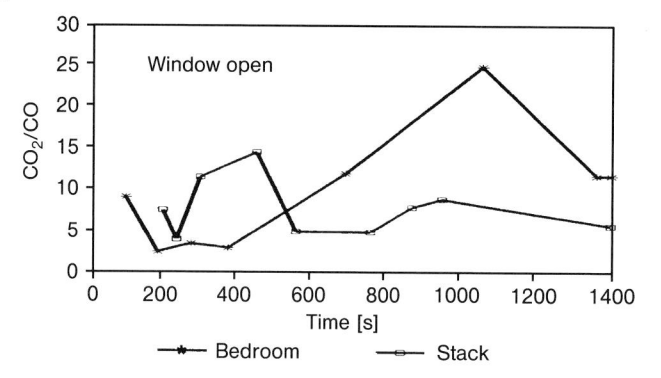

Figure 22a NBS 1:1 Test – CO_2/CO ratio in the flue-gas stream

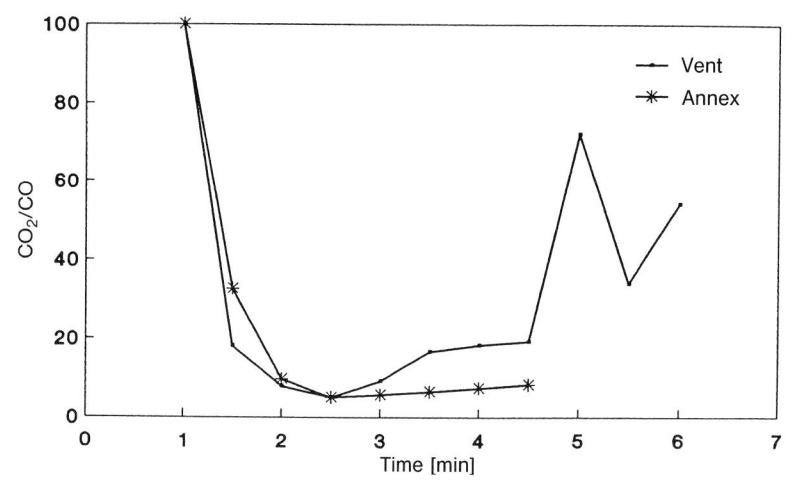

Figure 22b Simulation of the Star Dust Fire

In a fully developed fire, flashover may occur, where all the combustibles of the fire room will contribute to the fire. As demonstrated by the simulation of the Star Dust Fire by the Fire Research Station (FRS) the mass of smoke is characterised by a low CO_2/CO ratio [106] (see Fig. 22b).

The smoke movement will be initiated by the pressure increase in the fire room. The mass of smoke will be transported from the fire room to the neighboring areas and will thermally attack the wall and ceiling linings, ducts and cavities and suspended ceilings as well. Due to the lack of oxygen, decomposition products with a high level of CO will be released. These effluents may lead, after mixing with fresh air, to an instantaneous spread of flames. According to Emmons [107], such effects may cause an extremely fast flame spread. Due to the low oxygen content, an extremely high acute toxicity of the smoke layer has to be expected. The lethality of such smoke layers, which contain a high CO concentration, is reported in the relevant literature for fires in mines and underground fires like Kings Cross [6]. Figure 23a demonstrates this kind of danger in the railway field. The large-scale test, with an airbus cabinet demonstrated that a fire initiated by the kerosene burner [108, 109] impingement duration 45 s, lead to a flashover situation after 1.5 min. Analogous results were gained in a series of large-scale tests with railway vehicles in the transportation field [80, 110]. Bußmann et al. [111] report on comparative studies concerning the CO_2/CO of various large-scale tests. Figure 23b illustrates the dangerous smoke production after flashover.

The mass of smoke, produced during the flashover situation has to be estimated on the basis of the fire load taking part in the fire. The mass of smoke can be estimated relative to the ventilation. The laboratory test results gained by extremely high

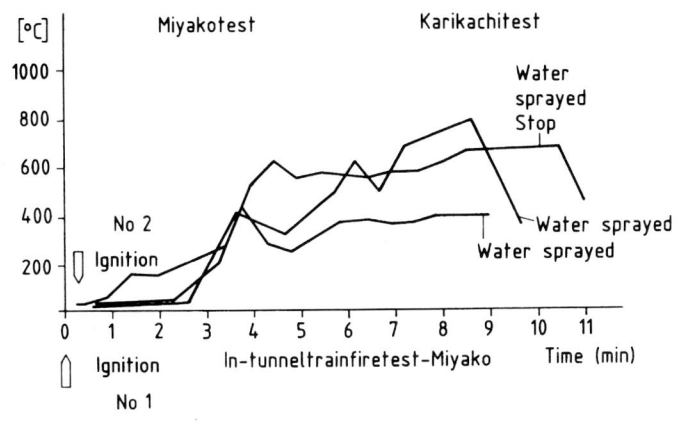

Figure 23a Temperature development by the burn out of a railway carriage

Figure 23b Flashover situation in the transportation field

ventilation conditions may lead to a corresponding faulty estimation. In the flashover situation 1/10 mm of the thermally stressed area will be instantly decomposed. Large areas will be covered by the fire in fully furnished rooms and large amounts of smoke will be released. The listings in Table 2 prove that material-related differences will be of minor importance.

Table 2 Fire Load Surfaces in Dwellings and in the Transportation Field

Fire Room	V (m^3)	Area (m^2)
Living Room	ca. 50	ca. 150
Cabin	ca. 12,5	ca. 40
Open-plan – carriage	ca. 150	ca. 450

In accordance with Smith [112] the fire load will not be the main risk. A comparison of the calorific potential alone will not be valid for a risk-related classification. If a flashover situation occurs, a dangerous situation as described above, might happen. If the smoke formation is a hindrance, the rescue actions will be, according to Emmons [113], especially influenced by the size of the thermally stressed surface area. The decomposing area will, e.g. as seen in the Düsseldorf airport fire [1], dominate the smoke formation in the flashover situation [114].

3
Research of Causes of Fires

3.1
General Experience of Fire Statistics

Detailed fire statistics are only able to provide information about

- which kind of ignition sources are reliable and
- of what importance the various ignition sources are
- which items are first ignited
- how large the proportion of residential fires might be, the numbers that spread to neighboring areas
- at what time of the day the majority of fires happen
- how many fatalities are recorded per 1000 fires and
- what is the extent of damage caused by the impact of flames and smoke.

Chandler provides in [115] an overview of the known experiences concerning residential fires. As well as the ignition-source problem, the aspects of a possible danger by smoke formation are reviewed. Social aspects and climate conditions (cold winters, etc.) seems to be linked to the high death rate. The increased number of fatalities during the "heating season" are confirmed by Jones [116]. Banks and Reardon reported in [117] that fire-statistics data of the World Health Organisation (WHO) show that the number of fatalities is decreasing worldwide. A comparison of the relevant national figures show quite differing records. The data sets of Fig. 24a show this effect very clearly, see also Fig. 24b. The comparison is based on the relevant data set for 1973, 1984 and some 3- and 5-year average values. The highest number of victims per 1 000 000 inhabitants are reported in the United States of America. These illustrations prove that for the comparison of such figures the basic conditions for the selection must be taken into account very carefully to prevent wrong conclusions. The graphical illustrations of Figs. 24b, c and d show which kinds of fire cause fatalities worldwide. The data set of Fig. 24e and f demonstrates that the majority of fires start in rubbish and in dwellings. The graphical representations show that the overwhelming majority of fire victims are killed in residential fires [118–123].

Polyurethane and Fire. Franz H. Prager and Helmut Rosteck
Copyright © 2006 WILEY-VCH Verlag GmbH & Co. KGaA, Weinheim
ISBN: 3-527-30805-9

3 Research of Causes of Fires

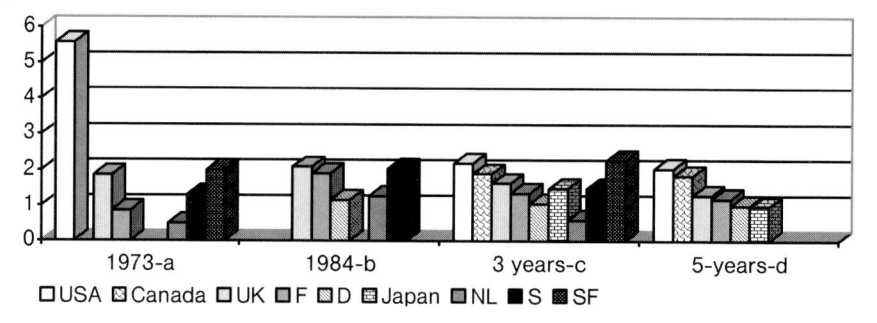

Figure 24a Fire victims worldwide per 100 000 inhabitants

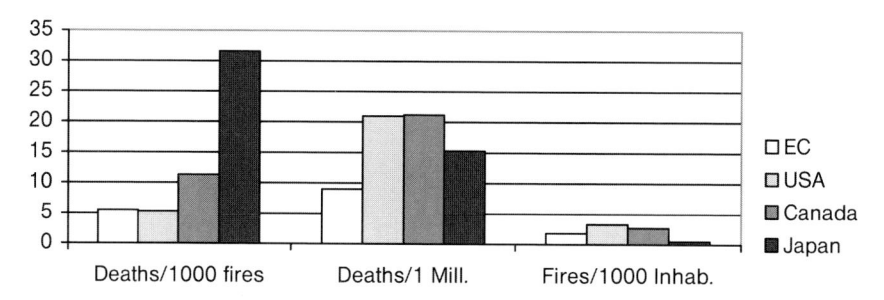

Figure 24b Fire victims in the EG per 100 000 inhabitants

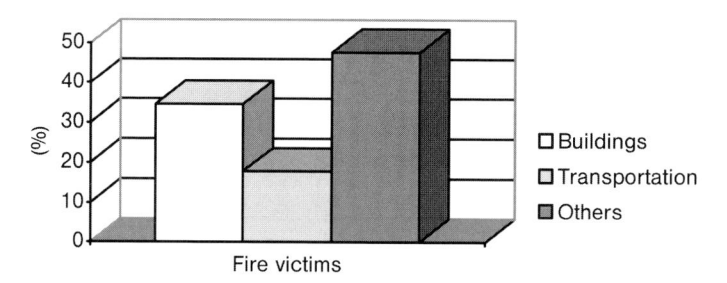

Figure 24c Fire victims worldwide – causes of fires

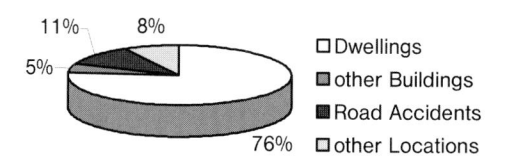

Figure 24d Fire statistics UK – fire development in 1998/99

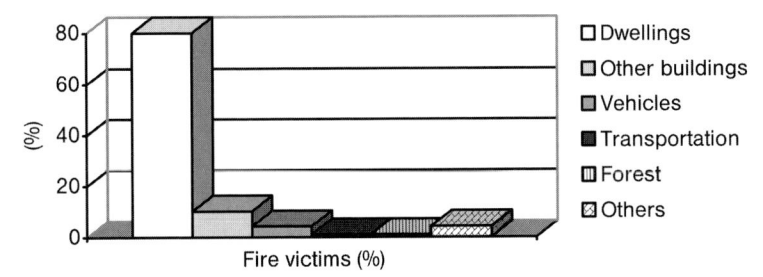

Figure 24 e Fire victims of different fire objects

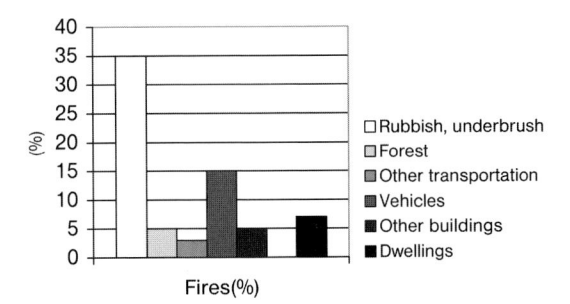

Figure 24 f First-ignited items

3.2
Knowledge of Fire Risks

Various studies in the United States of America, the United Kingdom, Japan and the EC must be interpreted in such a way that these figures concerning the number of fatalities per 1000 fires and per 1 million inhabitants underline the statements above about the marginal national divergencies (see Fig. 24a and b). The fire-statistics data of Figs. 25a, b and c illustrate the similarities figures for the United States of America, United Kingdom and Japan and demonstrate that the death rate for dwelling fires has been decreased significantly during the last few decades. The fire-statistics data of Fig. 25f demonstrate that the number of fire victims caused by furniture fires has continuously decreased in UK as well. Recent data of the EC for the period of 1982–1988 and 1996 to 2000 support this interpretation (see Figs. 25d and e). The fire statistics records of the Home Office according to Figs. 25c, d and f confirm this trend. The records of Fig. 25c illustrate the reduction of the number of fatalities during the years 1996 to 2000. These records document detailed information about how fires start and which items are first ignited [124–128].

Figure 26a illustrates, where residential fires generally start. Numerous studies have demonstrated that upholstered furniture in dwelling fires is the first-ignited item. Figure 26b provides information concerning dwelling fires in 1984 in the UK, while Fig. 26c illustrates the importance of dwelling fires for 1998/1999. The data set of Fig. 26a shows that cooking and smoking initiates most fires. Contrary to the

3 Research of Causes of Fires

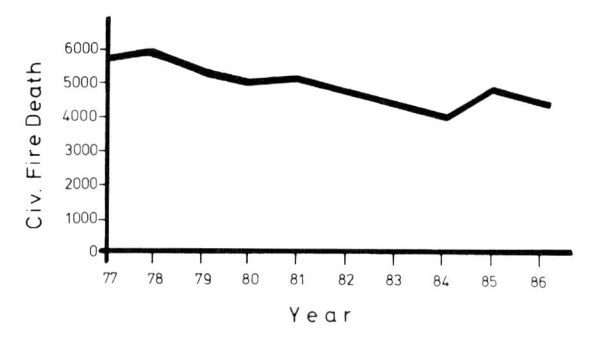

Figure 25 a Fire statistics USA – Fire victims during 1977–1986

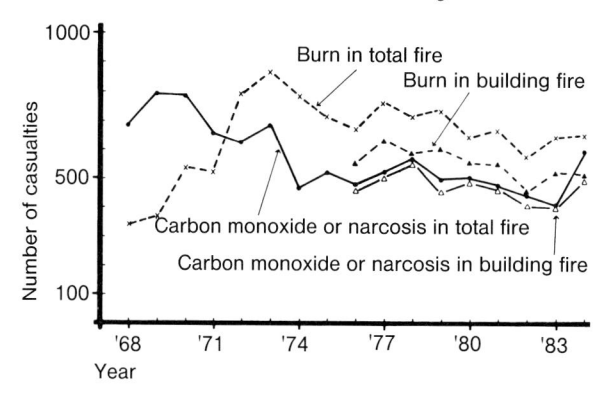

Figure 25 b Fire statistics Japan – Fire victims during 1969–1984

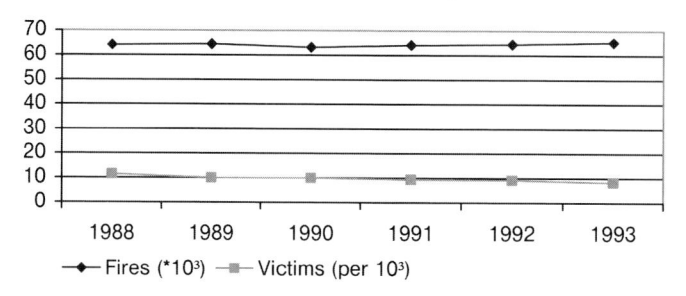

Figure 25 c Fire statistics UK – Fires (1000) and fatalities (per 1000 fires)

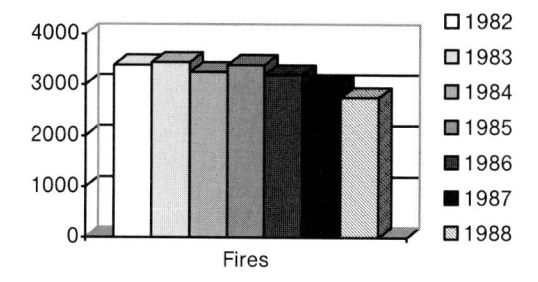

Figure 25 d Fire statistics EG – Fire victims

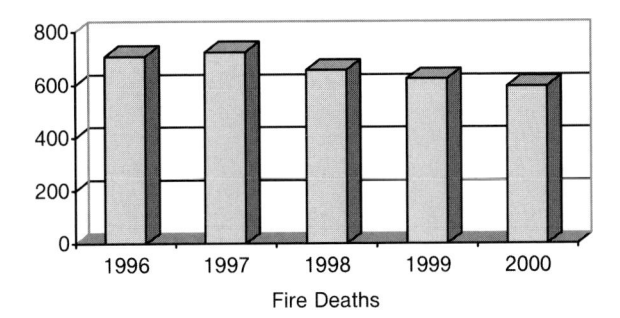

Figure 25e UK fire statistics – Estimated fire victims

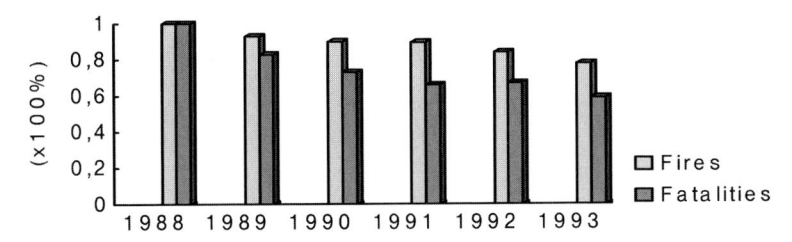

Figure 25f Decrease of fatalities of furniture fires

smokers fires only a small number of fire deaths were reported for fire initiated by cooking. While this trend seems to be relevant for the ignitions due to heating and matches, it seems to deviate from the Home Office report of 1984 for the ignition sources "cooking" and "smoking". It is possible that the increased use of smoke detectors may be responsible for the decreasing tendency. The major significance of early fire detection is therefore documented as well. According to the British fire-statistics data of 1984, there seems to be a constant level for both the ignition sources smoking and cooking. This is an additional indication that the intensified use of smoke devices is responsible for the break-off trend. The significance of an early fire detection was also documented [129–133].

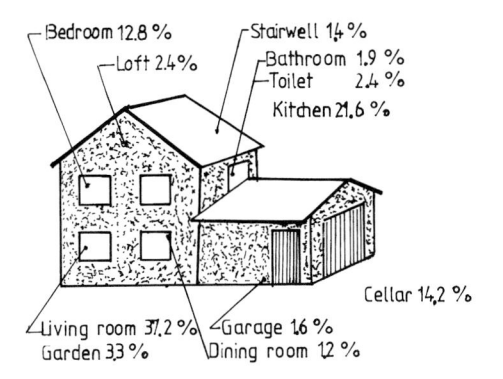

Figure 26a Fire statistics USA – Fire risk in residential homes

- Bedroom 12.8 %
- Loft 2.4 %
- Stairwell 1.4 %
- Bathroom 1.9 %
- Toilet 2.4 %
- Kitchen 21.6 %
- Cellar 14.2 %
- Living room 37.2 %
- Garden 3.3 %
- Garage 1.6 %
- Dining room 1.2 %

3 Research of Causes of Fires

UK - Fire statistics 1984

Fire and materials first ignited in dwellings		Fatalities
1. food	32%	3%
(1a. fat	26%)	(3%)
2. upholstery	22%	57%
3. electr. insul.	11%	5%
4. paper, carton	10%	6%
5. others	25%	29%
(unknown	15%)	(19%)
(structural	5%)	(1%)
(gases	3%)	(6%)
(liquids	2%)	(3%)

Figure 26b UK fire statistics – 1984

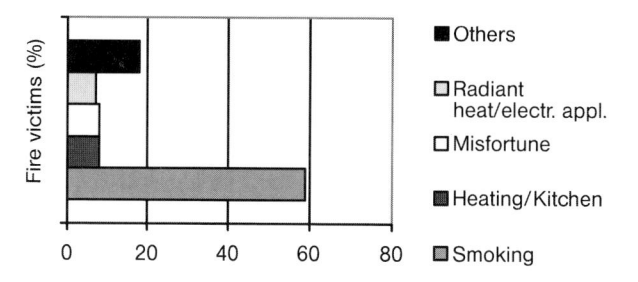

Figure 26c Scottish Fire deaths –1998/99

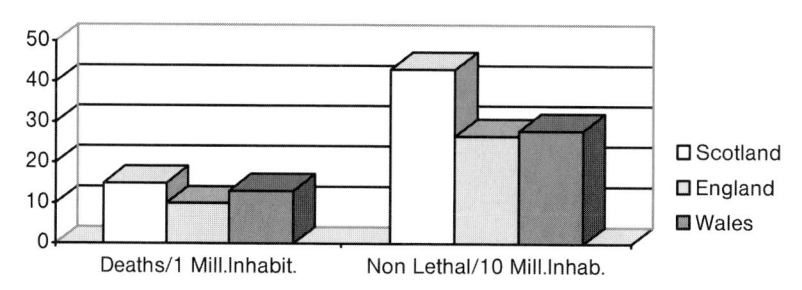

Figure 26d Fire victims in different fire scenarios

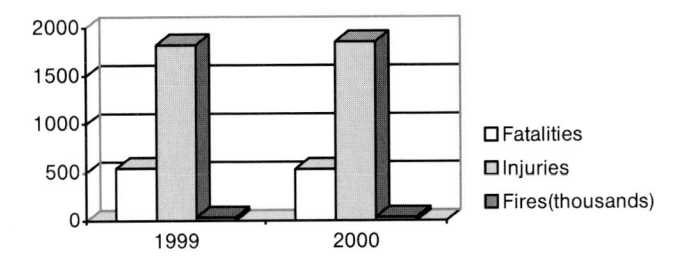

Figure 26 e Korean fire-statistics data

The data set of the excellent English fire statistics illustrate, as shown by Fig. 27a and b, that arson as a fire cause is becoming more important as far as the fire prevention is concerned. The graphical representations of Fig. 27a demonstrates the continuous increase of arson as the cause of fires for the period of 1961 to 1997. The data set of Fig. 27b reveal that the proportion of arson-initiated large fires has increased up to 50%. The data of 1991 and 1992 demonstrate (Fig. 27c) that the number of fires initiated by arson exceed those of careless ignition [134–136]. The data records of Fig. 27d provide an overview of the importance of arson for the risk of ignition worldwide in dwelling fires [136]. Arson as the cause of fires has to be observed very carefully, not least with respect to fatal fires like the fire at the Scandinavian Star on 06.04.1990 with 180 fatalities, as well as dance hall fires, because these fires, which only happen from time to time, are always linked with a high number of fatalities. As shown by Fig. 27d, the majority of the fires start during the night where naturally the probability of fire control is very small. Chandler reports data for 1992 that the majority of residential fires, which are already responsible for 70% of the fires victims, are caused by cooking and smokers. People cooking cause about 45% of these fires, but they are only responsible for approx. 10% of the fatalities. Contrary to this, smokers cause 14% of these fires, but these are responsible for approx. 40% of the fire victims (see also Figs. 27a and b). The reason for this differentiation is without doubt the usually early detection of the fire [137, 138].

The data sets of the Home Office (see Figs. 28a and b) underline the dominant role of the ignition source "cooking people" and "smokers". The fatalities will be significantly dominated by those fires initiated by smokers. A confirmation of these data

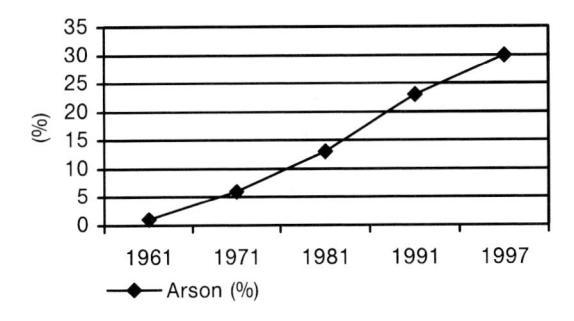

Figure 27 a UK fire statistics – Data of 1961–1997

3 Research of Causes of Fires

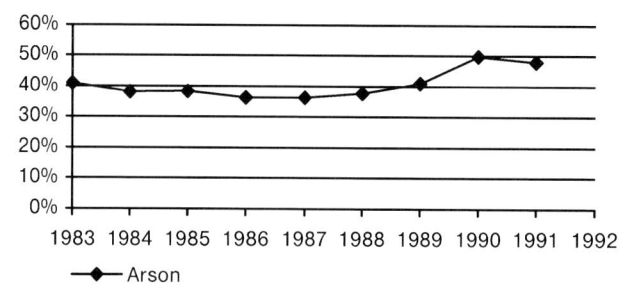

Figure 27 b UK fire statistics – Arson as ignition source 1983–1991

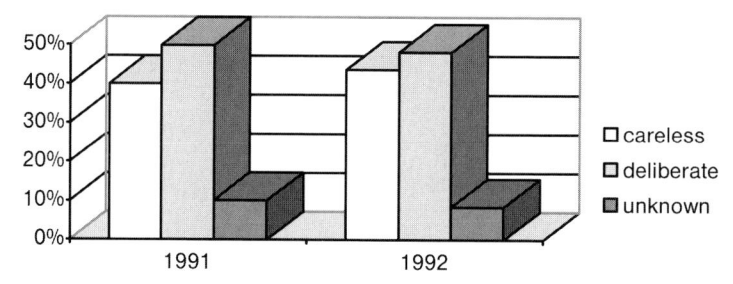

Figure 27 c Arson – Large fires 1991/1992

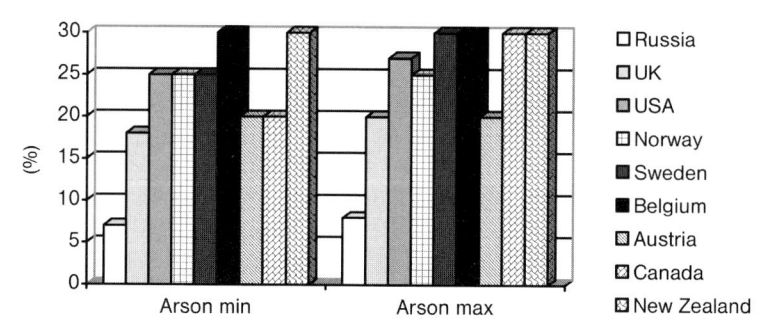

Figure 27 d Arson in residential fires

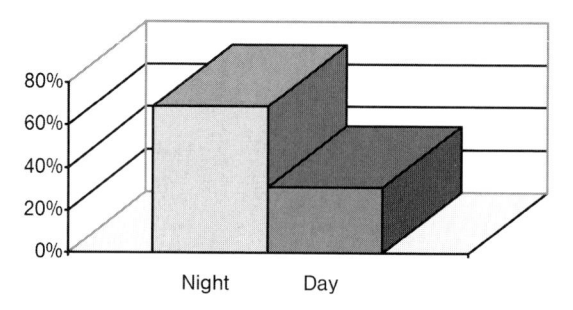

Figure 27 e Fire risk in relation to the time of the day

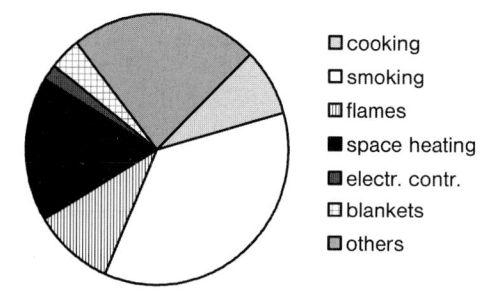

Figure 28a UK fire statistics 1984 – Fires as cause of death

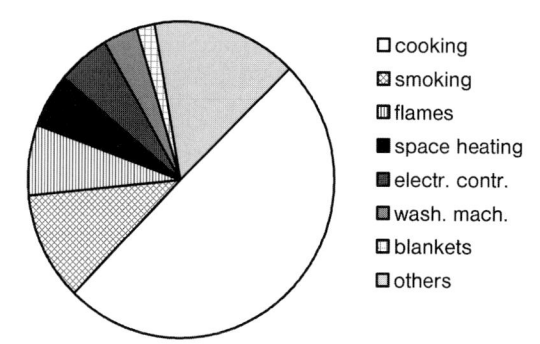

Figure 28b UK fire statistics 1984 – Ignition sources and fire victims

are the records of Figs. 29a and b on behalf of the NFPA year book, data about ignition sources and the first-ignited items. NFPA data provides general American records. These records describe information concerning ignition sources and items first ignited. These conclusions have been confirmed by the inquiries of the NFPA. The detailed fire-statistics data of the NFPA (see Figs. 29a and b) explain that smoker materials play an important role as the cause of fires. This statement is backed by the data set of average values of five years. The data set of Figs. 29a and b demonstrate this fact effectively. These data are average values of the years 1983 to 1987 [139–142].

These fires are 80% residential fires and are caused in the majority by smokers.

In [143] the experiences of Danish fire statistics are reported (see Fig. 30a). The data set with respect to 1980 to 1988 can be interpreted in such a way that smoker's materials are the dominant ignition source. Finnish records are given by Fig. 30b.

Christian [144] interprets fire-statistics data in relation to smoke assessment and fire prevention activity. Lundberg and Petersen [145] reports Norwegian experiences. The data set of the years 1970 to 1979 underlines the significant influence of the climate. 60% of the fires are caused during the "heating period" and about 30% were initiated in the bedroom. These figures are based on investigation reports of the police. According to the interpretation of the authors, 50% of the fires are caused by

3 Research of Causes of Fires

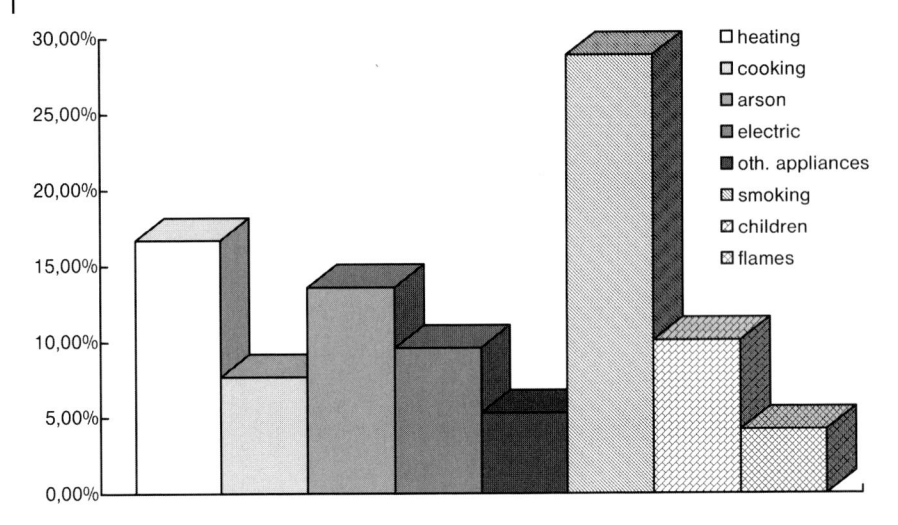

Figure 29a Fire statistics USA – Furniture Fires

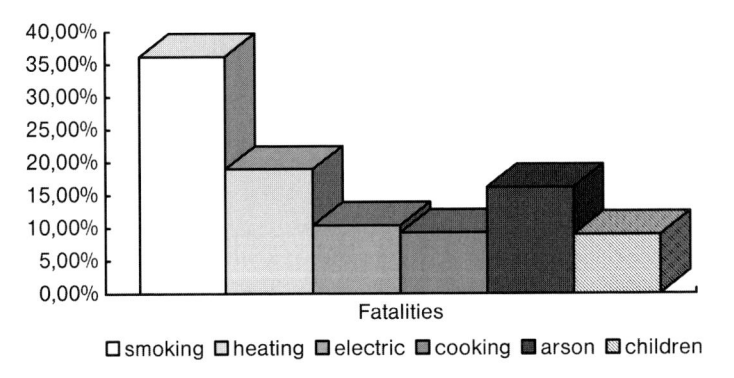

Figure 29b Fire statistics USA – Influence of ignition sources 1983–1987

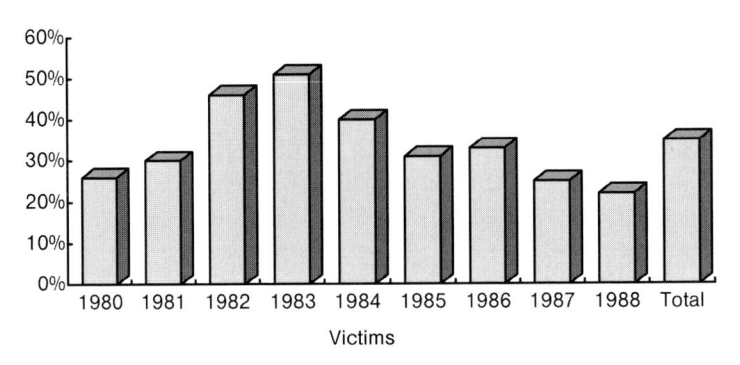

Figure 30a Fire statistics Denmark – Smokers and fire victims

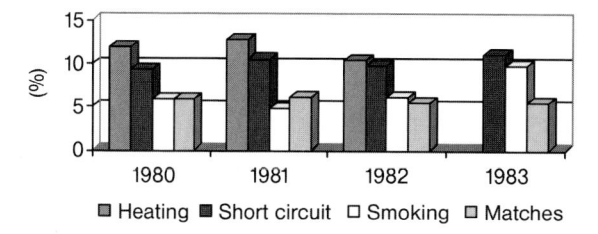

Figure 30b Fire-statistics data Finland (2562)

carelessness. 70% of the fires happen in dwellings with 1 or 2 storeys. These data confirm that in the majority of the fire incidents (87%) one person only was effected. Figures 31a and b underline further national experiences. About 42% of the investigated fires started in the period between midnight and 7 a.m. The graphical representations of Figs. 31c to e provide an overview of the number of residential fires and fire victims in Norway during 1990 to 1999. The number of fires stayed constant while the number of fire victims decreased slightly. The graphical representations of Fig. 31c illustrates the number of multifamily residential fires in Norway [146].

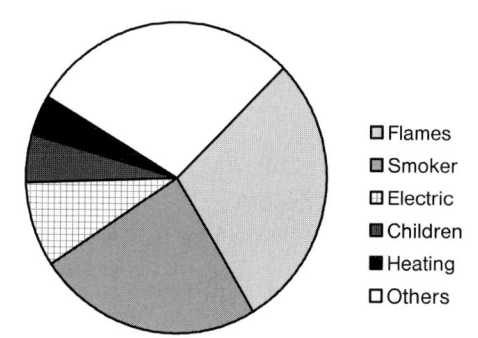

Figure 31a Fire statistics Norway – Ignition sources

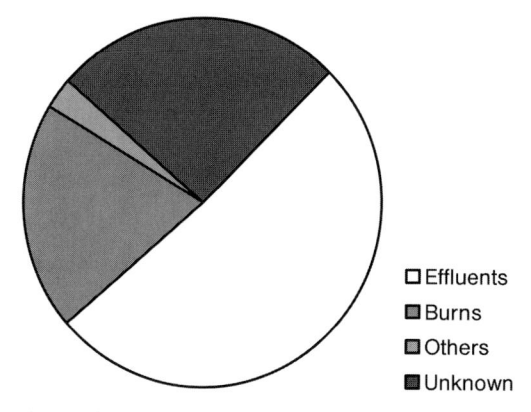

Figure 31b Fire statistics Norway – Fires in residential homes

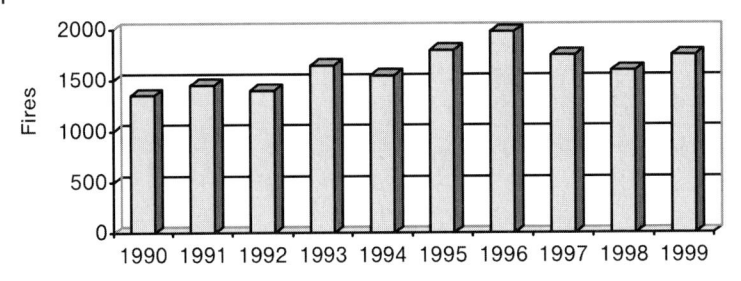

Figure 31 c Fire statistics in Norway – Fires in residential homes

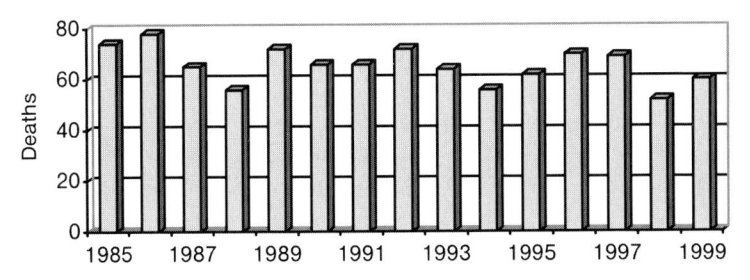

Figure 31 d Fire statistics in Norway – Victims in residential homes

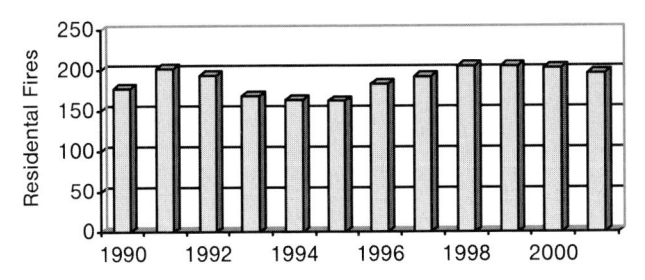

Figure 31 e Fire statistics in Norway – Fires in multifamily houses

3.3
Experience with Fire Spread – Fire-Detection Units and Sprinklers

A detailed study of the fire-statistics data demonstrates very clearly that in many cases interpretations are not meeting the real-fire risks. The data set of the years 1988 to 1993 confirm former findings (see Figs. 32a and b) [147].

A further explanation for the divergencies of fire spread to neighboring areas between different countries is due to important differences in the style of architecture. For lightweight or wooden construction building a higher fire-load contribution has to be expected. The fire-statistics data concerning the ignition sources and fires with fatalities during the 1970s and 1980s have to be interpreted as follows:

3.3 Experience with Fire Spread – Fire-Detection Units and Sprinklers | 47

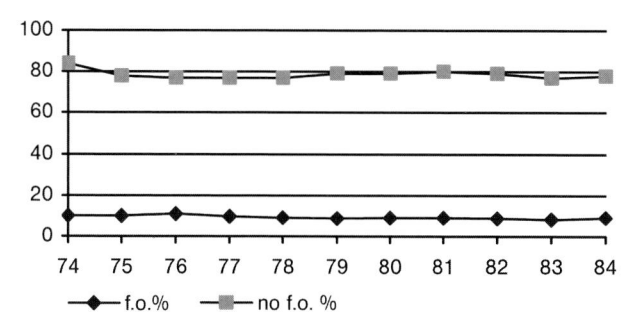

Figure 32a Fire statistics UK – Spreading of fires and fire victims

Figure 32b Fire statistics UK – Spreading of fires

For the UK a significant reduction of fires with fatalities has to be noted.

The large influence of the figures during 1985/86 may be caused by a modification of the data recording, because the retrograde trend started for the second half of the 1980s as well. The graphs of Figs. 32a and b demonstrate that in the UK only 10% of the fires spread to neighboring areas. On the contrary, in the USA most fires spread to neighboring rooms. This contribution, repeatedly mentioned [148], does not correspond, e.g. with the outcome of the Maryland investigations. These results can be interpreted according to Figs. 33a and b to show that the location of fatalities corresponds with the room of origin. The findings of the Maryland study correspond to the results of the Harland investigation [149–151]. The autopsy results of Harland (see Fig. 33c) clearly demonstrate that the majority of fatalities, in the present case (64%), have to be correlated to the room of origin.

Recent findings underline, as illustrated by Fig. 33d, that the overwhelming number of victims are located near the place of origin. Anderson et al. reported in [152] that the majority of fire victims are killed in the room of origin. A minority of the victims have been killed, as shown by Fig. 33e, on the same floor and above (see also Fig. 33f).

According to the English fire statistics (see Figs. 32a and b), 10% of dwelling fires only spread to the neighboring rooms, contrary to this only 1% spread in the hospital

3 Research of Causes of Fires

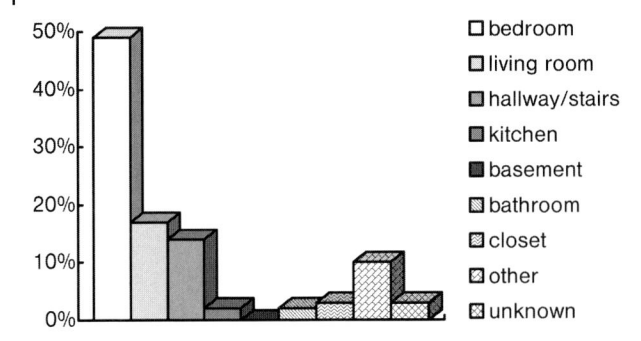

Figure 33a Fire statistics USA – Location of victims

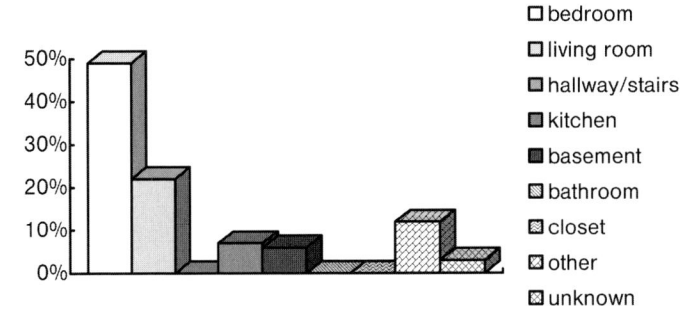

Figure 33b Fire statistics USA – Cause of fires

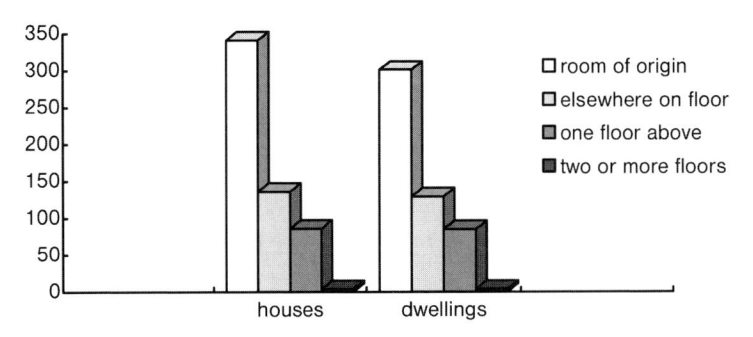

Figure 33c Harland study – Location of fire victims

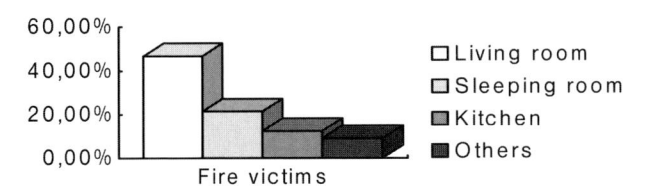

Figure 33d Fire statistics UK – Location of fire victims at residential fires

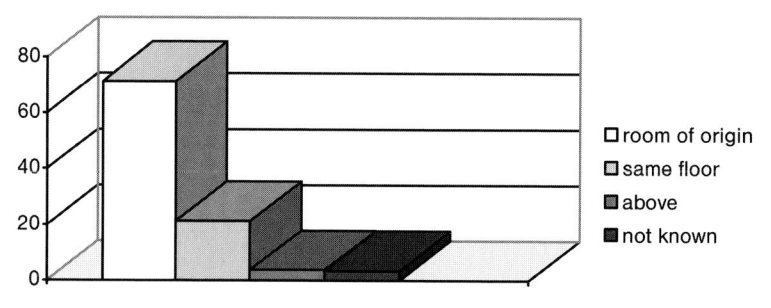

Figure 33e Location of victims

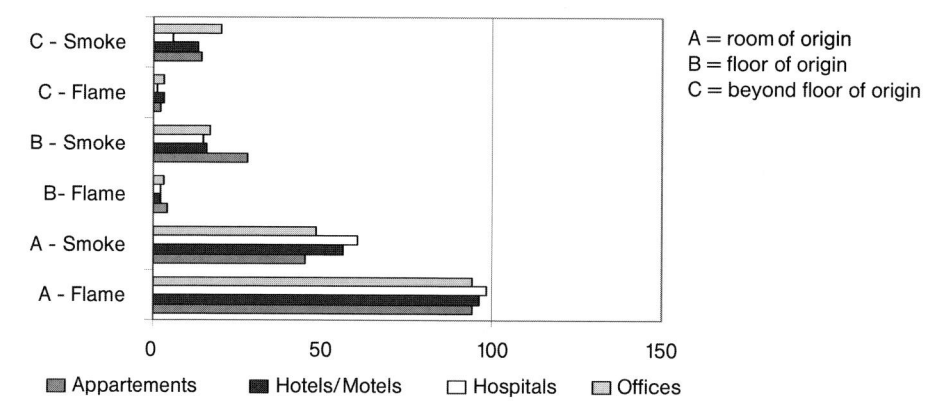

Figure 33f Fire statistics USA – Average extent of the smoke and fire damage – High-rise buildings

field. In 70% of these fires, the first-ignited items will only contribute to the fire. Recent findings of the British fire statistics for the period of time 1981 to 1991 are shown by Fig. 34a. The domination of smokers materials as ignition sources for fires with fatalities was confirmed. Comparing the results of hospital fires and large fires in general the outcome (Figs. 34b and c) can be interpreted in such a way that because of the controlling systems in hospitals, the fire will be limited to the item first ignited or the room of origin. Spreading to adjacent areas occurs in only ca. 2% of cases [153–156].

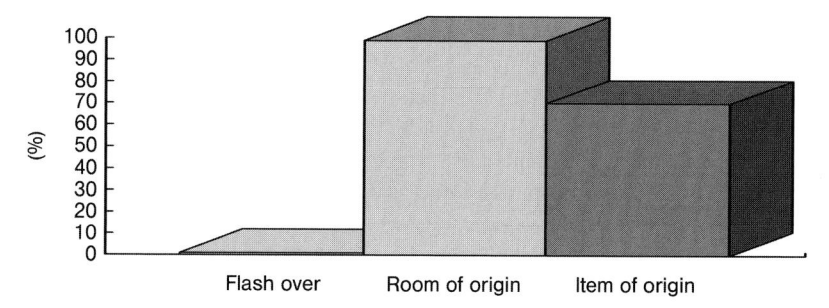

Figure 34a Fire statistics UK – Hospital fires

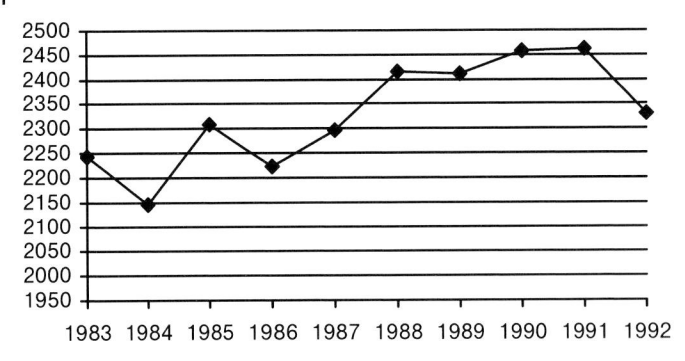

Figure 34b Fire statistics UK – Frequency of hospital fires

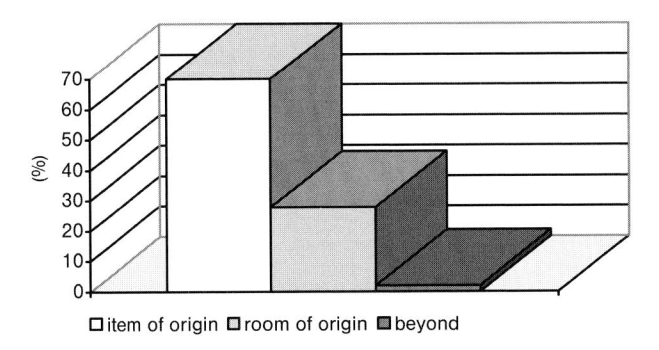

Figure 34c Temporal course of hospital fires

Based on British fire statistics, Christian [157] reports on the available safe egress time (ASET). The data set of Fig. 35a illustrate the differences between the key parameter times of fire detection. For the majority of fires this time will be larger than 30 min. The present knowledge of fire-statistics data has to be strictly interpreted and regional findings concerning fire detection can not be generalized. The findings of Christian [158] that dwelling fires as a rule will be detected after 30 min, have been confirmed. The use of smoke detectors has ensured (Table 3) a significant improvement in the detection of fires. Many more fires will be detected with a shorter time, e.g. within 5 min. The number of victims as well as the amount of the fire load consumed will be significantly reduced.

The prerequisite of fast detection is the multiplied use of smoke detectors [159–162]. The latest findings in the Strathclyde region for July 1996 to June 1998 show this trend according to Doherty [163] (see Figs. 35b, c and d).

Clark and Hirschler [164] demonstrated on behalf of the Canadian fire statistics that the smoke components present only one parameter of the various hazard inputs in the case of a fire. Lowry et al. [165] demonstrate on behalf of fire-statistics data that the smoke components HCN and HCl are of minor importance in dwelling fires. The dominant role of the smoke component CO has been confirmed. The UK Fire statistics data of 1983 to 1993 [166], see Fig. 36a, can be interpreted as follows:

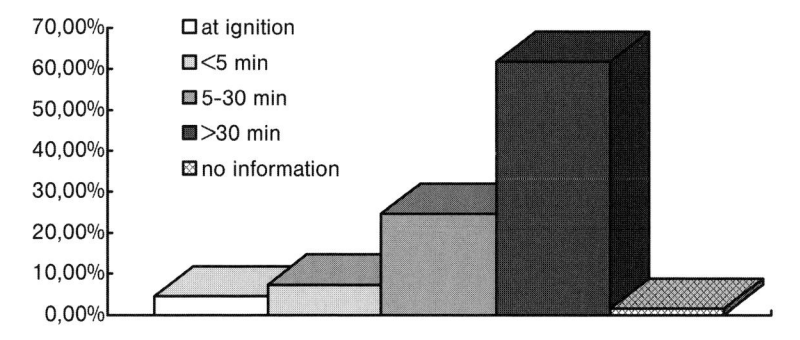

Figure 35 a Influence on the time of detection by smoke detectors

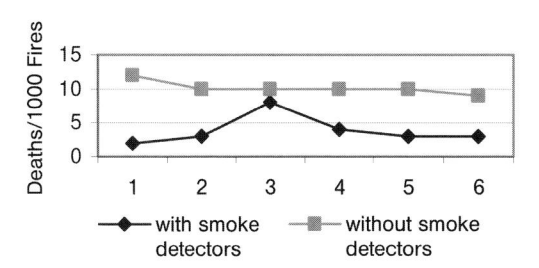

Figure 35 b Influence of smoke detectors on the number of victims

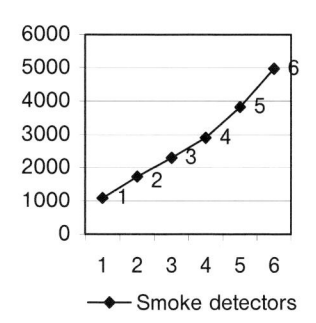

Figure 35 c Use of smoke detectors in residential homes

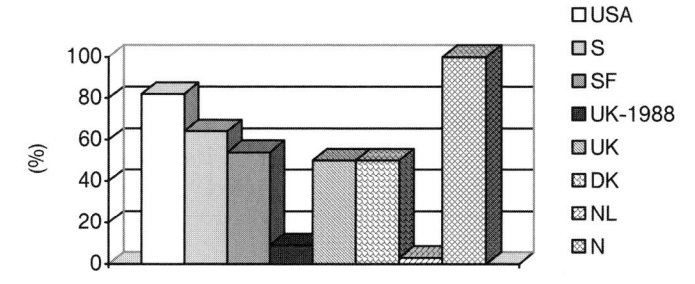

Figure 35 d Use of smoke detectors worldwide

Table 3 Influence of Smoke Detectors to Time Space

Time to detection	Percentage
At once till 5 min	64
5–30 min	33
30 min–2 h	3
> 2 h	0

- The overwhelming majority of fire victims is one person only.
- A general trend was established for fire with minor damage as well as by fires with multiple fatalities.

Comments concerning extremely high fatalities should be given only with respect to the general fire disaster scenario.

The data set of Fig. 36b explains the number of fire victims related to their location. The USA fire-statistics data according to Fig. 36c demonstrate that fire trends are not marked by a significant reduction of fatalities. The data set of the UK Fire statistics illustrate (Fig. 36d), where single or multiple-death fires occur [167].

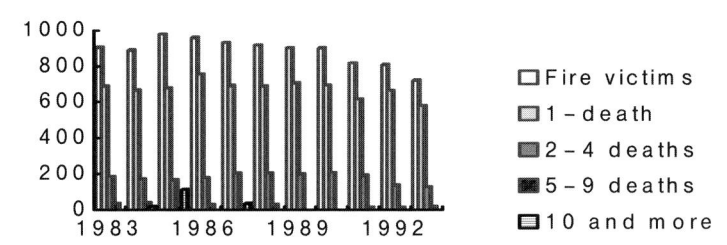

Figure 36a Fire statistics UK – Multiple death fires

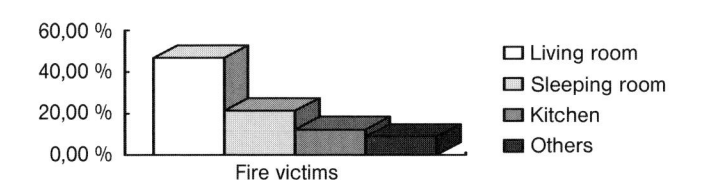

Figure 36b Fire statistics USA – Influence of ignition sources on multiple death fires

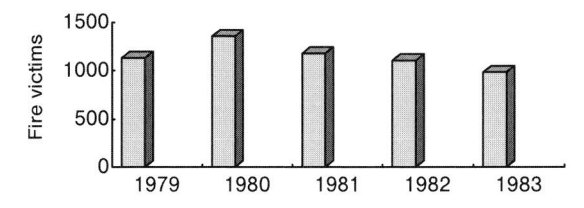

Figure 36c Fire statistics USA – Multiple death fires

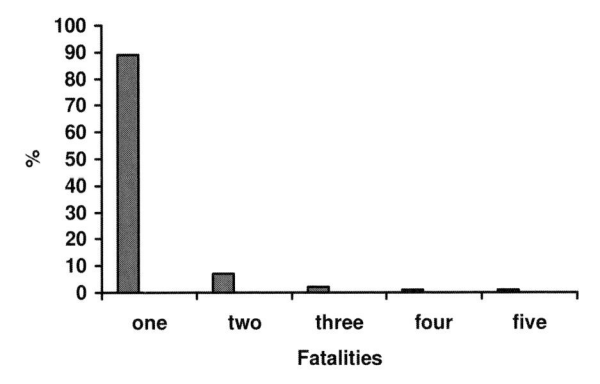

3.3 Experience with Fire Spread – Fire-Detection Units and Sprinklers | 53

Figure 36d Fire statistics UK – Multiple death fires

ignition source	Casualties	
	multiple death fire	single death fire
smoker material	113 / 250 (45%)	157 / 463 (36%)
trapped by smoke	95 / 179 (53%)	115 / 300 (38%)
overcame by smoke	134 / 173 (77%)	224 / 300 (74%)

Figure 36e Fire statistics UK 1984 – Multiple death fires

The graphs of Figs. 37a to f show a survey of the importance of the various ignition sources in the transportation field, which might lead to fire disasters [168, 169].

These figures explain the fire-statistics data of the transportation field and electro-mechanical appliances. The graphical representations of Figs. 37a to c, e.g. given by the Department of transport, local government and the region (DTLR) provide a view of the importance of road accidents involving fire victims [170, 171].

The data set of Figs. 37a and b provide details about ignition sources of car fires. The possible risk of fires in fast people movers are explained by Fig. 37c. Figures 37d and e demonstrate that wall fires and the failing of electrical wiring are the most important starting points of bus fires. The data set of Fig. 37f are to be interpreted that engine troubles are the most important ignition source in the maritime field. The compartmentation of the machinery room by fire-resistant walls is one of the most important actions in preventive fire protection. Figure 37f provides an overview of the importance of different ignition sources in the UK in 1991. Reference [172] provide information on the fire-statistics data of 1996. Failing electrical blankets caused the death of 19 people in 1996. 90% of the effected people were older than 65

3 Research of Causes of Fires

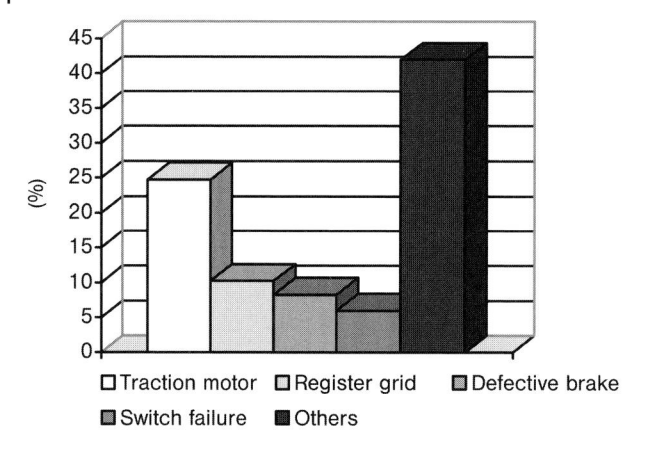

Figure 37a Ignition risks in the transportation field

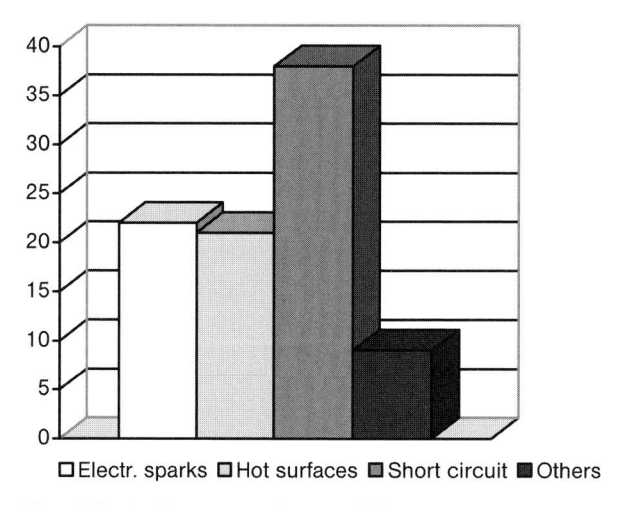

Figure 37b Ignition sources of motor vehicles

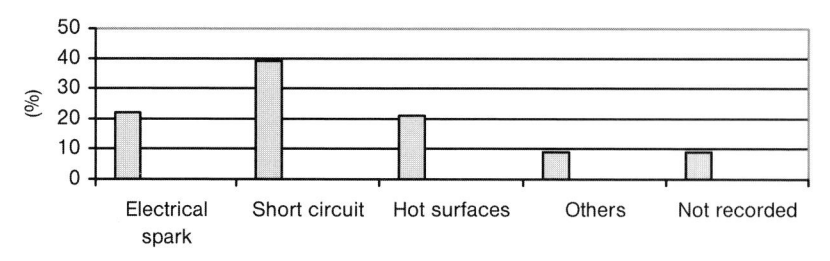

Figure 37c Causes of car fires

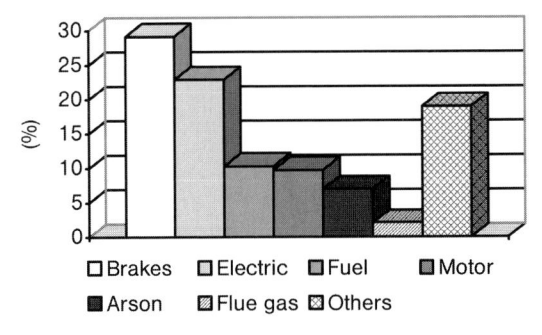

Figure 37d Ignition sources of bus fires

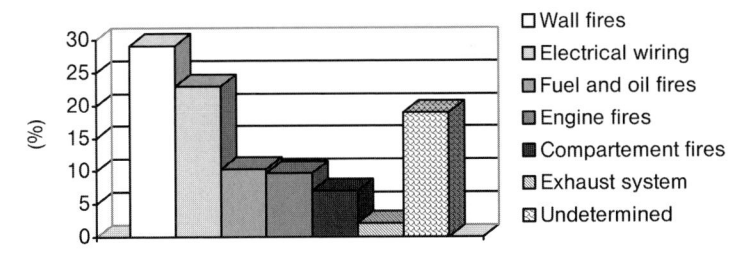

Figure 37e Bus fire scenario

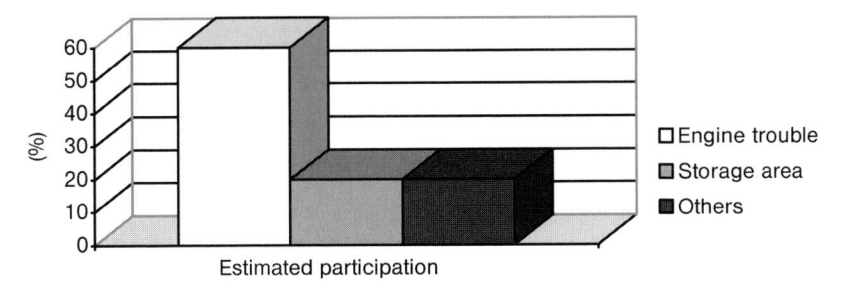

Figure 37f Ignition risks in the naval field

years. A detailed analysis of the results has demonstrated the overwhelming number of the ca. 5000 blankets are older than 10 years. The control sample of 931 blankets has shown that one in five blankets has faulty plugs and one of ten faulty fuses. Short-circuits, electric sparks and hot surfaces are obviously the most important ignition sources.

3.4
The Time of the Fire Initiation as a Classification Criterion

Experience has shown that a high percentage of fires develop in the evening or during the night. The graphs of Fig. 38 imply concern about the increased risk of smoke

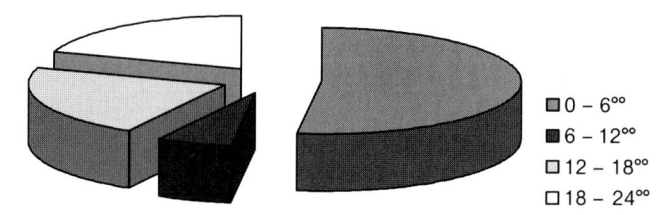

Figure 38 Day and night fires

- 0 – 6°°
- 6 – 12°°
- 12 – 18°°
- 18 – 24°°

inhalation effects in the evening and during the night [173]. The statistical findings have partly shown that environmental conditions like climate, town size and social conditions also may considerably influence the extent of damage. The way of life, e.g. smoking and drinking will be responsible for the high number of victims as well as the late detection of the fire incident. This interpretation will be backed by experience that indicates that the time of the year may play an important role. During the wintertime, when life takes place increasingly in the living room, there is a much higher probability that a fire occur. The restriction of the fire to the room of origin does not solve the second basic requirement "the spread of fire and smoke has to be prevented" even within the room. In general, this will occur but obviously fire victims have to be lamented in far away areas.

The mass of smoke produced during the fully developed fire and in the case when no flashover occurs, can create a smoke danger remote from the fire. If local flashover conditions produce a smoke layer of low oxygen content but that is then sufficient dilution in the peripheral area it may not put life at risk. The experience in the USA has been interpreted to show smoke-detection systems are responsible for this significant reduction and not specific material-related requirements. The fundamental findings of the fire statistics indicate that measures that enable an early detection and therefore rescue and extinguishing actions have resulted worldwide in lower death rates.

4
Preventive Fire Protection –
National Requirement and Classification Systems

"Charitable is the power of the fire, if it is controlled by" is the slogan of the poem "The bell" from Schiller. The fire that runs out of control will become a danger to all living persons. Since the beginning mankind has been trying to get effective protective measures. The aim of fire prevention has to be to stop fire incidents. An increased level of security has to be reached, e.g. by tighter legislation. According to the basic requirements of fire prevention the development of a fire has to be prevented. This can be achieved by different actions of the fire prevention, e.g.:

- The fire development can be effectively prevented by eliminating the relevant ignition sources, e.g. by banning smoking and by the observation of hot work.
- By effective material selection with resistance against relevant ignition sources under enduse conditions, e.g. by the use of safety wastepaper baskets constructed with incombustible materials and
- By using internal measures, e.g. smoke detectors, the fire can be detected in time and stationary fire extinguisher installations can be effective in fire fighting.

As far as the second basic requirement is concerned, the speed of fire and smoke has to be slowed and requires a series of actions. Structural and internal actions have to be used to guarantee that the fire will not spread uncontrolled beyond the place of origin. Spatial cutoffs with a sufficient fire resistance will ensure that the fire, smoke protection included, does not spread to the neighboring areas. As far as smoke distribution is concerned, experience has shown that it will be distributed by channels and openings to neighboring areas.

The fire risk of furniture and furnishings will be evaluated in relation to the environment. Differentiation is needed between the different risk situations of private and public sectors. In the relevant building codes a clear separation can be recognized. The extent of the possible dangers will be determined by the risk of ignition. Before any requirements are fixed and risk related, real defined classification criteria are needed and the following questions have to be clarified:

- What is the legislation?
- What kind of protection is sufficient?
- Does the intended measure lead to other uncertainties and dangers?

Polyurethane and Fire. Franz H. Prager and Helmut Rosteck
Copyright © 2006 WILEY-VCH Verlag GmbH & Co. KGaA, Weinheim
ISBN: 3-527-30805-9

- Is there a meaningful control possible?
- What is the necessary expenditure?

Before any new requirements are fixed, every current experience and relevant scientific knowledge concerning the connection of the different parameters of the fire performance has to be considered. Pure experience and test results without any scientific evidence of relevance have to be carefully considered with relation to the finalisation of requirements. Risk-related requirements may also consider human behavior in the case of a fire. The present trend in national standardisation with respect to so-called guidance papers must guarantee the relevance of test methods to be standardized and have a positive effect on requirements of the future. If, at least for safety-related fields of application, such a level of regulation can be achieved in the future, it would ensure that pseudoscientific-related standards, which have no relationship to reality, will have no chance to be accepted by responsible licensing authorities. It has to be constantly controlled, if the desired aim is to be achieved. This has to be unlimited and valid for planning, feasibility studies as well as the evaluation and the interpretation of the results. The guidelines for safety-related standardisation have to be absolutely accepted. The statement, given in [174], that the expert's opinion will be produced more and more by nonexperts, has to be recognized for the field of standardisation as well. The important parameters for the aim will, e.g. be changed with respect to the reproducibility. Risk-related requirements are obliged to consider the human behavior in the case of a fire. The human behavior, which varies, according to Fig. 39a, cannot be foreseen. Experience of fire disasters [106, 175] has demonstrated that the danger in the developing stage of a fire will be very often underestimated and therefore important parts of the egress time will be wasted. In particular, e.g. the desired increased level of safety for children in the case of a fire, which was the most important reason for the stringent UK furniture regulation, one had to be aware that measures to increase the period until rapid burning occurs, will be, in that case, as a rule useless. Children will usually hide and no advanced delayed ignition will be useful in this case. The evaluation of the reported findings [176, 177] show that the human behavior in a fire will be influenced by a series of factors, for example, curiosity and fear. The human behavior in escape and the rescue problem has been studied in various scientific investigations. As a rule, children do not flee. They believe they can escape by hiding. This means that they will neither try to escape earlier nor react differently to an earlier detection of the fire.

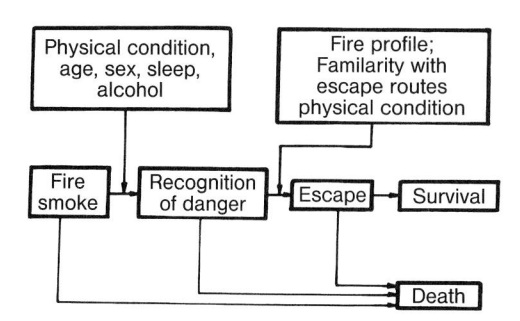

Figure 39a Human behavior in the case of a fire

3.4 The Time of the Fire Initiation as a Classification Criterion | 59

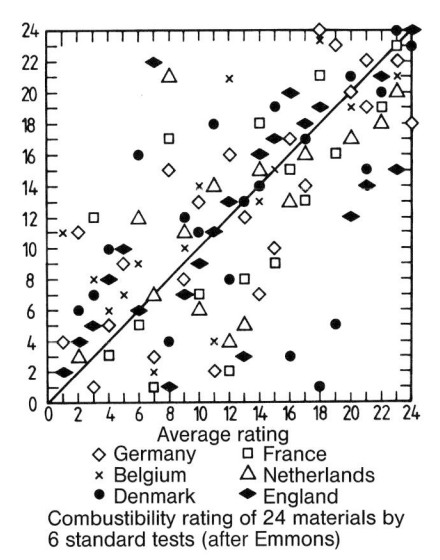

Figure 39b Comparison of different national classification systems.

Combustibility rating of 24 materials by 6 standard tests (after Emmons)

◇ Germany □ France
× Belgium △ Netherlands
● Denmark ◆ England

Factors essential for a positive human behavior in the case of a fire are therefore:

- Excellent knowledge of the surroundings. The familiarity with all possible escape routes increase significantly the chances of escape.
- With the aid of a guide who is familiar with the environment, the fleeing people will be about 1.5 times faster than inexperienced people [178]
- Excellent visibility [80].
- Good physical conditions. Young, trained humans need a significantly shorter time than elderly and old people, which might need the help of other persons. According to [176], men are more willing to go through black smoke than women.
- Avoid panic. Panic reactions are to be expected, when the hope of rescue for the endangered persons is decreasing [177]. Warnings in due time, will avoid the direct threat of panic.

The planned EC furniture directive, initiated by the UK, will not only set the risk of ignition in the main operational areas like dwellings, public buildings and high-safety areas like prisons but will guarantee that enough egress time is available for rescue activities in the case of a fire. To justify the first basic requirement, every possible ignition source has to be taken into account, which might include careless handling or arson as well as the failure in electrical parts of furniture. The typical ignition sources in the electrical field, which might become effective in the case of failure, like glowing wires, glowing contacts or self-ignition of the decomposition products and short-circuits have to be taken into consideration. In the case of a fire initiated by failure of an electrical component, the fire will not be limited to the ignited part, which will then become a secondary ignition source for the furniture item. Risk- and enduse-related actions are only possible on the basis of the knowledge of detailed fire statistics. It is desirable that regulations are assessed only after a detailed

Table 4 Robust Solution Philosophy

Large room
Small room
Walls and ceilings
Floorings
Facades
Roofs
Ducts and installations

---

study of fire-statistics data, to guarantee that only correct conclusions will be the basis of regulations. Detailed fire-statistics data are established in the USA and the UK. There is no doubt that upholstered furniture is a significant part of the first-ignited items. In general, the government takes care of adapting the requirement to the enduse conditions. At the Luxemburg Conference concerning the proposed Furniture Directive, the members of the national testing stations EGOLF required the enduse-related testing of furniture to be analogous to the building section in the sense of the robust solution philosophy. According to Table 4, a risk-related differentiation of the most important fire scenarios was required by the experts selected by the commission [179].

The robust solution was the aim of the European harmonisation of fire-reaction testing of building products. To solve the harmonisation problem, a group of experts was selected. It was agreed to lay down the key data in relation to the fire scenarios defined in Table 4. As far as furniture and furnishings are concerned, a similar differentiation of the scenarios, e.g. dwellings, public places, conference rooms and high-risk areas like prisons might be necessary. The basic requirements according to Table 4 had to be brought in line with the use of a building. The large divergencies of the regional requirements illustrated the extent to which these requirements are influenced by actual fire disasters. Like any other safety-related actions, most fire-performance requirements are based on the negative experiences of the community. The town fires of the middle ages were responsible for the development of the different national tests and classification systems of fire prevention. By comparing the different classification systems within the standardizing research work of TC 92, Fire Performance of Building Products, Emmons [180] documented the divergencies of these different methods, see Fig. 39b.

4.1
Building Section

A detailed description of the necessary different fire-safety problem as a function of the design and use of a building is given by the listing of Table 5 in relation to the Scottish Building Regulation. Although national test methods for the fire performance and reaction to fire testing and classification of combustible materials are different, the incombustibility furnace is an exception, while the fire-resistance requirements are, in principle, similar on a worldwide basis. The fire resistance is determined by the use of comparative, partial identical test protocols. Sections of building elements, as a rule 3000 mm × 3000 mm will be tested as a wall or ceiling element in the furnace, see Fig. 40a. The thermal stress will be imposed according to the time–temperature curve (ETK) of Fig. 40b.

Table 5 UK – Building Regulation – Differention of buildings with respect to use

I	Small redidential
II	Institutional
III	Other residential
IV	Office
V	Shop
VI	Factory
VII	Other place of assembly
VIII	Storage and general

Figure 40a Test furnace for building elements

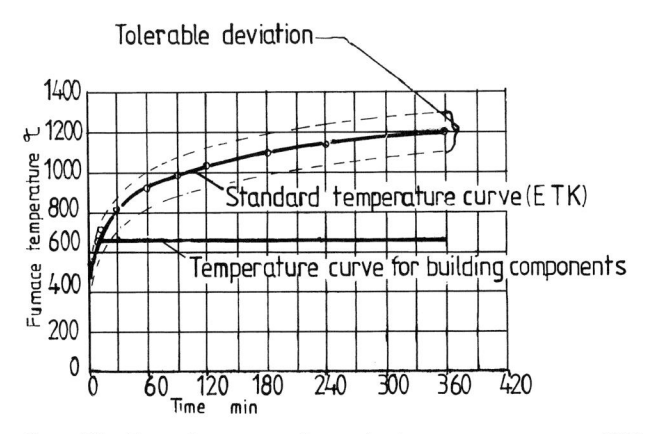

Figure 40b Thermal stress according to the time – temperature curve (ETK)

```
                      4
    ▲---▲
           ▲-▲
               2
                        3
  1
              Time  min
```

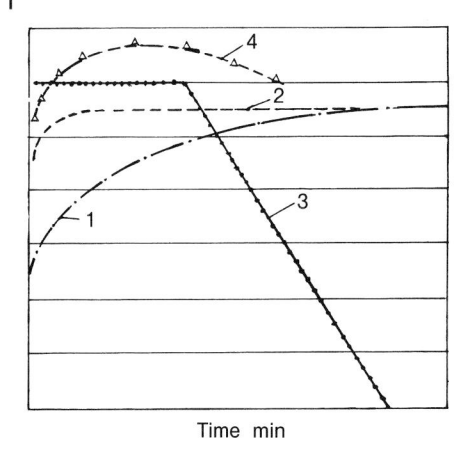

1 ETK per DIN 4102 ---
2 Hydro carbon curve (HC) ----
3 RATB curve ••••••
4 Rijkswaterstraat curve (RWS) -▲-▲-▲-

Figure 40c Variation of the thermal stress in the furnace test

The majority of the test furnaces are oil or gas fired. The possible variety of test results can reach an astonishing degree [180]. Variations of the thermal stress shown in Fig. 40c during testing have been agreed and are based on different fire scenarios. For facade elements, modified test and classification criteria are given in DIN 4102 part 1. The thermal stress of the outer facing is modified as shown in Fig. 40c. The pass/fail criteria are modified for the thermal stress of the inner part of the element. Due to research work, the fire performance of building elements can be estimated by mathematical calculation [77, 113].

Criteria determined in element testing are different time durations, where

- the integrity of the element is lost
- fire penetration happens
- an inadmissible temperature increase of the reverse surface happens and
- an ignitable mixture of smoke at the reverse surface has been proven.

The fire resistance of the tested elements will be classified according to the national regulation in relation to the determined test times, for example, 30, 60 or 90 min.

As far as the fire resistance of roof coverings is concerned, no generally accepted test procedure and classification criteria currently exist. In this connection the efforts of TC 92 (SC 1 and SC 2) did not reach an agreed solution.

4.1.1
United States of America

Preventive fire safety in the USA is defined in the different building codes, see Table 6a.

Table 6a Model Building Codes – USA.

Basic Building Code	BOCA 1981
Uniform Building Code	ICBO 1982
Standard Building Code	SBCCI 1982
Life Safety Building Code	NFPA 101

Table 6b Fire Performance Classification according to ASTM E84.

Spread	Flame spread – Index
Class A	0–25
Class B	26–75
Class C	76–200

BOCA is used in the majority of the Northeast and Midwest regions. IBCO will be used especially in the Western parts of the USA. SBCCA is the basis in the Southern regions and the Life Safety Code, developed by the NFPA (National Fire Protection Association) is used everywhere. This is part of the requirements of the HHS (US Department of Health and Human Services) and the JCHA (Joint Commission of the Accommodation of Hospitals and Homes for Elderly People).

The relevant chapters of the Building Codes are regulatory and permit the use of synthetic foam systems. Fire resistance is tested to ASTM Standard E 119, which corresponds with the IS 834. In relation to the enduse, construction elements are assembled into the wall or ceiling of the test furnace, see Fig. 40a. The wall or ceiling partitions are then subjected to the standard temperature–time exposure (ETK) for the time interval corresponding to the rating desired. Possible smoke hazard classifications to [181] will differ between high-rise buildings, covered mall buildings and atrium buildings.

The material-related classification of the building products is done in agreement with the most important Standards like ASTM – E 136, E 84 (see alsoo Table 6b). The noncombustible rating is given for the 750 °C furnace test, see Fig. 41a. Test details and classification criteria are given in the various standards like ASTM – E 136 and IS 1184 as well as DIN 4102 part 1. The cylindrical test piece, 45 mm diameter and 50 mm height and with a test volume of 80 cm^3, will be heated in the furnace for about 30 min. The tested building material will be classified as noncombustible, for example, class A 1 according to DIN 4102, if the material temperature does not increase by 50 K and the average flaming time of the decomposition products does not go beyond 20 s and the weight loss is not more then 50%.

The rating of combustible building materials is carried out to the ASTM-Standards E 84, E 162 and E 648. The resistance against ignition sources of low intensity will be tested to laboratory test methods like the Butler Chimney Test (ASTM-3014) or a cigarette test for loose filling materials [98]. The resistance against ignition sources of a higher intensity will be tested in the Steiner Tunnel UL 723, developed by the Underwriters Laboratories, which was standardized as ASTM – E 84 and NFPA 255 and is

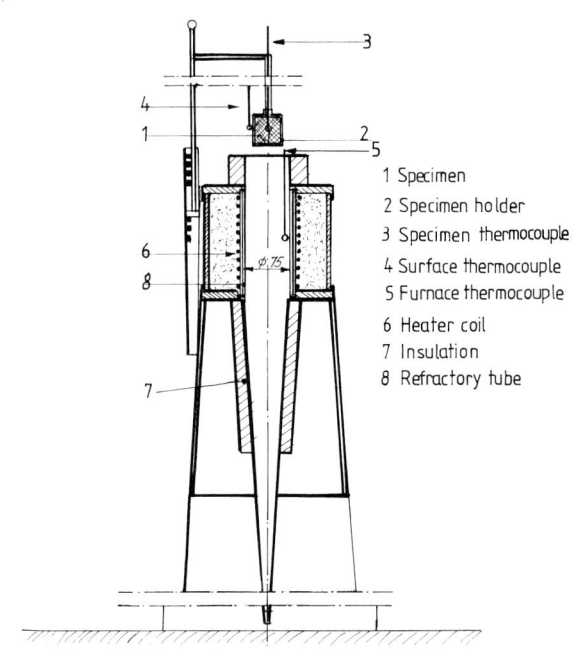

1 Specimen
2 Specimen holder
3 Specimen thermocouple
4 Surface thermocouple
5 Furnace thermocouple
6 Heater coil
7 Insulation
8 Refractory tube

Figure 41a Noncombustibility furnace according to ISO 1184

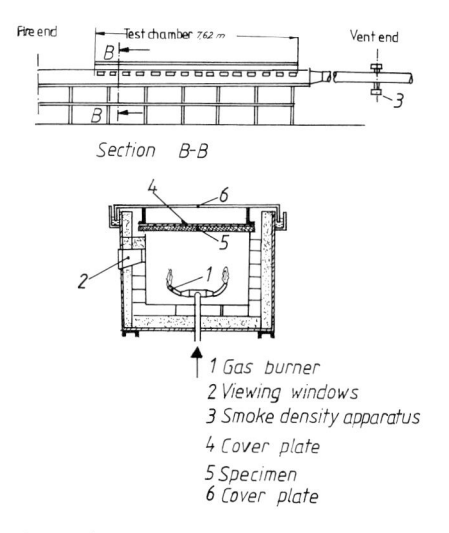

1 Gas burner
2 Viewing windows
3 Smoke density apparatus
4 Cover plate
5 Specimen
6 Cover plate

Figure 41b Steiner Tunnel Test according to ASTM-E84

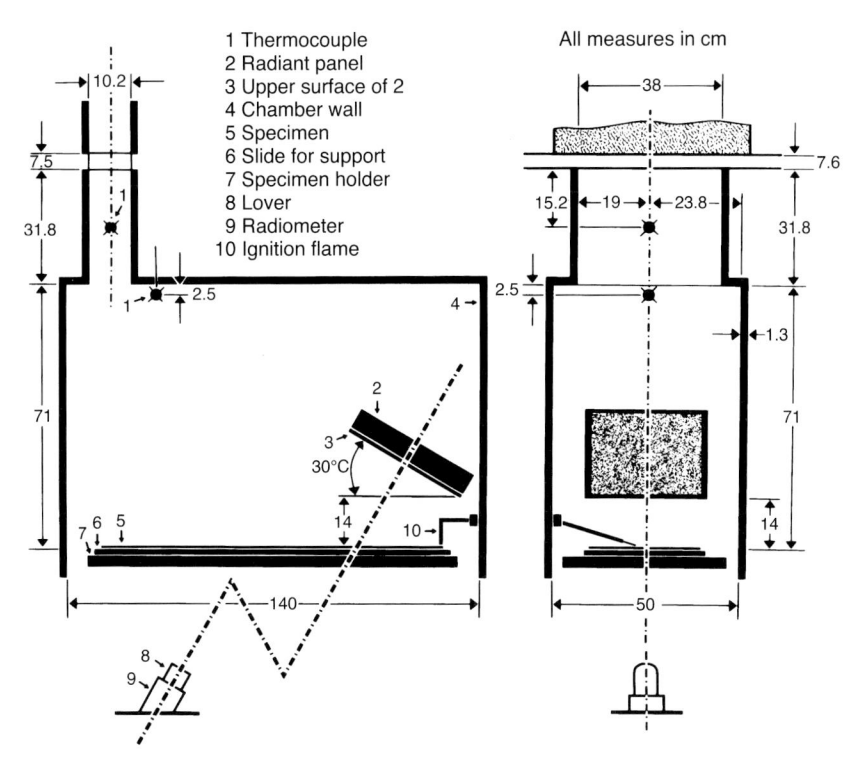

1 Thermocouple
2 Radiant panel
3 Upper surface of 2
4 Chamber wall
5 Specimen
6 Slide for support
7 Specimen holder
8 Lover
9 Radiometer
10 Ignition flame

All measures in cm

Figure 41c Critical Radiant Flux Test according to DIN 4102 et al.

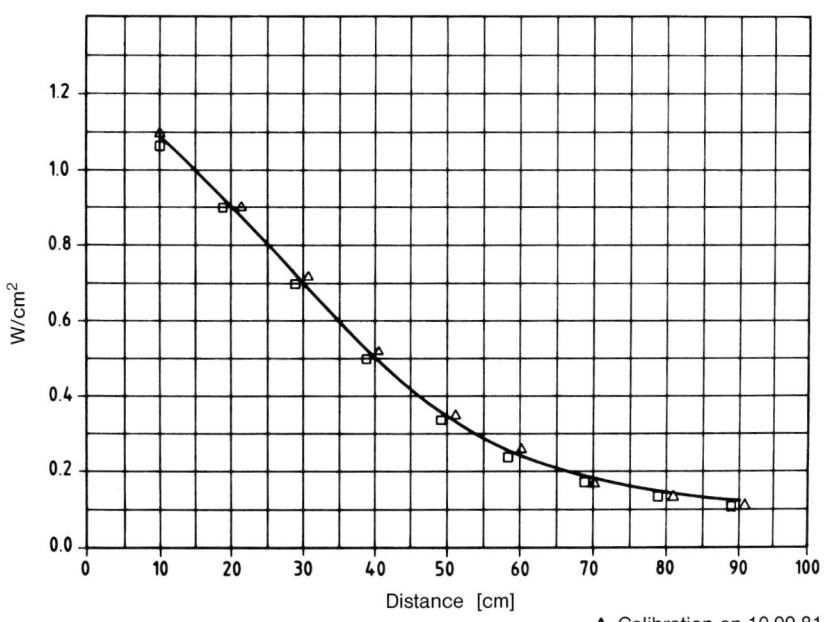

Figure 41d Thermal stress in the Critical Radiant Flux Test

△ Calibration on 10.09.81
☐ Calibration on 03.09.81

illustrated in Fig. 41b. The test pieces with outer dimensions of 0.51 m × 7.32 m × thickness of product are fitted as the ceiling lining. During the test period of 10 min, the test piece will be thermally attacked by two gas burners with 53 MJ/min. Red oak and asbestos cement boards are used for calibration. Asbestos cement was given the index number 0 and red oak the number 100. The spread of flame is observed, recorded in relation to the test time. If the resultant area below the spread–time distance curve is below <0.75 (ft min) the rating will be FSC (Flame Spread Classification) = 0.5 At.

At is the total area under the time–distance curve.

If this value increases above 97.5 ft min, the formula FSC = 4900/195–At will be valid. The building codes, e.g. the Life Safety Code and the UBC Code differentiate the combustible linings also according to Table 4.

In the USA floor coverings will be tested and classified as ceiling linings in this test. The Canadian test procedure is much more enduse related. Flooring materials will be tested there in the tunnel under flooring conditions. In the USA an alternative test method for floor coverings was standardized as ASTM – E 648 (see Fig. 41c). The test protocol has been developed at NIST, the former NBS, and is based on the test results of relevant large-scale corridor tests [182, 183]. The test piece with outer dimensions 1050 mm × 250 mm × thickness of product will be thermally stressed according to Fig. 41d. The extent of flame spread during the test period is the dominant criterion, which is the test intensity level at which the spread ceases.

The test and classification criteria of the smoke-density classification are given in ASTM – E 662 (see Fig. 42a). The validity of all the laboratory test methods has been questioned during the Federal Trade Commission (FTC) action caused by a complaint in Kansas City, which led to the Consent Order [14] that any classification has to be marked by a footnote such as "This numerical flame spread rating is not intended to reflect the hazards presented by this or any material under actual fire conditions". The consent order initiated a 5-million dollar program regarding every aspect of the fire performance of insulating materials. Because the classification terms "nonburning" and "self-extinguishing" according to ASTM-D 1692, illustrated by Fig. 42b played a major part in the FTC action, any laboratory fire-performance classification, which was not covered by risk-related large-scale testings have to be marked by such a warning label.

The SBCCA Building Code has in addition to the spread classification a limit set for the smoke release by values of 450 or 75 according to ASTM- E 84. The smoke measurement is done by using an optical measuring system with a light path of 914 ±102 mm. The absorption will be determined at the tunnel exit as a function of the testing time. The area value will be divided by the red oak result and multiplied by 100. An alternative test for the determination of the flame spread, especially in the building section, is the Radiant Panel Test according to ASTM – E 162. In this test method, illustrated by Fig. 42c, the test piece with outer dimensions of 150 mm × 450 mm × thickness is inclined, at 30° positioned to the radiator (300 cm × 460 cm, with a preset temperature of 670 °C). This gives a distance of 120 mm or 340 mm at the ends of the sample. The pilot flame with a length of 150 to 180 mm is positioned at a distance of 32 mm to the upper corner.

4.1 Building Section

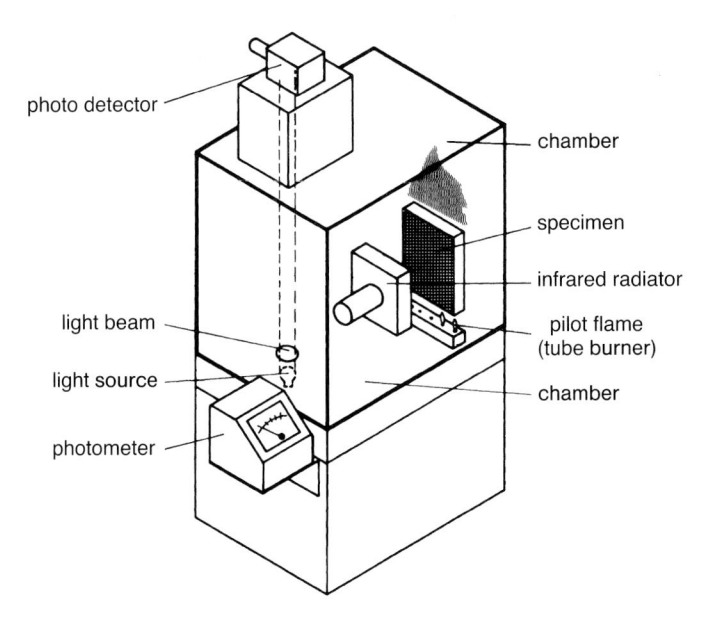

Figure 42 a Smoke-density measurement according to ASTM – E662

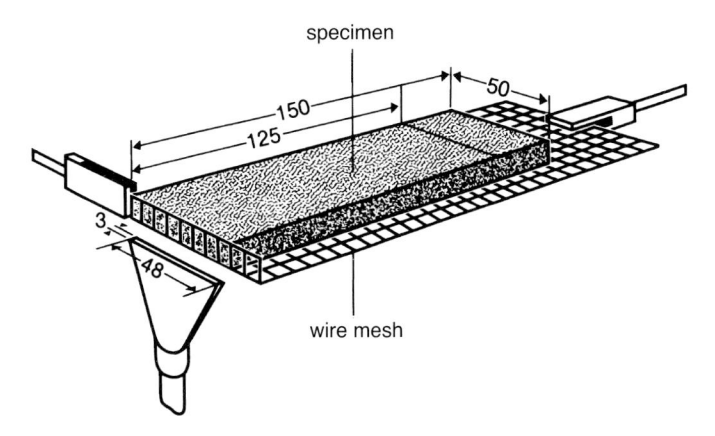

Figure 42 b Standard Method of Test for Flammability ASTM – D 1692

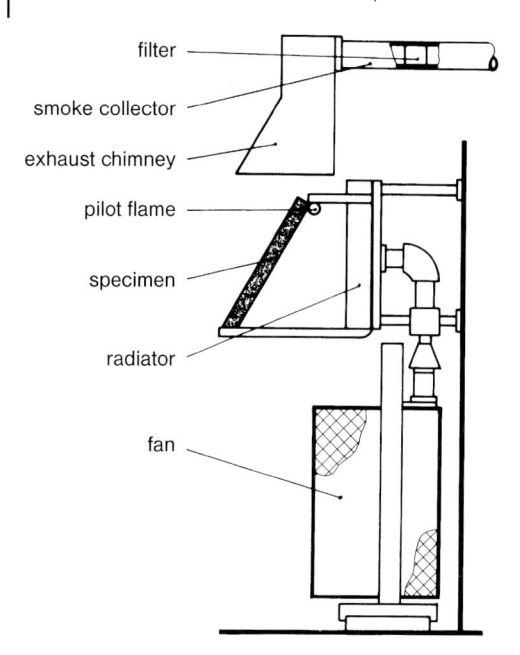

Figure 42c Standard Method of Test for Surface Flammability ASTM – E162

The rating of the contribution to the surface spread of flame has been carried out on the basis of an index

$Is = Fs \times Q$
$Q = cT/\beta$
$Fs = 1 + 1/t3 + 1/t6 - t3 + 1/t9 - t6 + 1/t12 - t9 + 1/t15 - t12$

where

$t3 - t15$ are the time space to the measuring mark of 3" to 15" have been reached.

$Q = cT/\beta$ correspond to the standard value 0 of the apparatus with an asbestos sample.

Bioassay tests have been widely used for the evaluation of the toxicity of fire effluents for material-related classifications. The Potts Pot method is based on the studies of Pott at Dow Chemical [184] and was taken over by the National Bureau of Standards (NBS) and international standardisation. Detailed advice and extensive test results are given especially by Birky et al. [185]. This method requires the decomposition of ca. 25 mg/l of a substance under defined conditions of up to 1000 °C. For the test volume of 200 l, the mass of material is about 5 g. The smoke release shall take place at 25 °C below and above the decomposition temperature. The sample weight will be measured continuously by a weighing device. The test animals, white rats, were exposed, as shown by Fig. 43a, by nose exposure. Due to the closed system, afterburn processes may happen during the recirculation of smoke. Clark and Hirschler [164] and Benjamin [186] explain the test-method-related effects for the Teflon-test problem.

4.1 Building Section | 69

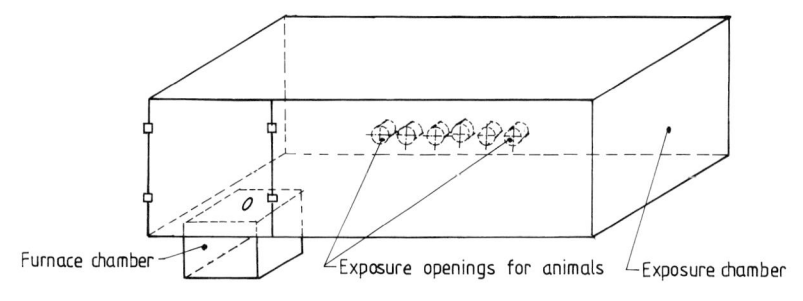

Figure 43 a Illustration of the Potts Pot Procedure 1053

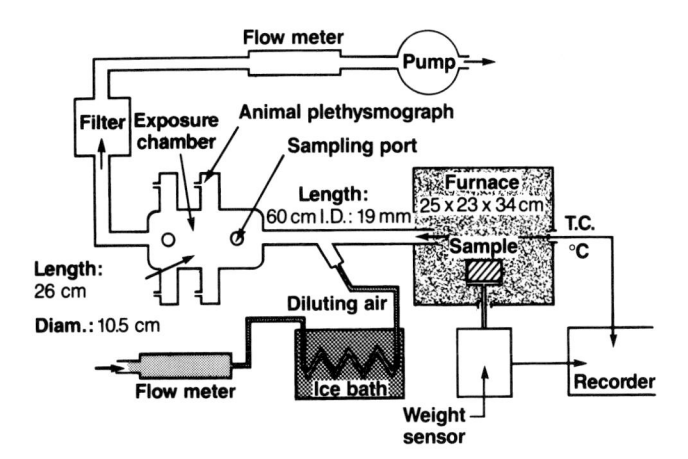

Figure 43 b U-Pitt decomposition model

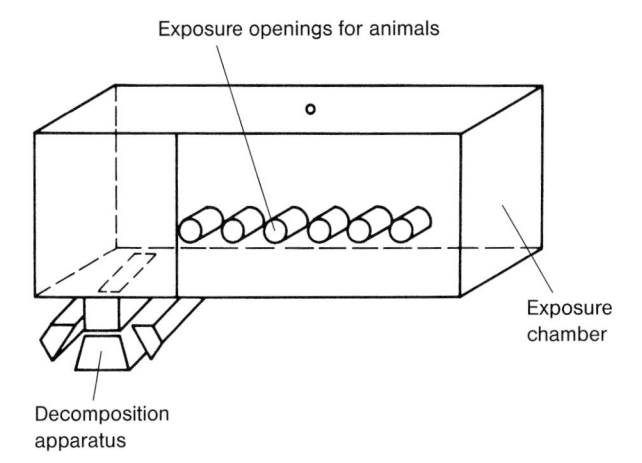

Figure 43 c US Radiant / NIBS Test Protocol

Within the PRC investigation [23] a series of decomposition models have been studied using the Potts Pot method, which was one of the selected models of the ISO TC 92 – SC3 research work. Of the numerous American decomposition models, the U-Pitt method, developed at the University of Pittsburgh (another method of the PRC program), was chosen as the second American method by SC3 of TC92. This protocol acquired legal status and was embodied in a law by the state of New York [187]. Test results gained by the U-Pitt method are the basis for registration of toxic-potency data [188]. Alarie and Barrow describe this test protocol in detail [92, 189]. The effluents are produced in an air stream of 9 l/h by a temperature ramp of 20 °C/min. The exposure of the test animals, white mice, is started after a significant weight loss of 1% of the heated material is recorded. Figure 43b illustrates the test procedure. The test animals will be exposed by nose exposure with a diluted stream of 20 l/h of the decomposition products. The evaluation is based on LC_{50} data. The behavior of the exposed animals is assessed from the toxocological point of view, e.g. breathing frequency.

The expert group of SC3 of ISO TC 92 have chosen the US Radiant protocol as a further American test method. The US delegation is recently supporting the modified US Radiant – NIBS protocol (see Fig. 43c), which is described in detail in [190, 191]. The test sample with dimensions of 72 mm × 72 mm is located horizontally in a glass cylinder and thermally stressed by external radiators. Contrary to the old version, there is now no contact of the effluents with the quartz radiators. The effluents are produced with an intensity level of 50 kW/m^2, and as a rule, LC_{50} data are determined. This level of intensity guarantees that ignition of the decomposition products will occur. If this does not happen, the stream of effluents will be ignited by a spark igniter. The loss of weight will be measured continuously. Basic investigations have been conducted in cooperation with SWRI [192].

The cone calorimeter, whose schematic is shown in Fig. 21, has been chosen by the expert group of SC3 of TC 92 with reservation, because the requested sufficient experience with animal testing was not available. Since the majority of the experts were convinced that this method will become dominant in the fire-performance classification of building products, it became part of TR 9122 selection according to Fig. 44. Proof that one of these selected methods will be qualified to simulate the decomposition products of the various, defined fire scenarios, is not part of TR 9122. This proof has to be given for each model separately. The cone calorimeter, see Fig. 21, produces a stream of fire gases by an external heat stress of up to 100 kW/m^2 with continuous flaming. The effluents are produced under flaming conditions only in an air stream of 0.035 m^3/s. The weight loss of the horizontally arranged samples with the dimensions 100 mm × 100 mm will be continuously measured and recorded.

The usefulness of some of the selected protocols of SC3 are compared with others, not selected by ISO and their relevance will be discussed in detail. Especially in the United States of America other models have been used for bioassay testing and are well documented. The FRC of Utah in Salt Lake City gathered a lot of experience with the NBS Smoke Chamber. Kolman et al. [193] investigated the toxic potency of effluents given off by rigid polyurethanes. The response behavior of the exposed animals, white rats, was noted and classified from the toxicological point of view. Smith

"Box" Furnaces
  NBS Cup Furnace
  UPITT Box Furnace

Tube Furnace Models
  DIN 53436 Tube Furnace

Radiant Heat Fire Models
  U.S.-Radiant Furnace (Modified)
  Cone Calorimeter
  Japanese Cone Furnaces
      BRI (Building Research Institute)
      Cone Furnace
      RIPT (Research Institute for Polymers
      and Textiles) Cone Furnace

  Japanese Ministry of Construction Fire Model

Figure 44 Decomposition devices according to TR 9122 – SC3 Toxic Hazard in Fire of ISO TC 92

et al. [194] reported details of the CAMI protocol, which has been developed at the Civil Aeromedical Institute (CAMI) of the FAA (Federal Airway Administration). Figure 45a illustrates this test protocol. The decomposition products are released at temperature levels of 600 °C and 800 °C. The test animals, white rats, are exposed to the effluents within a 2.61-l chamber and recirculation conditions. The test results demonstrate that the concentration of the decomposition components CO and HCN are, at the test temperature of 800 °C, significantly reduced. Skornik et al. [195, 196] determined the effects of smoke inhalation as well as under the conditions of recirculation, see Fig. 45b. The decomposition of the tested material and the exposure of the test animals happened in a stream of 100 l/min. The concentration of the effluents will be determined on behalf of the weight loss. The authors evaluated the outcome of the test in comparison to the reference material gypsum board.

A Furnace
B Exposure chamber
C TUF chamber

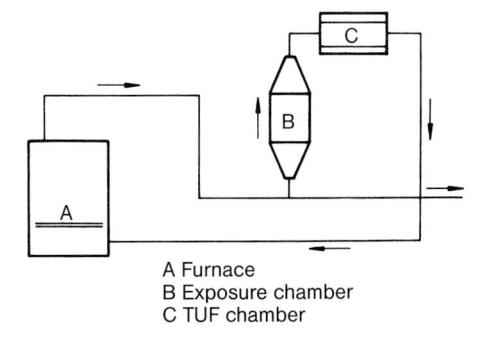

Figure 45a CAMI decomposition test equipment

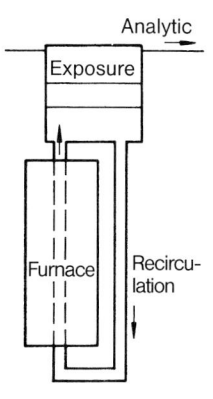

Figure 45 b Decomposition model Skornik

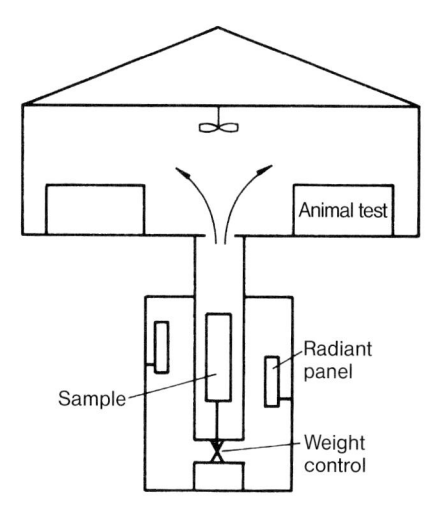

Figure 45 c SRI – decomposition apparatus

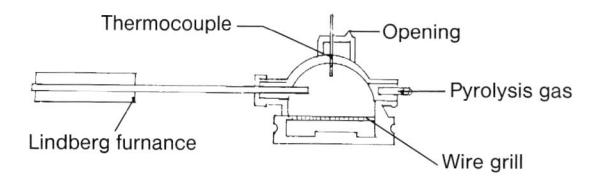

Figure 45 d USF decomposition model and NASA exposition chamber

In the model of the Stanford Research Institute (SRI) [76, 197], see Fig. 45c, the test sample is thermally stressed by an external heat flux. The effluents will be transported by natural ventilation to the exposure chamber above the furnace. The concentration of the most important components, the smoke density and the bioassay results, lethality and reduced activity (sense of sight, sense of touch and sense of hearing) are to be determined and recorded. In the USF method, developed at the University of San Francisco, the effluents are produced in an electrically heated furnace at constant temperature levels of 600 °C and 800 °C and at variable air stream or by a constant temperature increase of 40 °C/min between 600 and 800 °C. The exposure of the test animals, usually white rats, will be done in the NASA chamber according to Fig. 45d. The time to death will be determined and recorded [198, 199].

The modified U-Pitt method, U-Pitt 2, is a further device under investigation. The stream of effluents is produced by a heat stress up to 50 kW/m^2 by a conical radiator with an air stream of 1 l/min to 60 l/min. Details of the test procedure are given by Alarie [200] and Alarie and Caldwell in [201].

4.1.2
European Requirement and Classification Profiles

Germany

Preventive fire-protection measures in Germany are given in the relevant chapters of regional building regulation (see Table 7a) [202–204]. The relevant decrees apply to assembly rooms, public houses/restaurants or garages and guidelines, e.g. guidelines governing the use of combustible building materials in building construction and other standards like DIN 4102 implemented by building supervisory authorities [205, 206].

Table 7a LBO – Relevant Paragraphs of Fire Prevention

§ 1 Para.1	Fire Prevention of Building Products
§ 6	Space
§ 17	Basic Requirements
§ 28–30	Staircase
§ 34	Corridors as Rescue Ways
§ 38	Firing Station

Table 7b Fire-reaction classification of building materials and elements DIN 4102

Fire-Reaction Classification of Building Products	
incombustible	Class A1
incombustible	Class A2
combustible	
difficult to ignite	Class B1
normal to ignite	Class B2
easily to ignite	Class B3
building construction	Fire Resistance
fire resistant	F30, F60
fire resistant	F90
resistant against flying brands and radiation	hard roofing

4 Preventive Fire Protection – National Requirement and Classification Systems

For the most part, the requirements are only stated in a destructive, rather than an actual form. The real definitions of the various building-material and building-component classifications are covered by DIN 4102. The guidelines of combustible materials regulates when and where combustibles like polyurethane and polyisocyanurate foams are used. The 2nd article of §18 of the regional building regulation [203] states "The use of easily flammable building materials is not permissible, this does not apply to building materials when they are used in combination with other building materials and are no longer easily flammable". The court case of the Düsseldorf Airport Fire [1] demonstrated that the responsible regulatory people should be aware of the historical course of the relevant rule. This special advice, given by the Technical Commission building Supervision, finds expression in the relevant test and certification criteria of DIN 4102. Table 7b offers an overview of the classification system of DIN 4102.

The testing certificate has to be provided as follows: The result of the fire-resistance test under enduse conditions has to be produced in the furnace test according to DIN 4102 respectively IS 834 illustrated by Fig. 41a.

Details concerning the test-classification criteria as well as the failure criteria are given in the relevant standards. The proof for the resistance against flying brands and radiated heat of roof structures have to be provided in relation to part 7 of DIN 4102 (see Fig. 50a). A part of the roof structure with outer dimensions of 3000 mm × 3000 mm has to be positioned at an inclination of 15° and 45°. The thermal stress is given by 600 g burning wood wool shown by Fig. 46. The test is passed if no penetration of flames occurs and no inadmissible spread is observed, the burned or charred surface area will be not larger than $0.25\,\text{m}^2$ (single value smaller than $0.30\,\text{m}^2$) and the extent of the burning cover will be not more than 0.5 m (single value smaller than $0.30\,\text{m}^2$).

The proof of noncombustibility, e.g. class A1 of DIN 4102, must be gained in the furnace test at 750 °C according to Fig. 86. The classification of class A2 requires, see Table 8, a positive result in the chimney test (Brandschacht) according to Fig. 47a certificate concerning the smoke density and the acute toxicity of the decomposition products as well as a limited heat release or a reduced result in the 750 °C furnace test.

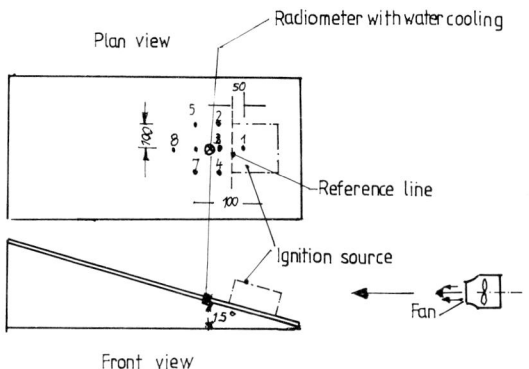

Figure 46 Resistance against flying brands and radiated heat

Table 8 Fire Reaction Testing – A2 Classification according to DIN 4102

Building material classification	
Furnace test 750 °C	Alternative testing Calorific value and heat-release rate
a) Total burning time <20 s b) Temperature rise <30 grd	a) < 600 kcal / kg b) < 1000 kcal / m^2 -30 min
Chimney test	Expert evaluation
c) undestroyed length > 35 cm d) fire-gas temperature <125 °C e) no ignition on the back side of the sample	c) Smoke density d) Toxic effects of decomposition products (DIN 53436)

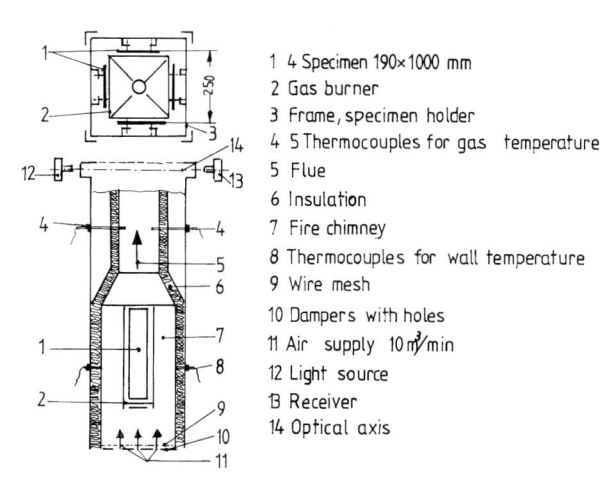

1 4 Specimen 190×1000 mm
2 Gas burner
3 Frame, specimen holder
4 5 Thermocouples for gas temperature
5 Flue
6 Insulation
7 Fire chimney
8 Thermocouples for wall temperature
9 Wire mesh
10 Dampers with holes
11 Air supply 10 m^3/min
12 Light source
13 Receiver
14 Optical axis

Figure 47a Chimney test according to DIN 4102

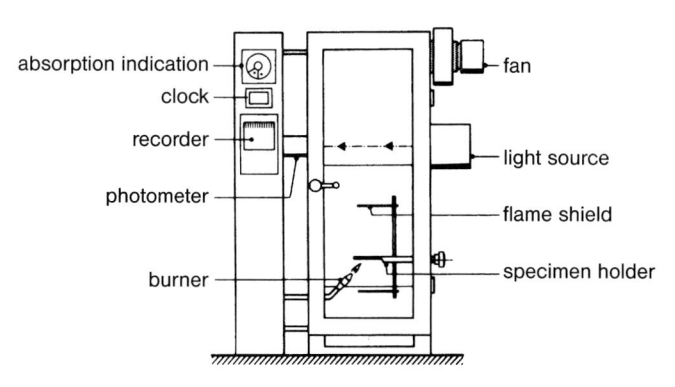

Figure 47b Smoke-density test method according to ASTM-D 2843 – XP2 Chamber

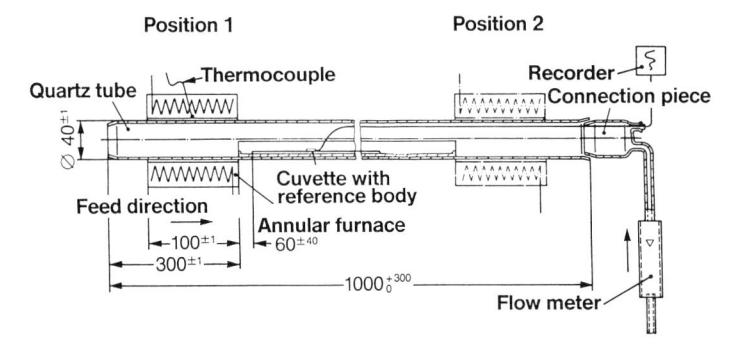

4 Preventive Fire Protection – National Requirement and Classification Systems

Figure 47 c Measuring device DIN 53436/37 – Decomposition device

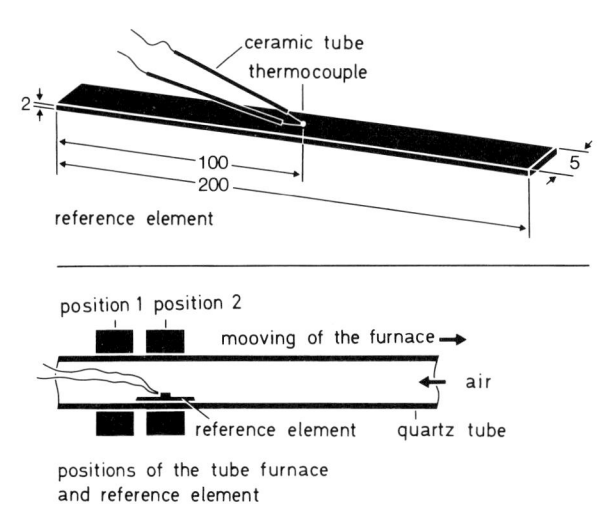

Figure 47 d Calibration body

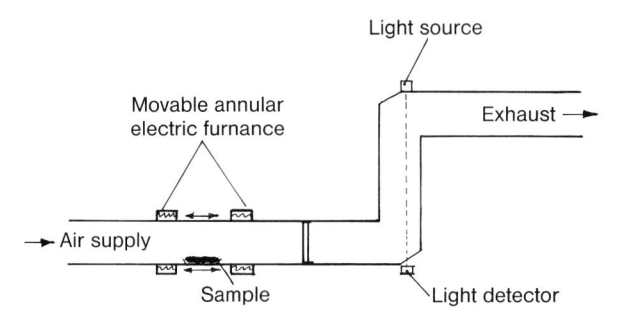

Figure 47 e Smoke-density measurement according to DIN 53436/37 – measuring device

4.1 Building Section

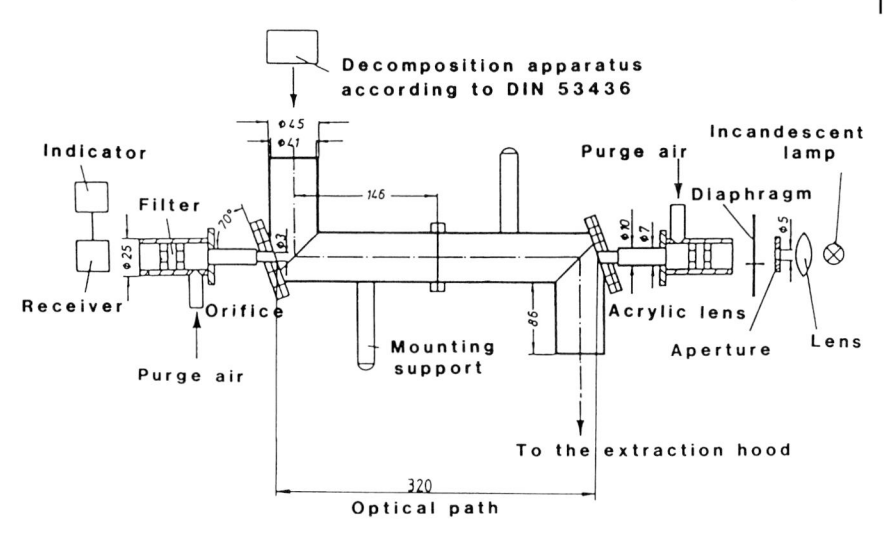

Figure 47f Decomposition apparatus according to DIN 53436

In the chimney test, the four samples, outer dimensions of 1000 mm × 190 mm × thickness mm (max. 80 mm), are arranged vertically, see Fig. 47a, and exposed to 150 to 250 mm high flames of a ring burner that simulates a burning wastepaper basket with a power of 21 kW. In the flame area, 100 mm above the lower end of the sample, the flames of the burner produce a radiant loading of well in excess of 40 kW/m². The undestroyed length of the tested samples has to be larger than 350 mm, the single value must be below 200 mm. The temperature level of the effluents must not rise above 125 °C and no flame penetration is allowed. The smoke density of the effluents is limited also under flaming as under smoldering conditions. The smoke-density values under flaming conditions are tested in the XP2-chamber according to ASTM – D 2843. In the XP2-chamber test, shown in Fig. 47b, the horizontally arranged test piece will be thermally stressed by a defined propane-gas burner for 4 min. In the upper half of the test chamber the reduction of the intensity of the horizontal light path due to the decomposition products will be measured and recorded. The effect of smoke layering is ignored. The XP2-chamber is the official test protocol for classifying the smoke density of building products in Germany (DIN 4102 class A2) and in Austria (Ö Norm B 3800) as well as in Switzerland [208].

The DIN furnace protocol, see Figs. 47c to d, is used to clarify the smoke density of A2-classified materials under smoldering conditions. The test furnace according to DIN 53436/37 has been developed by a working group of the DIN Plastic Division. Part of the expert group consisted of fire-prevention as well as fire-fighting experts. The tube furnace according to DIN 53436 essentially consists of a quarz tube, an annular furnace and a reference body for calibrating the test temperature. The furnace, 1000 mm long, travels at a standardized speed of 10 ±0.5 mm/min in the direction opposite to the air stream in the tube. The stream of effluents is produced in the temperature range of 300 °C to 550 °C with an air stream of 300 l/h. The reference body, as shown by Fig. 47e is a soldered metal strip with a special sheath thermocouple. Furnace speed and air-stream direction are opposing to ensure that no preheating of the undestroyed area of the sample occurs. The transmission of the light path will be determined, see Fig. 47e, with respect to the test temperature.

The limiting values of transmission for the A2 classification are max. 15% or 5%. The expert's evaluation of the toxic potency of fire effluents of A2-classified materials is done via bioassay testing with effluents produced at reference-body temperatures of 300°C and 400°C. The exposure of the test animals, white rats, over a period of 60 min must not lead to death, the CO-Hb value in the blood must not increase above 25% and from the smoke components, other than CO, like HCN or NO_x must not give rise to reservations from the toxicological point of view. The stream of effluents is constant during the testing period of 60 min. For the majority of tests performed so far, use has been made of 300-mm long samples of equal mass (e.g. 5 g) or equal volume (e.g. 3.6 g – 0.12 g/cm length). Tests with combustible materials have been run for 30 min only. The certification of class A1, A2 and B1 is given on the basis of positive test results of official test stations by the Institute for Building in Berlin. The B1 certificate is obtained when the test by an official testing station has shown that the chimney test requirements are fullfilled, e.g.

- an undestroyed medium length of 150 mm
- no test sample completely destroyed and
- the temperature of the smoke stream does not increase above 200°C

Burning debris and afterburning have to be observed and recorded. For pipe insulation the chimney test has to be run with cut samples as well as with complete pipes according to Fig. 48c [209, 210]. The testing of composites with core materials, e.g. steel sandwich elements with steel facings, need special preparation. According to the test regulation of the DIBt (Deutsches Institut für Bautechnik) test samples with protected edges, or samples with no edge protection, see Fig. 48a have to be protected by the standardized frame developed by the union of official test laboratories (ABM) ABM Nr. 18. Febr. 1993, see Fig. 48b, must be tested with the joint in the middle of the sample. The historical background of this testing frame had the aim of simplifying the production of the test samples of PUR steel sandwich panels. Beginning with the closed, riveted test case, the sample was later cut and the steel was folded to protect the edges and then riveted. Finally, the cut samples were stabilized by a massive frame to ensure that the protection of the metal facing was effective and was similar to that of real-fire situations established by large-scale testing and fire disasters [211, 212]. The equivalent stability of the samples with the middle joint, could not be achieved by the lightweight work of the official frame. Figures 48e to l demonstrate the extent of the damage of the PUR core of tested steel sandwich elements.

Combustible floor coverings can be classified class B1, if the Radiant Panel Test according to DIN 4102 pt. 16, shown by Fig. 41c is passed. The test samples with outer dimensions of 230 mm × 1060 mm × thickness mm will be thermally stressed, see Fig. 41d, at a radiation level of 1.0 to 0.1 kW/m^2 combined with a pilot flame at the hot end. The test requirements are passed if the burning extent is below 450 mm and the smoke density of the stream of effluents does not increase above 300% min.

Figure 48 a Testing of composites with cut edges

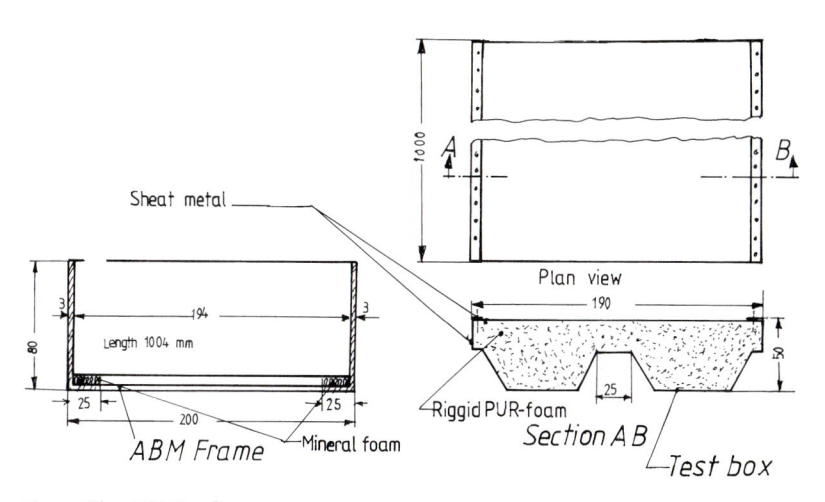

Figure 48 b ABM-Test frame

Arrangement A   Arrangement B

A Outside diameter < 60 mm
B Outside diameter 60-90 mm
C Outside diameter > 90mm

Arrangement C

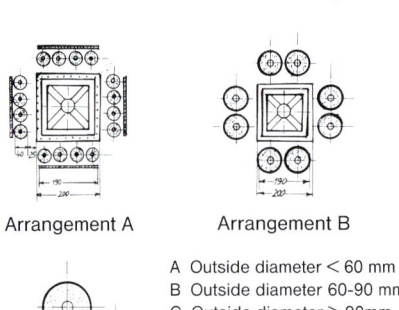

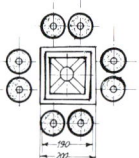

Figure 48c Test frame for insulated pipe samples

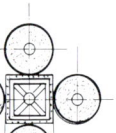

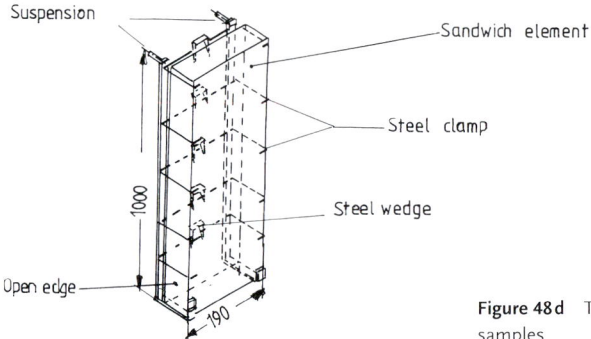

Figure 48d Test configuration of composite samples

Figure 48e Tested composite samples

4.1 Building Section | 81

Figure 48f to h Tested sandwich element samples

Fig. 48f

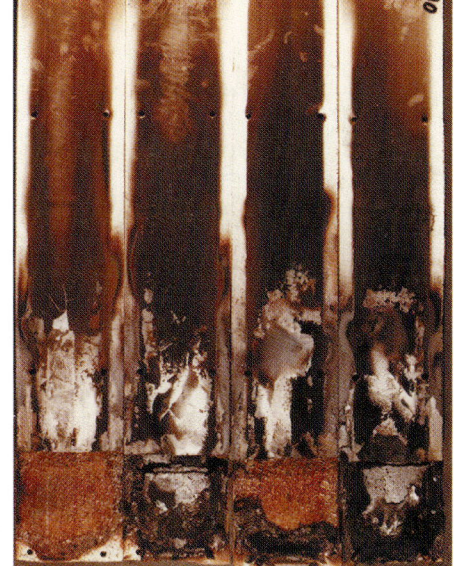

Fig. 48g

Fig. 48h

82 | *4 Preventive Fire Protection – National Requirement and Classification Systems*

Fig. 48i

Fig. 48j

Fig. 48k

Fig. 48l

Figure 48i–l PUR steel-sandwich element probes

The classification of class B2, normally flammable, can be achieved by passing the small-burner test, DIN 4102 pt.1, which simulates a match-type test, see Fig. 49a.

The sample has outer dimensions of 190 mm × 90 mm or 230 mm × 90 mm × thickness mm. A 20-mm long propane-gas flame will impinge on the lower end of the vertical sample for 15 s or at the surface, 20 mm above the lower edge. The flame of the ignited sample must not reach the mark 150 mm above the impinging point. The test period is 20 s that includes the 15 s impinging time. Figure 49b illustrates how the core of sandwich panels has to be tested and classified.

The small-burner test procedure is well standardized for evaluating the burning dripping of B2-classified materials. The tested materials will be classified as "burning dripping" if within the test period of 20 s, which includes 15 s impinging time, the filter paper arranged below the test equipment is ignited by burning droplets.

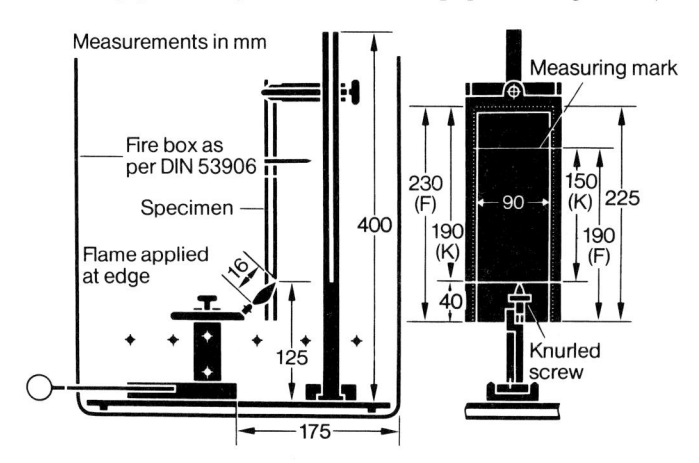

Figure 49 a Small-burner test according to DIN 4102

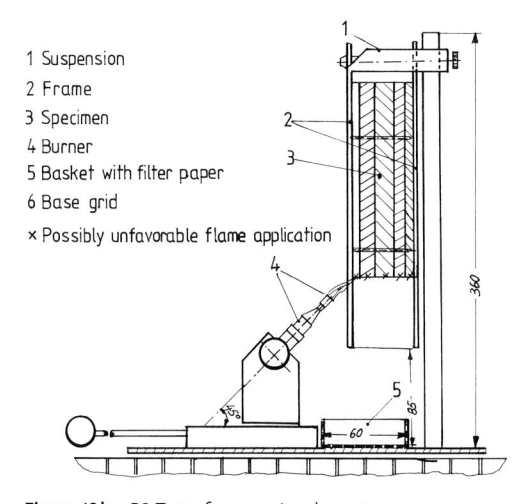

1 Suspension
2 Frame
3 Specimen
4 Burner
5 Basket with filter paper
6 Base grid
× Possibly unfavorable flame application

Figure 49 b B2-Test of composite elements

Great Britain

The requirements concerning building materials and constructions stated in the Building Regulations of England and Wales, Scotland and Northern Ireland, are specified in detail in the different parts of BS 476. Fire-performance results for building products according to part 5 of BS 476 are identical to the test and classification criteria of IS 1182 and DIN 4102. The fire-performance testing of building material is based on three different laboratory test methods.

The resistance against ignition sources of low intensity will be tested according to BS 476 pt. 5 as well as to the standardised test protocols within BS 2782. The surface spread of flame will be determined according to BS 476 pt. 7. As shown by Fig. 50a the test samples with outer dimensions of 885 mm × 270 mm × max. 50 mm will be arranged vertically and perpendicular to the gas-fired radiator. The thermal stress is applied for 10 min. The pilot flame for ignition of the decomposition products is located at the hot end of the sample. The classification in the four classes is specified in Table 9, where the extent of spread after 1.5 min and 10 min is documented.

In the fire-propagation test, illustrated by Fig. 50b, a sample of outer dimensions of 225 mm × 225 mm × max. 50 mm will be applied as with a wall lining. The combined thermal stress is produced by propane-gas-fired burners of 530 J/s and two electrical heated elements of 1800 W that become effective after 2 min 45 s and that are reduced to 1500 W after 5 min. The test period is 20 min. Based on the continuously recorded smoke temperature, the classification is given in comparison to an asbestos cement sample. Up to the 3rd minute the test temperature is measured every 30 s, until the 10th minute at every minute and until the end every 2 min. Using these test results, the classification is calculated by taking the formula

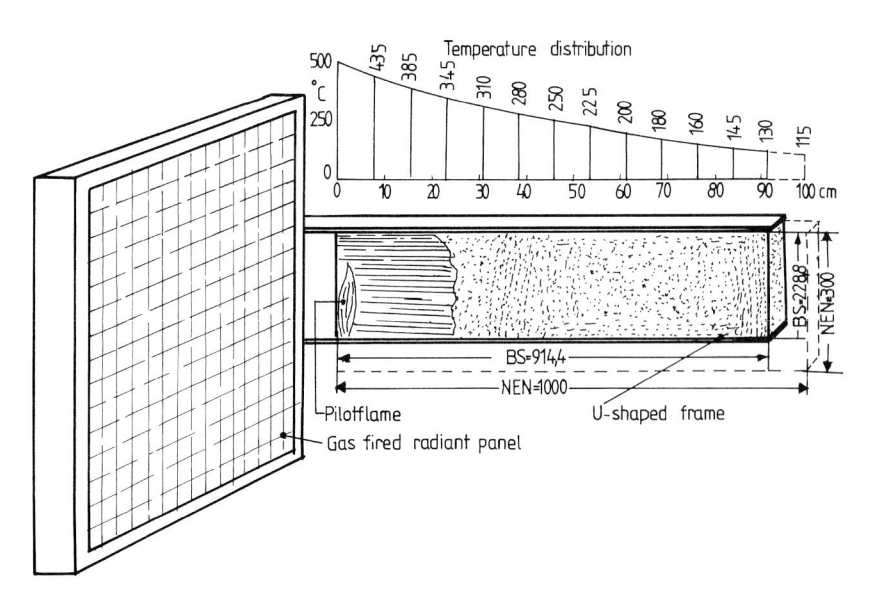

Figure 50a Surface Spread of Flame Test according to BS 476 pt.7

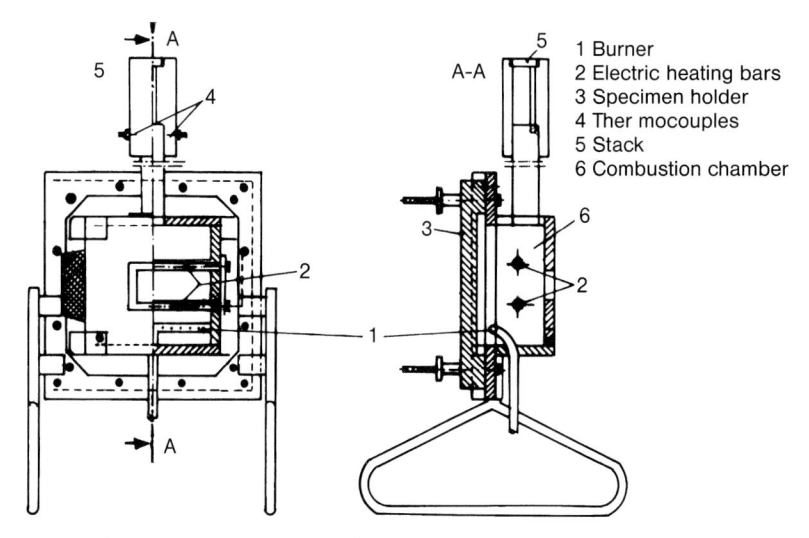

Figure 50b Fire propagation test according to BS 476 p.6

1 Burner
2 Electric heating bars
3 Specimen holder
4 Thermocouples
5 Stack
6 Combustion chamber

Table 9 Classification according to BS 476 tl.7

Class	Spread after 1.5 min (mm)	Tolerance for 1 sample (mm)	Total spread 10 min (mm)	Tolerance for 1 sample (mm).
1	165	25	165	25
2	215	25	455	45
3	265	25	710	75
4 Failing of the class 3 criteria.				

$$l = i_2 + i_2 + i_3,$$

$$i1 = \sum_{1\ 2}^{3} \frac{Qm - Qc}{10\,t}$$

where i_1, i_2 and i_3 are the subindices after 3, 10 or 20 min, t is the test time in min, Q is the smoke

$$i2 = \sum_{20}^{10} \frac{Qn - Qc}{}$$

$$i3 = \sum_{12}^{4} \frac{Qo\,10 - t\,Qc}{10\,t}$$

temperature level corresponding to the reference material at time t.

The fire-propagation test is used to upgrade class 1 material. This can be reached, e.g. by using noncombustible coverings or noncombustible substrates.

Austria

Preventive fire protection in Austria is overwhelmingly based on regional requirements, e.g. the conditions of the town council authorities of Vienna [213]. The fire-performance classification of building products/materials is determined by Ö Norm B 3800. Part 1 of this standard sets the testing of building materials and parts 2 to 4 are relevant to building constructions. Part 2 of this standard is responsible for the fire-reaction testing including side effects like burning dripping and smoke density. The noncombustibility is certificated analogous to IS 1182 with the 750 °C furnace test. Combustible materials are classified as class B1 – difficult to ignite, class B2 – normal flammable and class B3 – easily flammable. The classification of class B1 is done by the Schlyter test, see Fig. 51, where two vertically arranged samples were stressed by a multijet burner. There is a distance of 50 mm between the two samples with outer dimensions of 800 mm × 300 mm × thickness mm. The one sample is 50 mm lower and is thermally stressed for 15 min. The power of the burner is 21 kW. The classification of B1 is achieved if the flamed specimen is not ignited or the afterburn time is limited to 1 min, the afterglow time to 5 min and the undestroyed length of the sample is min 400 mm and no flashover to the second specimen occurs.

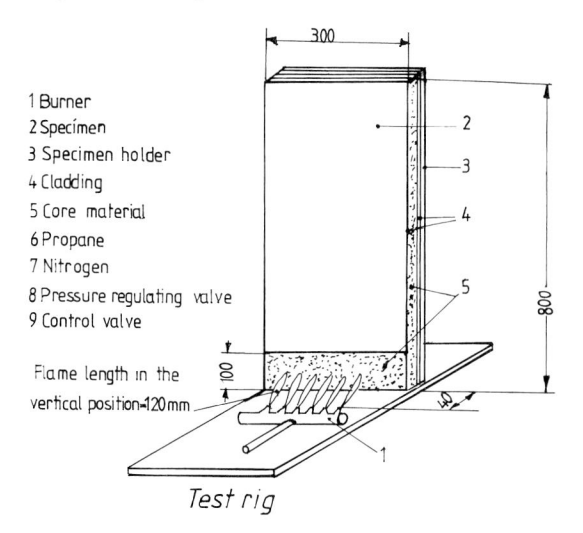

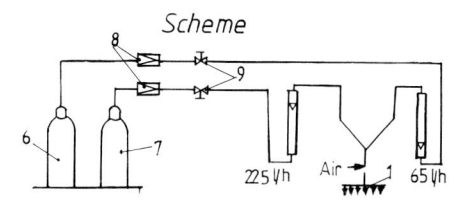

Figure 51 Schlyter –Test according to ÖNorm B 3800

If building materials like wood with a thickness above 2 mm and other products according to Sect. 3.1.3 of Ö Norm B 3800 are not classified as B2, they have to be tested by the small-burner test, which is similar to that of DIN 4102 (see Fig. 49a). The tested material is classified as class B2 if the small-burner test is passed. Like DIN 4102 the vertically arranged test piece will be thermally stressed at the lower edge or 15 mm above by the small burner. No flame spread above the 150-mm mark must occur within 20 s, which include the 15 s impinging time of the match-type flame. The burning dripping will be assessed by the same test procedure.

For B1-classified materials the impinging time is 10 min and for the B2-classified ones it is 10 s. The differentiation is shown in Table 10.

Table 10 Classification of the dripping behaviour – Ö Norm B3800

Classe	Evaluation
Tr 1	no dripping
Tr 2	dripping, filter paper will not be ignited
Tr 3	burning dripping, filter paper will be ignited

The B1 classifications of flooring materials are determined analogous to DIN 4102 pt. 16 in the radiant panel test according to ASTM – E648. The test sample with outer dimensions of 800 mm × 200 mm, is located horizontally and will be thermally stressed by a gas-fired radiator similar to DIN 4102. The inclination of the radiator, 300 mm × 360 mm has been modified to about 60°. The radiation intensity varies from 1.5 to 17 kW/m^2. The pilot-flame impinges on the hot end for 2 min. The test is finished after 20 min, if the flames are not extinguished before. The classification will be done according to Table 11.

Table 11 Classification of floorings according to Ö Norm 3800

Class B1 – damaged extent > 4 kW/m^2 on the surface of the test sample
Class B2 – damaged extent ≤ 4 to 3 kW/m^2 on the surface of the test sample
Class B3 – damaged extent < 3 kW/m^2 on the surface of the test sample

Switzerland

In Switzerland safety requirements are regulated by federal law. Preventive fire safety is basically defined by the regions (cantons). The general fire-performance regulations (Association of Cantonal Fire Insuranceces – VKF) [208] and Fire Prevention Service for Industry and Commerce – BVD) incorporate test and classification criteria. The material-related fire-performance requirements for insulating materials are fixed according to the type and use of building and include the side effects like smoke release. In special enduse areas, for example materials with good fire performance and medium smoke may be interchanged with materials with better fire performance but low smoke. According to the Swiss regulation, the burning behavior is differentiated into 6 values. An increasing number means a better classification as shown by Table 12.

4 Preventive Fire Protection – National Requirement and Classification Systems

Table 12 Fire reaction classification according to the Swiss requirements

Classification – basic test – edge test	
II easily ignitable – rapid burning	Test sample ignites and burns rapidly ≤ 5 s
III easily burning	Test sample ignites and burns fast ≤ 20 s
IV normally burning	Test sample ignites, burning time > 20 s
V little burning	Flames extinguish before the 150 mark, burning time ≤ 20 s
VI non burning	No flames, no carbonisation
Classification – basic test – area test	
III easily burning	Test sample ignites, burning time ≤ 40 s
IV normally burning	Test sample ignites and burns > 40 s, burning time > 50 s
V little burning	Test sample ignites and extinguishes ≤ 5 s, burning time ≤ 50 s

The test protocol corresponds to the small-burner test, see Fig. 49a. The test samples with outer dimensions of 160 mm × 60 mm × 4 mm, or 160 mm × 60 mm × 6 mm for foam materials, will be arranged vertically and thermally stressed by a 20-mm long propane/butane-gas flame (900 °C) at the lower edge or 25 mm above. The testing time is 15 s or 45 s [214]. To gain a risk-related classification with respect to actual fire conditions, an additional small-burner test was agreed at an elevated temperature level of 200 °C [215]. The fire-reaction classification of flooring materials, class I and class II is also achieved by the small-burner test. The differentiation of class III and IV is based on the radiant panel test of DIN 4102 with a slightly modified smoke-density determination. The large-scale test arrangement according to Fig. 52 offers the opportunity to test facade elements under the thermal stress of room fires using a special gas burner [215, 216].

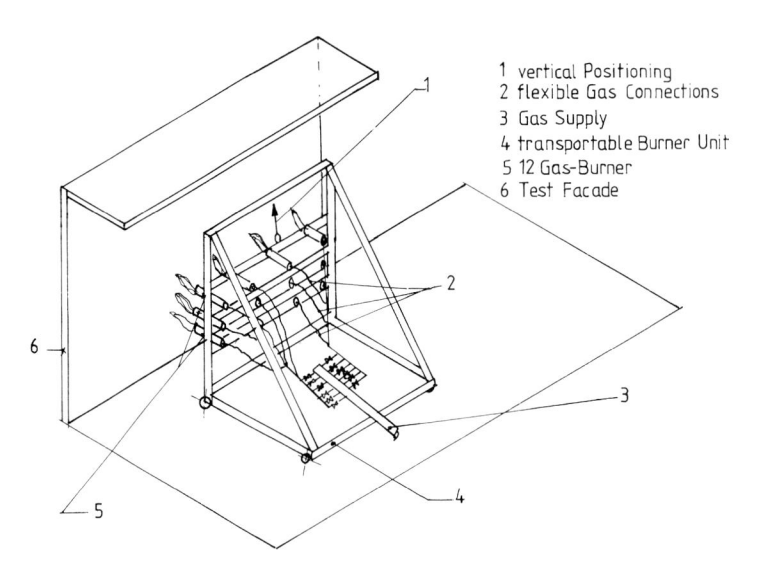

Figure 52 Swiss Facade Testing Station

1 vertical Positioning
2 flexible Gas Connections
3 Gas Supply
4 transportable Burner Unit
5 12 Gas-Burner
6 Test Facade

The usual smoke-density measurement is based on the XP2-chamber procedure. The characterisation is given by the percentage decrease of the horizontally arranged light transmission measuring system. The decomposition products of the test piece, 39 mm × 30 mm × 4 mm (solids) and 60 mm × 60 mm × 25 mm (foamed materials) will be produced by the impact of a special gas burner (the so-called Lüscher Burner) with a flame length of 150 mm. The length of the light path of the optical measuring system is 308 mm. The classification is given in Table 13.

Table 13 Smoke density classification according to the Swiss requirements

Classification	max light absorption (%)
class 1 strong	> 90
class 2 moderate	> 50 – ≤ 90
class 3 light	0 – < 50

If this kind of differentiation was not sufficient, e.g. in conflicts, additional tests (gas analysis) have been necessary. The effluents were produced by smoldering. The test material will be placed on a steel plate (1.2 mm thick, 120 mm diameter) heated to 450 °C with a bunsen burner.

Italy

The Italian fire-performance tests and classification component requirements according to the decree of the 26.06.1984 are similar to IS 834. The noncombustibility test is identical to the IS 1182 furnace test at 750 °C. The classification of combustibility is based on a combination of three different test methods.

CSE – RF – 1/75/A small-burner test, time= 12 s
CSE – RF – 2/75/A small-burner test modified time = 30 s
CSE – RF – 3/77 radiant panel test according to ISO DP 5658 of ISO TC 92 SC1 – 20 min.

The vertically arranged test pieces with outer dimensions of 390 mm × 104 mm × thickness mm are ignited by the propane-gas burner with a flame length of 20 mm. The classification will be related to the product of categories and weighting factor. The category is formed by the classification. The weighting factor takes into consideration the different classification criteria. The classification of the 12-s test is illustrated by Table 14.

The resistance to larger ignition sources is measured by the radiant panel test according to CSE – RF – 3/77, which has been standardised in line with ISO DP 5658, see Fig. 53a. The test samples with the dimensions of 800 mm × 155 mm × max. 40 mm are exposed to a thermal stress, Fig. 53b, which illustrates the different intensity levels for wall and ceiling linings and flooring application. The decomposition products will be ignited by a pilot flame located at the hot end of the specimen. Figure 53b illustrates the level of intensity.

The classification system according to ISO/CSE – RF – 3/77 is shown by Table 15.

The selection is one of the four classes of building materials is done according to the listing of Table 16 in relation to these differing test methods.

4 Preventive Fire Protection – National Requirement and Classification Systems

Table 14 Classification scheme in the small burner test according to CSE

Grade	Afterburn time (s)	Afterglow time (s)	Damaged extent (mm)	Afterburn time of drippings (s)
1	< 5	≤ 10	< 150	non burning
2	> 5 – ≤ 60	> 10 – ≤ 60	> 150 – ≤ 200	≤ 3
3	> 60	> 60	≥ 200	> 3

Evaluation criteria	Factor	Category	Evaluation
Afterburn time	2	I	6–8
Afterglow time	1	II	9–12
Extent of damage	2	III	13–15
Dripping behavior	1	IV	16–18

Table 15 Classification scheme of the spread test

Grade	Spread (mm/min)	damaged extent (mm)	Afterglow time (s)	Afterburn time of drippings (s)
1	not measurable	≤ 300	≤ 180	not burning
2	≤ 30	≥ 300 – ≤ 600	> 180 – ≤ 360	≤ 3
3	> 30	≥ 650	> 360	> 3

Classification

Parameter	Importance Factor
Spread	2
Extent of damage	2
Afterglow	1
Dripping – flooring	0
Dripping – wall	1
Dripping – ceiling	2

Class	Product from grade and factor		
	Flooring	Wall	Ceiling
I	5–7	6–8	7–9
II	8–10	9–12	10–13
III	11–13	13–15	14–17
IV	14–15	16–18	18–21

Table 16 Fire reaction testing of building products in Italy

CSE RF 1/75A or 2/75A	I	1
CSE 3/77	I	
CSE RF 1/75A or 2/75A	II I	2
CSE 3/77	I II	
CSE RF 1/75A or 2/75A	III II I III II	3
CSE 3/77	II III III I II	
CSE RF 1/75A or 2/75A	IV III III IV II IV I	4
CSE 3/77	III IV III II IV I IV	
CSE RF 1/75A or 2/75A	IV	5
CSE 3/77	IV	

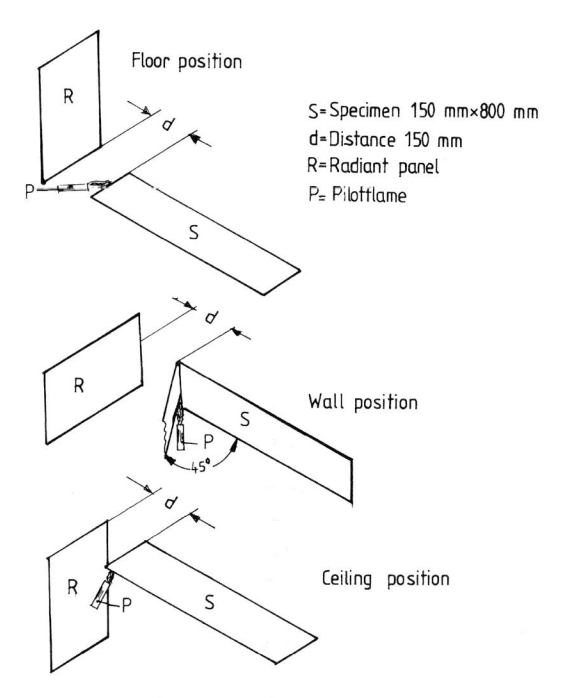

Figure 53a Surface Spread of Flame Test according to CSE – RF 3/77

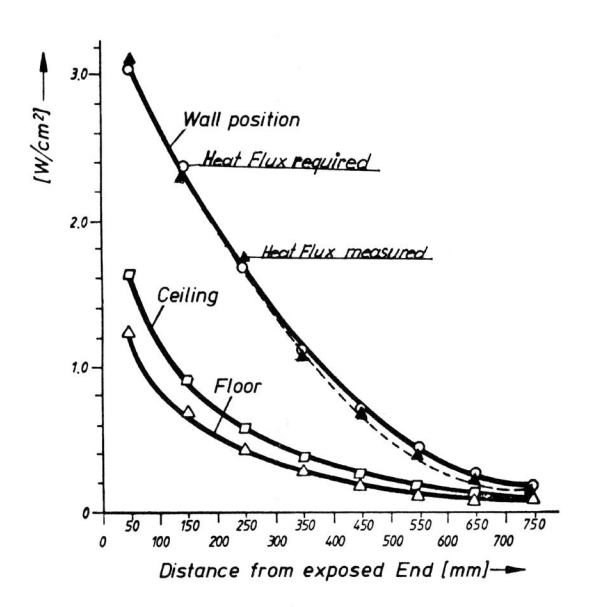

Floor position
S = Specimen 150 mm × 800 mm
d = Distance 150 mm
R = Radiant panel
P = Pilotflame

Wall position

Ceiling position

Tolerances ± 0.1 W/cm² above 1 W/cm²
± 0.05 W/cm² below 1 W/cm²

Figure 53b Thermal stress of the surface area of the test sample

France

The type and use of buildings determines the fire-performance requirements decisively in France as well. The most important requirements of the regulation are given in the Journal Officielle De La Republique Francaise (19.11.1991 p. 15 040 to 15 047) published on 28.08.1991. The fire performance of building products is given by the classes M0, M1, M2, M3 and M4. These test methods, standardised in NF-P 92501 to P-92510 form the basis of the fire-reaction testing of building materials in Belgium, Greece, Spain and Portugal. The noncombustibility, class M0, is determined by means of a bomb calorimeter. This test method corresponds to IS 1716. Provided that the upper calorimetric result does not exceed 2500 kJ/kg, the tested material will be classified as class M0. Due to the material performance, two different key test methods are used for the classification of combustible materials. Rigid materials of arbitrary thickness and flexible materials with a thickness above 5 mm are tested and classified in the cabin test, with the epiradiateur, in accordance to NF – P 92-501 (NBN – S21, UNE 23 – 721-81). In this primary test method, see Fig. 54a, the plate-shaped specimen is held at an angle of 45°. An electrically heated radiator provides a heat flux of 30 kW/m^2. Two butane-gas flames provide the ignition source for the decomposition products. The parameters for the fire reaction to fire classification are the following: Time to ignition with a burning time less than 5 s, afterburn time of flames above the test piece, maximum values of temperatures in

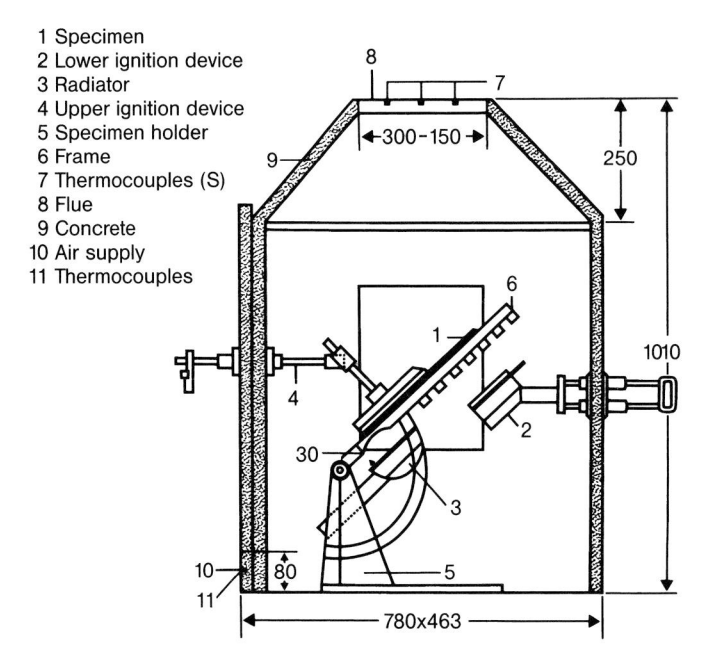

1 Specimen
2 Lower ignition device
3 Radiator
4 Upper ignition device
5 Specimen holder
6 Frame
7 Thermocouples (S)
8 Flue
9 Concrete
10 Air supply
11 Thermocouples

Figure 54a Epiradiateur Test according to NF-P 92-501

4.1 Building Section | 93

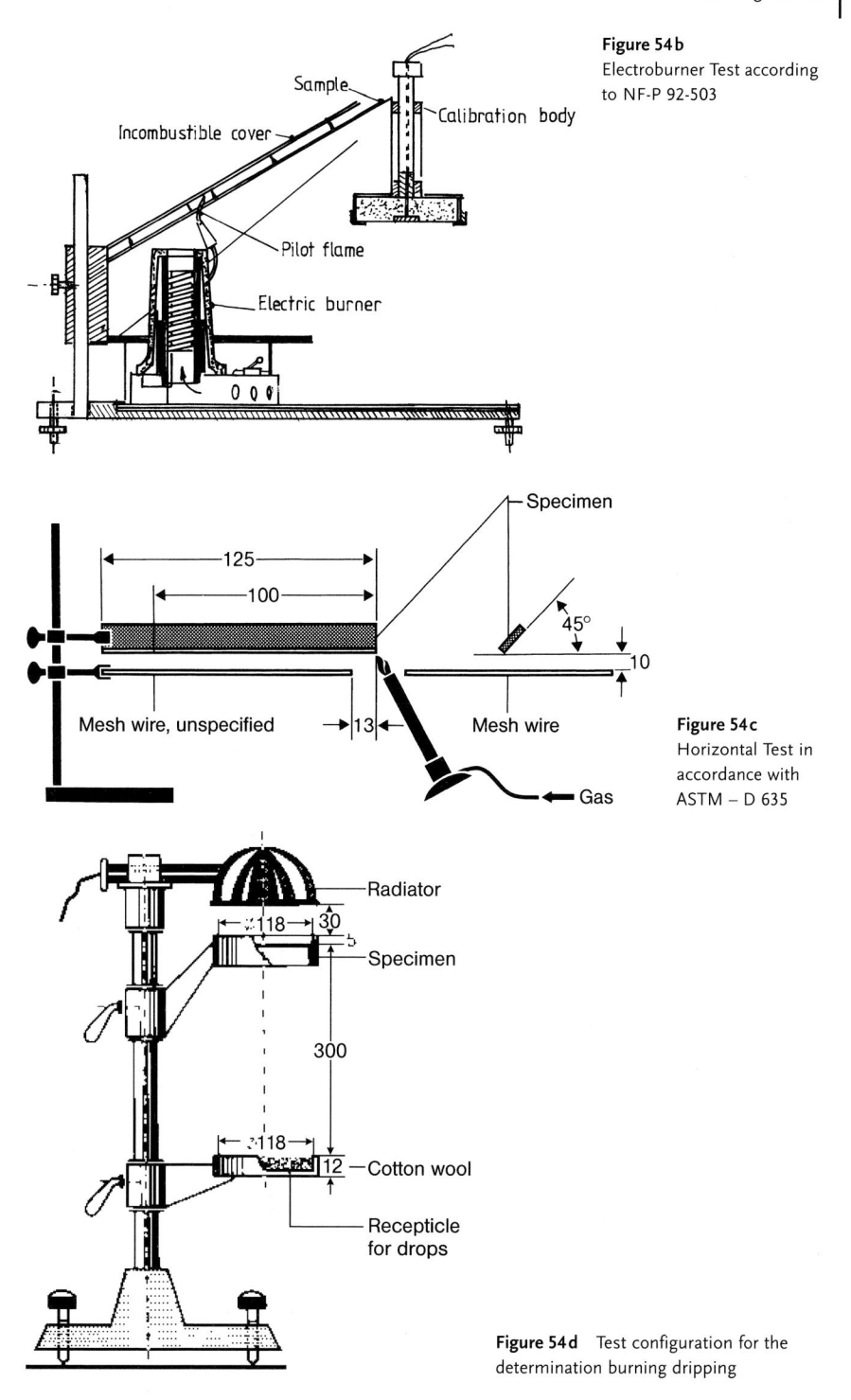

Figure 54b
Electroburner Test according to NF-P 92-503

Figure 54c
Horizontal Test in accordance with ASTM – D 635

Figure 54d Test configuration for the determination burning dripping

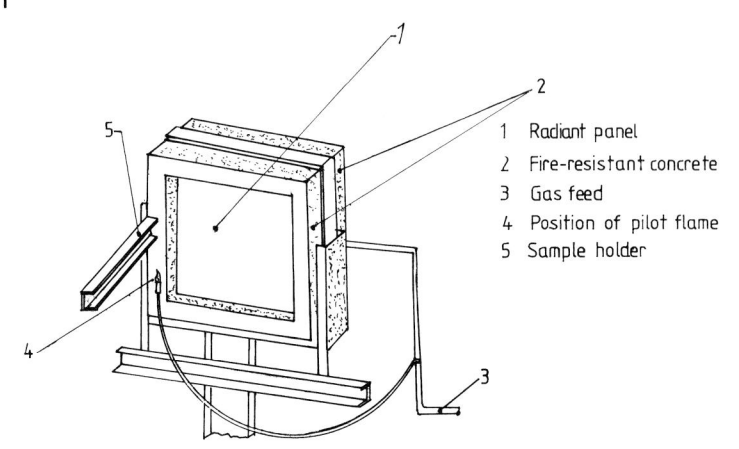

Figure 54e Radiant Panel Test according to NF-P 92-506

1 Radiant panel
2 Fire-resistant concrete
3 Gas feed
4 Position of pilot flame
5 Sample holder

intervals of 30 s, burning droplets and side effects like melting and dripping. Composite materials with an incombustible exterior layer, e.g. steel-sandwich elements are tested in the cabin test with a slot of 3 mm × 18 mm. The slot is positioned from the lower edge. The outer edges of such test pieces are to be protected by Al foils. The ranking (see Table 17) is based on the classification formula:

Table 17 Classification in the Epiradiateurtest

M1	M2	M3	M4
q < 2,5	q < 15	q < 50	M3 not obtained

$$q = \frac{100 \sum h}{t_i \sqrt{\Delta t}} \qquad q = \frac{\sum g}{n}$$

h (cm) = sum of flame height
t_i (s) = time of ignition
t (s) = flaming time, where the flames go above the diameter of the radiator

Flexible products with a thickness < 5 mm are classified by means of the so-called electric-torch test, see Fig. 54b. The test piece with dimensions of 600 mm × 180 mm × >5 mm is arranged at an angle of 30°. The thermal stress generated by the electro-torch is 30 kW/m² and is applied for 5 min. The thermal decomposition products are ignited by pilot flames applied at 20 s, 45 s and in addition after every 30 s. The test duration is 5 min or until the flames of the test pieces are extinguished. After burn time, the dripping behavior and the size of the damage area are the basis for ranking materials in classes 1 to 4 (see Table 18).

If the tested material is classified in the main test as class M4, an additional flame-spread test with a test piece of 400 mm × 35 mm × thickness mm similar to ASTM – D635 has to be conducted. A small-burner flame has to be applied 10 times for

Table 18 Classification in the Electroburner – Test

Classes	M1	M2	M3	M4
Afterburn time r < 5 s	x	x		
no spread	x	x		
damaged extent <350 mm		x	x	
damaged extent <600 mm			x	
damaged width <90 mm			x	
200–450–600 mm				
burning dripping		x	x	
M3 failing				x

x) if the sample is full of holes, than spread

xx) M4, if the cellucotton ignites, no other rating obtainable and the spread test at an impingement of 30 s a burning rate of < 2 mm/s results.

5 s to the horizontally fixed sample. Afterburn time and dripping behavior are the basis for ranking according to the listing of Table 19.

Table 19 M – Classification in the supplementary test

Class	M1	M2	M3	M4
Afterburning	0	≤ 5 s	≤ 5 s	M3
Burning dripping	–	–	x	failing
Spread (mm)	–	< 25 mm	< 250 mm	failing

A special test according to Fig. 54d has to be conducted for melting materials. The test piece with outer dimensions of 70 mm × 70 mm with a minimum weight of 2 g is positioned horizontally on a special metal grid at a position 30 mm below. During the thermal loading of 30 kW/m^2, heat stress is applied for 10 min, the flammability and dripping behavior are observed. If the sample ignites within the first 5 min, the radiator will be removed until the flaming ceases. During the following 5 min the thermal stress will be continuously maintained. If the cellulose cotton wool, 300 mm below the grid is ignited by burning droplets in at least one test of the four, the tested material will be classified as "burning dripping". This behavior means that the classification drops to M4.

The fire performance of flooring is evaluated in the Radiant Panel Test according to Fig. 54e. Test details are given in NF – P 92-506. The 400 mm × 95 mm × max. 55 mm size test piece will be upright in the longitudinal axis. The thermal stress is produced by the gas-fired radiator. The stress duration is 10 min. The thermal decomposition products will be ignited by a small burner at the hot end of the test piece. The burn distance determines the ranking classes M3 and M4.

In realistic full-scale tests, see Fig. 55, the possible contribution of cladding materials to fire spread can be examined. The facade element to be evaluated is exposed to the thermal stress of a room fire. The room fire is generated by a fire load of wooden cribs in the basement of the two-storey test house of approximately 25 kg/m^2. The facade material is thermally stressed by the hot gas stream emerging from the win-

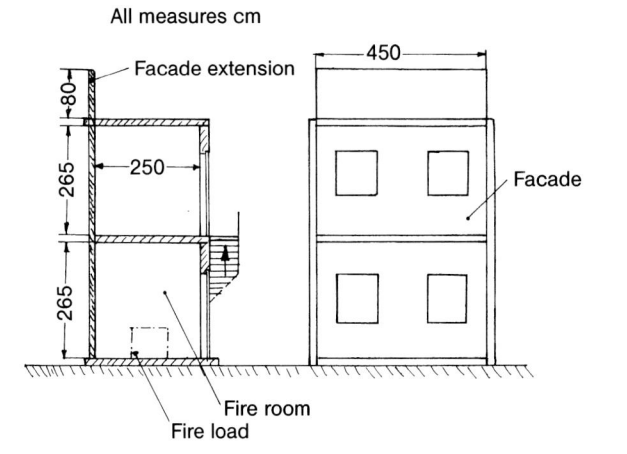

Figure 55 Facade Testing

dow of the fire room. The contribution to the flame spread across the facade lining as well as the possible ignition of the curtain material in the upper-floor window by radiation is recorded. The test arrangement and procedure of the test run have been regulated. The use of synthetic materials in public buildings has been regulated by the Arrete of 4.11.1975 and the renewed version Nr. 1540 II of 1.12.1976. [217]. The use of nitrogen- and chlorine-containing synthetic materials is limited to 5 g/m^3 and 25 g/m^3, if the ranking of M0 or M1 is not achieved. Consideration of special risk and enduse-oriented situation is carried out in accordance with the listing given in Table 20.

Table 20 Reglementation – nitrogen and chlorine containing products

Classification	Ceilings	Floorings	Miscellaneous
M0 or M1	0	0	0
M2 or M3			
density $> 0,02 \text{ g/m}^3$	4.bP/3	0	b.P
density $< 0,02 \text{ g/m}^3$	16.bP/9	0	4. bP/3
M4 density $>0,02 \text{ g/m}^3$	–	6.P/5	b.P
M4 density $< 0,02 \text{ g/m}^3$	–	–	4:bP/3

Authorisized test stations are CSTB, LNE and SNPE. Extensive investigations were conducted at the University of Paris, using the decomposition apparatus similar to DIN 53436 [218, 219]. Further decomposition models for evaluating effluents have been developed in France. Capron [220] describe a decomposition model, where the effluents will be produced by radiation within an air stream of 300 l/h. Two radiators (Epiradiateurs) are used to generate a radiation of 25 kW/m^2. Effluents were produced for an exposure time of 20 min. The volume of the measuring chamber is 64 l. The evaluation was done on behalf the damage using test animals and a physiogramm given in the work of Jouany et al. [218, 219]. Daniel et al. [221] report on a specific decomposition model for electro-isolating liquids. The test material is used to produce an aerosol mixture in an air stream, which will be ignited by a heating plate. Analytically determined test results and bioassay test results are used for the evaluation and classification criteria.

Netherlands
The permitted use of building products depends on the determinations of the building regulation. Both material-related fire-reaction rankings and the effectiveness of construction elements are to be verified. The element examination requires a fire-resistance test according to NEN 3883/IS 1182 using wall-, ceiling- or flooring-sections and orientations. Tests for resistance against flying brands and radiated heat for roofing elements will be conducted in accordance with the wood-wool basket test of DIN 4102 pt. 7. The acceptance of building materials is regulated by the ranking according to both standards of NEN 6066, the Vlamuitbreiding Test illustrated and the Vlamoverslag Test. The Vlamuitbreiding Test corresponds to BS 476 pt. 7, see Fig. 50a. Modifications are made to the tolerances of the radiation profile as well as for the classification criteria listed in Table 21.

Table 21 Classification according to NEN 3883

Class	Spread after 1.5 min (mm)	Tolerance for 1 sample (mm)	Total spread 10 min (mm)	Tolerance for 1 sample (mm)
1			175	25
2	250	50	550	50
3	350	50	750	100
4	500	50	Failing of the class 3 criteria	

In the Vlamoverslag test rig, illustrated by Fig. 56a, test pieces of outer dimensions of 220 mm × 230 mm × (thickness) mm, used as inner surface linings will be subjected to a combined radiation/flame thermal stress. The ranking of the tested material depends on the level of radiation, where a flashover to the second test sample during the testing period of 20 min occurs. Table 22 provides an overview of the classification criteria.

Table 22 Classification in the Vlam Overslag Test Protocoll

Classification	Heat stress E5	Heat stress E15
class 1	> 1875	> 1500
class 2	< 1875 > 1125	> 1500 < 1500
class 3	< 1125 > 565	> 750 < 750
class 4	< 565 —	> 190 < 190

98 | 4 Preventive Fire Protection – National Requirement and Classification Systems

The application of the building material determines the fire-performance requirement profile including the smoke-density ranking according to NEN 6066. The test samples, composite samples as well, will be horizontally arranged according to Fig. 56b. The area distribution of the thermal stress produced by an electrically heated radiator, is illustrated by Fig. 56c. This two-chamber system, developed by the SC1 of ISO TC 92. TR 5956 provides a detailed test protocol, which offers a variation of the irradiance level of 10 to 50 kW/m^2. The reduction of visibility caused by the stream of effluents will be measured by means of a photoelectric system. A fan installation ensures that smoke layers are not built-up in the dual-chamber box. Based on the recorded maximum transmission data material-related data of the optical density D are calculated.

Belgium
Preventive fire protection in Belgium is mainly based both on the French and the Dutch test and evaluation criteria. The examination of components are listed to the

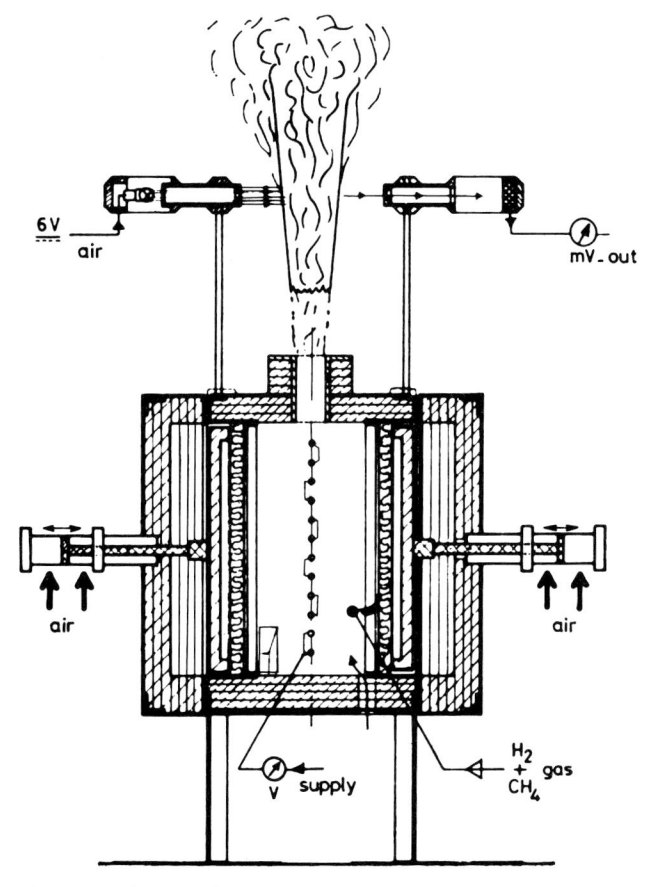

Figure 56a Vlam Overslag Test

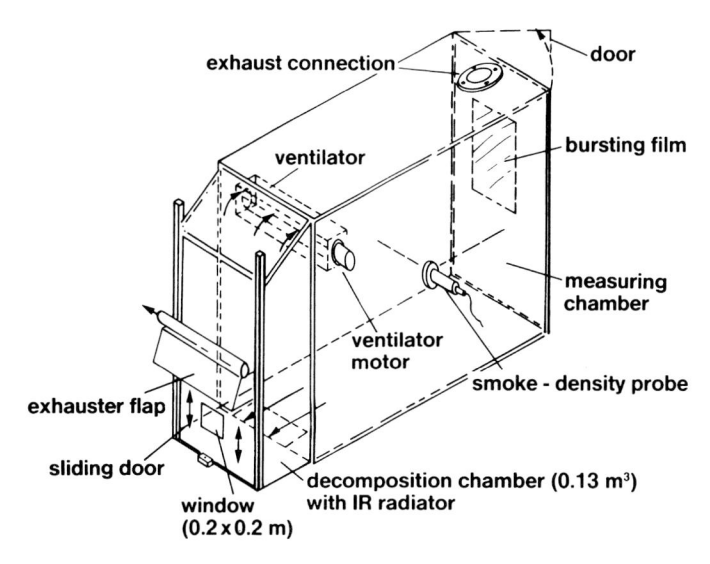

Figure 56b ISO Smoke Box –Dual Chamber Box according to NEN 6066 1075

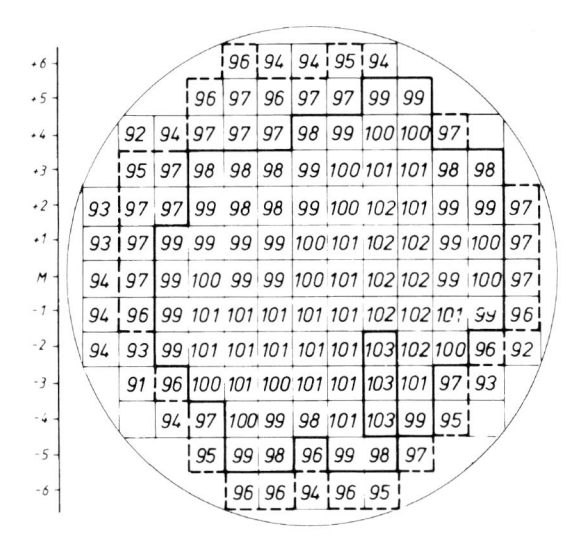

Figure 56c Radiation distribution at the test sample surface

valid ISO standards. Both the French and the Dutch fire-reaction test methods are accepted by the Belgian building authorities. In the beginning of the 1970s Herpol developed at the University of Gent a special, universal test protocol, which has been reported by Minne in detail [222]. Figure 57 illustrates the test equipment. The thermal stress was generated by radiation and flame impingement. The objective of the test protocol was a classification system based on indices for ignitability, spread of flame, heat-release rate and smoke production as illustrated by the following formula

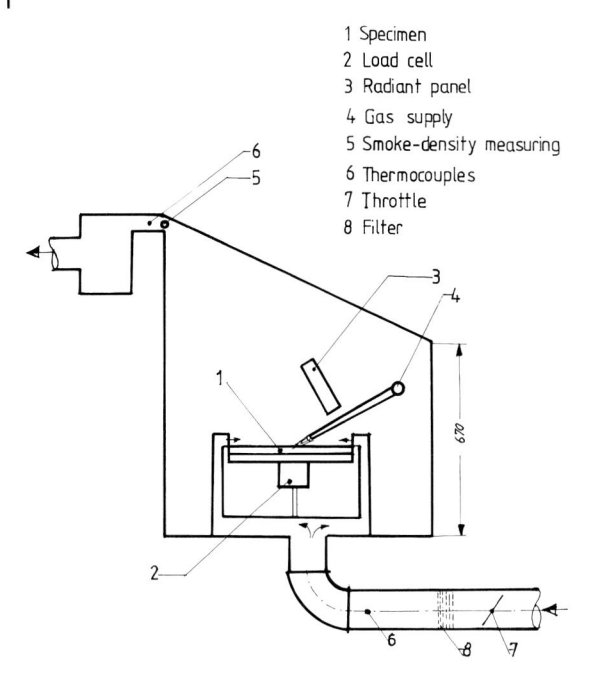

```
              1 Specimen
              2 Load cell
              3 Radiant panel
              4 Gas supply
              5 Smoke-density measuring
              6 Thermocouples
              7 Throttle
              8 Filter
```

Figure 57 Herpol Test Procedure

$$Irf = \frac{K1C + K2/1 + K3/2 + K4P + K5Os + K6Ot}{K1 + K2 + K3 + K4 + K5 + K6}$$

where $K1$ to $K6$ are dimensionless factors. C the net release value, Ir and Ip parameters of ignitability, P the parameters of surface spread and Is and It the parameters of smoke release.

Scandinavia
In Scandinavia, the different national fire-reaction testing and evaluation criteria have been harmonized. Where the building regulations include specific requirements, the national test protocols are still relevant components of the national standardisation. The fire-resistance test protocol according to NT – 005, similar to IS 834 is widely accepted like the fire-reaction test procedures NT – 002, the Schlyter test, NT – 004, the Brown'sche box, NT – 007, the flooring test protocol and NT – 006, the roofing test procedure. The NT – 004 and the NT – 007 test protocols provide, in addition to the ignition-risk values, material-related smoke-release data. The ignition risk of building materials in Scandinavia will be determined by means of the Schlyter test and the Brown'sche box. In the Schlyter test, illustrated in Fig. 58a, one of the two test pieces with outer dimensions of 800 mm × 300 mm × thickness mm will be directly effected by the flames of a special burner. The contribution to the surface spread will be determined and registered. Failure occurs, if the flame front spreads to the second sample. In the Brown'sche box procedure (see Fig. 58b), samples with

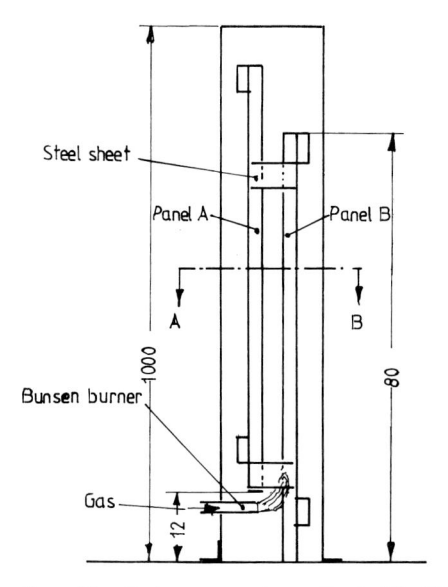

Figure 58a Test arrangement according to NT – 002

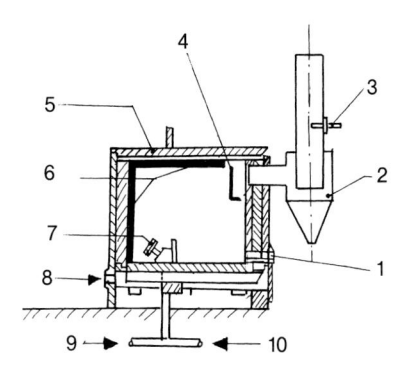

1. View port
2. Cyclone
3. Thermocouple
4. Flame guard
5. Lid
6. Specimen
7. Burner
8. Secondary air
9. Primary air
10. Gas

Figure 58b Test arrangement according to NT – 004 Brown'sche Box

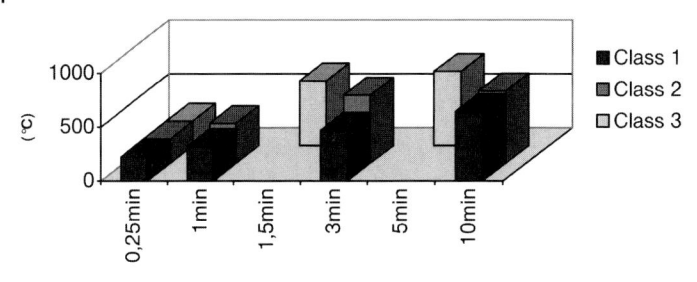

Figure 58c Classification scheme of NT – 004

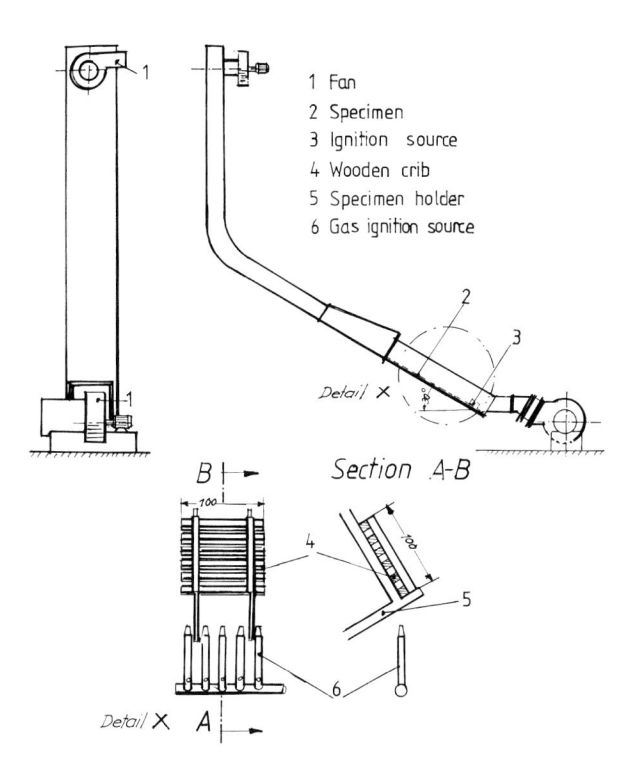

1 Fan
2 Specimen
3 Ignition source
4 Wooden crib
5 Specimen holder
6 Gas ignition source

Figure 58d Test arrangement according to NT – 006 and 007

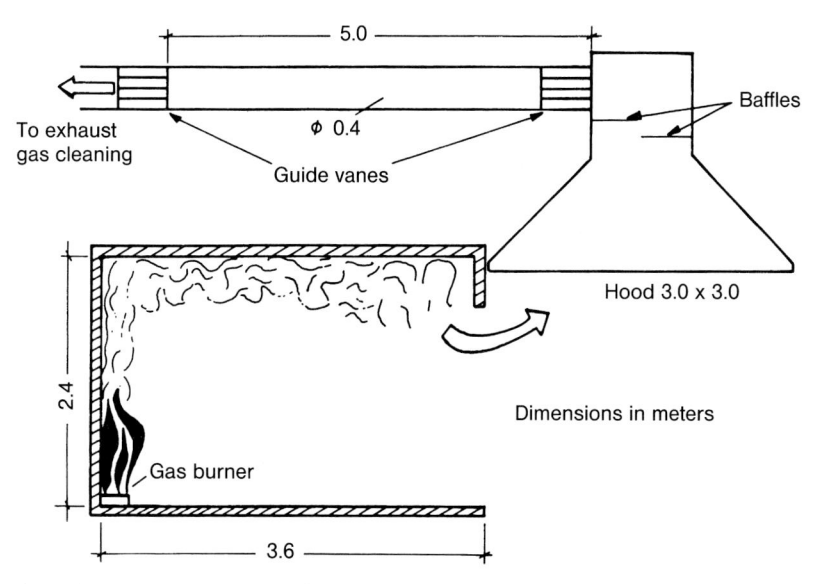

Figure 58e Room-Corner Test Configuration ISO 9705

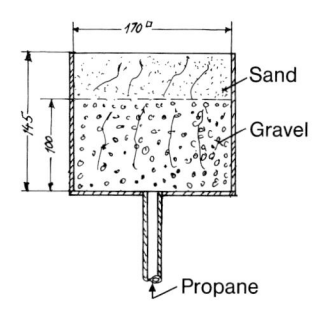

Figure 58f Sand-bed burner

outer dimensions of 228 mm × 228 mm × 4 mm are arranged as wall and ceiling linings. The samples will be effected by flames of a propane-gas burner (2.4 l/min propane 24 l/min air) for 5 and/or 10 min. The ranking is based on the temperature development of the effluents according to Fig. 58c. The smoke-release rate will be determined by an optical measuring system. The course of the light absorption will be registered. The test sample of the flooring test NT – 007 will be exposed to a special ignition source, see Fig. 58d within an airstream of 2 m/s. This test is passed if the flame front is not exceeding the 550-mm bench mark and the light absorption of the effluents within the first 5 min stay below 30% and later below 10%. The same test procedure is standardised for determination of the resistance against flying brands and radiated heat. Smoke-release rate is to be assessed.

Scandinavian testing institutions have an essential part in the relevant standardisation and research work of SC1 of TC 92 and introduced the cone-calorimeter test, illustrated by Fig. 21, and the room-corner test procedure, see Fig. 58e, in connection with the Eurific Program [223, 224]. The cone-calorimeter test equipment consists of a conical electrically heated radiator, a sample holder, a weight-loss measuring system and an exhaust hood. The weight-measuring system permits a continuous weight-loss determination. The radiator generates a variable thermal stress of the horizontally or vertically test sample with outer dimensions of 100 mm × 100 mm of 10 to 100 kW/m^2. The thermal decomposition products will be ignited by a spark igniter. The heat-release rate will be determined by thermocouple recordings in the exhaust hood or by the oxygen consumption technique. The latter is based on the fact that the oxygen consumption is directly proportional to the heat of combustion. Standardisation of the room-corner test according to IS 9705 was initiated at the ISO TC 92 session 1976 at Sydney, in order to offer an appropriate, material-related test protocol of acceptable size in relation to the standardisation activities of TC 61 Plastics for expanded insulating materials [225]. The material to be tested will be applied as wall and ceiling lining of the test room according to Fig. 58e. The outer dimensions for the test room are 3.6 m × 2.4 m × 2.4 m. The ignition will be initiated by a special propane-gas burner (170 mm × 170 mm), see Fig. 58f. The thermal stress is limited for the first 10 min to 100 kW. If flashover does not occur, the heat stress is increased for the second 10 min to 300 kW. Smoke-density measurement and the analytically determined concentrations of the most important components are conducted in the fume hood situated above the doorway. The heat-release rate is determined by means of the oxygen consumption according to the general equation $XC_nH_n + YO_2 = ZCO_2 + H_2O + kcal$.

4.1.3
Eastern Europe and Far East

Eastern Europe
Hildebrand discusses the relevance of those test and classification systems of the former DDR. In [226] he gives a survey of the common test procedures of the DDR including their validation by risk-related large-scale tests [227–229]. The fire-reaction testing in the former DDR was regulated mainly by TGL 10658. The main test procedures were:

- the fire-resistance testing of building constructions analogous to IS 834
- the determination of the combustibility of building products by TGL 10685, a 750 °C furnace test
- the determination of the contribution to the surface spread of flame by TGL 1068, a chimney-test procedure similar to DIN 4102
- the contribution of facade and roofing partitions to the surface spread of flames under realistic enduse conditions.

A risk-oriented facade test, similar to a SNIP procedure of Russia, could also be conducted in the fire-test house of Laue/Leipzig. The risk of fire spread to the next floor was assessed with a defined gap between the ceiling and the facade test element. The smoke problem was mainly determined by the procedures standardized by GOST. The schematic drawings of Figs. 59a and b illustrate the test procedures, which are used for the smoke-density determination and the bioassay testing and evaluation of the effluents. In the two-chamber system of GOST [230, 231], the samples with outer dimensions of 40 mm × 40 mm × max. 8 mm are tested under flaming as well as smoldering conditions. The irradiance level of the radiator was varied from

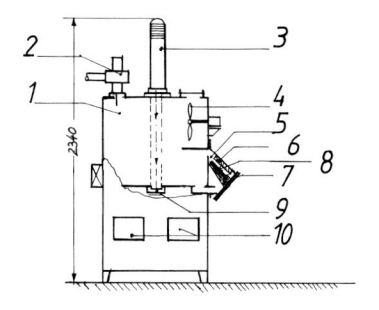

1 Smoke chamber
2 Ventilation
3 Light source
4 Fan
5 Combustion chamber
6 Radiator
7 Specimen holder
8 Specimen
9 Photocell
10 Measuring instruments

Figure 59a Smoke-density measurement according to Gost 12.1012 1052

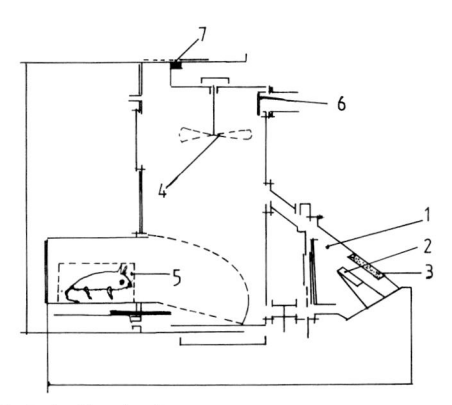

1 Combustion chamber
2 Specimen holder
3 Electric radiating panel
4 Mixing fan
5 Cage for experimental animals
6 Purging valve
7 Inflammable rubber gasket

Figure 59b Toxic potency of effluents according to Gost 12.1012 ???

96 kW/m² (ca. 750 °C) to 25 kW/m² (ca. 400 °C). The size of the combustion chamber is 0.003 m³. The standardized test duration is 15 min. The decomposition products are guided into the measuring or exposure chamber, whose volume is similar to the NBS Chamber, 0.51 m³.

Australia

In Australia the design specifications are based on the Australian Model Building Code. The fire-reaction test and evaluation criteria are standardized in the different parts of AS 1530, as there are:

- part 1 – Combustibility of Materials
- part 2 – Test for Flammability of Materials
- part 3 – Test for Early Fire Hazard Properties of Materials
- part 4 – Fire-resistance test on Element of Building Construction

The test procedure for building elements correspond to the framework of IS 834, see Fig. 40a. Basic scientific investigations have been carried out especially by Harmathy [232, 233]. Noncombustibility is determined in a similar way to the 750 °C furnace test illustrated by Fig. 41a. The framework of the fire-reaction testing of building materials is given in part 3. The Early Fire Hazard test equipment, which was developed from realistic large-scale testing [234], tests samples with the dimension 600 mm × 450 mm × thickness mm that are vertically positioned on a movable platform parallel to the gas-fired panel. The distance between sample and the radiator with outer dimensions of 300 mm × 300 mm is reduced during the test run from 850 mm to 175 mm. Two pilot flames, each 12 mm long, are positioned in front of the sample surface. Movement of the sample holder is stopped as soon as the decomposition products are ignited. Ignition is defined as flaming for more than 10 s (see Fig. 60a).

Figure 60a Early Fire Hazard Test according to AS 1530 1028

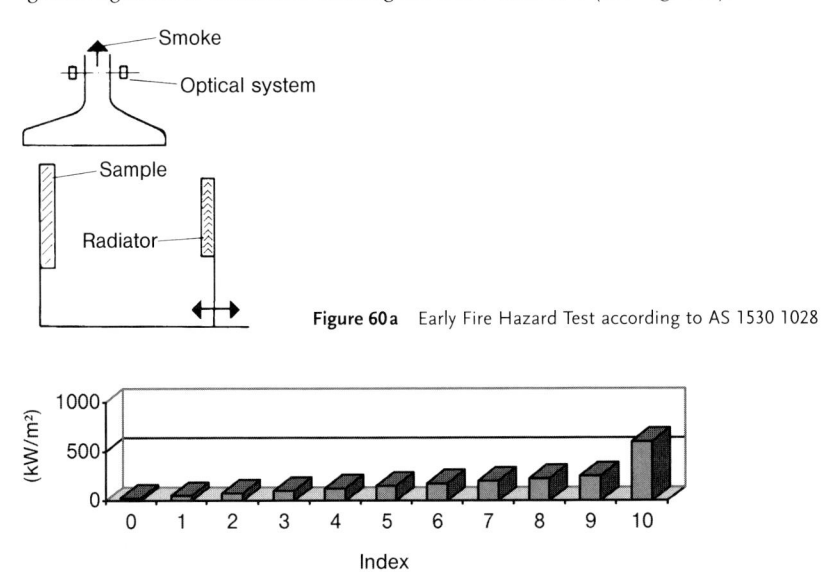

Figure 60b Classification of the heat-release rate according to AS 1530

Table 23 Ignitability – Index evaluation according to AS 1530

Index	1.33x medium time space till the increase of 1,4 kW/m² (s)
0	≤ 270
1	240 – < 270
2	210 – < 240
3	180 – < 210
4	150 – < 180
5	120 – < 150
6	90 – < 120
7	60 – < 90
8	30 – < 60
9	10 – < 30
10	< 10

The classification of AS 1530 is based on four indices for ignitability, flame spread, heat release and smoke evolution. The irradiance level of the radiator is recorded during the test run every 3 s. The ignitability index is calculated as the difference from 20 and the time up to ignition. Provided that less than three of the nine specimen are ignited, the index 0 will be achieved. Classification is according to the listing of Table 23. Data will be collected and used for calculation as soon as the radiation of the panel increases more than 1.4 kW/m² within 203 s after ignition. If this increase does not occur, the index 0 is valid.

The material-related ranking of the heat-release rate in (kJ/m²) is determined by means of integration over a period of 2 min after ignition took place. The data set of Fig. 60b provides an overview. As soon as 5 or more samples do not ignite, the index value 0 is valid.

The smoke-density measurement according to AS 1530 pt. 3 1982 is carried out by means of an optical system with a light path of 305 mm within an exhaust chimney. The test data are measured and recorded every 3 min. The determination of the optical density D according to the mathematical relationship

$$D = 1/L \log. 10 \times 100/100R$$

at what

D = optical density (1/m)

L = light path (305 mm)

R = maximum reduction of the transmission during the measuring space of 1 min.

The index values will be calculated also for smoldering conditions as flaming conditions with respect to Table 24.

Japan

The measures of preventive fire protection are based on the definitions given in the Building Standards Law, which became effective in 1950 (CIB Szuzuka) The testing of fire resistance of building constructions is analogous to IS 834. The noncombustibility testing of building materials is carried out according to IS 1182, the 750 °C fur-

Table 24 Smoke density evaluation according to AS 1530.

Index number			Mean optical density			
	[m⁻¹]					[m⁻¹]
0	<K					< 0,0082
1	K	–	< 2K	0,0082	–	< 0,0164
2	2K	–	< 2²K	0,0164	–	< 0,0328
3	2²K	–	< 2³K	0,0328	–	< 0,0656
4	2³K	–	< 2⁴K	0,0656	–	< 0,131
5	2⁴K	–	< 2⁵K	0,131	–	< 0,261
6	2⁵K	–	< 2⁶K	0,261	–	< 0,525
7	2⁶K	–	< 2⁷K	0,525	–	< 1,05
8	2⁷K	–	< 2⁸K	1,05	–	< 2,10
9	2⁸K	–	< 2⁹K	2,10	–	< 4,20
10		≥ 2			≥ 4,20	

K = 0,0082

nace test illustrated in Fig. 41a. The acceptability of building materials is regulated by the conditions of:

- Notification Nr. 1828 for noncombustible materials.
- Notification Nr. 1231 for seminoncombustible and fire-retardant materials
- Notification Nr. 101 for fire-retardant plastics for roofing.

In the notification of 1231 also the aspects of heat release, smoke density and smoke toxicity are regulated together with combustibility.

Combustible and seminoncombustible materials have to be classified according to JIS A1231 Testing Method for Incombustibility and the so-called Face Test. Samples of lining materials with outer dimensions of 220 mm × 220 mm × ≤ 15 mm are positioned at a distance of 35 mm from the radiating elements in the test procedure similar to BS 476 pt. 6, the fire-propagation test illustrated by Fig. 50b. The electrical heating elements are switched on 3 min after the flaming impingement. It is calibrated using an asbestos cement board. The time–temperature curve is given in Fig. 61a. None of the test samples must be penetrated by melting or cracks nor critically deformed. The afterburn time is limited to 90 s. The deviation of the time–temperature curve of the noncombustible materials must not increase more than 50°C above the calibration curve. As far as seminoncombustible and the fire-

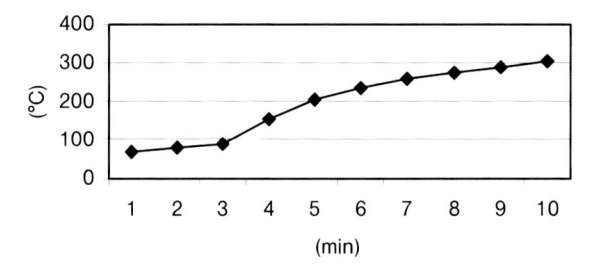

Figure 61a Calibration curve according to JIS A 1231

retardant materials are concerned the integral of the smoke temperature curve must not deviate after 3 min more than 100 °C min and/or 350 °C min of the standard curve. The classification is illustrated by Table 25.

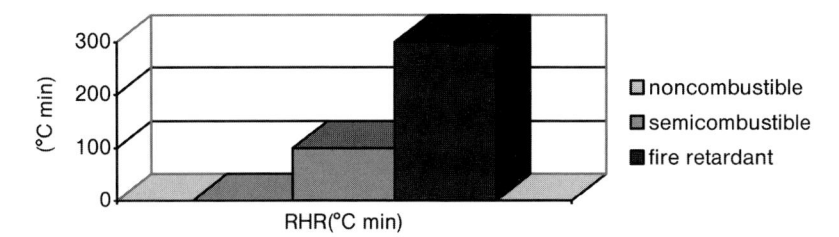

Figure 61 b Classification of building products

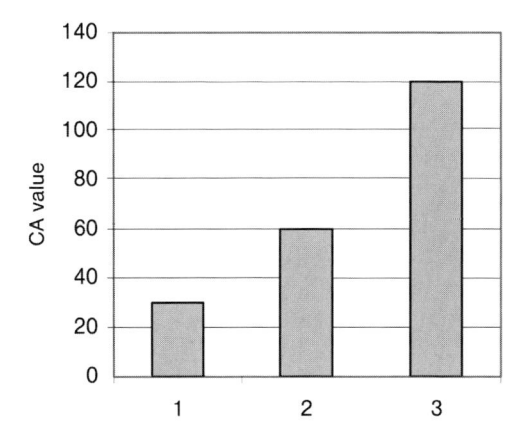

Figure 61 c Smoke formation according to JIS A 1231

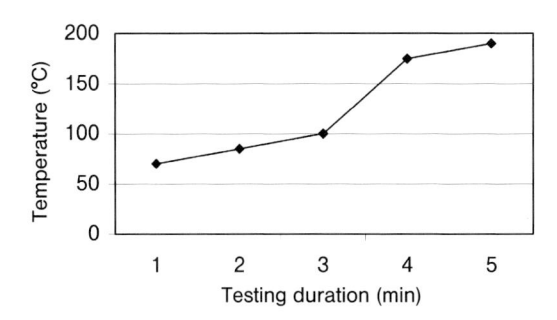

Figure 61 d Thermal Stress in the Face Test

Table 25 Fire reaction classification according to JIS A 1231

Classification	Heating time of burner (min)	Heating time of radiator (min)	Total heating time (min)
Non-combustible	3	7	10
Semi-non-combustible	3	7	10
fire retardant	3	3	6

By notification Nr. 1231, smoke density and smoke toxicity are regulated by the Ministry of Construction. The required test procedure, the so-called Hole Test is similar to the combustibility test method. The test pieces with outer dimensions of 180 mm × 180 mm will be perforated by additional holes (25 mm Ø). The afterburn time is limited to 90 s and the heat release must not increase above 100 °C min. The smoke-density classification has to be carried out according to Fig. 61c. The determination of the material data has to be calculated according to the formula $CA = 240 \log. 10 \times I_0/I$.

In the case of the flue-gas toxicity the bioassay experimental results will be compared with the results of the reference material Lauran. The decomposition products will be produced with samples (220 mm × 220 mm or 68 mm × 68 mm × thickness mm) in the Face Test (JIS A1231). The primary air stream is 3 l/min and the secondary air stream is 100 l/min. The decomposition products are produced by a combined thermal stress of radiation and flame impingement (see Fig. 61d). The flame impingement is limited to the first 3 min of the total test duration of 15 min. For the second 3 min of the test an additional thermal stress is applied by radiation. The resultant stream of effluents is guided into the mixing chamber above the combustion device. The CO_2/CO ratio of the effluents stays, as a rule, below 10. The assessment of the test results uses the criteria of lethality and reduced activity, for example, with the rotating wheels (see Fig. 62a). The second decomposition model, the RTB protocol shown in Fig. 62b, is based on the cone-calorimeter test method with the widely used test protocol for heat-release measurements. The thermal stress upon the horizontal test sample will be produced by the conical, electrically heated radiator, the ISO ignitability apparatus as used in the ISO Dual Chamber Box. The thermal stress can be varied between 10 and 50 kW/m² and the air stream between 5 l/min and 20 l/min. The loss of weight will be continuously measured and recorded.

The relation to actual fires is the goal of the Box Procedure developed by Saito. Testing details of this method, see Fig. 62c, are given in [235, 236]. The test samples can be used as wall linings (145 mm × 145 mm × 50 mm) and ceiling linings (120 mm × 120 mm × 50 mm). The thermal stress is produced by a burning crib (cypress wood 8.8 to 9.2 g, single pieces 5.8 mm × 5.8 mm × 40 mm). The airflow of 12 l/min is similar to a five-fold air change. The smoke release is significantly reduced after 6 min burning to avoid a dilution. The time until a significant reduction of the activity of the test animals on the rotating wheels occurs is recorded as the test result.

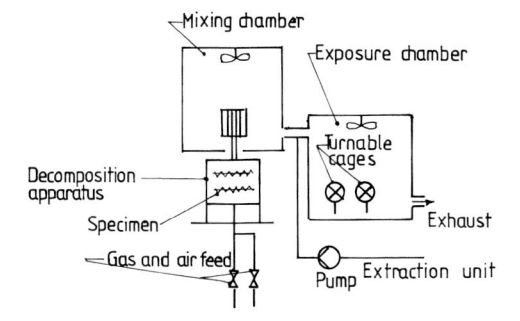

Figure 62a BRI-Decomposition Model

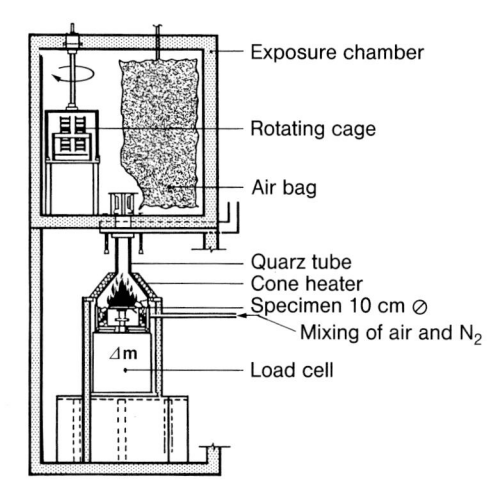

Figure 62b RTB-Decomposition Model

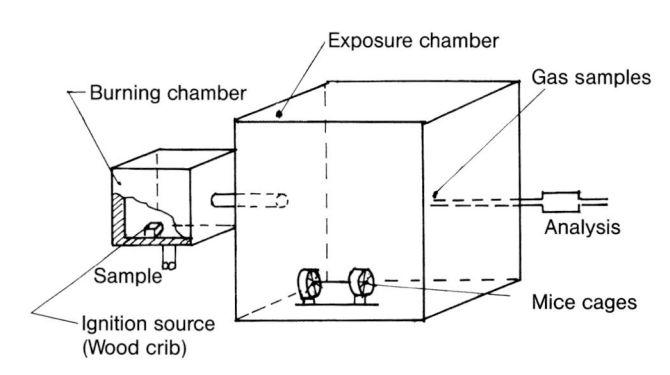

Figure 62c Decomposition Protocol – Box Method

People's Republic of China

The test and evaluating criteria of the fire-reaction testing are laid down in the National Standard GB 86 24-88 are analogous to DIN 4102, the German test specifications. The noncombustibility test corresponds to ISO 1182-83. The test procedures for classifying the combustible materials are the chimney test according to Fig. 47a and the small-burner test according to Fig. 49a. The smoke density of fire effluents are tested according to the A classification, if necessary under flaming conditions only by using the XP2 chamber.

4.1.4
International Standardisation – European Harmonisation

At the beginning of the 1970s Herpol tried to develop a universal European fire-reaction test method at the University of Gent. The goal of that investigation was a material-related classification. Figure 57 illustrates the test procedure. The test specimen is tested under the impact of external radiation and indices concerning the ignitability, spread of flame, heat and smoke release were to be determined. Within the Nordtest project 34-75 Minne [222] discussed this test protocol in detail. Within the EC the CE mark symbolizes the free-trade market. The attempts to harmonize the different national fire-reaction testing systems have underlined the findings of Emmons [180]. Emmons demonstrated, see Fig. 39b, the difficulty in harmonizing the various classification systems. The Blachere report [237] initiated by the European Community, came to the same conclusions. No chance existed to compare the different ranking systems by calculation. The first attempt to get a solution by a) the three-sister model, a combination of the three important European fire-reaction tests, the chimney test according to DIN 4102 pt. 15/16, the Epiradiateur according to NF-P 92501 and the Surface Spread of Flame according to BS 476 pt. 7 and b) the Eurofic Program [238], a combination of the Cone-calorimeter test according to IS 5660 and the Corner Test according to IS 9705 failed. This latter combination was especially propagated by the Scandinavian representatives, which claimed it was a risk-oriented solution but neglected the fact that those pure material-related data do not correlate with the enduse conditions required by the robust solution according to Table 3. This test combination provides realistic, practice-oriented test conditions for a very special fire scenario only. The test parameters correlate with a fire scenario of an empty room, a small cabin, e.g. an empty kitchen, where the fire has been initiated by an ignition source of high intensity, similar to a broken gas pipe. In the cone-calorimeter test, the test sample with outer dimensions of 100 mm × 100 mm × max. 50 mm is horizontally or if necessary vertically orientated. The thermal load of 10 to 100 kW/m^2 will be produced by a conical radiator, shown by Fig. 21. The distribution of the irradiance across the surface of the samples is illustrated by Fig. 50b. The common test period is 20 min. The thermal decomposition products will be ignited by a defined ignition source (pilot flame, spark ignition). Smoke gas temperature and the concentrations of the most important smoke components will be measured in the exhaust duct. The reduction of visibility by the effluents will be assessed by an optical measuring system. In the room-corner test, ignition will be by a sand-bed burner, see Fig. 58f, which is located in one corner of the room producing the thermal stress for 20 min. The thermal power, which is limited for the first 10 min to 100 kW will be increased to 300 kW for the second 10 min. The evaluation criterion is the time of flashover situation, which, e.g. is defined by the first escape of flames. This criterion is obviously effected by the material properties, the room geometry and the ventilation conditions. The determined data records are not transferable without special evidence. The initiators of the Eurific-program suggest that this reflects natural fires. The verified correlation of the cone-calorimeter and the room-corner test prove, in their opinion that the Cone-Calorimeter method is risk related and justify its choice as the basic test procedure for mathematical modelling [239, 240]. Risk-related estimations of this programme suggest that the required increased fire-protection safety can be defined by means of cone-calorimeter test data. The data records of Table 26 clarify the proposed classification scheme of the Eurific program.

Table 26 Classification scheme of smoke density requirements

classification	test (min)	melting, cracks defraction	afterburn time (s)	time – temperature area (°C* min)	smoke (CA)
non combustible	10	not permitted	≤ 30	0	≤ 30
semi non combustible	10	not permitted	≤ 30	≤ 100	≤ 30
fire retardent	6	not permitted	≤ 30	≤ 300	≤ 120

For the future robust solution, see Table 4, the European harmonisation of the various national fire-performance requirement profiles all indicate, currently, a reliance on relevant ISO procedures and a new central test protocol, the Single Burning Item (SBI) test [241]. The listing in Fig. 63a provides an overview of the connections of possible fire development with the relevant parameters of the selected test methods.

Noncombustibility will be alternatively determined with the IS 1184 and the bomb calorimeter standardized in NF-P 92510 similar to IS 1716. Released heat potential must not be greater than 2500 MJ/kg.

The resistance against ignition sources of low intensity levels will be tested by a small-burner test, see Fig. 49a. Resistance against ignition sources of raised intensity has to be assessed by the new test method, the SBI test, whose schematic is illus-

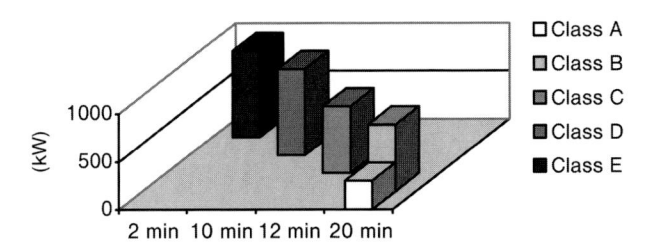

Figure 63a EG-Classification Scheme

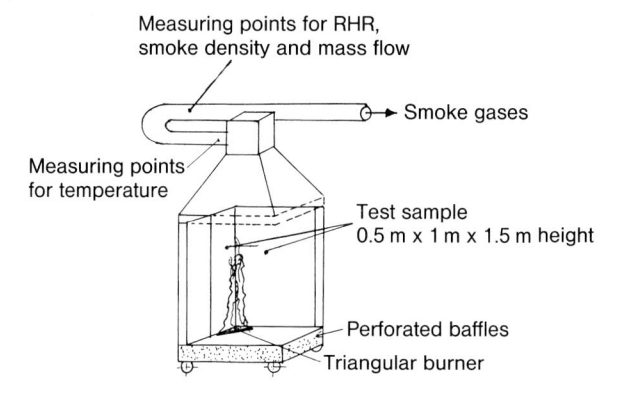

Figure 63 b Single Burning Item (SBI) Test

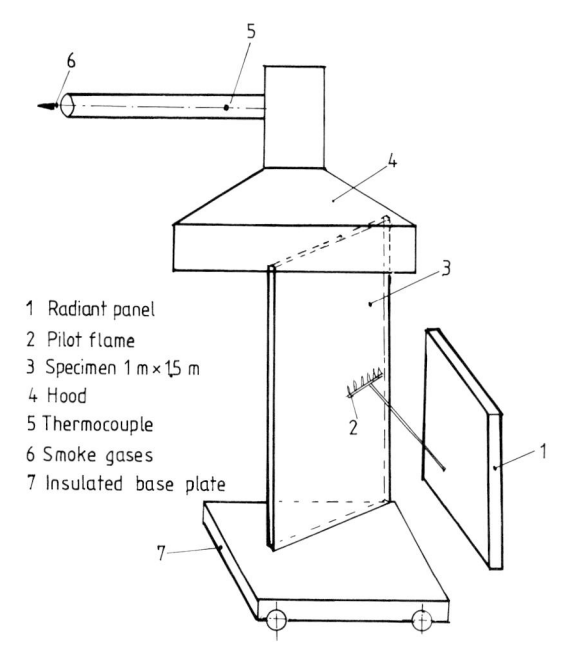

1 Radiant panel
2 Pilot flame
3 Specimen 1 m × 1.5 m
4 Hood
5 Thermocouple
6 Smoke gases
7 Insulated base plate

Figure 63 c Roland Test Configuration 1063

trated by Fig. 63b. Evaluation criteria are the contribution to the surface spread of flame and the side effects burning dripping and the release of heat and smoke. In the SBI test two test pieces with outer dimensions of 1000 mm × 1500 mm and 485 mm × 1500 mm with a maximum thickness of 200 mm will be arranged in a corner configuration and thermally stressed by a special triangular sand-bed gas burner 200 mm × 250 mm for 20 min. Two different layers of sand are used for the sand-bed torch, 60 mm layer of 4–8 mm and a layer of 2–4 mm granules. The gas flow of the torch is 0.5–0.6 l/min. The thermal stress with the torch flame (30.7 kW) takes place in the corner, where a maximum contribution to the surface spread of flame and heat and flue-gas liberation is to be to expected. The time to ignition has to be observed and registered. Surface spread of flame and burning dripping are to be visually judged while heat and smoke release have to be measured and noted. The material to be tested is to be mounted on a calcium silicate board. Application-oriented free-hanging specimens with and without an airgap may be used. Composites, for example, steel-sandwich elements, have to be tested with an additional real life joint, arranged at a distance of 200 mm in the larger test piece to simulate the end use.

The most important evaluation criteria are agreed as:

THR 600 The total released energy within the first 600 s.
LSF   Edge spread of flame to the lateral edge.
TSP 600 The total formation of smoke within the first 600 s.
FDP   Burning dripping

The main burner will be ignited within the time period of 1–120 s. The observation time is 1200 s. At $t = 120$ s, the auxiliary torch with 647 ±10 mg/s propane gas will be ignited. At $t = 300$ s, the gas flow to the auxiliary burner is coupled with the main torch. The end of the testing period is at $t_{end} = 1560$ s.

Burning dripping will be judged during the first 600 s only. It has to be confirmed, if

- the RHR value exceeds 350 kW or for more than 30 s the average value exceeds 300 kW
- the temperature in the exhaust pipe reaches 400 °C or for more than 30 s the average value of 300 °C.
- material falls on the burner

As illustrated by the listings in Tables 27 to 29 an additional classification of the smoke release is used for classes C and D. The data records of these tables provide an overview of the material-related characteristics to be considered as classification parameters. A further differentiation of the ranking includes fire-related characteristics like burning dripping and smoke release. Other results are to be determined including average $RHRav(t)$, total value $THR(t)$ for 1000 $RHRav(t)/(t-300)$ for $0 \leq 1 \leq 1500$ s. Figra, the fire growth values is the total of the first 600 s of THR 600 s, $RSPav(t)$, $TSP(t)$ and 1000 $RHRav(t)/(t-300)$

SMOG RA – Smoke growth ratio index
Figra 0.2 MJ – Figra 0.4 MJ for both the border values of THR

In a parallel research work, the using of the so-called Roland test (see Fig. 63c), the similar heat stress is generated by radiant heat.

4 Preventive Fire Protection – National Requirement and Classification Systems

Table 27a Fire reaction classification according to the Eurific Program

Class	Testing time (min)	RHR[1] (kW)	RHR[2] (kW)	RHR[3] (kW)	Smoke[4] (m²/s)	Smoke[5] (m²/s)
A	20	300	0600	050	10	3
B	20	700	1000	100	70	5
C	12	700	1000	100	70	5
D	10	900	1000	100	70	5
E	2	900	1000	–	70	–

1) Maximum value without burner
2) Maximum value with burner
3) Medium value
4) Maximum smoke density
5) Smoke density – medium value

Table 27b Aspired fire reaction classification of building products

Fire Scenario	EURO-Building Materials class	Contribution to fire	Test method
Fully developed fire $I > 60\ KW/m^2$	A	No contribution	Furnace test-CEN TC 127 Calorific potential CEN TC 127
	B	Very limited contribution	Furnace test-CEN TC 127 Calorific potential CEN TC 127 and SBI –Test *
Burning Item $I < 40\ KW/m^2$	C	Limited contribution	SBI Test* CEN TC 127 and Small flame **
	D	Acceptable contribution	SBI Test * CEN TC 127 and Small flame**
Match	E	Acceptable burning behavior	CEN TC 127 Small flame ** DIS 11925–2
	F	No requirement	

* Flame spread including smoke production and burning dripping
** Flame spread and burning dripping

Table 27c Aspired fire reaction classification of flooring materials

Fire scenario	Class	Fire contribution	Criteria
Fully developed fire $I > 60\ kW/m^2$	Afl	No contribution	Very limited calorific potential, very limited heat- release rate, limited mass loss, no flames
	Bfl	Very limited contribution	Very limited calorific potential and limited mass loss, no spread, very limited smoke production
Fully developed fire in the neighboring room $I <10\ kW/m^2$	Cfl	limited contribution	Spread and smoke production very limited
	Dfl	Acceptable contribution	Limited contribution to spread and smoke production
Heat stress by small ignition sources f.e. cigarette	Efl	Acceptable burning behavior	Acceptable ignitability
	Ffl		No requirement

Table 27d Recommanded classification in the SBI Test

A2	Tign < 120 W/s HR600 $< 7,5$ MJ	+ Furnace test ISO 1182 or Calorific Value ISO 1116 Flames do not reach border
B	Tign < 120 W/s THR600 $< 7,5$ MJ	+ Small burner ISO 11925-2 / 30 s Flames do not reach border
C	Tign < 250 W/s THR600 < 13 MJ	+ Small burner ISO 11925-2 / 30 s Flames do not reach border
D	Tign < 750 W/s	+ Small burner ISO 11925-2 / 30 s

Table 28 Smoke density classification in the SBI Test

S1	Smogra $< 30\,\text{m}^2/\text{s}^2$	TSR 600 $< 50\,\text{m}^2$
S2	Smogra $< 180\,\text{m}^2/\text{s}^2$	TSR 600 $< 200\,\text{m}^2$
S3	not S1 or S2	

Table 29 Classification of burning dripping in the SBI Test

D0	No burning dropping (dripping, particles) within the first 600 s
D1	No burning dropping (dripping, particles) with an afterburn time > 10s within the first 600 s
D2	Not D0 or D1 burning dripping in the standard procedure pren ISO 11925-2 part 1

Façade testing

These test methods are based on full-scale trials and on the research work of CIB. The real-scale testing of facades has been given by the working group CIB WG14, part of the predominant national and international committees. The goal of these investigations is the improvement of the resistance of the facade lining against the impact of a room fire. The schematic of Fig. 64 illustrates the usual thermal loading of the facade lining in the full-scale experiment. In two workshops of the WG14 of CIB in Leipzig and Lund [242], the different national approaches were compared. One common effect was the possible spread of the room fire due to the stream of effluents emerging from the windows. The national test houses, e.g. in France (Paris, CSTB see Fig. 55), in Germany (Krefeld-Uerdingen/Bayer AG, Dortmund and Leipzig), Sweden (Boras and Lund), UK, Hungaria, USA and Canada were discussed in detail, see Table 30. The majority of test houses consist of 2 or 3 and more storeys.

The Bayer AG® experimental test house was similar to that of Paris and allows investigations of facade elements. Most of the tests were run to get approval for facade and balustrade elements in a single building project. These test methods were intended also to confirm laboratory test procedures. Wood cribs served in most cases as fire load. The absolute fire load was mainly 25 to 50 kg/m². Partial oil and gas burners were also used to simulate the fire load [230, 243].

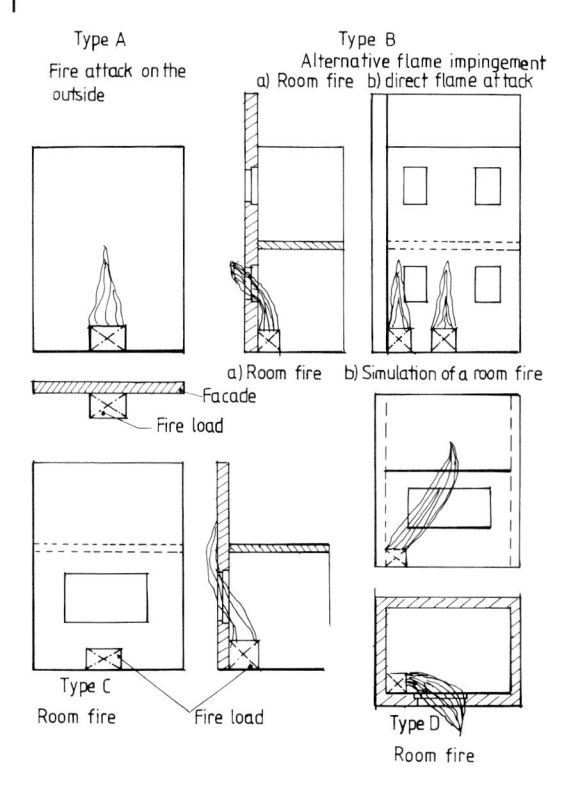

Figure 64 Overview of various test concepts

Table 30 National test arrangements for facade testing

Country/Town	Façade area (m²)	Number of floors	Fire room area	Fire load (kg/m²)
Canada	69,0	3	29,7	25
Germany				
Krefeld	38,2	2 (3)	20,6	50 (25)
Leipzig	88,2	3	21,8	Variable
Paris	29,9	2	12,0	50
Sweden				
Boras	87,2	3	4,2	60l Heptane
Lund	49,2	3	9,2	310 kg
Hungary	77,0	3	18,4	Variable
South Africa	66,2	4	19,1	20 (50)
UK	39,5	4	8,0	400 kg
USA	88,0	2	20,9	580 kg

Flame-spread tests of roof constructions conducted at TNO.

To safeguard against possible fire spread across roofing constructions different full-scale investigations were carried out. Figures 65a to c show in schematics the possible fire scenarios that were evaluated. The resistance against a specific fire load

4.1 Building Section | 119

Figure 65 a Fire risk situation – Fire attack from outside

Figure 65 b Fire risk situation – Fire attack from inside

Figure 65c Fire risk situation – Fire penetration

was investigated in the many relevant studies. In the test series at TNO Delft [244] and a similar test at FRS/UK the influence of a room fire to the spread across the roof was studied (see Fig. 66a). The fire load used was 25 kg/m^2. Careful selection of the ventilation conditions showed that the fire load produced a thermal stress shown in Fig. 66b. Figure 66b show the fire-gas temperatures in the ceiling field.

Large-scale testing test building

Arrangement of fire load and measuring points

• = Temperature
▲ = Smoke
■ = Radiation

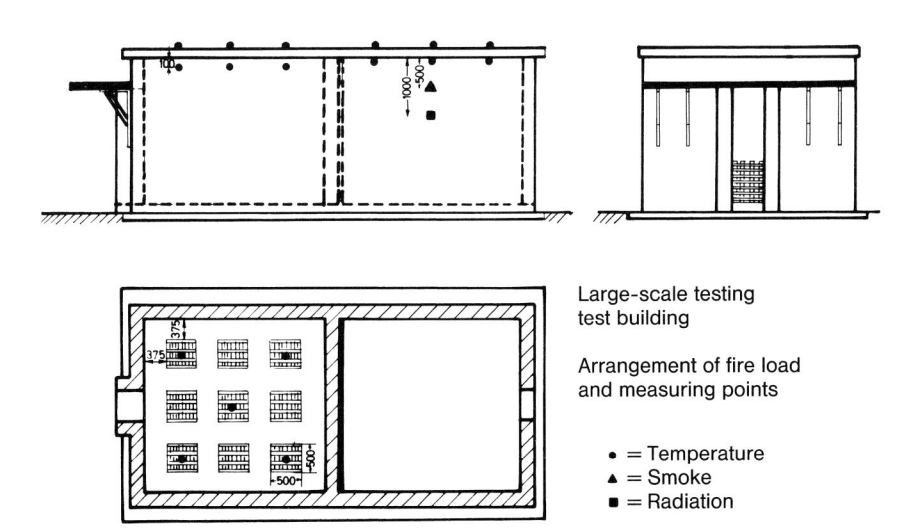

Figure 66a TNO Test House – Fire-load distribution

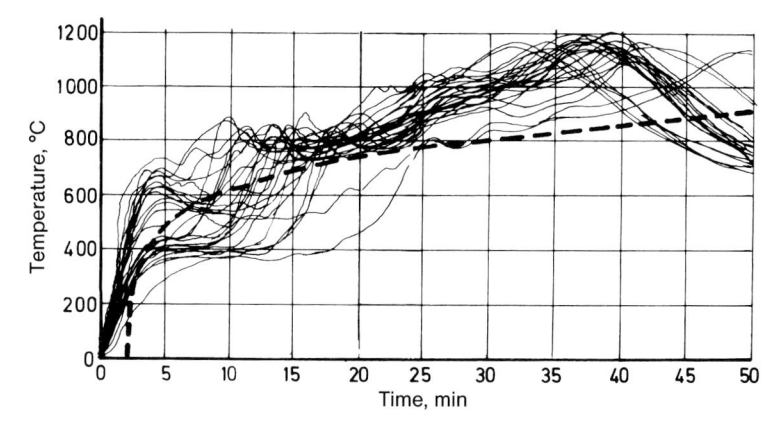

Figure 66 b TNO Roof Tests – Smoke-layer temperature and ETK

Harmonisation of European Fire Reaction Testing

The harmonisation of fire reaction testing and evaluation performed in the building section by CEN TC 127 is aspired as well as in the furniture and furnishing field (CEN TC 207) as well as in the transportation area (CEN TC 256 and CENELEC TC 9X). In the furniture and furnishing field agreement was reached for ignition sources of low intensity with the cigarette test (DIN EN 1021-1) and the match test (DIN EN 1021-2). The impact of larger ignition sources is investigated by the furniture calorimeter (EN 45545-2). EN 455 45 provide information about the fire reaction test parameters in the railway section. These requirements are usually based on ISO standards (ISO 5656-2 spread, ISO 5659-2 Smoke Generation, 5660-1 Reaction to Fire, EN ISO 9239-1 Flooring Test and EN 45545-2 Furniture Calorimeter). The national test orders are layed down e.g. in DIN 5510, NF-F16-10 or BS 6853. The European Directive 95/28/EG requires e.g. an additional dripping test for ceiling linings in motor vehicles, which is covered by NF-P902-505. A test simulating vandalism (with an ignition source of higher intensity 8100g paper cushion) has been standardized by DIN 66084 Classification of the burning behaviour of upholstery compounds.

4.2
Furniture and Furnishing

4.2.1
American Testing and Evaluation Criteria

National and regional fire-performance requirements have been adapted in the United States of America. Based on the Flammable Fabric Act several minimum standards were defined. At the investigation of the Consumer Safety Commission (CPSC) testing and evaluation criteria of a cigarette test were developed for mattresses [245]. The schematic drawing in Fig. 68a clarifies the construction of this test procedure. A glowing cigarette will be applied to a mattress section. The influence of a smoldering cover, possibly present in a real fire, was taken into account. The standard requires the examination of the complete mattress construction, as demonstrated by Fig. 68a, with two cotton sheets. The glowing cigarette has to be placed on various positions on the bare mattress as well as with the cotton sheets. The mattress meets the specified performance criterion, if no charring/smoldering extends for more than 5.1 cm in any direction. A similar cigarette test has been standardized in Canada. While a cigarette test has been adapted for the mattress section as a federal regulation, a similar test for upholstered seating has so far only reached the draft stage. The draft federal cigarette test development at NIST, the former NBS, is still not standardized. Figure 67b illustrates the test procedure.

The proposed standard later became the basis for the UFAC cigarette test protocol for self-certification of the relevant industry [246]. This self-certification offered by

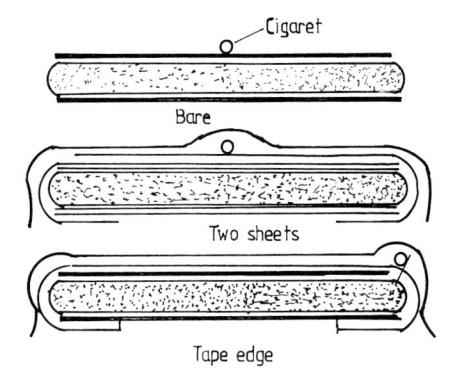

Figure 67a Mattress – cigarette test

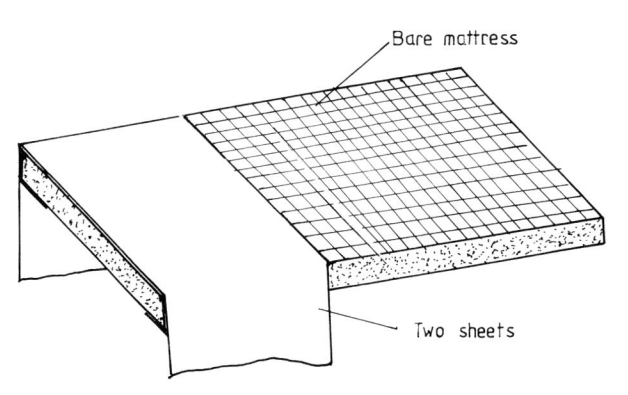

Figure 67b Upholstered furniture – cigarette test

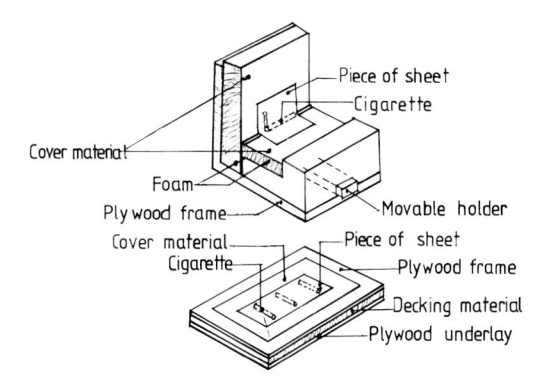

Figure 67 c UFAC Regulation

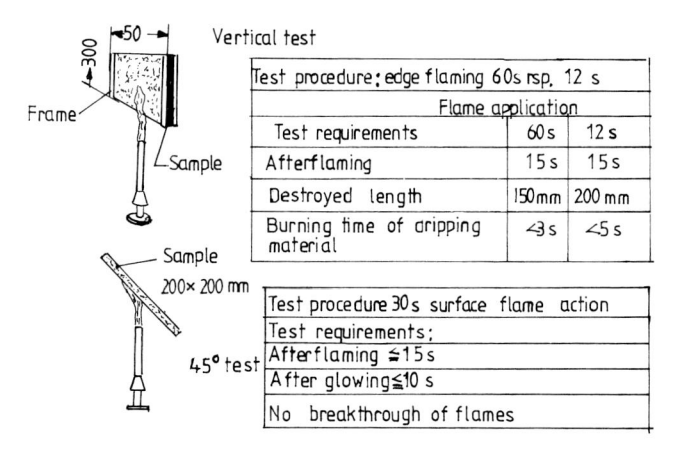

Figure 67 d FAR vertical test configuration

the upholstered furniture industry requires tests according to Fig. 67c. The UFAC marking guarantees that consumers have a real chance to select furniture with respect to safety reasons. Figure 67c demonstrates that only those components where proof is available that there is no risk of a smoldering fire are used in the production. The validity of the self-certification organized by UFAC, is checked by the CPSC. The most stringent regional requirements have been adopted in California on the basis of the Home Furnishing Act of California. A detailed overview is given in [247] and [248]. Pure material-related requirements as well as enduse-related compound testing have been defined. The vertical test according to the FAA (Federal Aviation Administration) regulation, which is shown by Fig. 68c and that should result in flame retardant upholstery components, and the cigarette test according to Fig. 67d have been standardized.

The ignition requirements in relation to Fig. 67 are met if on average the destroyed length does not exceed 152 mm. Single values must remain smaller than 203 mm. The average afterburn time is limited to 5 s. Single values must stay below 10 s. The

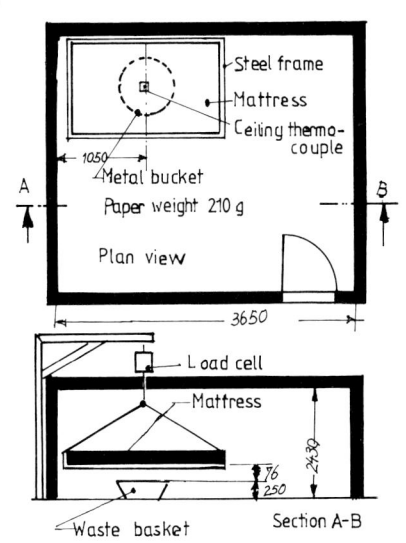

Figure 68a Mattress test according to TB 121

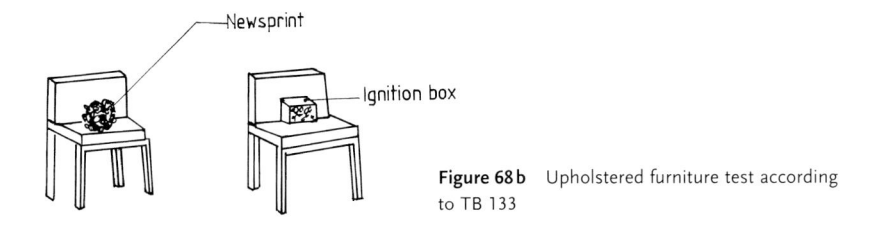

Figure 68b Upholstered furniture test according to TB 133

permitted weight loss of 3% of upholstered furniture with shredded fillings will be tested by a 12-s flame impingement. By methenamine tablet impact the weight loss of upholstery furniture with EPS filling must be below 5%.

The resistance against ignition sources of higher intensities has been regulated, to ensure sufficient safety in application areas of high risk, like hospitals or prisons. Figure 68a illustrates the test configuration, which is described in detail by the Technical Bulletin 121 [247]. Under the heat stress of a burning wastepaper basket neither the flue-gas temperature nor the weight loss and the critical effluent component CO must increase above the permitted values. The relevant test protocol for upholstered furniture is regulated by TB 133. A detailed survey is given among other things in [247, 248]. The test configuration, shown by Fig. 68b, requires the resistance against 106 g of burning newspaper. The permitted weight loss is limited on 10%. To gain a better reproducibility of test data, the newspaper ignition source is supposed to be replaced by a special torch burner developed at NIST [38]. Comparison with various upholstery combinations demonstrated a good agreement. A modification of the NIST burner became part of the EC research program for larger ignition sources within a harmonized furniture regulation.

In the United States of America some additional regional fire-performance requirements for upholstered furniture are regulated. After the disastrous BOAC ter-

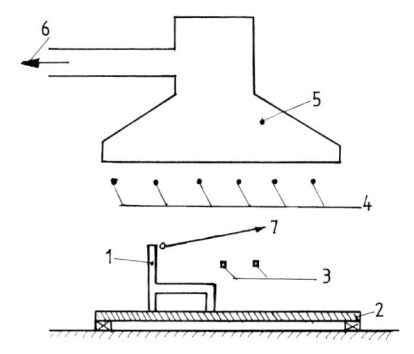

1 Test specimen
2 Weighing platform
3 Radiometer
4 Thermocouples
5 Hood
6 To gas analysis
7 To HCN analysis

Figure 68c Furniture Calorimeter 1061

minal fire [249] testing and evaluation criteria have been stipulated by the New York Port Authority. Material-related requirements are published on the basis of FAR part 25 [250], see Fig. 67d, and the horizontal flammability test according to ASTM- D635 or UL Subject 94, see Fig. 54c. More stringent requirements for upholstery fillings has been enacted to reduce the contribution to the surface spread of flame. The Radiant Panel Test according to ASTM – E 162, which was developed for the preselection of building materials, must not give results above admissible values. The certification can be achieved by the upholstered combinations. A modification of ASTM- E 162 has been standardized (ASTM – D 3675) for upholstery fillings. Realistic test results are aspired to by the furniture calorimeter test, shown in Fig. 68c. The fire performance of whole items of upholstered furniture are investigated in this large-scale test procedure. Increased ventilation ensures that the released thermal decomposition products pass into the exhaust duct.

4.2.2
European Test and Evaluation Criteria

In Europe, disregarding the United Kingdom and Ireland, there are currently no fire-performance regulations in the private area. As far as furniture and furnishing are concerned, fire performance has been set in the contract sector, for buildings of specified use. These are applied in connection with building-material-related regulations, especially for the public sector.

United Kingdom and Ireland

In Great Britain, fire-protection regulations for the domestic furniture were supported by detailed information of the excellent fire-statistics data of the Home Office as well as suggested by the findings of extensive, risk-relevant large-scale tests [64, 251]. The Furniture Regulation of 1980 required the resistance of upholstered furniture to glowing cigarettes, see Fig. 69a.

The testing protocol of BS 5852 was the basis of international standardisation as well as numerous national standards. Figure 69b illustrates the relevant test method, which has been standardized in BS 5852. The 1988 regulation introduces the match test. The flame impingement time of 20 s may initiate an afterburn time not more than 120 s. This test is applied to all covering fabrics that are tested over a standard

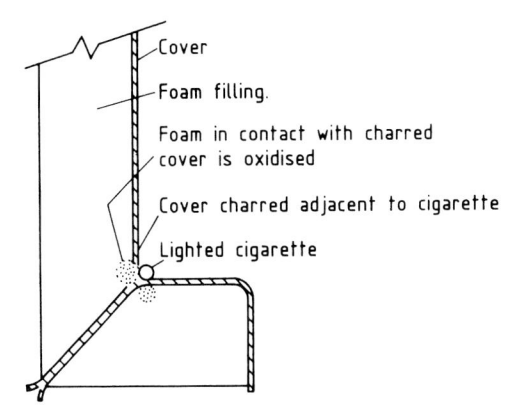

Figure 69a Cigarette test according to BS 5852

Figure 69b Match test according to BS 5852

Figure 69c Peeling effect of upholstered combinations

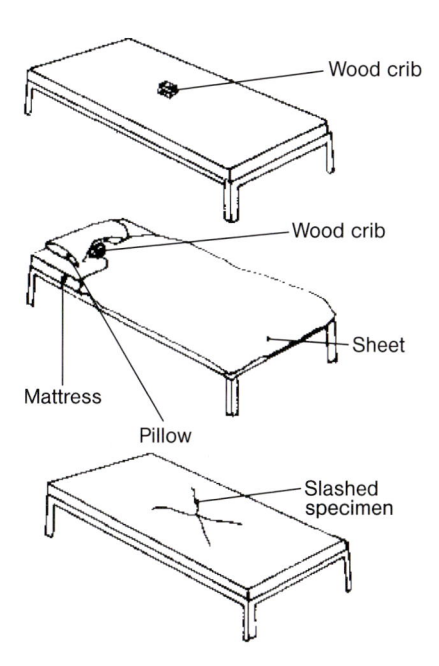

Figure 69d PSA Test set up of the regulation nr.5

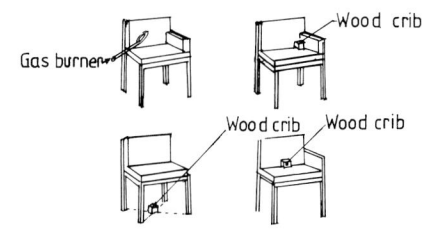

Figure 69e PSA Test set up of the regulation nr.6

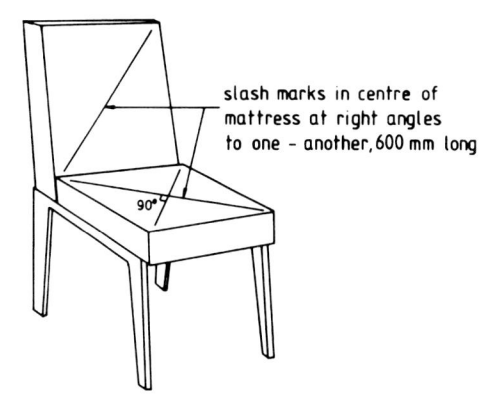

Figure 69f Test arrangement under vandalism conditions

PUR foam. The peeling effect according to Fig. 69c plays an important role. All fillings are tested with an FR polyester cover using larger ignition sources. Block, flexible PUR foam is tested with the Nr. 5 wood crib with burn time, smoldering and weight loss criteria. All other fillings are tested with gas flame 2. The testing of fabrics with a standard filling and of fillings with a standard cover is intended to allow interchangeability of materials to reduce the test burden to industry. The adverse effect is that results may not comply with the request that tests should reflect real-life situations.

A series of test protocols were published by the Property Service Agency (PSA) that no longer operates. The following list gives a view of the relevant selected ignition sources, which are part of the PSA requirements for furniture and furnishings in the public sector.

- Nr. 3 Fire Barrier Standard for Upholstery (Seating and Bedding, Screening Test)
- Nr. 4 Composite Upholstery Ignition Standard (Seating and Bedding, Screening Test)
- Nr. 5 Flammability of Beds and Bedding (Test Room $30\,m^3$)
- Nr. 6 Ignitability Standard for Seating (Test Room $30\,m^3$)

Figures 69d and e demonstrate the test set up of the specifications Nr. 5 and Nr. 6. Figure 69f shows the examination of cushion units and beds under vandalism conditions in the scheme.

In order to hold the necessary testing expenditure as small as possible, a prediction on behalf of testing of components of the supplier industry with the aid of a grid system is a general goal.

Germany

In the Federal Republic of Germany the acceptability of combustible furniture materials in general is controlled on the basis of the regional building regulation (LBO) as well as the Department Store and Assembly Place Regulation [252]. The detailed regulation devolves in a multiple way on the latitude of the lower construction supervision authorities and/or the responsible members of the preventive fire-protection organisation. The requirements for B1- and B2-classified textiles for curtains, decorations and upholstered furniture in accordance with the Assembly Place Regulation of Hessia (VO) [253] are exemplary. Both classes are defined by DIN 4102, which provides a detailed description of the test procedures, illustrated in Figs. 47a and 49b. Risk-related testing and evaluation criteria for upholstered furniture compounds has been developed within the framework of the relevant standardisation committee Agt5 (Committee for Consumer Suitability). The relevant test methods for classifications of upholstered composites are

- the cigarette test shown in Fig. 69a
- the match test in accordance with Fig. 69b and
- the paper-pillow test according to the UIC guidelines illustrated in Fig. 70a.

These realistic test methods have now been adopted in the latest version of the Assembly Place Regulation [254]. With the UIC test the fire-caused damages initiated by vandalism in mass-transport systems are supposed to be simulated in particular. The test-piece concept corresponds to the commitments of the match test as shown by Fig. 69b.

A positive rating presupposes that

- the surface spread of flames
- the afterburn time of the upholstery compound and
- the weight loss

are limited. A general regulation is currently prescribed only for smoker's boxes in cinemas. The present requirement profile is based on the cigarette test.

Western Europe

In Austria the use of easily ignitable materials, classified class B3, is unacceptable in assembly places such as cinemas. The test piece is, in accordance with BS 5852/BS EN1021-2, exposed to the flame of a small burner (flame height 25 mm) for 30 s. Class B1 requires no flame spread and the afterburn time is limited on 10 s and/or 60 s. The flame-spread velocity is limited to 3.6 cm/s for class 2. The test piece must not be completely consumed. Failure results in the rating of class 3. Figure 70b shows in the schematic the experimental setup of the Austrian Standard ÖNorm B3825.

In France the regulations are based mainly on the fire-reaction tests of building materials. Depending on the fire-risk situations, classes M2 and M3 may be necessary. The relevant test configurations are illustrated by Figs. 54a and b. Provided that there are no requirements for class M1, furniture in public buildings, open for the public, must met the Nitrogen-Chlorine Law. The general conditions of this law,

Figure 70a UIC paper cushion test

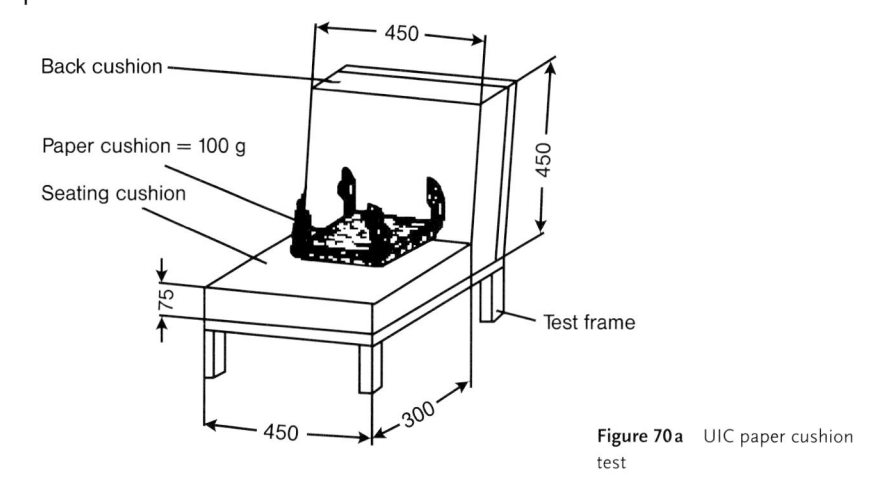

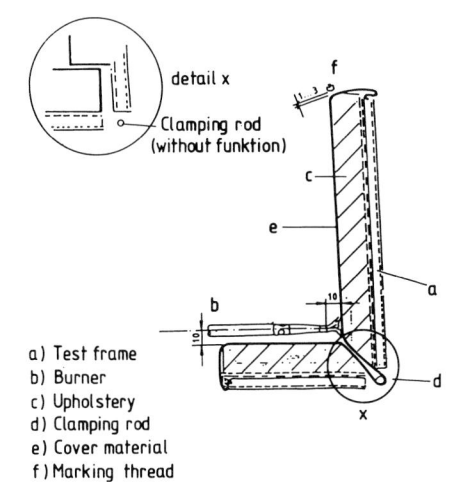

a) Test frame
b) Burner
c) Upholstery
d) Clamping rod
e) Cover material
f) Marking thread

Figure 70b Test configuration according to ÖNorm B 3825

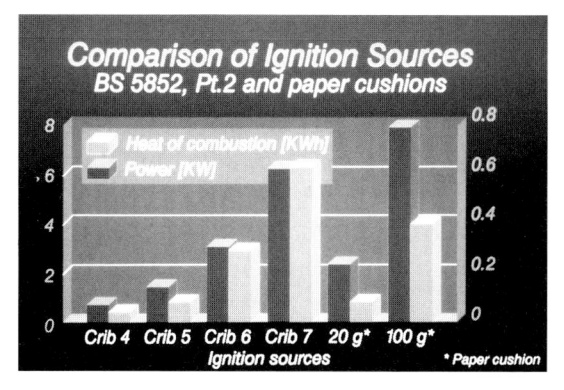

Figure 70c Characterization of the selected EG ignition sources

Table 31 Spain – upholstery furniture evaluation.

Class	flame impingement (s)
class 1/ M	20
class 2/ M	80
class 3/ M	140

which was set in force in 1971 following the two fire disasters, the dance hall fire of Saint Laurent du Pont and the Varig Aircraft Crash in Paris require the assessment of the chemical nature of the material [129, 255]. A detailed description is given in Table 20. The nitrogen and/or chlorine content is to be determined at a temperature level of 700 °C. Regulations were determined for public buildings open for the public but different requirements were set for different buildings.

In the province of Barcelona, the demands on upholstered furniture in public buildings are regulated analogous to the French building-material fire-reaction test methods. General regulations are based of numerous, recent fire disasters [129, 256].

In Italy minimum standards were set after the cinema fire in Turin [48] for upholstered furniture in assembly places. A rating in three classes is carried out by means of a modified small-burner test according to Fig. 49a as illustrated by Table 31. The propane-gas flame reaches a height of 40 mm with a gas flow of 45 ml/min. The classification depends on the duration of the flame impingement.

Since the outer covering as a rule can not suppress the flame attack, the filling material itself must resist the flame impingement. In the Scandinavian countries standards based on BS 5852, e.g. cigarette and match-type tests were developed and standardized. Regulations were set exclusively for public buildings only.

EC Draft Directive
The European draft directive for furniture was initiated by England and Ireland. The acceptance of regulations for furniture in the private sector is still controversial within Europe [257, 258]. Next to resistance against ignition sources of different intensities, limitations of the possible fire-gas dangers has been discussed. Figure 70 clarifies the effectiveness of individual typical ignition sources of the relevant test methods. For the primary danger level, the living area, the relevant standards, e.g. the cigarette and the match test were developed on the basis of ISO documents by the relevant CEN committee. The standards concerning ignition sources of increased power are widely discussed [9, 179]. At the request of the commission, risk-relevant ignition sources of higher intensity, e.g. 20 g and 100 g paper had to be developed for the three risk fields:

- dwellings
- assembly rooms, e.g. discos and cinemas and
- areas of high risk like prisons

Figure 70c illustrates the intensity level of the selected EC ignition sources.

The reduction of the possible danger will be achieved by minimising the risk of ignition, the first essential requirement of the furniture directive. The second essen-

Table 32 Calculated mass of smoke (stoichiometric combustion)

Material	Density (kg/m³)	Calorific value (MJ/kg)	Calorific value (MJ/m³)	Smoke (m³/kg)	Smoke (m³/m³)	Smoke (m³/m²) 4 mm
Particle board	450	17,0	7650	5,3	2385	2,4
Wood	800	16,5	13200	5,2	4160	4,2
Cork	80–500	27,1	2170–13550	7,4	592–3700	0,6–3,7
PUR	50	26,1	795	7,2	360	0,36
EPS	15–30	38,9	585–1165	9,9	148,5–297	0,15–0,3
Cotton	20–60	15,7	315–940	5,0	100–300	0,1–0,3
Wool (sheep)	20–80	20,5	410–1640	7,3	146–584	0,15–0,6
Cellulose	35–75	17,5	610–1310	5,4	189–405	0,19–0,4

tial requirement concerns fire gases and escape. There is no current generally acceptable solution.

The following listing of Table 32 gives a view over maximally possible effluent amounts that arise at a stoichiometric burnup. The data records of Table 32 show very vividly that in the case of the arrangement of mass- and volume-related boundary values other risk-estimation result.

Because the smoke release is as a rule not related to mass but rather to the volume and/or the surface contribution, data for expanded plastics should be striven for in technical information sheets.

In France, a legal arrangement is being worked on currently for upholstered furniture as well as bedding for the private and public sections like hotels, pensions and official buildings. Also vacation apartments are affected among other things, including caravans. Next to the resistance against ignition sources of low intensity like cigarettes and matches also regulation of the possible danger through smoke inhalation is striven for. The mass loss is taken as the decisive evaluation criterion. The test-related details for this specification are being developed currently.

4.2.3
Far East, e.g. Australia

Both Australia and New Zealand have procedures for the evaluation of the ignition and burning behavior of upholstered furniture. The English Furniture Regulation formed the basis of these standard activities. In order to get a better reproducibility in the cigarette test, Australia tried to replace the glowing cigarette by a heating element. The schematic representation of Fig. 71a shows the electrically heated element and the test configuration [DR 80123, DR 80124].

The resistance against ignition sources of higher intensity was evaluated by burning wooden cribs, see Fig. 71b. [259].

In New Zealand, the English cigarette protocol has been generally standardized. As far as the crib test is concerned, the BS 5852 wooden cribs were replaced by balsa cribs [260].

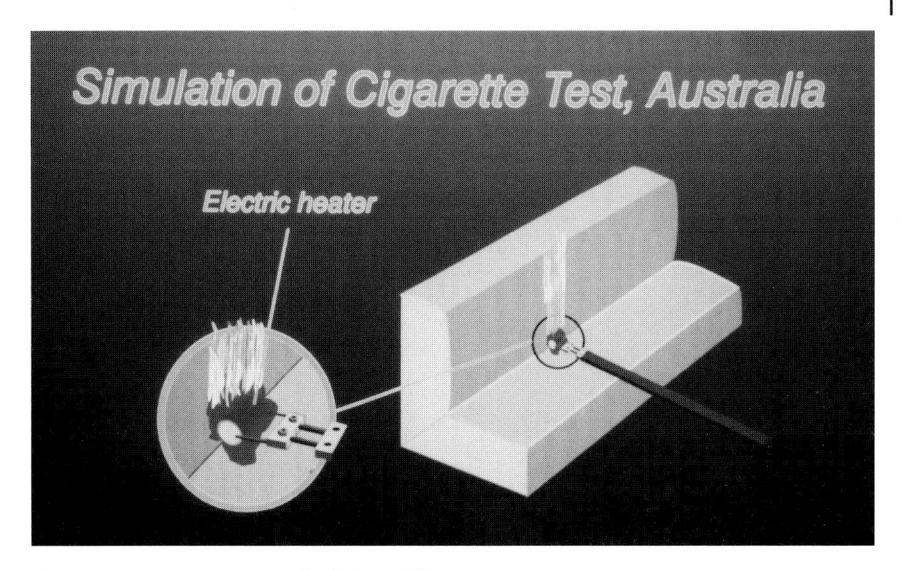

Figure 71a AS – Cigarette test of upholstered furniture

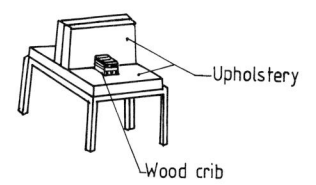

Figure 71b AS – Crib test of upholstered furniture

4.3 Transportation

4.3.1 Road Traffic

The determination of the causes of fires in motor vehicles is difficult. It can be assessed that the ignition of fuel, e.g. caused by electric short-circuits, through flying sparks or through hot exhaust units is the starting point. Carbureter fires or glowing cigarettes are further possible ignition sources. The results of current fire investigations in England (see Figs. 37c–e) indicate a high percentage is initiated by arson. Only in individual cases is it a question of attempted insurance fraud. A purposeful fire-performance preselection of the material of the motor vehicle for use in the car interior is supposed to guarantee that no rapid spread of fire occurs due to the used materials. Regulations based on the Motor Vehicle Safety Standard (FMVSS 302) are implemented worldwide. The test and evaluation criteria of the Federal Highway Administration (FHA) in the United States of America have been developed by

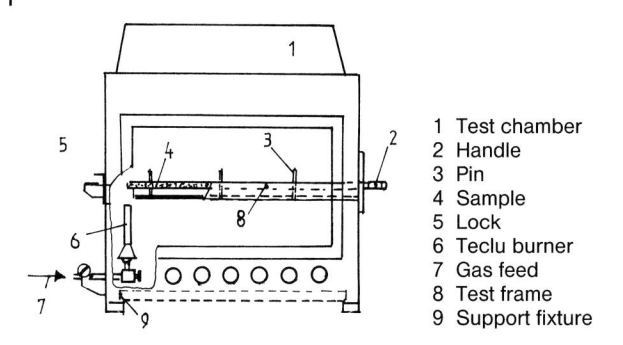

Figure 72 Motor Vehicle Safety Standard (MVSS) 302

1 Test chamber
2 Handle
3 Pin
4 Sample
5 Lock
6 Teclu burner
7 Gas feed
8 Test frame
9 Support fixture

extensive studies with different laboratory test methods. Test and evaluation criteria were incorporated in the national and international standardisation work [261].

As demonstrated by Fig. 72, the strip-shaped test piece with outer dimensions of 350 mm × 10 mm × thickness mm (max. 20 mm) is horizontally arranged and thermally stressed by means of a defined burner for the duration of 15 s. By modifying the requirement profiles [262] also end-application-oriented influences became among other things, e.g. by lamination of filling material by adhesion or flame bonding, considered. The admissibility of the tested materials will be decided by the burning rate measured between the at 38 mm and 292 mm attached markers. The permissible boundary value is fixed by 101.6 mm/min (4 inch/min). Altogether more unfavorable boundary values were determined by the companies exporting in the United States of America in their specifications. [263–265] in order to guarantee that also outliners fulfil the American admissibility criteria even after long exterior storage.

In the Federal Republic of Germany the permitted limits are given in §30 Str. VO [266] as 110 mm/min. In common with the ISO standard 3795 of TC 22/SC16 the German standard DIN 75200 does not contain any requirements. The guidelines of the Str. VO also regulate the fire-reaction testing and evaluation of body units. For this area of application the rating F1 in the small-burner test to DIN 53438 part 3 is required. The test piece, arranged vertically (see Fig. 49a) will be exposed to a match-type burner flame for the duration of 15 s. The flame must not reach the 150 mm above the flame-impinging point. Therefore it is guaranteed that vehicles with bodies constructed with plastics will not be ignited by ignition sources of low intensities, e.g. burning matches.

4.3.2
Aviation Sector

Also in the aeroplane sector the regulation of the relevant American Authority, the Federal Aviation Administration (FAA) [267] is used worldwide. The different requirement profiles for the material use in the interior of aircraft are application oriented and defined in FAR §25. 853, appendix F [268–270]. Test and evaluation criteria are based mainly on material-related laboratory test methods. There are Federal Standards Nr. 191 method 5903, ASTM-E 162 or D 3675, ASTM -E 648, ASTM-E 662 and ASTM E 906 as well as ASTM – E 119. The risk of ignition is evaluated in the FAR test, see Fig. 67c, with the test sample upright or horizontal and/or at a slope 45° or 60° and with outer dimensions of 325 mm × 50 mm × thickness mm will a heat stress by a defined bunsen burner for about 12 s and/or 60 s. The afterburn time of the sample (15 s/15 s) and of the possibly dripping material (5 s/3 s) as well as the burned length (200 mm/150 mm) are the evaluation criteria. In supplementary rules the requirements of the risk of ignition for upholstered seatings and the heat and smoke released are determined. A real seat, consisting of one (508 mm × 457 mm × 102 mm) seat part and one (635 mm × 457 mm × 54 mm) back part are thermally stressed by a defined kerosene burner (T = 1043 °C), see Fig. 73a. The defined temperature and radiation cause within the testing time of 2 min burning for a limited time and over a limited area The permitted loss of weight must not exceed 10%. Burning dripping is not permitted.

The contribution of large-area units of the interior lining to the heat release in the case of a fire are regulated by means of the OSU Chamber (see Fig. 73b) in accordance with ASTM – E 906. The heat output of the 150 mm × 150 mm × thickness mm (max. 45 mm) large test samples must not exceed the limiting values of the heat-release rate 100 kW/m^2 and 100 kW/m^2 min for an external thermal heat stress of 35 kW/m^2 for within the stress duration of 5 min. These limiting values of the 1988 were revised in 1990 and defined as 65 kW/m^2 and 65 kW/m^2 min.

The kerosene burner of the seat test is used also to determine the ease of penetration of floors and partitions. A 150 mm × 150 mm or a 250 mm × 250 mm-sized sample will be thermally stressed by the kerosene burner. If no penetration by the

Figure 73a Kerosene burner test

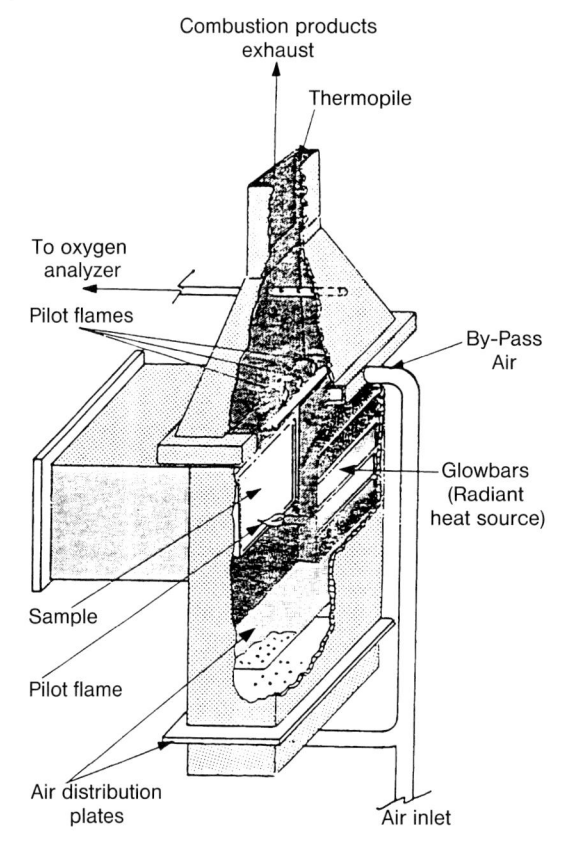

Figure 73b OSU Chamber test

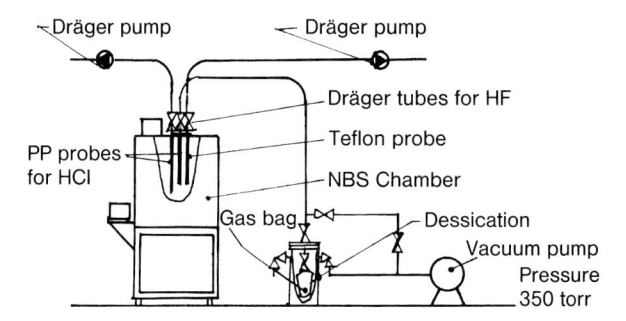

Figure 73c Determination of the concentration of the main fire-gas components

flames occur within 5 min or 15 min, the part is rated as resistant or fire proof. In the aeroplane sector the effluents to be examined will be produced in the NBS Chamber, see Fig. 48, under a thermal stress of 25 kW/m² with and without flame impingement. The limits of the smoke release are based on the internal requirements of Airbus and Boeing [271, 272]. The smoke-density regulations are based on measurements with the NBS Chamber according to ASTM-E 662. Figure 42a clarifies the requirement profiles and the test protocol. The basis of the examination and evaluation of the smoke toxicity is a NBS Chamber test procedure as well. The concentrations of the essential effluent components are determined, see Fig. 73c.

4.3.3
Shipbuilding Sector

The examination of the fire retardance of wall and ceiling constructions will be controlled by means of resolution A163 of IMO (Intergovernmental Maritime Consulting Organisation) [273] in accordance with the building components tests. The thermal loading of the wall and ceiling sections is carried out in the relevant test furnaces in accordance with the ETK. For the evaluation of the possible contribution to the fire spread the relevant IMO committee is proposing the ISO Surface Spread of Flame test procedure in modified mode [274], see Fig. 53a. For sea-going vessels, the international agreement for the protection of human life at sea, SOLAS – Safety of Life at Sea is given validity in all states of the world. Different IMO recommendations and regulations, national safety measures as well as requirements of classification associations as, e.g. Norske Veritas have to be considered. The fire-performance requirements of IMO and the national fire-safety requirements of SSVO (Schiffssicherheits Verordnung) [275] are valid for sea-going vessels under the German flag. The federal responsibility is taken over by the shipping section within the Ministry of Transport. Control is handled by the Seeberufsgenossenschaft (SeeBG). According to Sect. II – 2, regulation – 3 of SOLAS (Safety of Life at Sea) noncombustible materials should not ignite or evolve combustible gases when heated to 750 °C, see Fig. 41a. While the use of noncombustible materials the responsibility is regulated by the IMO Resolution A270c, the permission to use combustible materials is given to responsible national organisations by IMO resolution A 166 [276].

The national classification system as a rule is based on the respective fire-performance test certificates of building products. The responsible trade organisation (SeeBG) requires that for sea-going vessels under the German flag combustible insulation materials with limited use has to be classified as B1 according to DIN 4102 under enduse conditions with metal linings. The requirement includes an expert judgement of the acute toxicity of the thermal decomposition products by means of bioassay test results on the basis of DIN 53436 in relation to reference materials such as cork and wood. In order to restrict fire spread as much as possible, the combustible insulating materials must be covered by a metal sheet of minimum 1 mm thickness. The general conditions for fire protection for furniture and furnishings are regulated by IMO in the SOLAS Convention. The relevant technical regulations are required for passenger ships with more than 36 persons. Furniture and furnishings must achieve a limited contribution to the danger in the case of a fire. The thickness of veneers and coatings is limited in order to guarantee that there is only a limited contribution to the possible surface spread of flame test. The procedure, which has been developed in the relevant IMO committee [277], is based on the former ISO draft of SC1 of TC 92 and the official Italian test protocol, see Fig. 56a. The modified version allows next to the evaluation of the surface spread of flames the judgement of the heat-release and smoke-release rates, which depend on the progress of the fire and fire load. In the IMO test procedure, which is illustrated by Fig. 53a, an increased radiation intensity is produced by a modified gas-fired radiator.

The smoke test for current provisional procedures use a modified NBS Chamber [278, 279]. Smoke-release criteria have been modified following various fire disasters. The modified NBS test protocol, see Fig. 74a, was stipulated. The basic research work of SC1 of TC 92 Fire Reaction Test ISO DP 5960 was carried on by SC4 of TC 61 Plastics. The horizontally arranged test piece, with outer dimensions of 74 mm × 74 mm, as illustrated by Fig. 74b, is thermally stressed by an electrically heated conical-shaped radiator, see Fig. 74c, by a variable heat stress of 10–50 kW/m². The disadvantages due to melting effects, which occur with the FAA regulation, will be ruled out by the horizontal orientation. As with the NBS Chamber test, an additional pilot flame can be used. The released stream of smoke is assessed by means of a perpendicular optical measuring system. Analogous to the NBS Chamber system, material-related data of the optical density D and the specific optical density D_s are determined. Earlier studies with a prototype of the modified NBS Chamber were conducted by Hovde [280]. The extent to which the material-specific mass optical

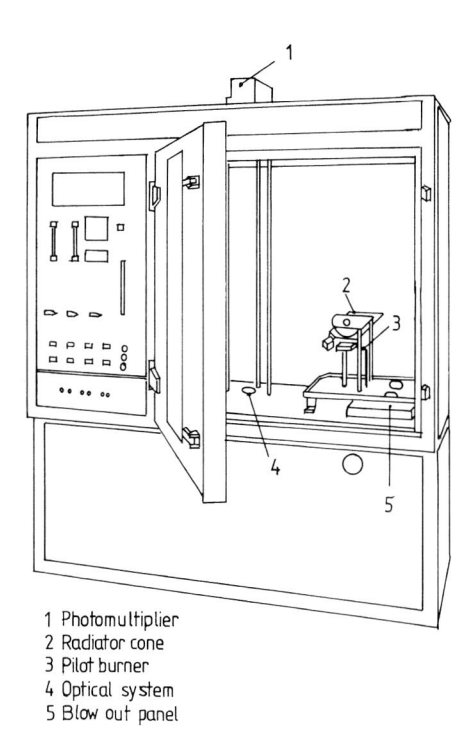

1 Photomultiplier
2 Radiator cone
3 Pilot burner
4 Optical system
5 Blow out panel

Figure 74a Smoke-density measurement in the single chamber box

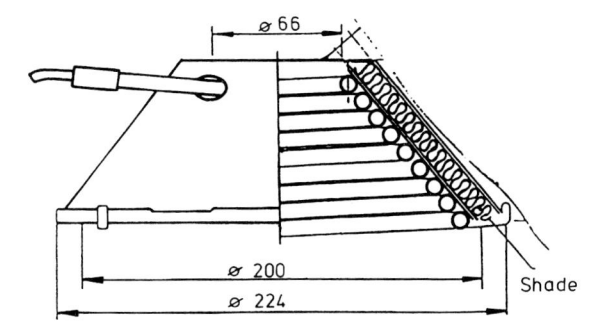

Figure 74b Conical-shaped radiator

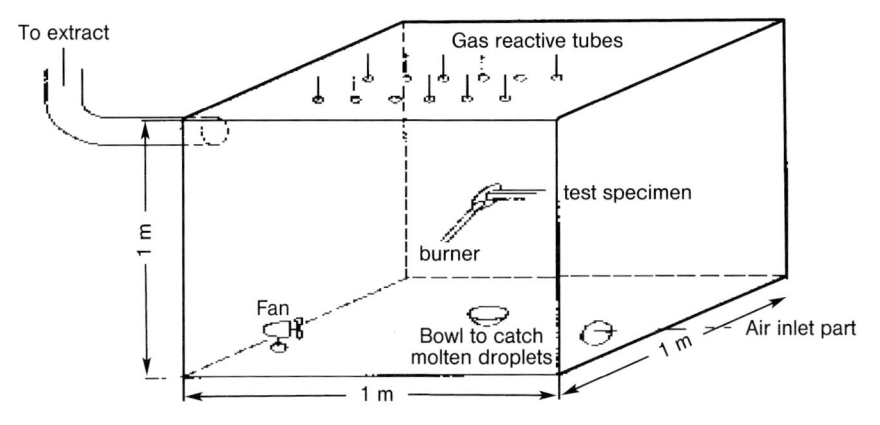

Figure 74c NES 713 – Classification of fire-gas components

density (MOD) results are risk related must be doubted. Basic investigations on this specific subject were carried out at the Fire Research Center (FRC) Salt Lake City [281].

The Naval Engineering Standard (NES 713) is used for naval applications. In this, the substance to be evaluated is ignited by means of a bunsen burner, see Fig. 74c. Provided no ignition occurs, the decomposition process is forced by a continuous action of the bunsen burner. According to the chemical nature of the product, 1 g of the material to be tested will be thermally exposed to flames for the test duration. The determined concentration data are calculated for $100\,g/m^3$.

The acute toxicity of effluents will be assessed in Denmark on behalf of the Scandinavian Star Fire as well by the Cone-Calorimeter Test Method shown by Fig. 21 and the Room-Corner Test illustrated by Fig. 60a and in the, up to now, valid prescribed decomposition model according to DIN 53436 shown by Fig. 57c. After the fire on the Scandinavian Star modification of the current test and classification criteria according to DIN 53436 has been required by the responsible authorities. The DIN tube system was used under the conditions of the A2 classification (DIN 4102) of building products.

4.3.4
Railway Sector

The various material-related fire-safety requirements are based as a rule on the relevant fire-reaction criteria. The different national classification criteria were reworked and redefined in connection with spectacular fire disasters [282–285]. In the United States of America the fire-performance requirements are based on the ASTM and NFPA Standards. A comprehensive view of the entire problem including the various national requirements is given by the final report DOT/ORD 93/23 Dec. 1993, which also reports the demands concerning the different European test methods. The material-specific fire-protection classification contains as a rule an evaluation of the possible contribution to the surface spread of flame and the smoke release. Specified test methods are ASTM-E 162, Surface flammability using a radiant Heat Energy Source (solid materials) and/or ASTM – D 1354 (expanded materials) analogous to Fig. 42c, the Radiant Flooring Test according to ASTM – E 648 as shown by Fig. 41c and the smoke-density measurement in accordance with ASTM – E 662 illustrated by Fig. 42a basis for the admissibility. The legal arrangement for the building and the operation of trams inclusive of underground and overhead railroads as well as cable-railroads is given in the Federal Republic of Germany by the street-car building and operation arrangement BO Strab [286]. The German Railway AG is responsible for the safety of their entire operation including buildings according to §38 of the Federal Railway Law. The different parts of DIN 5510 as well as DIN 54341are application oriented and are now valid as controls instead of the former regulation [287, 288]. Harmonisation of the differing national fire-performance testing and evaluation criteria is proposed within the framework of EN 45504 for the railway-car production. The listing of Table 33 clarifies the different national points of view. Since the BO Strab does not include any well-defined requirements, the requirement profiles of DIN 5510 Fire Protection in Railway Vehicles form the basis for regulation.

Table 33 Fire safety requirements in the railway field

DIN 5510	
Part 1	General rules
Part 2	Burning behavior and side effects of materials and components Classification, requirements and test procedures.
Part 3	Requirements of fire resistance for barriers and partitions
Part 4	Construction of carriages – Safety requirements.
Part 5	Fire-safety requirements for the electrical equipment
Part 8	Fire-reaction test procedures

The test samples, arranged vertically in the test chimney according to Fig. 75a DIN 54341, will be thermally stressed by a special bunsen burner (DIN 54837) for 3 min. Combustibility, smoke release and dripping behavior will be assessed. The extent of the destroyed surface area is decisive for the fire-performance rating. Table 34 gives the criteria of DIN 5510.

Table 34 Fire performance classification DIN 5510.

Classes			
Class B1	easily ignitable	damaged area	90–100%
Class B2	combustible	damaged area	76–90%
Class B3	difficult to ignite	damaged area	<75%
Class B4	incombustible	no damage	

The rating with respect to burning dripping will be done by observation in the chimney test. Differentiation between the classes is based on visual observation by T1, burning dripping to T4, no visible deformation/no softening.

The smoke release during the chimney test will be assessed by means of the reduced transmission values. Two classes, $SR1 \leq 10\%$ min and $SR2 \leq 50\%$ min. are used. Within the framework of international cooperation "UIC (Union International de Chemins de Fer)" a risk-relevant examination of the upholstery seating was agreed, see Fig. 70a. By means of the DIN standards 5510 and 54341 an upholstered seat will be thermally stressed by a burning paper pillow (100 g) with outer dimensions of 360 mm × 270 mm, conditioned for 24 h according to DIN 50014. The test

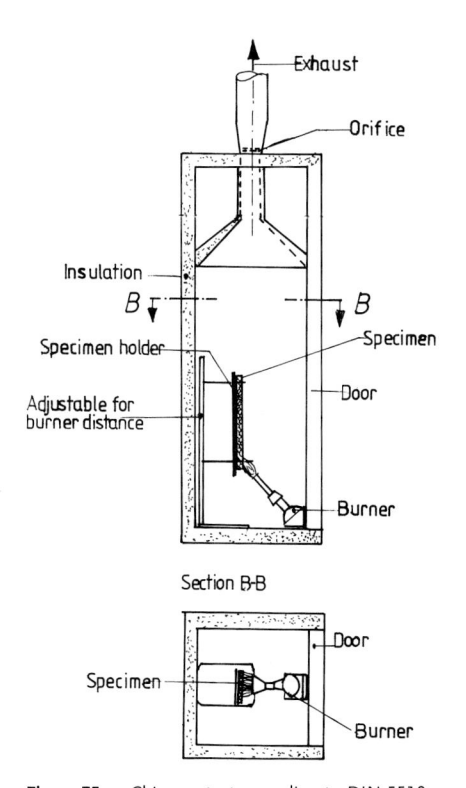

Figure 75a Chimney test according to DIN 5510

is passed if neither a total burn up nor burning dripping occurs and flames extinguish within 10 min (18 min). The testing and evaluation of the smoke release, smoke density and smoke toxicity are part of the same test report. The smoke release of cables will be tested according to IEC 15 (IEC TC 89). The test in the 27-m³ box is carried out with different ignition sources. The thermal loading of the horizontally arranged cable test pieces (1000 mm × 45–90 mm) occurs from below by means of burning alcohol (1000 ml). The extent of the surface spread can significantly influence the obscuration of a photoelectrical system, see Fig. 75b. Upholstery test pieces, two L-shaped samples with outer dimensions of 400 mm × 400 mm × 25 mm, are tested with 0.5 kg of alcohol-impregnated charcoal. The material-related data determined in both scenarios have validity only for the examined general conditions and cannot be transferred without special evidence. Transformation to other fire scenarios, e.g. onto the smoldering fires caused by overheated wiring in the cable field, is not possible. Analogous evaluations are possible only by means of a relevant investigation (see Fig. 76). This statement is valid, e.g. for the smoldering fire initiated by the heat load of smoke layers in suspending ceilings.

In France the present rules of the SNCF and the RATP for railway vehicles are based on comparative investigations including laboratory, bench and large-scale tests. The fire-performance classification contains the ignitability and the contribution to the surface spread of flame including the burning dripping and the smoke release. There is a different set of classifications for large parts of interior fittings and furnishings, the cable classification and the I-Classification for interior fittings of railway vehicles (see Table 35). The I-classification based on the two test procedures of the glow-wire test according to IEC 695, shown by Fig. 77a and the Oxygen Index Test according to IS 1716 illustrated by Fig. 77b. The M-classification is based on the test procedures standardized for building products like NF -P 92 – 501 (Epiradiateur)

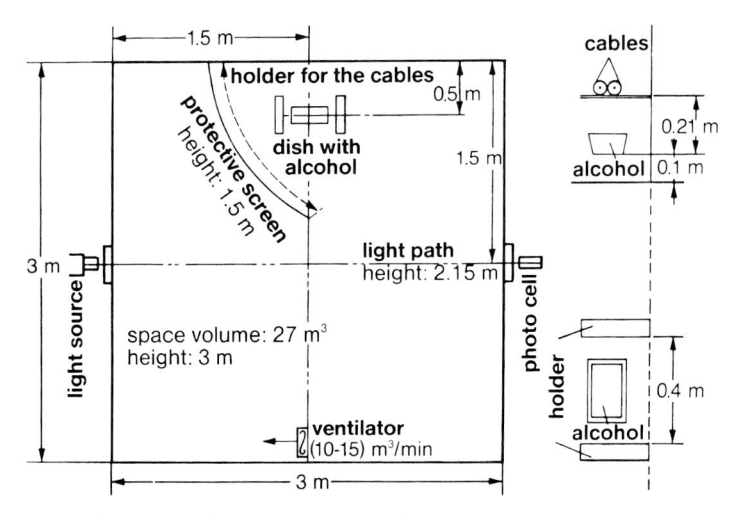

Figure 75b 3×3×3 m³ smoke-density protocol

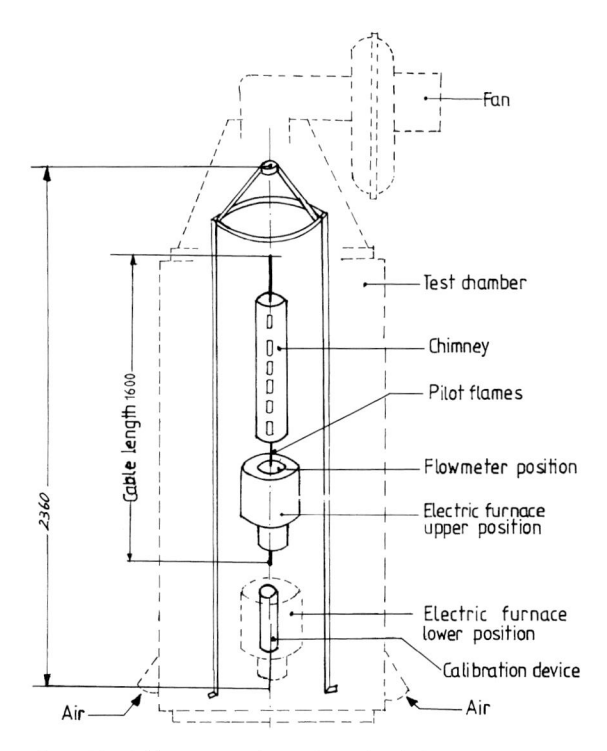

Figure 76 Cable test according to NF – C 32-070

or/and NF-P92 – 503 (Bruleur Electric), see Figs. 54a and b. Table 35 provides a view of the factors decisive for the I-classification. The cable classification is based on the standard NF- C32-070 whose testing scheme is clarified in Fig. 76.

Table 35 I – Classification in the glow wire test

Class	O2 – Index NF-T51-071	Glow wire test NF-C20-455
I0	≥ 70	960 °C +)
I1	≥ 45	960 °C +)
I2	≥ 32	850 °C +)
I3	≥ 28	Afterburn time ≤ 2 s
I4	≥ 20	Afterburn time ≤ 2 s
NC	< 20	Afterburn time ≤ 2 s

+) no ignition   NC – Not Classified

In the glow-wire test according to Fig. 77a the ease of ignition is determined under the impact of the glow wire heated up of 600 °C to 950 °C. The test temperature is stipulated depending on the electromechanical product and enduse application. The oxygen-index test to ASTM-D 2863 and/or NT-Fire 013 was developed primarily for

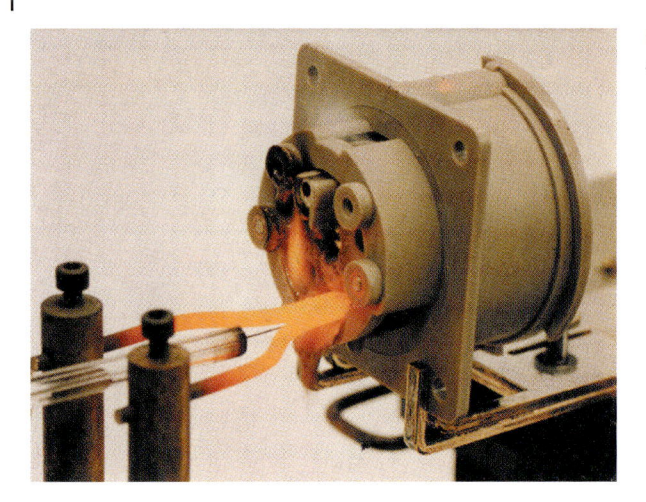

Figure 77 a Glow-wire test according to IEC 695

the aerospace field. The aerospace industry was working until the fire disaster with an oxygen atmosphere of more than 21% O_2. In this laboratory test method, which has gained great popularity due to excellent repeatability/reproducibility, the oxygen content of the air mixture is varied. The rod-shaped test piece with the external dimensions of 150 mm × 6 mm × 3 mm is arranged in a vertical position and ignited at the upper end of the test sample and burns down like a candle. The oxygen index is determined by means of the formula $n = 100 \times O_2/O_2 + N_2$ (%).

The smoke evaluation according to NF-F16 is using two different test methods. In one test smoke-density measurements are carried out in the NBS Chamber according to NF-X10 (see Fig. 42a) and in the other test smoke-concentration data are determined in the tube furnace according to NF- T51 as illustrated by Fig. 77c in which a 1-g mass of the material to be tested is decomposed in the temperature range of 600 °C to 800 °C. The decomposition products, generated by the thermal stress for 20 min are gathered within the measuring bag of 40 dm³. The concentration of the main components is analysed and recorded.

In the NBS Chamber test (see Fig. 42a and Fig. 77d) the two characteristics are maximum optical density D_{max} and the smoke characteristic value VOF_4 during the first four test minutes that is calculated according to the formula $VOF_4 = D_{S1} + D_{S2} + D_{S3} + D_{S4}/2$.

In the tube furnace in accordance with NF – T51, whose testing-scheme is illustrated by Fig. 77c, about 1 g of the material is decomposed by pyrolysis in the stream of air (120 l/h). The furnace at 600 °C to 800 °C for the duration of 20 min is contrary to DIN 53436 not moved. The effluents are collected in the approx. 40-dm³ large measuring room and analysed. The acute toxicity of effluents is evaluated by means of the main smoke components assuming an additive-effect mechanism. The individual components are related to bioassay test results, e.g. the ILDF-20min data that guarantee that no adverse effects occur within the exposure duration of 15 min. The material-related classification is according to the mathematical relationship

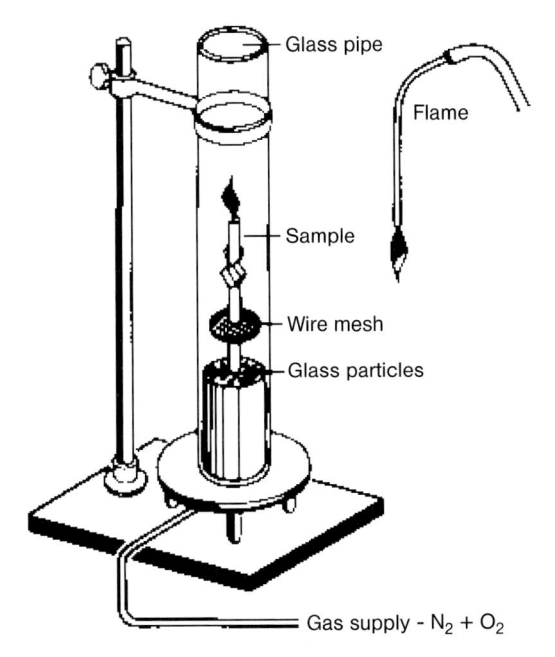

Figure 77b O₂-index Test according to ASTM – D 2863

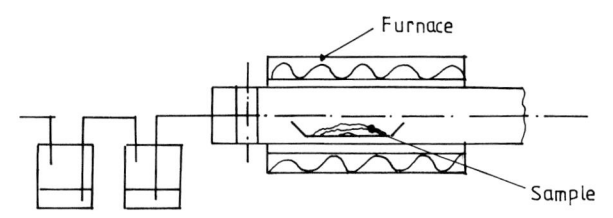

Figure 77c Decomposition model according to NF-T 51
à

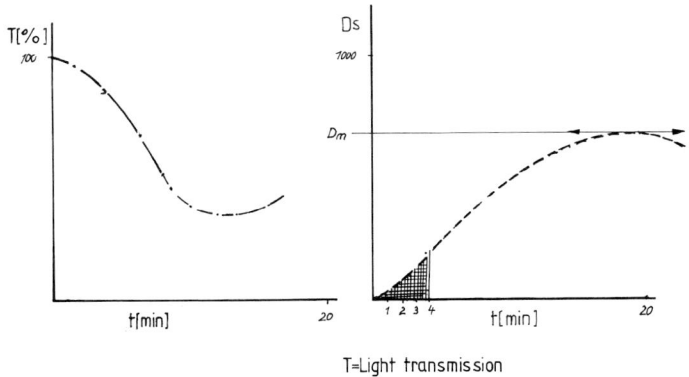

T = Light transmission
Ds = specific optical density

Figure 77d Scheme – Determination of smoke-density potential values

$ITC = 100 \Sigma\, t_i/cc_i$

where t_i = concentration of smoke components in (mg/g) and
cc_i = critical concentration ILDH 15 min in (mg/g).
The evaluation is carried out according to $NF = ITC/2 + VoF_4/20 + D_m/100$
The IF data are basis of the F-classification according to Table 36.

Table 36 Smoke density evaluation according to NF – T51

F0		IF	≤ 05
F1	05<	IF	≤ 20
F2	20<	IF	≤ 40
F3	40<	IF	≤ 80
F4	80<	IF	≤ 120
F5	10<	IF	

The smoke requirements for the classes M0/I0/M1/I1/A are determined by means of a grid system in connection with the M, I and C classes depending on the vehicle category and the enduse conditions.

A decomposition model providing the possibility to produce effluents depending on the decomposition temperature with a relatively low CO_2/CO ratio is standardized by GOST, see Figs. 59a and b. Detailed descriptions are given by various publications [230, 231]. In the small decomposition chamber, the test material is thermally stressed for 20 min and the decomposition products are produced at 400 °C (18 kW/m²), 600 °C (38 kW/m²) and 750 °C (65 kW/m²). The mixture of effluents is guided into the exposure chamber (V = 0.1 and/or 0.2 m³). Determined data are the optical density D and the smoke toxicity. The test animals, white rats, are exposed for the duration of 30 min. The assessment is carried out by the mortality data (LC_{50} values) and the CO-Hb results. The test methods were used in the different application areas, e.g. building and transportation field, naval vessels and railway vehicles.

4.4
Electrical Engineering

The aim of the preventive fire protection of electrical products is to guarantee that in the case of failure these devices do not become an ignition source for the surrounding environment. Detailed fire statistics provide knowledge about possible fire risks of electromechanical and electronic products and so it makes it possible to take measures for the extensive minimization of the fire risk. Table 37 gives a view of usual ignition sources.

Table 37 Ignition sources of the electrical field.

Ignition sources in the electrotechnical and the electronic field
Hot and glowing wire
Glowing contacts
Small and large flames (tracking, short – circuit)

The admissibility of electromechanical products is controlled by regulatory rules [289, 290]. These safety aspects are based in the Federal Republic of Germany on the relevant VDE (Association of German Electrotechnic) Rules. The VDE label guarantees compliance to the legal safety conditions. A specific label for the fire-performance classification does not exist. The necessary testing conditions are developed by the DKE (German Commission for Electrotechnic), a joined commission established by DIN and VDE. The international standardisation work is carried out in the European (CENELEC) and international (IEC) standardisation committees. In the United States of America the safety of electromechanical products and installations is guaranteed by legal regulations. These are based on the realistic procedures and listings of the Underwriters Laboratories, a nonprofit organisation that was set up in 1884 with the aim to carry out technical safety investigations for the protection of people and properties. The profits of UL must be reinvented for the further development of the safety standards. The specifications of UL form the basis of the regulation [291]. The UL system contains three sectors: the listing, the classification and the recognition. The listing procedure has a quality marking similar to the VDE sign. The classification covers only part of the qualities of the final product. The essential feature of the UL Approval System is the possibility of the examination of components and materials with the aim of certification. The certification of electromechanical products is considerably simplified, when accepted components and materials are used. In the follow-up system it is guaranteed through regular inspections that current production corresponds to the tested and certificated test samples. The material-related and application-oriented regulations that guarantee that these products do not ignite in the case of failure and the fire spread to neighboring areas, are based on material- and product-related test methods. In the international standardisation two different testing philosophies exist. A risk-oriented finished part test correspond according to Finger [292] to the knowledge of modern fire-protection engineering, because the application-oriented component test under enduse conditions corresponds to a risk-relevant classification. Screening examinations of material-related data of standardized test samples are required as decision aids for designers as a alternative. The latter also forms the basis for technical insurance-requirement profiles. The standards of UL are examples of material-specific characteristics that facilitate both the specification of material-specific values as acceptance criteria and the fixing of a tolerance threshold for control examinations. The agreement with the different classification criteria is striven for in the different committees for finished components in national and international standardisation institutions. Finger states that in the IEC the test philosophy of the risk-related testing of finished parts is favored. The conflict between the Underwriters Laboratories, whose bonus system is decisively defined by material-related test results and the IEC test philosophy of enduse conditions caused the attempts of UL to get the ISO status for the UL standards via the SC4 of TC 61 Plastics. To what extent the advantages of the pure material-related testing, e.g. direct material comparison and production control, balances the possible disadvantages of the different parameters, e.g. material thickness, material compound/composites, sandwich or shaping is still controversial in the field of standardisation. The diverse influences on the fire performance of finished parts can not be assessed from pure material testing either as single effects or in combination. Indeed ignition and environmental conditions as well as the test-sample configuration in the field of pure material testing are modified and continuously advanced, to gain a significantly approximation to final application. A general approximation cannot be achieved because of the complexity of the physics and also in the case of material identity and geometrical similarity no dimensionless key numbers can be developed.

4.4.2
Proof of the Resistance against Overheated (Hot and Glowing) Wiring

All the relevant procedures such as the Hot Mandrel Test, the Glow Wire Test, the Hot Wire Ignition Test and the Bad Connection Test can be carried out with defined test pieces and finished parts. Test methods like the determination of the decomposition temperature allow preselective testing of the different materials. The possible release of flammable decomposition products due to thermal stress on electromechanical insulating materials is one aspect of the possible danger of the failure. Flash- and self-ignition temperatures of the decomposition products are comparable to the flash point of flammable liquids. A defined material quantity will be treated by pyrolysis in the electrically heated crucible, see Fig. 78a. The thermal decomposition products are ignited by means of a pilot flame or through the hot wall. These temperature levels are not absolute figures. They can be significantly changed by spatial circumstances, the kind and amount of the material, to be tested and in particular by mixing with air. The Setchkin apparatus according to ASTM-D 1927 was incorporated in national and international systems of regulations. The temperature levels under standardized test conditions that lead to ignition, are determined in a similar way to VDE 0345, see Fig. 78b. The determined temperature levels will be called flash-ignition (FI) and self-ignition (SI) temperatures. Figure 78b illustrates the construction of the test apparatus. 3 g of the material to be tested will be heated in stages.

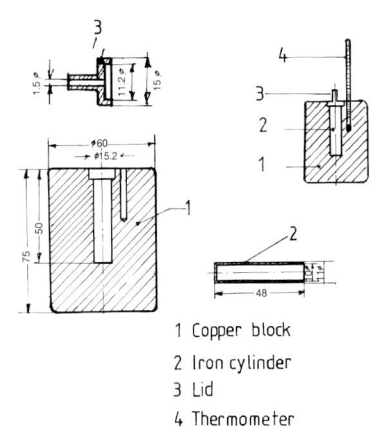

1 Copper block
2 Iron cylinder
3 Lid
4 Thermometer

Figure 78a Decomposition temperature according to VDE 0472

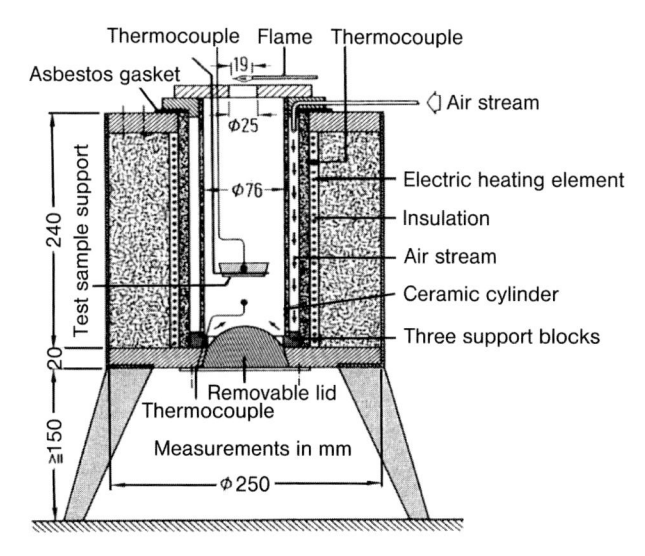

Figure 78 b Determination of the decomposition temperature

That temperature level of the stream of effluents will be determined when ignition of the gas stream occurs initiated by a pilot flame and/or by the heated furnace wall.

During the hot mandrel test according to VDE 0470, illustrated by Fig. 79a, a conical hole is drilled into the test piece. The conically shaped and electrically heated mandrel is pressed into the hole. The released thermal decomposition products must not be ignited by sparking of a high-frequency generator. The weakness of this procedure is demonstrated during the testing of thermoplastic materials. The tracking of the thorn in the case of the melting material did not produces satisfactory results. The responsible VDE commission limited the depth of penetration of the thorn as the problem solution. In the glow bar testing, see Fig. 79b, the former glow resistance test according to Schramm-Zebrowski, the rating is assessed as the product of weight loss and the burn length due to the heat stress of an electrically heated (910 °C) silicon carbide rod.

The glow bar is withdrawn after 3 min and any flames must be extinguished. The three classes are

- BH1 no visible flames
- BH2 burn length < 100 mm
- BH3 burn length > 100 mm

The contact with the test piece plays a significant role in the glow-wire test, see Fig. 77a, which was developed by CEE [293]. During this test procedure the defined glow wire coil, made of chrome-nickel alloy with a build-in miniature thermocouple, will be heated to various test temperatures, e.g. 550 K, 650 K and 750 ±10 K and 850 K and 950 ±15 K. The test piece is fixed on a trolley and is pressed onto the glow wire for 30 s with a force of 1 N. The depth of the penetration of the coil is limited to 7 mm. Thermoplastic material can lose contact with the coil. The test is passed if no

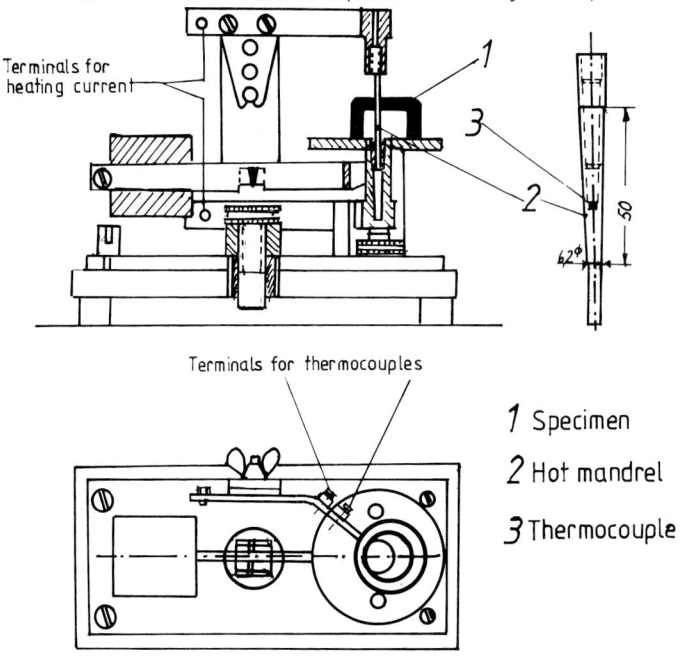

Figure 79a Hot Mandrel Test

flaming and no glowing occurs and flaming and glowing within the test piece, in the area surrounding the coil and on the support situated 20 mm below are extinguished within 30 s after the coil has been removed. If the component or a part of the component is tested the paper wrapping of the support must not be ignited by dripping of glowing or burning parts of the specimen and the wooden support must not be charred. A similar function has the hot wire ignition test of UL 746 see Fig. 79c. The aim is the simulation of contact with power cables. The wire temperature is limited

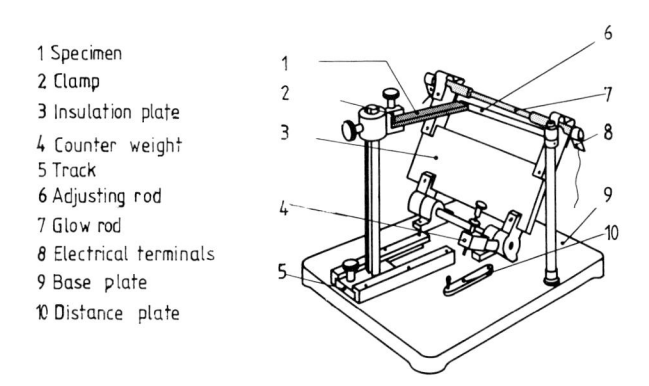

1 Specimen
2 Clamp
3 Insulation plate
4 Counter weight
5 Track
6 Adjusting rod
7 Glow rod
8 Electrical terminals
9 Base plate
10 Distance plate

Figure 79b Glow-Bar Test

Hot Wire Ignition Test

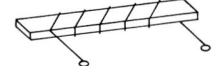

Figure 79c Hot-Wire Ignition Test

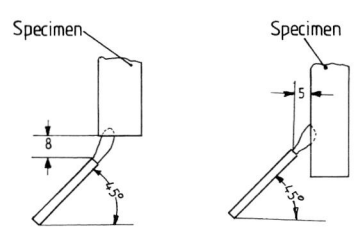

Figure 79d Impingement by the needle flame

to 930 °C. The time to ignition, max. 120 s, has to be determined. The thickness of the specimen shall be similar to the enduse conditions.

Two different aspects have to be followed for the risk of ignition by direct flame impingement. First, ignition due to the failure of the electrical component and secondly by flames initiated by tracking, e.g. in washing machines. Whether tracking currents occur, depends not only on the tracking resistance but also on other factors. These include the distance between live components, the nature of residues and the weatherability. The risk of ignition by direct flame impingement is tested and classified by various laboratory test methods.

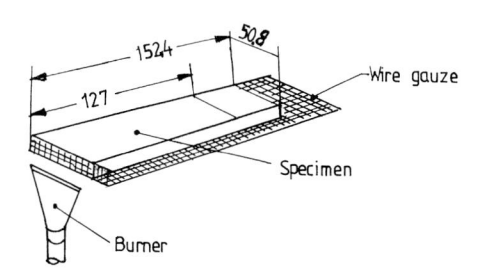

A Horizontal test for foam materials (ASTM D1692; ISO)

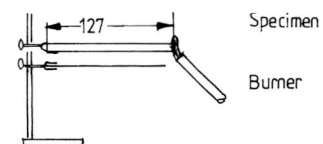

B Horizontal test for rigid materials (ASTM D635; IEC)

Figure 80a UL – Test procedure

4.4.3
Thermal Stress with Flames

The goal of the flame impingement with test flame Nr. 1 according to VDE 0471 part 2. 2 is the imitation of flame that will be formed by failing electromechanical components, e.g. television sets. Figures 79d and 80a illustrate these tests. The requirements are fixed in the different standardisation committees, e.g. finished parts or backs of television sets. Different impingement times of 5, 10, 20, 30 and 60 s have been agreed. The test will be passed if after the agreed impingement

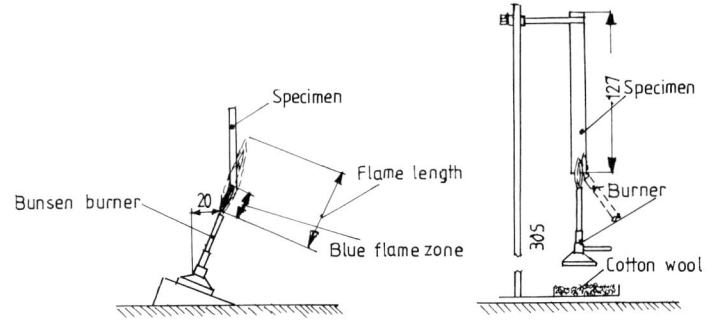

C Testing according to $94V_0$, $94V_1$, $94V_2$

Figure 80b UL – Test procedure

Criteria		Method		
		Vertical test		Horizontal test for foams
		Solid materials	Film test	
Flame length (mm)		19 mm	19 mm	38 mm
Exposure time (s)		2 x10 s	2 x 3 s	60 s
		Class 94 V-0	Class VTM-0	Class HBF
Extent of combustion s (mm)		s ≤ 127 mm	s ≤ 127 mm	s ≤ 127 mm or
Rate of combustion V (mm/min)		–	–	V ≤ 38 mm/min
Afterflame time t_x (s)		≤ 10 s; $\sum t_x$ ≤ 50 s; x=10	≤ 10 s; $\sum t_x$ ≤ 50 s; x=5	–
Afterglow time (s)		≤ 30 s	≤ 30 s	–
Burning droplets		no	no	–
		Class 94 V-1	Class VTM-1	Class HF-1
Extent of combustion s (mm)		s ≤ 127 mm	s ≤ 127 mm	s ≤ 57 mm
Rate of combustion V (mm/min) Afterflame time t_x (s)		– ≤ 30 s; $\sum t_x$ ≤ 250 s; x=10	– ≤ 30 s; $\sum t_x$ ≤ 250 s; x=5	– ≤ 2 s; (1 Probe ≤ 10 s)
Afterglow time (s) Burning droplets		≤ 60 s no	≤ 60 s no	≤ 30 s no
		Class 94 V-2	Class VTM-2	Class HF-2
Burning droplets		yes, otherwise as 94 V-1	yes, otherwise as VTM-1	yes, otherwise as HF-1

Figure 80c Classification according to UL Subject 94

- no flames or glow occur,
- the afterflame time stays below 30 s
- the destroyed length s will not be exceeded and
- flames in the form of burning dripping do not become effective in the neighboring environment.

To produce thermal stress with larger flames, e.g. test methods like the small-burner test (flame height 20 mm) and bunsen-burner tests (flame height 100 mm) are used.

The well-known test methods of Underwriters Laboratories are in use in this field of application. UL 94 is one of the most important standards and applies for many various areas of applications except building products. For the use of plastic materials in electromechanical products UL classifications are of greater importance, because favorable ratings of the materials and the positive listing by UL are often the basis for their acceptability. The different flame tests of UL 94 are to be characterised as pure material tests, which have been standardized by various national and international standards. The drawings of Fig. 80b illustrate these different test principles. The horizontal test procedure provides information concerning the spread of the flame front and the afterburn time. The vertical test procedure offers detailed information about the afterburn time of the test piece and eventually burning dripping. Figure 80c provides an overview concerning the important test and evaluation criteria.

In general, for the fire-performance testing of cables a cable section will be thermally stressed by a defined direct flame impingement. The evaluation is based on test data concerning the extent of burning. According to VDE 0472 part 804, the cable section will be thermally stressed by a propane-gas burner with a flame length of 125 mm and/or 175 mm. The test is passed if no flaming occur or the flaming does not reach the upper end and extinguishes after the burner was taken away. In the large-scale railway cable test according to VDE 0472 part 804c, the possible contribution to the fire spreading will be determined according to Fig. 81. The test sam-

Figure 81 Testing of cable tracks

ples will be exposed to the flames of a special burner for 20 min according to American (IEEE 883) and Swedish (SS 4241 475) requirements for cables used in atomic power stations. In France various test protocols for cables have also been standardized by the Union Technique de l'Electricite as: UTE C20-452, UTE-C20 -453 and UTE-C20-454.

4.4.4
Side Effects of a Fire

For preventive fire protection of electromechanical products, the aspects of the fire-related characteristics have to be taken into account. Requirement profiles exist both with regard to the burning dripping and the possible visibility reducing, noxious and corrosive effects of the thermal decomposition products. The effect of burning dripping will be of greater importance with respect to the postulated spread of fire. Burning dripping with respect to the external flame impingement becomes a dangerous ignition source, where

- the intensity (strength) of the flames of the dripping material is similar to the intensity of the primary ignition source and
- the surroundings are not resistant against the ignition-source intensity of the burning droplets.

Only in this case will an increased danger of surface spread of flame occur. Flaming dripping from electromechanical products like lamps, lights, power points, etc. that might become ignited in the case of failure has to be categorized as a secondary ignition source and has to be suppressed by effective fire-prevention measures. Corresponding evaluation criteria are to be regulated as effective measures of the preventive fire protection of electromechanical products. The test-relevant evaluation of the thermal decomposition flue gases of electromechanical products is based on various laboratory test methods. As well as possibly corrosive, sight-reducing and noxious/health-endangering effects are assessed and classified. A greater risk-relevant aspect of the side effects in the electromechanical field is the possible corrosive effect of the thermal decomposition products. These material-related regulations in this respect are based on different test philosophies. According to DIN/VDE 0266 part 85 cables are judged with regard to a reduced corrosive danger. According to VDE 0472 part 813c thermal decomposition products are produced with cable sections and guided through gas-absorbing bottles. The pH values of the washing water is supposed to stay below 3.5 and the conductivity shall not increase above 100 µS/cm.

The pH measurement is also the basis of the relevant French Standard NF-C20-453. Jann. 85. According to the Italian Standard UTE-C20-453 a mixture of 300 mg of cable granules and 100 mg polyethylene will be combusted in an enclosed cabinet at 50 °C. The electric resistance as a function of time (2 h – 55 °C, 2 h – 40 °C) is determined with a copper wire of 0.06 mm diameter, 1 m long and "acitfree". The development of risk-related test procedures has been initiated within the framework of international standardization (SC4 of TC 61) and in agreement with TC 89 of IEC. Smoke

4.4 Electrical Engineering | 155

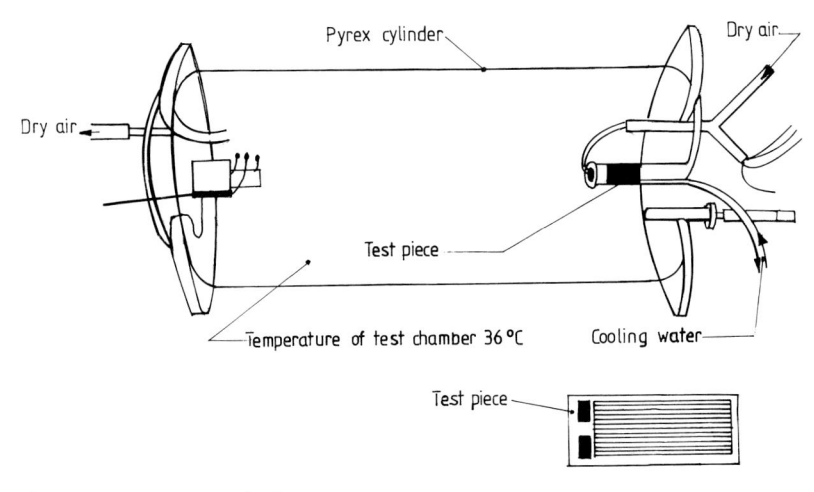

Figure 82a CNET Test Method

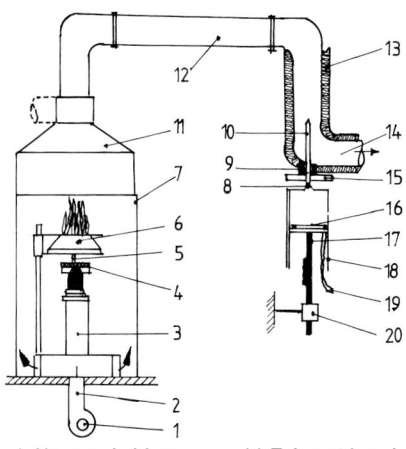

1 Air-supply blower
2 Temperature and RH measurement
3 Load cell
4 Specimen
5 Igniter
6 Conical heater
7 Pyrex enclosure
8 Disconnect valve
9 PTFE union
10 Glass sampling tube
11 Exhaust hood
12 Transport section
13 Temperature-controlled section
14 Outlet
15 Stainless steel tube
16 Target
17 Plunger disconnect
18 PMMA syringe; 5 Liter
19 Electric test connections
20 Stopper motor

Figure 82b Corrosimeter Test Method

release per unit time, exposure conditions and spatial parameters are the dominant test criteria, to ensure that risk-related smoke concentrations are the basis of the assessment. Basic investigations are carried out with the static test procedure, the CNET-test, see Fig. 82a, and dynamic systems like the DIN tube furnace test, see Fig. 47c and the modified cone-calorimeter procedure, a draft standard, see Fig. 82b [295].

Examination and evaluation of the smoke-release rate were investigated by TC 89 of IEC. The smoke density problem is handled on the basis of the NBS smoke-density chamber. As far as the cable section is concerned, studies are also carried out with the 3-m cube test (see Fig. 75b). The problem of the acute toxic potency of effluents is assessed using a tube furnace system similar to DIN 53436 and NF-X70. Taking into account the previous studies of SC3 of TC 92, the stream of effluents is produced at 600°C and 850°C [296]. It is intended to determine the concentrations of the most important components and to evaluate the toxic potency by mathematical modelling corresponding to SC3 of TC 92.

5
Material-specific Fire-Performance Characteristics of PUR

The entire field of polyurethane chemistry including production and application is treated in detail in the Polyurethane Handbook of Oertel [297].

5.1
Polyurethane Production

The formation of the polyurethanes occurs according to the polyaddition procedure discovered by Otto Bayer 1937. Polyurethanes gained acceptance due to their universal qualities in almost all relevant fields. They are the most important representatives of the reaction-expanded plastics. The main reaction is the conversion of polyisocyanates with polyhydroxyl combinations to produce a polyurethane. The urethane group, a relatively constant carbodiimid acid group emerges from an isocyanate and a hydroxyl group. Polyethers and polyesters are used as the preferable polyhydroxyl compound. Mineral oil is the basis of the two main components, Polyol and Polyisocyanat. TDI (toluylene-diisocyanate) and MDI (diphenylmethane – diisocyanate) are the most commonly used within the group of isocyanates (Figs. 83a and b). TDI is mainly used in flexible-foam production and MDI in rigid-polyurethane production.

Different recipes form the basis of flexible and rigid foams, semihard systems, elastomers, coatings and binders. The range of properties can be substantially changed by catalysts, emulsifying agents, inhibitors, crosslinks/chain extenders, stabilizers, antioxidants, dyes or special additives, for example, additives for flame retardancy. The foaming reaction is either a chemical reaction due to the release of carbon dioxide formed by the reaction of isocyanates with water or a physical nature due to vaporisation of added low-boiling liquids. The formation of CO_2 as a result of the reaction of isocyanates and water (NCO/H_2O) produces the oldest and for the flexible-foam production also the most customary foaming agent for expanded products with an overwhelming open-cell structure. The physically effective blowing agents for low-boiling liquids, such as for example the most commonly used FCkWs (chlorofluorohydrocarbons), e.g. R11, will be judged by their thermal conductivity. Such blowing agents have been suspected to contribute to the depletion of the stratospheric ozone layer. Therefore the environmental organisation of the United Nations ratified the Montreal protocol on Sept. 16 1987. The goal of this protocol was to reduce and eliminate the production and use of substances depleting the ozone

Polyurethane and Fire. Franz H. Prager and Helmut Rosteck
Copyright © 2006 WILEY-VCH Verlag GmbH & Co. KGaA, Weinheim
ISBN: 3-527-30805-9

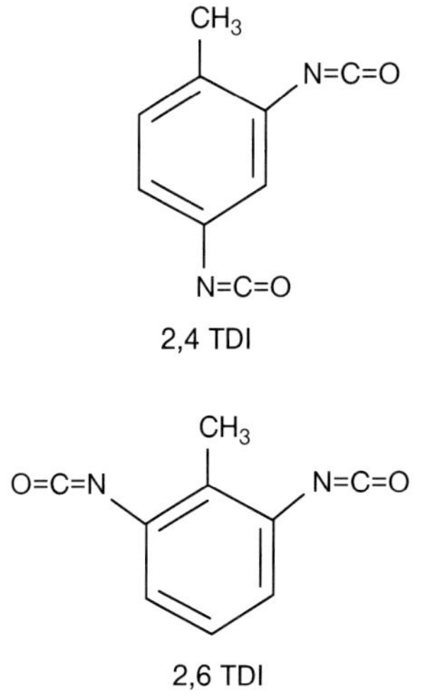

Figure 83a Structure formula TDI

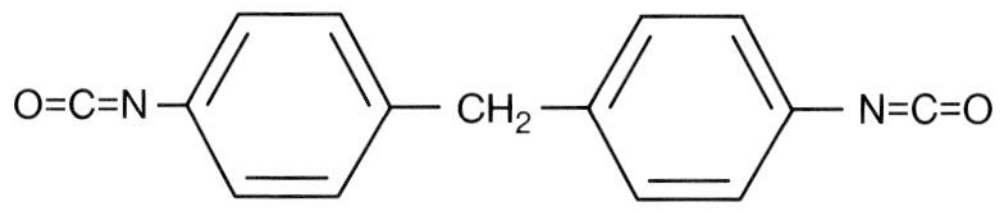

Figure 83b Structure formula MDI

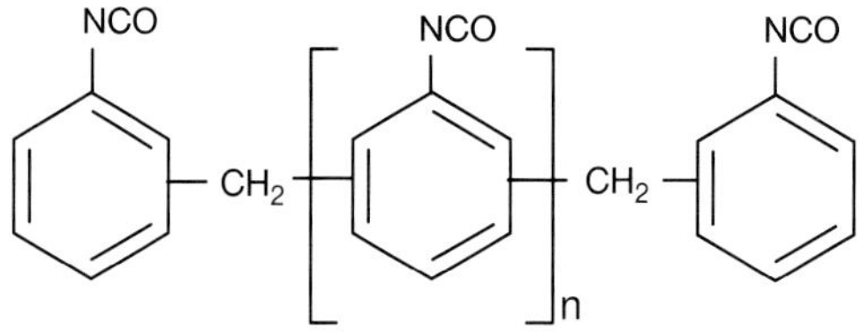

Figure 83c Structure formula crude MDI

layer. FCkWs were excluded from the market because of environmental conditions and were replaced by pentane. These foaming agents evaporate due to the heat of reaction, but remain within the closed cells of the expanded plastics and then decisively influence the thermal conductivity of the product. The extremely small thermal conductivity of rigid polyurethane foam is determined to a large extent by the cell gas. The injection of air in the mixture forms a further variant of the foaming process. For rigid foams, an increasing trimerisation leads to polyisocyanurate systems, which have an increased tendency to carbonize under flame impingement. Mainly halogen–phosphorus combinations are used as flame-retardant additives. The effectiveness is based on intervention into the vapor state and/or development of protective isolating layers. Both chemically incorporated additives and nonincorporated substances are used. The aluminum hydroxide, $Al(OH)_3$ changes at approximately 180–200 °C via release of water from alumina that forms a protective layer according to the formula 2 Al(OH)$_3$ + Q = Al$_2$O$_3$ + 3 H$_2$O. Heat is absorbed from the fire-reaction zone and furthermore by the endothermic reaction. The resulting steam dilutes the released stream of effluents. A number of manufacturing processes were developed for the production of polyurethane foams. DIN 24450 defines the technical terms of foaming and reactive-liquid molding. The production occurs either in the one-shot process or in two stages, the prepolymer or the semiprepolymer process. With the one-step or one-shot procedure all reactive chemicals are brought simultaneously to the reaction, which slowly or quickly progresses exothermally. In the case of the two-shot procedure part of the polyol component reacts with a surplus of the prepolymer isocyanate component and this reacts, still fluid, with the component. Raw-material manufacturers supply the chemical batches including the necessary formula. Equipment and machinery producers offer the necessary instrumentation and the processing industry produces these different polyurethane products. The mixing head is provided with the raw-material components by means of different pump systems. The reaction mixture is poured by the self-cleaning mixing head. During block production, the mixture streams into a U-shaped paper former, see Fig. 84a. The flowing lid produces a flat surface of the ascending foam. Continuous block production is the most economic production for large volumes. The block material is cut onto the desired sizes, e.g. for mattresses and upholstered furniture.

In the double-conveyor-belt process, the reaction mixture is spread through a mixing head, see Fig. 84b, onto the lower facing. The rising mixture adheres to both the lower and the upper facing. The boards are cut to the desired length. The rigid-foam products for standard insulating boards have facings on both sides. Examples of these are mineral fleece, alumina foil and multilayer foils. Special products may have wood, plywood, chipboard or plasterboard facings. Sandwich elements with rigid and flexible facings or half-pipes can also be made in discontinuous processes. The composites with flexible polyurethane foam can be laminated by use of adhesives and by flame lamination. Sandwich elements with a core of rigid PUR foam and elements in particular with metallic facings have been preassembled into large building elements. These elements have been used for industrial buildings, warehouses and exhibition centers. A high degree of prefabrication within the industrial manufacturing process guarantees a good accuracy and quality. Construction is based on a tongue-groove joint system.

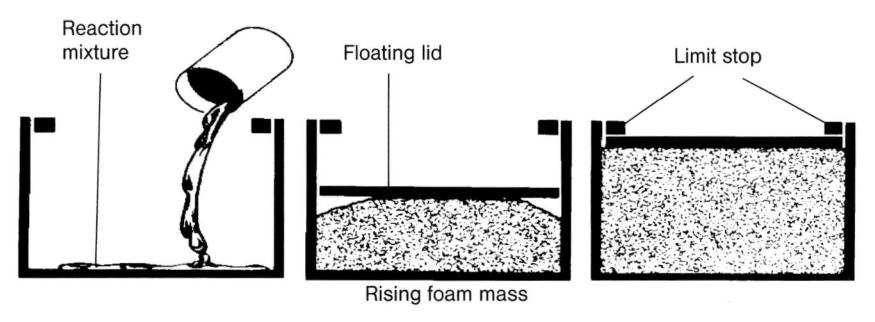

Figure 84a PUR slabstock production

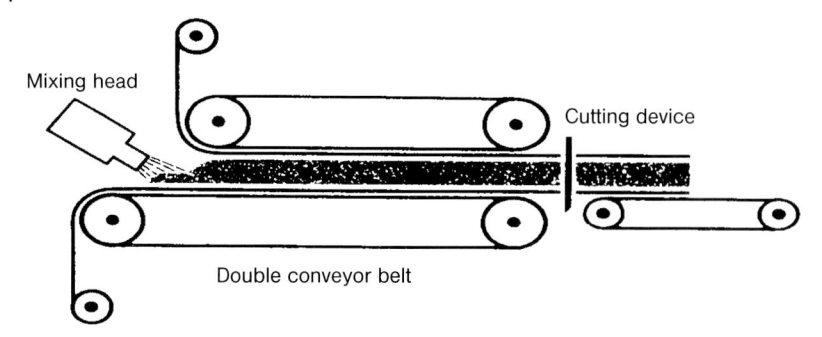

Figure 84b Double-convenor-belt process

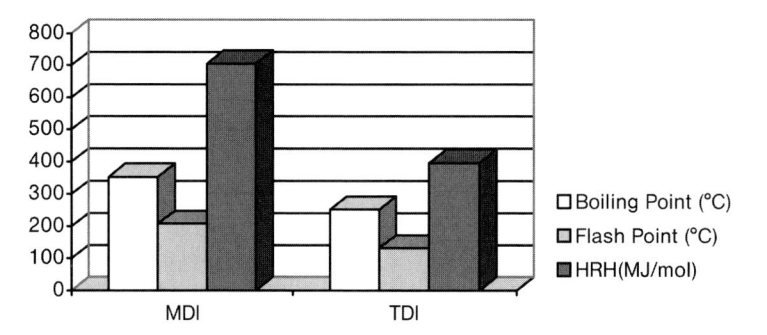

Figure 84c Physical properties

Different qualities of molded foam can be gained through a careful selection of the formulation. In the case of the low-pressure technique the mixing of the components is guaranteed by means of a special mixer. Composites can be produced by pouring systems that are foaming behind outer coverings. An essential area of application for molded parts lies in the transportation field, in the furniture and furnishing sectors. With the spray-foam technique, insulating layers are manufactured on ceilings and roofs and can also be deposited on parking decks and underpasses. Polyisocyanates are used also for paints and coatings. There are a very large number of types of PUR paints that gives an abundance of production kinds. One distinguishes in detail, for example, between:

- Two-component systems, where the binding agent consists of poyol and polyisocyanate
- One-component systems, whose polyisocyanate binders dry with the humidity and harden.
- One-component systems, whose binders consist of polyol and polyisocyanate. The film-forming reaction occurred through application of heat.

DD (Desmodur®/Desmophen®) paints are PUR-reaction resins. Polyurethane as well as polyisocyanurate systems are produced by formula variation and by containing biuret groups. Curing occurs by means of polyaddition reactions. Due to modern research work in surface finishes, the formerly usual a splashing procedure, the classical modes of painting (brush or rollers) are declining and spray modes are dominating.

5.2
Risk of Ignition in the Production and Storage Area

5.2.1
PUR Raw Materials – Basic Characteristics

Fire disasters like the Seweso disaster or the Sandoz Fire led the attention of the public to possible dangers of chemicals in fires [298, 299]. During the blast in the ICMESA fabrication, e.g. TCDD (tetra chlorodibenzo-dioxin) was released. Due to this disaster, research programmes have been initiated and financed by the EG to clarify the possible danger in the case of a fire with chemicals [300]. Petrochemical and chemical industry fires must be differentiated by the characterisation of possible fire scenarios between the pool-fire situation and the torch-like fire generated in the case of possible leakage due to negligent, arson and possible sabotage acts. The differing general conditions of a manufacturing plant and the storage area for raw materials, e.g. drums and tanks have to be considered, in order to limit such possible dangers by preventive fire protection. Kourtides et al.[301] and Koseki et al. [302] report on pool-fire investigations with heptane.

About 70% of the heat released was attributed to the convective part and 30% was effected by radiation. The influence of the radiation part increased due to the smoke and soot formation with the pool diameter. The 0.15 heptane pool fire produced in a tunnel a four times greater RHR according to Carvel [303]. Carvel described combustion pool-fire tests with a 1-m^2 heptane pool. He reported that the inside tests (Ineris Tunnel) reaches 180% of the equivalent fire in the Norwegian tunnel fire (Eureka project [304]). The methanol pool fires (0.1–0.25 m diameter released about 1.0, 1.1, 1.4 and 2.7 times greater RHR values than in the open air. Einfeld et al. [305] reported on pool-fire investigations conducted at the Scandia laboratories. These tests with a pool size of 0.1 to 0.15 and 0.36 m^2 were run within the test room (7 m × 8 m floor area). The hydrocarbon fires with a pool size of 17 m^2 and 170 m^2 were run outside. Fiala [306]report about similar tests. Due to these fire disasters, the national requirements of the "Störfall VO" as well as the international Seweso directive had to be considered [307–310]. A careful analysis of accessible fire records did not show any indications of a significant contribution of PUR raw materials to such fire disasters. Neither polyisocyanates nor polyols were obviously part of the main fire load of large fires. Despite their positive findings basic studies were initiated and financed by the affected industry. These basic investigations contain among other things pool-fire conditions under laboratory and bench-scale conditions as well as large-scale testing including real drum fires. These studies were carried out and/or initiated by the industry and used predominantly in the two important material groups, the

80%/20% 2.4/2.6 TDI and the polymeric MDI (PMDI) containing 30–32% NCO. The essential aspects of the investigative programs were demonstrated in particular by the three publications of Malair et al. [311, 312] and Sand et al. [313]. A comprehensive representation of this problem including all available, relevant findings is given in Chapter 6 "Combustibility and Related Fire Hazard" in the book "TDI and MDI Health and Environmental Sciences" John Wiley, London UK [314]. The present findings can not be transmitted without restrictions onto arbitrary standard isocyanate products, such as, e.g. 100% TDI or polyether-based TDI Prepolymer-MDI products. In particular solvent-containing mixtures will show a different combustion behavior. Mixtures can be characterized by DSC/TGA measurements. Investigation results, see Table 38, clearly show that, e.g. 80/20 TDI is vaporizing and decomposing in the temperature range of 250°C to 300°C, while the analogous temperature range for PMDI is at approximately 350°C to 400°C. If these vapors are ignited, then the decomposition products consist mainly of CO_2, CO and steam as well as small amounts of nitrogenous components as NO_x and HCN. The respective concentration data depends decisively on the composition conditions and the availability of oxygen. Due to the chemical reactions, PMDI carbonises very rapidly, which suppresses the flaming combustion and initiates a smoldering process. 80/20 TDI is, see Fig. 85a, a

1 Thermocouples
2 Sample
3 Wood crib of 10 kg
4 Ignition sticks

All measures in mm

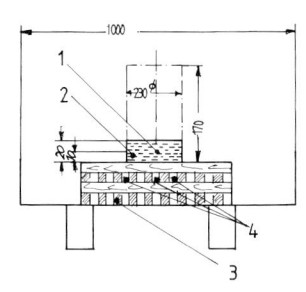

Figure 85a Arrangement of the ignition source in the drum area

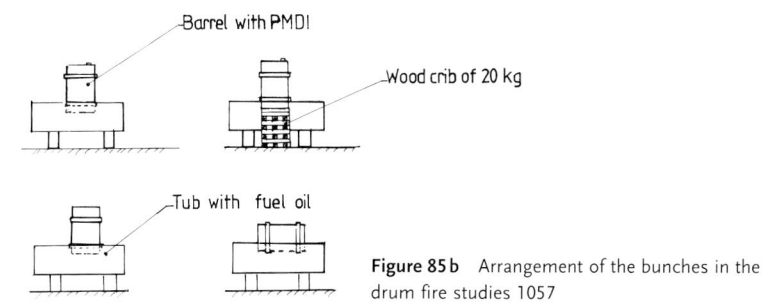

Figure 85b Arrangement of the bunches in the drum fire studies 1057

mixture of 2.4 and 2.6 isomers in relationship of 80:20 toluene-diisocyanate isomer. At the normal processing temperatures (20–30 °C) 80/20 TDI is a liquid of low viscosity. As illustrated by Fig. 85b the main component of PMDI is about 45% diphenylmethane 4,4' diisocyanate (MDI). The rest of the dark-yellow fluid has an isocyanate content (NCO) of 30–32% and exists as isomers and oligomers of MDI and polymeric MDI ($n = 0.3$). The fire-safety-relevant physical qualities of these isocyanates are compared in Table 38.

Table 38 Physical properties of the tested Isocyanates

Properties	80 / 20 TDI	PMDI
Melting point (°C)	10	>
boiling point (°C)	253	> 300
(°C)	132	208
Self – ignition – temperature (°C)	> 595	> 600
No explosion DSC / TGA proof	not explosive	not explosive

5.2.2
Laboratory Research Work with the Tewarson Apparatus

The results gained by investigations of Rhone Poulenc [315, 316] with the Tewarson apparatus, see Fig. 9 are compared in the listings of Tables 39–43. In the stream of air with an oxygen content of 21% as well as with a reduced content of 15% in relation to the product hexamethylene diamine (HMD) that has a comparable chemical construction with a hydrocarbon core with two hydrocarbon reactive groups, however, no isocyanates are thermally stressed up to 30 kW/m². Relevant characteristics, the risk of ignition and the heat as well as the smoke release are determined. These characteristics, see Tables 39–43, can be interpreted in such a way that PMDI cannot be ignited as easily as HMD. TDI ignites more quickly with increasing external radiation under the test conditions than HMD. The heat-release rate is significantly reduced in the sequence HMD, TDI and PMDI. The reduced oxygen availability leads for all tested products to reduced heat-release rates. During the fire-gas formation the product HMD produces much smaller smoke densities with both test conditions. The thermal decomposition products of both isocyanates produce a significantly larger reduction in visibility, see Table 36. With a reduced oxygen availability slightly smaller values are measured. The analytically determined concentration data of the essential smoke components show similar circumstances to the smoke-density data. The comparison ISSeP cone-calorimeter studies, financed by the EG, were carried out within the Mistral project 3085 [314]. Results of these investigation confirm basically the values of the Tewarson studies. 36.4 g TDI (30 ml) with a pool area of 0.0113 m² was exposed to a heat flux with an intensity level of 15 kW/m². The amount of heat released within the fire duration of 21–22.5 min has an average of 250–328 (max. 351–445) kW/m². During the smoke release max. concentration levels of 1680–1760 g/kg CO_2, 85–89 g/m² and 35–44 g/kg are produced.

Table 39 Time till ignition of pool area (min / s)

Ext.Radiation Material	0 (KW/m²)	5	7,5	10	20	30	Air O₂ (%)
PMDI	a	a	a	6,33	2,30	1,50	21
TDI	a	7,10 (min/s)	–	2,03	0,51	0,41	21
HMD	a	6,48	–	2,27	1,43	1,10	21
PMDI	a	a	a	9,20	3,10	2,90	15
TDI	a	11,45	–	3,30	1,45	0,57	15
HMD	a	8,35	–	3,45	2,00	1,35	15

a = no ignition

Table 40 Total heat release (kW/m²)

External radiation	5 (kW/m²)	10 (kW/m²)	30 (kW/m²)	Air O₂ (%)
PMDI	a	660	1037	21
TDI	735	1116	1432	21
HMD	1097	1860	3057	21
PMDI	a	177	614	15
TDI	242	353	1320	15
HMD	446	902	1557	15

a= no ignition

Table 41 Smoke-release – optical density D

External radiation	5 (kW/m²)	10 (kW/m²)	30 (kW/m²)	Air O₂ (%)
PMDI	a	0,95	2,07	21
TDI	0,62	0,83	1,66	21
HMD	0,03	0,06	0,08	21
PMDI	a	0,32	1,00	15
TDI	0,5	0,28	1,14	15
HMD	0,17	0,02	0,03	15

a = no ignition

Table 42 Maximum concentration of main fire gas components

Material	Airstream O_2 (%)	I (kW/m²)	O_2 (%)	CO_2 (%)	CO (ppm)	NO_x (ppm)	HC total (ppm)
PMDI	21	10	0,57	0,53	250	43	<2
TDI	21	10	1,0	0,9	400	58	10
HMD	21	10	1,60	1,0	80	76	<2
PMDI	15	10	0,17	0,15	67	13	<2
TDI	15	10	0,32	0,28	100	28	<2
HMD	15	10	0,80	0,50	50	55	<2
PMDI	21	30	0,90	0,82	530	25	40
TDI	21	30	1,32	1,3	1060	50	30
HMD	21	30	2,80	1,85	450	260	30
PMDI	15	30	0,63	0,50	350	35	10
TDI	15	30	1,20	1,13	1600	43	35
HMD	15	30	2,20	1,47	450	168	80

Table 43 Quantities of fire-gas components per gram of sample

	O_2 LEVEL (%)	I (kW/m²)	PMDI (mg/g)	TDI (mg/g)	O_2 (g/g)	CO_2 (g/g)	CO (mg/g)	NO_x (mg/g)	HCN (mg/g)	HC total (mg/g)
PMDI	21	10	6,7	–	0,89	1,14	27	0,65	2,0	<0,5
TDI	21	10	–	2,0	1,43	1,87	38	9,8	7,4	1,7
HMD	21	10	–	–	2,68	2,35	9,6	14,4	1,0	<0,5
PMDI	15	10	1,8	–	0,60	0,85	18,7	0,5	2,4	<0,5
TDI	15	10	–	0,9	1,43	1,90	41,6	13,0	5,1	<1,0
HMD	15	10	–	.	2,34	2,19	1,0	18,8	0,4	<0,5
PMDI	21	30	5,8	–	1,07	1,29	36	6,0	5,0	3,9
TDI	21	30	–	5,0	1,33	1,68	80	4,7	15,8	4,4
HMD	21	30	–	.	2,54	2,28	29	17,5	9,6	3,6
PMDI	15	30	2,2	–	1,03	1,29	48,4	7,2	4,8	1,9
TDI	15	30	–	4,1	1,47	1,81	92,5	5,7	16,7	3,3
HMD	15	30	–	–	2,57	2,20	35	12,2	18,2	11,9

5.2.3
Drum Tests and Supplementary Pool-Fire Experiments

The investigation project of Bayer AG pursued two different goals [317]. The first was to determine if the attack of a primary fire causes the drum content to escape at weak points and ignites. On the other hand, it has to be clarified whether the drum content that emerges in fluid or gaseous form achieves a substantial contribution to the possible surface spread of flame. As ignition sources, primary fires of 10 kg of heavy

Figure 86 Tray construction

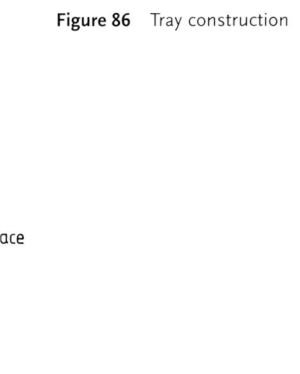

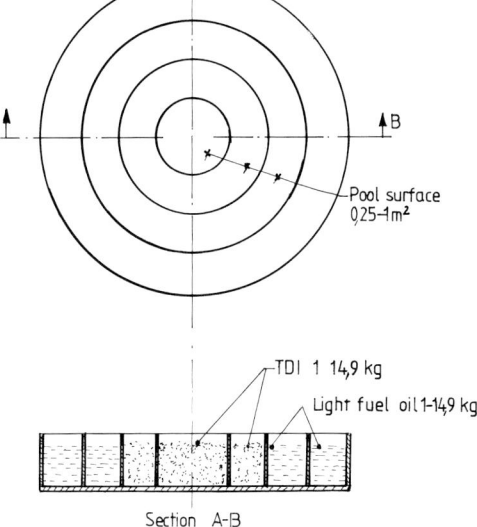

Pool surface 0.25–1 m²

TDI 1 14.9 kg
Light fuel oil 1–14.9 kg

Section A-B

wood cribs as well as mineral-oil pool fires (50 to 80 l) were used. The thermal loading of the isocyanate containers, drums with 10 and/or 250 kg content, is shown in Figs. 85a and b. As illustrated by Fig. 86 mineral oil, used as the primary ignition source in the pool-fire investigations, was poured into the tray. The obvious resistance of TDI against ignition sources of low intensity underlines the necessity of sufficient preheating in order to guarantee a self-supporting combustion process. The ignition and the self-supporting combustion process could be initiated with both isocyanate products by a sufficiently intensive preheating by means of a primary fire. As shown by Tables 46 and 47, isocyanate lots of 10 to a maximum of 250 kg were examined. Metal drums were mainly used. Polyethylene drums were investigated for comparison.

The listings of Tables 44–50 provide an overview of the ignition and burning behavior of the thermally stressed TDI and PMDI drums. The metal containers showed different behavior. As a rule, the temperature rise in the container interior combined with the increase in pressure, led to the splitting of the container at a weak point of the welds. Separation of the lid also occurred sometimes. Continual heating with gradual deformation caused after a longer test time by particularly stable containers without serious weak points and distortion explosion as listed in Tables 44 and 45. Depending on the preheating of the drums as well as on the geometry of the crack, the content of the drums escaped in fluid as well as in gaseous form. If spillage occurs by a nozzle-shaped aperture the gaseous decomposition products produce a torch flame with temperatures above 1000°C. At the drum tests with PMDI initiated the thermal stress of the primary fire a flaming process of the drum content. The splitting of the container during the experiment could occur without explosion due to the foaming reaction inside the drum initiated by the thermal stress. As

Table 44 Performance characteristics – TDI drum tests

Test parameter	Unit	1	2	3	3	4	4	5
Quantity of TDI	kg	10	60	60	60	30	30	250
Drum position								
Fuel oil	l	40	30	10	10	10	10	80
Container deformation	min	3	5	5	–	5	5	7
Contents gas release	min	3	8	4	4	2	–	37
Liquid	min	–	18	–	11	2	–	39
Burning liquid	min	–	18	–	–	–	–	–
Torching burst out	min	–	–	–	6	–	3	44
Burning first observed	min	4	8	28	6	6	3	39
Max. flame length	m							
Vertical		>5	–	–	3	–	4	Fireball
Engled		–	–	–	4	–	4	–
Torch		–	–	–	3	–	3	4 to 5
HCN	ppm	–	–	–	>10	–	–	–
Drum explosion		no	no	yes	no	yes	no	no

Table 45 Performance characteristics – PMDI – drum tests

Test Nr..	1	2	3	4	5	
PMDI (kg)	1 × 10	1 × 30	1 × 30	1 × 60	2 × 30	
drum position	3 b	3 a	3 a	3 a	3 c	
					left	right
Fire load	–	2 × 5	20	30	10	10
fuel oil						
wood crib (kg)	2 × 10	–	–	–	–	–
drum rupture (min)	32	17	4	3	–	–
release (min)						
gas	32	17	5	5	2	5
liquid	–	–	6	–	–	21
burning liquid	–	–	12	14	–	11
drum closure blown	–	19	–	–	28	–
torching	–	–	5	4	–	3
first flame	–	17	4	4	3	3
max. flame (m)						
horizontal	–	–	5 – 6	–	–	–
vertical	>5	2	4	5 – 6	0,3	3
angled	–	3 – 4	4	4 – 5	0,2	–
drum explosion	no	no	no	no	no	no
foaming reaction	–	yes	yes	yes	yes	yes

Table 46 Gas analysis – TDI drum tests

Test Nr.	1	2	3		4	5
Time of test (min)		7	25	26	18	30
CO_2– max (%)	–	2,1	3,6	2,3	11	7
CO –max (%)	–	6	0,95	0,16	>1	2,8
O_2–min (%)	–	16	17,5	18,8	5	13
CO_2 / CO	–	13	4	14	11	3
HCN (ppm)	–	–	–	>10 (78 min)	–	–

Table 47 Gas analysis – PMDI outdoor drum tests

Test nr	1	2	2	2	3	3	3	4	4	4	5
t (min)	–	4	18	29	6	10	12	2	9	14	6,5
CO_2max (%)	–	0,7	1,0	0,01	1,6	2,2	1,5	2,0	2,8	2,7	0,3
CO max (%)	–	–	–	0,118	0,08	0,14	0,49	0,07	0,27	0,40	–
O_2min (%)	–	19,8	18,5	–	17,0	16,3	17,3	16,0	14,5	15,5	20,5
CO_2/CO	–	–	–	0,1	20	16	3,1	28,6	10,4	6,8	–
HCN (ppm)	2,5–3,0	–	–	–	–	–	–	–	–	–	–

Table 48 Performance characteristics – PMDI indoor drum tests

Test Nr.	6	7
Quantity of PMDI (kg)	15	10
Pool area (m^2)	0,25	0,25
Drum material	Steel	PE
Drum location Fig.	3a	3a
Ignition source fuel oil (l)	10	7,5
Contents release (min)	4	0,25
Flames first observed (min)	4	0,5
Torching (min)	7	–
Drum lid forced off (min)	10	–
max. flame length (m)	3	2,5
Drum explosion	no	no
Foaming reaction	yes	yes

Table 49 Gas analysis – PMDI indoor drum tests

Test Nr	Measuring point	6	7
O_2 min (%)	G1	19,7	20,1
	G2	20,7	20,4
	G3	20,6	20,6
CO_2 max (ppm)	–	16 500	13 800
Measuring time (min)	–	19	5
CO max (ppm)	–	>500	445
Measuring time (min)	–	19	12
CO_2 / CO	–	21	31

Table 50 Gas analysis – TDI indoor drum fires

Test Nr.	1	2
Medium values		
CO_2 (%)	0,21	1,00
CO (ppm)	65	350
NO_x (ppm)	10 – 12	10 – 12
HCN (ppm)	(2)	(40)
TDI (mg / m^3)	10 – 15	8–10
Total values		
CO_2 (mg/g)	2090	1620
CO (mg/g)	41,1	36,0
NOx (mg/g) (als NO2)	10,4–12,5	1,7–2,0
HCN (mg/g)	(1,2)	4
TDI	5–7,5	0,7–0,9

demonstrated by Fig. 87a gases escaped after 4 min of the thermal insult for the 30-kg bunch. The torch flame was created after 5 min. The foaming process resulted in a temperature rise in the drum to 450°C. For the resultant highly porous carbonized expanded material a core temperature of approximately 500°C could be verified after an observation time of 90 min. The foaming reaction led mainly to the rapid sealing of the leaks. The reduction of gaseous decomposition products occurred correspondingly with the reduction in pressure. The ignition of these decomposition products produced a torch-type flame with extremely high flame temperatures of more than 1000°C, see Table 44. The generation of such flames causes a great fire risk for neighboring bunches in the storeroom due to the high temperatures.

The results of the fire-gas analysis of the drum- and pool-fire tests carried out in the open-air show very vividly according to Fig. 88a that, e.g. wind effects can significantly vary the value of the information The spatial variations of the flue-gas column and the extremely different dilution effects combined with that, complicate risk-relevant statements. The complementary experiments carried out in the interior of a test tower confirm this statement. The schematic of Fig. 88b shows the CO_2/CO ratio of the released stream of effluents produced in relevant studies. Figure 88c gives information about the arrangement of the measuring point in the tower room. The CO_2/CO ratio clarifies the different burning behavior of the two investigated isocyanates TDI and MDI.

The graphical representation of Fig. 87e underlines this statement. After ignition occurs, TDI burns uniformly to completeness, whereas the PMDI pool fire is changing into a smoldering fire. Unambiguous judgements from a toxicological point of view suppose a detailed knowledge about the temporal course of the fire and effluent concentration.

Because of defined ventilation conditions the test tunnel of INERIS allows more risk-relevant judgements of the smoke release during the thermal loading of isocyanates. The test tunnel, see Fig. 88a, consists of a 50-m tunnel with a sectional area of 10 m^2 and a two-part tower section. In the lower half of the tower are the take-off measuring points for the flue-gas analysis and the smoke-density measuring system.

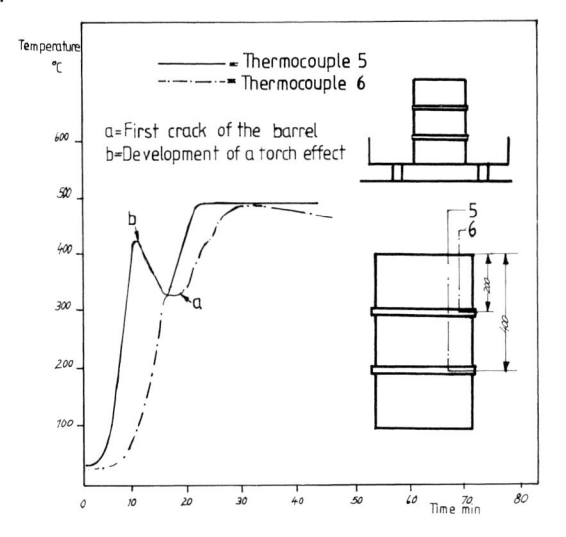

170 | *5 Material-specific Fire-Performance Characteristics of PUR*

Figure 87a Temperature increase within the drum during the preheating process

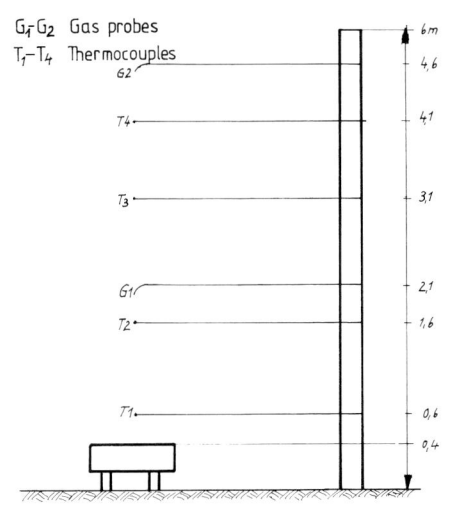

Figure 87b Arrangement of measuring points in the open air test

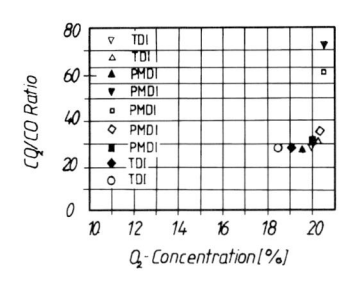

Figure 87c CO_2/CO ratios in the stream of effluents

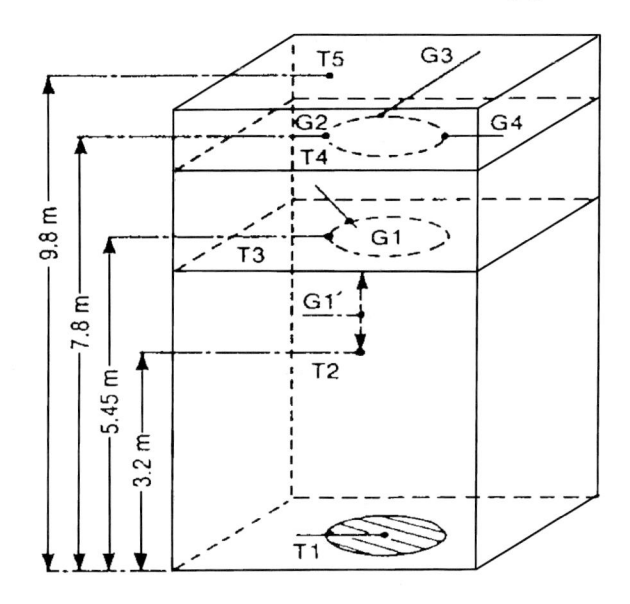

	CO_2	CO	O_2	HCN	Isocyanate
G1				x	x
G1'	●	x	x		
G2			x	x	x
G3			x	x	x
G4			x	x	

G1–G4: Gas sampling positions
T1–T5: Thermocouples

Figure 87 d Arrangement of the measuring points in the tower

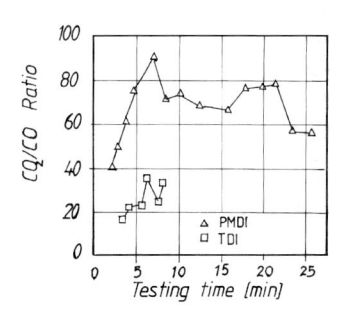

Figure 87 e CO_2/CO ratio in the flue-gas stream

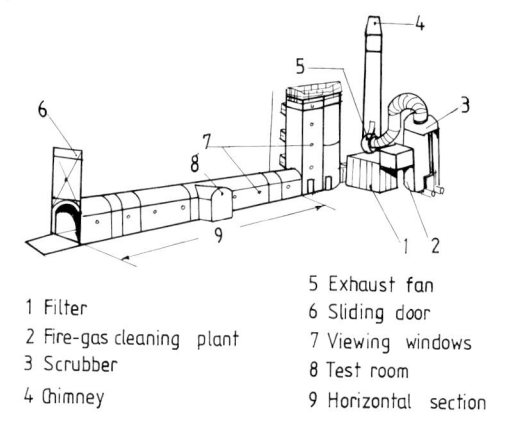

5 Material-specific Fire-Performance Characteristics of PUR

1 Filter
2 Fire-gas cleaning plant
3 Scrubber
4 Chimney
5 Exhaust fan
6 Sliding door
7 Viewing windows
8 Test room
9 Horizontal section

Figure 88a Ineris Test tunnel

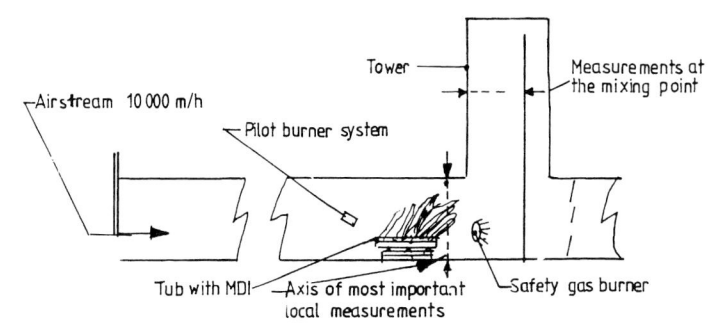

Figure 88b Test setup for the pool-fire experiments

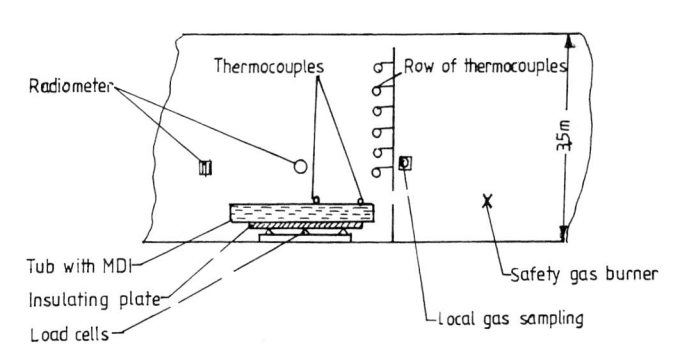

Figure 88c Arrangement of the measuring points

The upper half of the tower consists of the smoke-scrubbing system and the two measuring points for estimation of smoke particles. The defined ventilation system guaranteed a somewhat constant air supply to the scene of the fire as well as the extraction of the combustion products. For the investigations with 25 kg and 75 kg TDI as well as 7 and 35.5 kg PMDI a ventilation of 10 000 m^3/h was used. During the experiment with 35 kg PMDI within the framework of the separate Ineris program, a different ventilation of 30 000 m^3/h was employed [314].

The material is tested in a cylindrical tray, which was arranged on a weight-loss platform at a distance of approx. 2.5 m from the test tower, that became ignited and combusted with the aid of a primary fire (see Fig. 88b). The selected areas of PMDI pool fires of the Ineris test run were 0.25 and 1.00 m^2 as well as 2.25 m^2. The pool area of the TDI test run was always 1.00 m^2. In the PMDI pretest with 7.5 kg, the pool surface was ignited by means of a pilot flame applied for 3.5 min with a simultaneous thermal stress generated by a radiator. In the case of the remaining investigation, ignition and combustion were initiated by means of a primary fire of 1l hexane. After ignition, the TDI burned regularly without passing through the different stages of burning observed for PMDI. In the first phase, between the 1st and the 6th minute a steady-state pool fire occurred. In the second phase between the 6th and up to the 12th minute a smoldering fire occurred due to the foaming reaction that lead to a strongly reduced flaming and to a smoldering process. Figure 88c illustrates the arrangement of the radiometers, the thermocouples as well as the smoke-density measuring equipment. The schematic representations provide information to the washing system as well as the two impingers for the determination of the smoke particles.

Tables 51 to 57 also give a view of the chosen test and evaluation criteria. Ignition occurred at a temperature level of about 135 °C during the application of the igniting flame to the TDI pool surface. Thus is unambiguously higher than that of the hydrocarbons. With a burning rate of 2.1 kg/m^2 min for the bigger pool area of 1 m^2, TDI obviously lies at the lower end of the scale of flammable liquids. For Hexene the comparable value is 4.8 kg/m^2 min. Also the heat-release rate for TDI with approx. 560 kW/m^2 is below the level of hydrocarbons that is listed in [305] with values of 3000 kW/m^2. As far as the possible reduction of visibility is concerned, the results gained for TDI are relatively unfavorable as judged by characteristic data of soot deposits determined for the 7.5 kg pool with 0.3 kg soot/kg and for PMDI with 0.1 to 0.2 kg soot/kg carbon. Measured values illustrate that with increasing pool size an intensified heat-release rate occurs. Maximum data are reached with hydrocarbon fires by a pool size of 1 m^2. Except for the two components TDI and MDI, the effluent concentrations of the fire-gas components are similar to that of the aliphatic hydrocarbons.

Table 51 Pool fire conditions in the Ineris Fire Test Gallery

Test Nr.	1	2	Pretest	1	2b
TDI quantity (kg)	25	75	–	–	–
PMDI quantity (kg)	–	–	7	35,5	35
Pool area (m^2)	0,25	1,00	0,25	1,0	2,25
Ventilation (m^3/h)	10470	9930	10590	10590	33000
Propane – flow					
Safety burner (m^3/h)	1,45	1,44	1,124	1,124	–
Total amount propane (m^3)	2,03	1,24	–	1,124 (60 min)	1,84 (35 min)
Atmospheric conditions			–		
External temperature (°C)	10,5	9,0		11	12
External pressure (mm Hg)	770	747	–	769	744
External humidity (%)	72,3	82	–	–	–

Table 52 Summary of test results – Pool fire tests in the gallery

Test Nr.	TDI 1	TDI 2	PMDI-pretest	PMDI 1	PMDI 2 b
Burning rate ($kg/m^2\,min$)	1,38	2,0	0,16 – 0,2	0,9	–
Burning duration – Pool (min)	75	46	3,30–6,0	1,3–6,0	–
Transitional			6,0–20,0	6,0–12,0	
Smoldering				12–90,0	
Carbonisation			5,0	3,30–60	
Maximum temperature (°C)					
Pool surface	700	800			
Fire room ceiling	60	160			
Decomposition products	35	90			
Heat release rate (kW)	122	560			
Heat flux					
Lateral (kW/m^2)	1,15–2,0	5 – 6	–	1,0	–
Downstream (kW/m^2)	7 – 9	15	–	2,0	–
Gas concentration					
CO_2 total (%)	0,25	1,05	0,13	0,6	0,48
CO (ppm)	70 – 80	350	72	1275	248
Smoke density D					
Light path 3m	1–1,5	3 – 5	–	4,5	4,8
Residue (kg)			5,5	29,5	25,8
Residue (%)			80	83	74
CO_2/CO local			17	17–25	19
CO_2/CO at tower				21–26	19

Table 53 Gas analysis – PMDI -pool fire tests in the gallery

Test Nr.	PMDI – 1 Medium value	PMDI – 2 Medium value	PMDI- 1 Total amount (mg/g)
CO_2 (%)	0,47	0,12	1800
CO (ppm)	180	44	44
NO_x (ppm)	30	4,3	12
HCN (ppm)	15	22,9	3,5
MDI (mg/m^3)	18	474	3,5
CO_2/CO	21 – 26	27	–

Table 54 Comparing evaluation of laboratory and large scale test results

Observation test results	Tewarson – apparatus	Tewarson – apparatus	Ineris – fire test gallery
Pool area (m^2)	0,054	0,054	1,12
Heat flux (kW/m^2)	10	30	–
Burning rate (kg / m^2 min)	1,7	–	0,9
Heat – release rate (kW / m^2)	660	1037	280
Gas analysis			
CO2 (g/kg)	1140	1290	1800
CO (g/kg)	27	36	44
NOx (g/kg)	0,65	6,0	12
HCN (g/kg)	2,0	5,0	3,8
MDI (g/kg)	3,2	5,8	3,5
Residue (%)	32	20	83

Table 55 Fire effluents

Fire – gas components	part (%)
CO_2	10
CO	–
NO_x	19
HCN	11
MDI	60

Table 56 Observations and test results of the TDI pool test.

Test Nr	1	2	3	4	5	6	7	8
Quantity TDI (kg)	1,0	1,0	14,7	2,63	2,43	9,5	20,5	14,9
Pool area (m^2)	0,04		0,25	0,25	0,25	0,25	0,50	1,0
Ignition sources								
Wooden crib (kg)	10	10	–	–	–	–	–	–
Light oil (kg)	–	–	4,3	3,1	3,1	11	9	5
Flames first observed (min)	–	–	7	4	8	3	4	3
Max. flame length (m)	3,0	3,0	3,0	2,0	2,5	2,0	2,5	3,0
Combustion rate (kg/min)	–	–	0,8	0,6	0,6	0,8	1,8	3,4
HCN (ppm)	2	8	10	2	2	10	–	–
Residue (kg)	0	0	0	0	0	0	0	0

Table 57 Records of TDI pool fire test in the Ineris Test Gallery.

Test Nr.	Measuring point	3	4	5
O$_2$ Minimum (%)	G1	20,7	19,7	20,0
	G2	20,5	20,6	20,6
	G3	–	–	–
	G4	20,7	20,6	20,7
HCN medium value (mg/m^3)	G1	–	–	–
	G2	–	–	–
	G3	–	–	–
	G4	–	–	–
Free TDI (mg/m^3)	G1	–	–	–
	G2	–	–	–
	G3	–	–	–
	G4	–	–	–
CO$_2$ max (ppm)		–	–	6900 1 min
CO max (ppm)		–	–	< 500 19 min
CO$_2$ / CO medium value		–	–	30

5.3
Polyurethanes

The fire disaster of the Sandoz Company increased the attention paid to large fires in industry. Of special interest are the dangers to the environment, as well as the danger of fire gases and the fire debris. Störfall VO regulate the release of specified gases from plants and storage facilities. If these gases are generated by fires, then they need to comply with these regulations. During the estimation of the danger it has to be assumed that, e.g. all of the nitrogen content of a product will be transformed into HCN and this will be released in the smoke plume. The regulation will therefore lead to limitations in the quantity of material in stock.

In order to be able to estimate the real danger degree of a flue-gas cloud in the case of a fire, the effluent concentration and the judgement from the toxicological point of view must be known. The estimating of the possible concentration assumes knowledge and understanding of

- the kind of storage
- the form of the storage
- the surface finish
- the spatial order/the arrangement
- the ventilation conditions
- the preheating effect and similar factors.

For the estimate of the possible danger through toxic performing gas clouds a number of computation models about the spreading danger were developed [318–320]. The surface area of the fire and the burning-rate density present, determine the scale of the released stream of effluents.

The igniting and combustion behavior is not a material property. These data records will be determined by the compatibility/interplay of all relevant parameters. The same conditions are valid for the fire-related characteristics. Pure material-related data are:

- the melting point
- the decomposition temperature and
- the net calorific value.

The amount of effluents released maximally at complete combustion as well as the nessesary air volume can be identified as material related constant factors in relation to the net calorific value.

The polyurethanes are last but not least due to the 5-million dollar program of the Product Research Committee (PRC) caused by the Federal Trade Commission (FTC) the best fire-safety-examined material group [14]. However, numerous studies also became independent of the PRC program in particular in the building section as well as in the furniture field, also in significant 1:1 test procedures carried out for the classification under end-application-oriented conditions [56, 244, 321, 322].

As standardized, e.g. in DIN 18230, the complex m-factor (see Table 58), categorizing the combustion behavior of stored materials, composites and endproducts submits advice for the calculation of the required fire resistance of elements and complete constructions of industrial buildings.

Table 58 m-factor records according to DIN 18320

Number	Material	m – factor
4.5	Polycarbodiimid (rigid foam)	0,2
4.9.1	Polyurethane rigid foam, grade B1	0,2
4.9.2	Polyurethane rigid foam, grade B2	0,3
4.3.1	Phenol formaldehyde foam, grade B 1	0.7
4.8.3	Polystyrene rigid foam, grade B1	0.4
4.8.4	Polystyrene rigid foam, grade B3	0.9
4.9.3	Polyurethane flexible foam	1.2
10.3	Leather	1,2
3.2.1	Cotton	0.4

5.3.1
Material-related Igniting Risks of Polyurethanes

During the risk assessment all the aspects of production, storage and the final application must be identified. The risk situation of the production and storage of the expanded-plastic producing and manufacturing industry is directly effected by the amount of the combustible products. In particular the production must differentiate between slabstock and molded parts. The question of a possible spontaneous combustion plays an unimportant role particularly with preventive fire protection required by the insurance companies. Because of the danger of self-ignition, curing is an especially critical phase in the slabstock foam production. Experience has shown that the danger of heating and self-ignition as the result of exothermic reactions only occurs in the slabstock production and then only in the case of polyether block production. This fact has been taken into account in the elaborated safety regulations [323] of the German Federal Republic in alliance with the Association of Insurers (VdS) and Association of the German Industry (BDI). Corresponding actions became established, e.g. from the H. M. Inspectorate in Great Britain as well as in the British Insurance Industry [324]. The fact that in the molded-foam production there is no risk of spontaneous combustion was accepted by the German Insurance Industry through a more favorable premium structure. In the manufacturing industry the usual safety regulations are valid, e.g. in the furniture industry [325]. Sand [326] discussed this issue at the Polyurethane World Congress 1987 in Aachen. For the building field the relevant determinations of the construction supervising authorities are yet to be considered. Figure 89 shows stored sandwich element samples which proved resistant against ignition sources of different intensities.

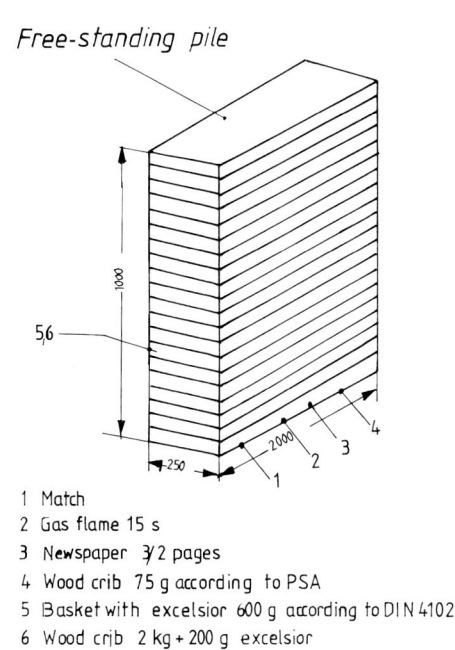

1 Match
2 Gas flame 15 s
3 Newspaper 3/2 pages
4 Wood crib 75 g according to PSA
5 Basket with excelsior 600 g according to DIN 4102
6 Wood crib 2 kg + 200 g excelsior

Figure 89 Ignition risk of stacked sandwich elements

Material-related ignition risks of PUR under enduse conditions

The relevant general conditions of the determination, appropriate storage and application state are to be duly considered during the assessment of material-related risks of ignition. The ignition and combustion behavior is not a constant. These characteristics are decisively influenced by the interplay of the relevant parameters. This reservation is valid also without limitation for all fire-related characteristics. The determination of application-related, material-particular characteristics of resistance against active ignition sources occurs in spite of the handicap of the different national test methods, such as, e.g. the Setchkin Test, see Fig. 78b. Since fire performance is not a material property, these material-specific PUR values have to be checked to assess whether they have already been able to be substantially modified by the different test parameters, or to be taken as a fixed value. The impact of an external heat source leads to a decomposition of the thermally stressed surface of nonmelting materials. The thermal impact initiates with melting products in the first stage a melting process. Additional thermal stress lead in a second stage to the decomposition of the surface of the molten material. With solid materials like wood, heat stress leads to carbonisation of the attacked surface and release of decomposition products. The pyrolysis mixture with the differing atmospheric oxygen will be ignited by a pilot flame if sufficient concentration is available or will self-ignite if heated to a sufficiently high temperature. The determined values can be modified to a limited extend by variations of the test-sample mass and the air supply only insignificantly. These data can be classified as material-related data [327], see Table 59.

Table 59 Decomposition Temperatures according to Setchkin

Material	FI (°C)	SI (°C)	Density (kg/dm³)
ABS	390	480	1,06
PMMA	300	450	1,18
PVC	390	455	1,40
PE	340	350	0,91
PUR rigid	310	415	0,05
PUR flexible	250–280	300	0,03
Wood	260	260	0,7
Paper	230	230	0,5
Cotton	210	400	300g/m²

The numerical values show that the decomposition temperature level does not offer any special features. The data records of Fig. 90 illustrate the risk of ignition of different products by means of relevant surface temperatures as well as the corresponding external intensity of radiation and distinguish considerably from the flash-ignition temperature of the Setchkin method. The comparison of the data levels clarifies the necessity to consider very clearly the definition and determination of the material-related ratings.

T_o – Surface Temperature (°C)
F_I – Flash-ignition temperature (°C)

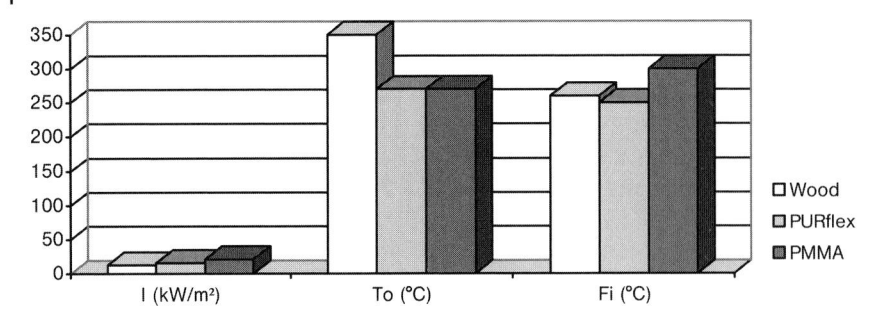

Figure 90 Risk of ignition by radiated heat

Ryan [328] demonstrates to what extent different test parameters are able to influence the characteristic size on the time to ignition. The risk of ignition of polyurethanes is mainly judged in the different fields by means of material-related tests. The relevant test and evaluation criteria as well as the admissibility of the values are reviewed in detail. The importance of material-related characteristics demonstrate, e.g. the requirements for insulants in the guidelines for the use of combustibles in the building construction in Germany [329] where, e.g. in Section 7.1 the demand of the construction supervision compartment. "Insulants tested on their own" with the statement "Insulants below the roofing felt must be tested on their own and classified at least normal ignitable (Class B2)" implemented in the Building Regulation. Independent of the different formulation variations, the resistance of polyurethane-foam materials against the impact of ignition sources of most different intensities has been studied in various research investigations. The open-cell structure and the air permeability allows a self-sustained smoldering process initiated by the exothermic reactions. Reference [330] fully reports the relevant studies within the framework of the PRC program that confirm the above statements. There is a correspondingly latent danger with mineral-fiber products and with phenolic-foam systems, which can be evaluated with the punking test according to BSI. With PUR and PIR rigid foams it has been demonstrated by research work carried out within the PRC program that self-sustained smoldering fires could only be initiated for dust-waste samples. Numerous studies verified that as clearly shown by Figs. 18b and 18c such smoldering fires, depending on the nature and concentration of the binder, will happen in noncombustible classified mineral-fiber products with smoldering temperature levels of 800 °C and more. For flexible polyurethane foams and because of the open-cell structure a strong tendency towards the self-sustained smoldering process has been verified, which does not occur for the closed-cell structure of rigid foams, in particular for PUR and PIR foam systems [331–335]. Damant, of the Flammability Research Laboratory (Bureau of Home Furnishings, Department of Consumer Affairs-State of California) has clearly shown, as illustrated by Fig. 91a that only the parallel use of several glowing cigarettes will lead to an acute risk of a smoldering fire for flexible PUR foam depending on the recipe/formula.

With the pure polyurethane system the smoldering process initiated by a cigarette will not propagate because heat will be absorbed from the smoldering process by the melting processes. Interliner made of flexible PUR can, according to [336, 337], suppress the smoldering fire risk of glowing cigarettes acting on mattresses. Figures 91b and c illustrate this behavior, which is similar to the damage of a cushion with a thermoplastic cover. The reported studies of Prager according to Tables 60a and b confirm this statement. Smoldering tobacco, in connection with outer coverings capable of smoldering, will initiate a self-sustained smoldering process [56, 103]. This smoldering process, initiated by glowing tobaccos, progresses through exothermic reactions in upholstery furniture. It achieves, with polyurethanes fillings, temperature levels up to 400–500 °C (see Fig. 17a).

Combinations, resistant against glowing cigarettes, show only an insignificant contribution to the combustion process. Where sufficient heat is withdrawn from the glowing process, e.g. by melting processes of the outer covering, the smolder process will be stopped. Provided that in the current fire case there is a similar behavior, there will be no danger from fire gases. The graphs of Fig. 19c show that a significant variation of the smoldering rate can occur depending on air permeability.

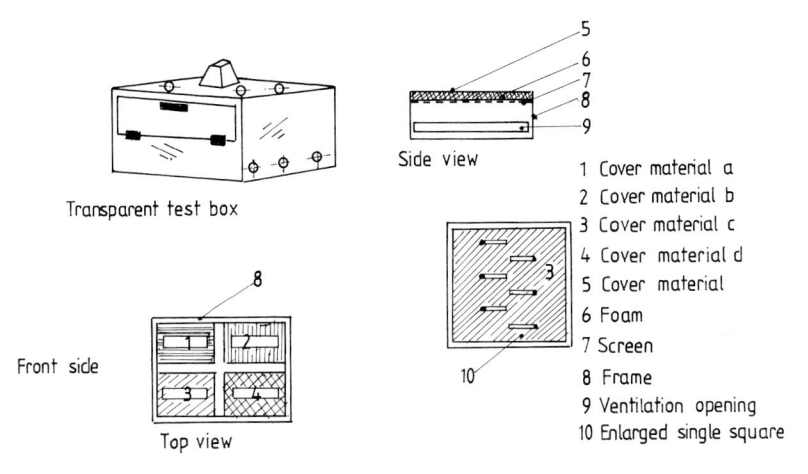

1 Cover material a
2 Cover material b
3 Cover material c
4 Cover material d
5 Cover material
6 Foam
7 Screen
8 Frame
9 Ventilation opening
10 Enlarged single square

Figure 91a PUR cigarette test with several glowing cigarettes

Figure 91b Cigarette test – Damaged area of upholstery composites

Figure 91c Cigarette test – Damaged area of PUR filling

Table 60a Cigarette test -Influence of the upholstery combination

Covering	Interlayer	Latex	PUR-NFR	PUR-FR	PUR-HR	PUR-Est
0	0	+	–	–	–	–
C/A	0	+	+	–	–	–
C/A	CFR	+	(–)	(–)	(–)	(–)
VIS	0	+	+	(–)	(–)	(–)
VIS	CFR	+	(–)	(-9	(–)	(–)
PUR/C	0	nt	–	–	–	nt
PVC/CFR	CFR	nt	–	–	–	nt
PP	0	nt	–	–	–	nt
PAN	0	nt	–	–	–	nt
MOD	0	nt	–	–	–	nt
PA/PI	0	nt	–	–	–	nt

+ = ignition (–) = no ignition, but strong smoldering
– = no ignition nt = not tested

Table 60b Cigarette test results

Outer covering	Cigarette test according to DOC FF 4-72					
	Influence of the cotton sheet and number of cigarettes					
	Filling material					
	NFR		HR		FR	
	1 cig.	2 cig.	1 cig.	2 cig.	1 cig.	2 cig.
C without sheet	+	+	−	−	−	−
C with sheet	+	+	−	+	−	+
C/A without sheet	+	+	−	−	−	−
C/A with sheet	+	+	−	+	−	+
VIS without sheet	+	+	−	−	−	−
VIS with sheet	+	+	−	−	−	−

+ = ignition proven
− = ignition not proven

Table 61 Methenamin tablet test results

Outer covering	Interlayers	Resistance against naked flames				
		Burning methenamine tablets				
		(According to ASTM − D 2859)				
				Filling materials		
		LAT	NFR	FR	HR	EST
C/A	0	+	+	+	+	+
C/A	CFR	+	+	+	+	+
PP	0	+	+	+	+	+
PP	CFR	+	+	+	+	+
MOD	0	n.t.	−	−	−	−
MOD	CFR	n.t.	−	−	−	−
PA	0	n.t.	−	−	+	+
PA	CFR	n.t,	−	−	−	−

+ = ignition proven
− = no ignition proven
n.t. = not tested

5 Material-specific Fire-Performance Characteristics of PUR

The possible risk of ignition by glowing particles, e.g. by welding beads, in the building industry, depend on environmental-related factors. The action of welding beads on flat surfaces of rigid PUR foams, shown by Fig. 92a, effects a limited burning. Figure 92b illustrates that ignition and a fast flame spread can occur where a slit in the surface leads to extremely preheating effects. The risk of ignition caused by the impact of welting beads is illustrated in [98]. The investigation results should be interpreted in such a way that the ignition temperature should be regarded not as a material but rather as a system property. Preheating effects, heating rate and heat built up due to spatial factors can have a decisive influence. After an extensive investigation, Jach [338] demonstrated that welding beads depending on environmental conditions can initiate even with melting products like rigid polystyrene foams and not only with products being capable of glowing, a smoldering fire and flaming combustion. The necessity of using protective measures for unprotected PUR-foam surfaces during hot works was documented in the technical information of the Bayer Company [339]. Ignition and a rapid surface spread must be expected due to spatial conditions that cause strong preheating effects of the surface of the foam, for example, if it is exposed to direct flame impingement. The possibility of a negligent ignition through burning matches was among other subjects of the extensive investigations by the Carpet Research Institute in Aachen [340]. Figures 92c and d show the scale of damage caused in the case of a cold-cure (high-resilience) PUR foam.

The degree of open-cell structure has little influence on the smoldering behavior and on the resistance to direct impingement of ignition sources of low intensity. Haas et al. [341] reported on investigations with conventional and HR PUR types and showed that the air permeability can change the resistance significantly. Compression, that is load of the test piece in daily use as seats, leads, see Table 62, to a signifi-

Table 62 Influence of the compression to the air permeability

	with compression	without compression
Air resistance	50	350
BS 5852 source 3	+	−

Figure 92a Welding tests – Area impingement

Figure 92 b Welding tests – Slit impingement

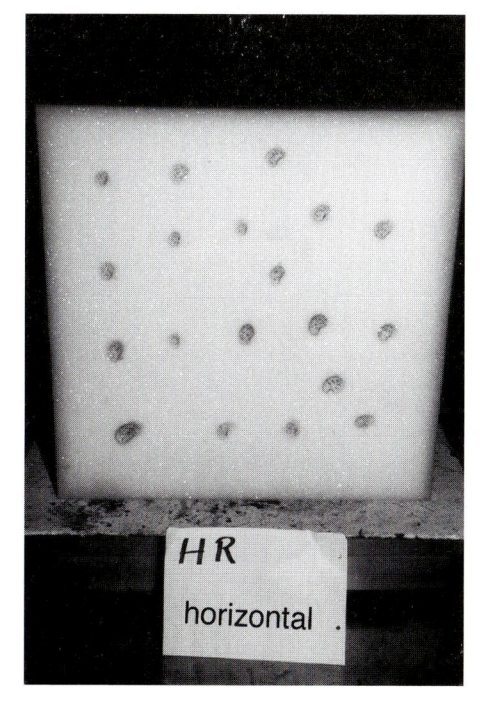

Figure 92 c Match type tests – Scale of damage at level test

Figure 92 d Match type tests – Scale of damage at sloping test

cant decrease of the air resistance. This changes the behavior of the test piece considerably and increases the failure probability. With increasing open-cell structure, the degree of resistance decreases and is possibly crucial.

In addition, Haas et al. fully report on the influence of the formulation variation on the degree of resistance against ignition sources of low intensity. These studies were carried out with conventional HR PUR foam grades with a weight per unit volume of 24 to 45 kg/m³. The formulation base on Desmophen ® 7135 (Conventional grade) and Desmophen® 7652 (HR and PMD Polyol) caused by variation of the water addition (2–4 parts) as well as the variation of the fire-retardant additives (TCEP and TCPP – 2 to 4 parts) a great variation of resistance against ignition sources. Table 63 provides a view of the individual ignition sources used.

Table 63 Divergencies of the intensity of small ignition sources

Ignition source	Flame height (mm)	Flaming time (s)
Cigarette	–	–
BS 5852 source1	40	20
BS 5852 source2	140	40
BS 5852 source3	240	70
CSE – RF – 4/83	40	20
CSE – RF – 4/83	40	80
CSE – RF – 4/83	40	140
FAR § 25.853 b	38	12

The addition of 7.5 to 10.0 parts of TCPP to the conventional PUR foam grade did not suppress the smoldering ability of the foam manufactured with 4 parts of water. These foam grades gained positive results according to Table 64 in the cigarette tests only. TCPP did not obviously improve the HR grades manufactured with 2 parts of water. With the extra addition of 4 parts water failure occurred only under the impingement of source 2 of BS5852. Table 64 shows the influence of the combined effect of water and fire-retardant additives. Failure occurs with the conventional PUR, manufactured with 2 parts of water and with fire-retardant additives up to 7.5 parts of TCPP. Only with the addition of 10 parts TCPP was the resistance against the ignition sources significantly increased and this grade ignited when exposed to flame 2 of BS 5852.

Table 64 Fire reaction testing in small laboratory test methods

Recipe	CAL-117A	CAL-117D	CAL 121	UFAC	O_2 Index
MRC with Al_2O_3	f.	n.v.a.	passing	f.	n.v.a.
ATH / HR	f.	passing	n.v.a.	passing	n.v.a.
CMHR	n.v.a.	n.v.a.	n.v.a.	n.v.a.	23–28,5

n.v.a. – no value available   f. – failing

The admissibility of combustible materials in Europe is mainly coupled with their resistance against (simulated) match-type flames. A limited surface spread of flame, as documented by Fig. 93 for three different rigid PUR foam grades, is not a guarantee for the proof of material class B2. Resistance against the simulated match flame of the small burner (see Fig. 49a) is obtained by the addition of special fire-retardant additives (e.g. Baymer DS1 ®) or an increased trimerisation, which means special PIR products. The fire performance of rigid PIR foam systems is due to their marked carbonisation tendency much more favorably classified than comparable PUR grades. The failure of rigid PUR foam systems to DIN 4102, see Figs. 49a and b, is not suppressed by the addition of fire-retardant additives, see. Fig. 93, when assessed by the visible flame tips in the stream of effluents. The limited surface spread of flame, which is documented by Fig. 94, does not change the standardized classification system.

Resistance against ignition sources of increased intensity was also part of extensive investigations with rigid PUR foam systems. With integral skin foams and RIM (reaction injection molding) the resistance against the flaming attack, similar to UL subject 94, had to be obtained for use in electromechanical products. The positive classification under laboratory test conditions was confirmed by realistic large-scale tests [342].

Szabat et al. discuss in [343] investigations with ignition sources of different intensity levels where the risk of ignition was evaluated for a direct flame impingement and a combined thermal stress of flaming and radiation. The tested materials were flexible PUR foams based on PHD with special fire-retardant additives. Basic classification were determined in the oxygen index test according to Fig. 77b, the resistance against direct flame impingement with respect to MVSS302, the California Bulletin

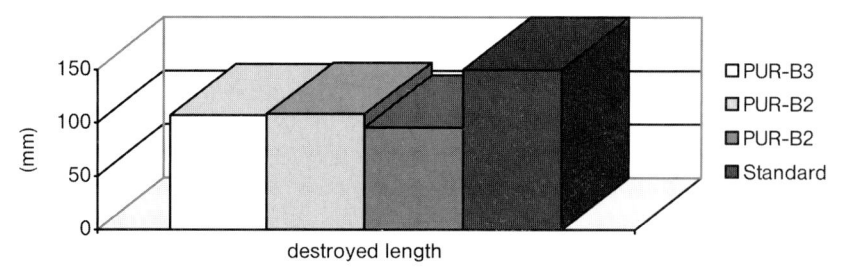

Figure 93 B2 classification of PUR rigid foam grade.

117 test, the FAR test requirement according to §25. 853 as well as the UFAC requirements and especially the contribution to the surface spread of flame according to ASTM-E 162/D 3675. The listing of Table 65 provides a view of the O_2-index values of various flexible PUR grades in comparison to a reference material. These material-related characteristics show the strong influence of the formulation variations. The authors refer both to the obvious weakness of the fire-performance classification and to the high probability of possible faulty interpretations. Table 66 shows the possible influence of the formulation variations, e.g. water and fire-retardant additives on the combustion behavior that is the possible influence on the Q and Fs characteristics of the radiant flammability test.

The graphical representation of Fig. 95a documents, according to the opinion of the authors and based on the test results of a special rigid PUR foam grade a sufficient agreement of the two standardized test procedures (ASTM-E 162 and ASTM-D 3675).

The risk of ignition of the upholstery filling polyurethane foam is demonstrated in [343] as well as in [14]. Madaj et al. show in [344] that the influence of the formulation to the test configuration is independent detectable in the experimental, in the lab tests and the full-scale trials. Madaj et al. [344] report the fire-performance classification of CMHR foam qualities based on MDI and TDI. The fire-reaction classification was based on O_2-index test results as well as that on material-related characteristics

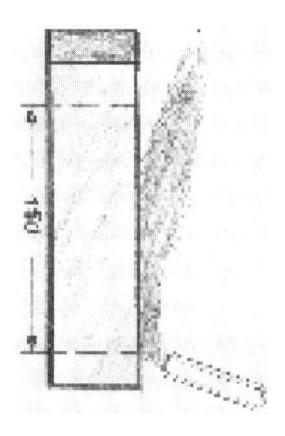

Figure 94 Scheme – Failure criteria in the B2 test procedure

5.3 Polyurethanes

Table 65a O$_2$ Index numbers

Material	O$_2$-Index
Neoprene LS200	51,9
CMHR I	53,5
CMHR II	33,1
highly filled PUR	30,7
no dripping Ester	28,8
Ester FR	30,8
Ether FR	24,1
Ester FR highly filled	36,5

Table 65b O$_2$ Index Numbers

Additive per weight (%)	Melamin O$_2$ Index	AL(OH)$_3$ O$_2$ Index
0	18	18
10	21	19
20	24	20,4
30	26,5	22
40	27,9	24
50	28,1	26,2

Table 66 Influence of the recipe to the spread

Material	Q	Fs
CMHR Variation		
CMHR I	15,0	93,4
CMHR II	10,0	60,0
CMHR III	6,0	6,0
Ester HR (35 kg/m^3)	6,0	374
Ester HR (40 kg/m^3)	7,0	428
Ether FR (28 kg/m^3)	9,3	568
Ether FR (30 kg/m^3)	6,8	413
Ester (30 kg/m^3)	12,1	736
Ester nondripping	11,9	724

Table 67 Reproducibility of Spread Data.

CMHR III	Is	Fs
E 162	10–31	2,3–5,46
D 3675	4 – 6	1

5 Material-specific Fire-Performance Characteristics of PUR

Q and Fs determined in the Radiant Flammability Test. The so-called flame spread index Fs was determined by means of the mathematical relationship:

$Is = Fs \times Q$

Under the flame spread factor Fs the summation

$Fs = 1 + 1/t3 + 1/t6 - t3 + 1/t9 - t6 + 1/t12 - t9 + 1/t15 - t12$

is understood. The graphical representation of Fig. 95b illustrates the views of Madaj et al. who carried out a statistical evaluation of the various test results. The polyol component causes a variation of 18% (Fs) and 49.1% (Q), while the addition of fire-retardant additives initiated deviations of 69.1% (Fs) and 39.1% (Q). Comparable deviations were measured in the case of the isocyanate component.

Table 67 provides a detailed view of the results with respect to flame action achieved with highly filled systems. Fire-retardant rigid PUR foams, e.g. spray system Baymer DS1® achieves the classification B2, normal flammable. Rigid polyisocyanurate foam grades are, due to their tendency for carbonisation, more favorably classified. Systems with special fire retardancy can achieve, see Table 68, the most favorable national classes for combustible materials. Among other things, specific PIR grades like Baymer® achieve class B1, difficult to ignite according to DIN 4102 as demonstrated by the certification PA III 2. 363, the classification M2 and M1

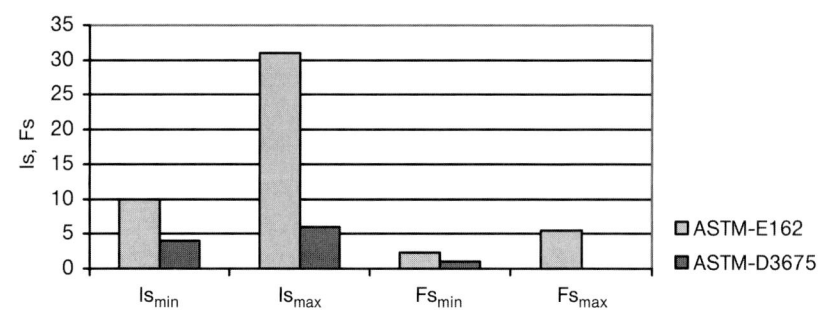

Figure 95a Reproducibility of spread values

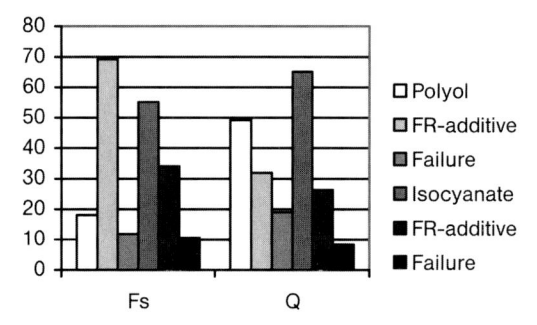

Figure 95b Influence of the recipe to the surface spread of flame

gained in the Epiradiateur test as well as classes II and I according to NEN 3883 and the boundary rating 25 in the tunnel test ASTM-E 84 in the United States. This information of the set of data refers to the former norm. The clip values refer, e.g. to special products. The influence of fire-retardant additives on the resistance of rigid PUR and PIR grades against ignition sources are reported in [345, 346]. Table 68 shows that by careful formulation a pure expanding system resistance against ignition sources of higher intensities is possible and favorable classifications can be achieved. The investigation of different rigid polyurethane grades showed the strong influence of the test configuration, see Fig. 96a and the graphical representations of Fig. 96b. These experimental results illustrate that the simulation of a duct leads to considerable flame spread.

Dowling and Feske [347] report on the fire-protection classification of eleven different MDI-based polyurethane foams. These tests were carried out in the Steiner Tunnel in accordance with ASTM – E 84 as well as in the cone-calorimeter procedure to IS 5660. The influence of variously chlorine-, bromine- and/or phosphorus-containing flame retardants on the contribution to the surface spread of flame as

Table 68 Fire reaction testing of building products

Country	PUR	PIR	PCD
Germany			
DIN 4102	B3(B2)	B2(B1)	B2(B1)
Belgium			
Epiradiateur	M4	M2(M1)	M1(M2)
Vlamoverslag	III	II (I)	I (II)
France			
Epiradiateur	M4	M1	M2(M1)
PPF Test	4	1	2
Great Britain			
Ignitability Test	X	P	P
Surface Spread of Flame	IV	II (I)	II (I)
Fire Propagation (I/i)	44/34	12-20/7-10	15/8
Netherlands			
Vlam uitbreiding	IV (III)	II (I)	I
Vlam overslag	III	II (I)	I
Swiss			
Small burner	II	V	V
Scandinavia			
Brown'sche Box	worse than wood	better than wood	better than wood
Paneltest (DK)			
Austria			
Schlytertest	flashover	no flashover	no flashover
USA , Canada			
ASTM – E84	rating > 500	rating 25 (< 25)	–

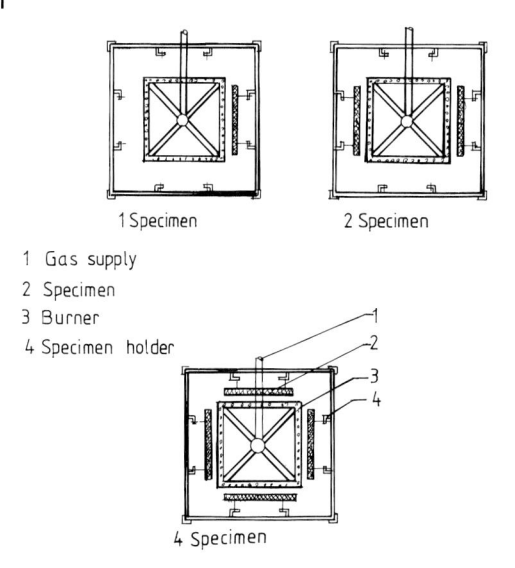

5 Material-specific Fire-Performance Characteristics of PUR

Figure 96a Different test configuration in the chimney test

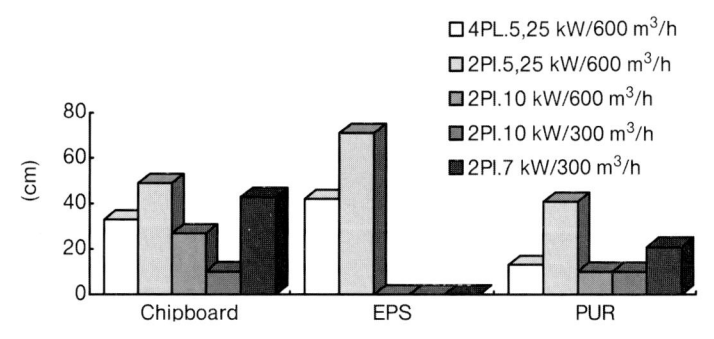

1 Gas supply
2 Specimen
3 Burner
4 Specimen holder

Figure 96b Influence of the test configuration on the test results

□ 4PL.5,25 kW/600 m^3/h
□ 2PI.5,25 kW/600 m^3/h
■ 2PI.10 kW/600 m^3/h
■ 2PI.10 kW/300 m^3/h
■ 2PI.7 kW/300 m^3/h

well as to the heat and fire-gas release rate. The results, see Table 69, show valid correlation of the determined test records. The authors think that the cone-calorimeter test results offer the ability to predict tunnel-test classifications.

Hanusa [348] reports on an investigation program initiated by the Urethane Safety Group (USG) and supported by the SPI (Society of Plastic Industry) with PIR and PUR systems in the room-corner test according to IS 5906. In this case, the contribution of wall and ceiling linings to the surface spread of flame was examined. The ignition source, an 18-kg wood crib, was arranged in the corner of the room. The influence of different linings and coating materials, e.g. metal sheet, Al-film, decorative chipboard as well as cementous coatings on the combustion/burning behavior was studied. The test results show a comparatively good agreement with cracking, flame spread and fire-gas temperature rise.

Table 69 Influence of the recipe to the fire performance

Test method E 84	THR 25 (kW/m²)	40 (kW/m²)	70 (kW/m²)	100 (kW/m²)
FSI				
20,3–29	1,24 – 3,41	4,74–8,39	9,24–12,95	10,93–14,54
20,2 – 29	6,7 – 18,85	20,6–55,3	51,8–82,2	63,5–97
Smoke Index	SEA (m²/kg)			
90–54	35,52–85,15	76,7–152,6	137,8–343,2	166,2–361
90–540	334,3–725,8	8,3–14,2	503,2–791,5	503,6–22,1

5.3.2
Material-specific Burning Behavior of Polyurethanes

Testing philosophy of the burning dripping

Flexible and semirigid PUR foams, due to their chemical compositions will possess the danger of a hot and burning molten material. The examination and evaluation of this property gives unfavorable classifications. Special systems, e.g. cold-cure systems, achieve, due to their melting behavior in particular the classification "does not drop burning" in numerous small laboratory test methods. For this quality, the melting occurs faster than the ignition of the molten mass of the tested material. Ignition of the liquefied surface has to be expected for any combustible melting material. The flowing and/or dripping of burning/melting are to be expected with relevant environmental conditions. The environmental conditions are decisive in the real-fire case with regard to ignition and to burning dripping, see Fig. 97a and/or Fig. 97b. Figure 97b shows that, e.g. B1-classified expanded polystyrene foam, qualified as not burning dripping due to the standard test procedure, though coatings can cause burning

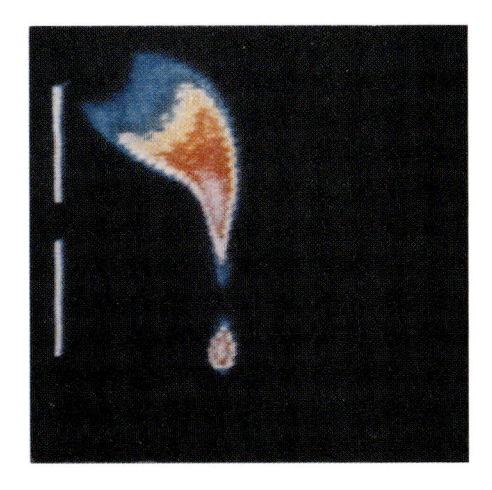

Figure 97a Burning dripping

Figure 97 b Burning rain

rain. The Düsseldorf airport fire verified that this standardized classification according to DIN 4102 "does not drop burning off" will be lost also possibly through the lamination with Al foil [1].

Rigid PUR and PIR foams in general duroplastic materials achieve the nonburning dripping classification. Realistic large-scale tests/full-scale field trials, see Figs. 98a and b, result in carbonisation of the thermally stressed surface and the release of decomposition products in a similar manner as wood.

Heat-release rate and material-related records
The characteristic value of heat release is taken, as a rule, as the key value for burning behavior. This material-related data is determined as part of national fire-protection classification systems in various specified laboratory test procedures like the OSU chamber, Tewarson apparatus or cone calorimeter. The possible influence of material and environmental conditions is to be considered for the evaluation of the results. The data records of Figs. 99a, b and c illustrate vividly the differences between volume- and mass-related characteristics.

The influence of the material- and environmental-related factors are to be considered as well as the possible contribution to the heat release. According to Tewarson [73, 349], the composition of the fire gases comes both from the chemical nature of the combustion and also from the decomposition conditions. These are responsible for the effectiveness of the fire process and with that also for the amount of heat released per unit time. The heat of combustion is one of the essential thermody-

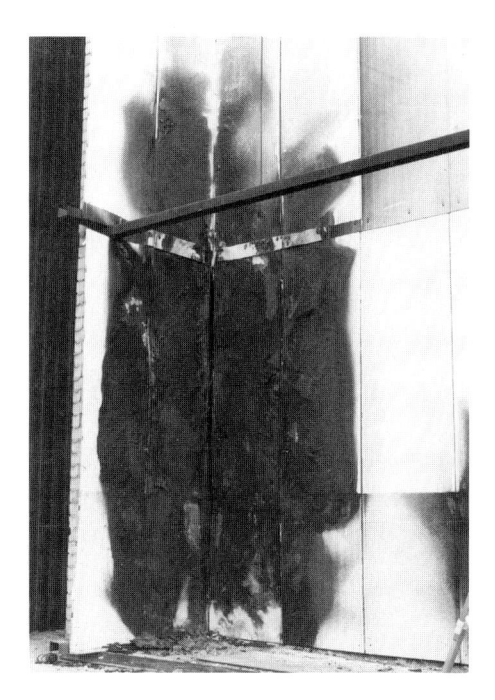

Figure 98a Carbonisation of rigid PUR foam in a realistic large-scale test

Figure 98b Carbonisation of rigid PIR foam in a realistic large-scale test

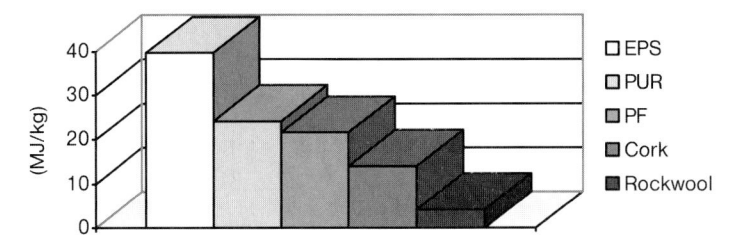

Figure 99a Volume based heat-release potential of various insulants

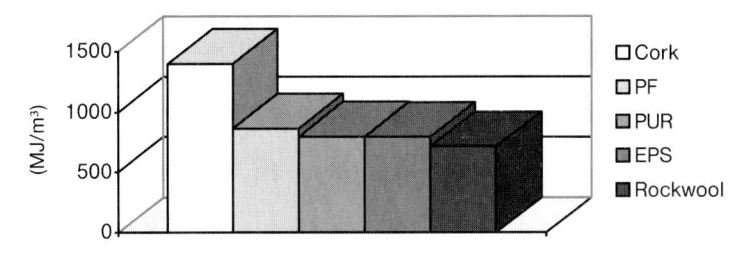

Figure 99b Mass-based heat-release potential of various insulants

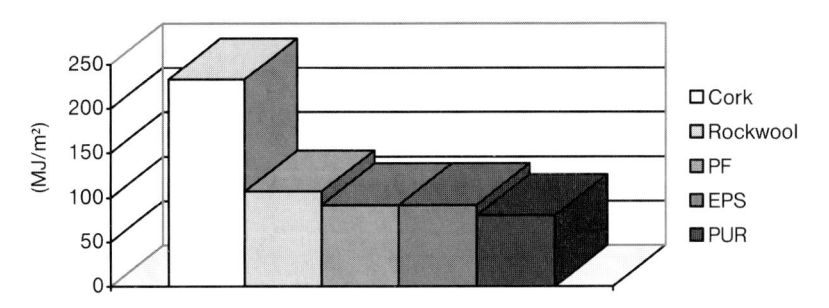

Figure 99c Area-based heat-release potential of various insulants

namic and/or calorific values of a combustible material. During a combustion process, it comes next to the formation of decomposition products also to the release of heat. This heat of combustion, the so-called net calorific value of a material, is an essential material-specific characteristic of a combustible material. The heat of combustion is the amount of heat that is released by a unit mass or volume of a combustible material in the case of a complete combustion. These distinguish the upper and the lower net calorific value. The upper net calorific value contains the heat of evaporation of the water of the combustible. In all calculations, one starts from the lower net calorific, that is the upper net calorific value is reduced by the heat of evaporation since the stream of water vapor arising during the combustion process disperses out into the atmosphere. The net calorific value is determined in the colorimeter bomb or is calculated by means of the elementary composition of the material in accordance with the calorific value formula

$H_u = 8100\,C + 29000\,(H - O/8) + 2500\,S - 600\,W$
at what H_u = lower net calorific value
C Carbon
H Hydrogen
O Oxygen
S Sulfur
W Water

The net calorific value is rated as the classical criterion for the classification of the possible heat release. With regard to the changing fire load it appears justified to use the net calorific values per kg as the basis for all considerations for the preventive fire protection. This definition has to be questioned as not risk relevant not only for the application to building materials. If one takes a closed room as a guideline, everything indicates for the need of volume- and/or area-related ratings. The data sets of Fig. 99a illustrate that without doubt the observation of the risk-related dimension would lead to a better safety aspect. The official report on the Düsseldorf airport fire required disclosure of the material characteristics from the viewpoint of their enduse application. Figures 99b and c show different insulants of supposedly identical performance actually have different performances when assessed on the basis of volume- or area-related properties. While with the mass-related records for mineral insulating boards, a much smaller heat potential value is achieved, results with volume- or area-related data, presupposed equivalent insulating records, give an approximately equivalent value. The listing of Fig. 100 provides a view of these material-related data.

These numerical values illustrate how important it is, especially for expanded materials, to use the realistic dimension. The volume-related characteristics demonstrate that the usually favored mass-related records supply data of comparison that do not correspond to the final application. The technical papers should supply both records at least for information [350, 351]. With increasing humidity a significant reduction of the burning rate occurs and with that the heat-release rate. Figure 101a demonstrates the degree/scale of the possible influence and Fig. 101b illustrates the

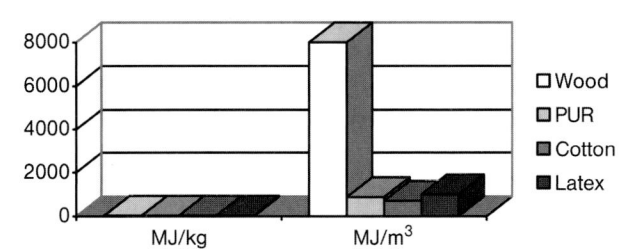

Figure 100 Net calorific values

influence of the humidity of the fire resistance of chipboard. Figure 101c clarifies the thermal decomposition caused by external thermal stress. Different physical and chemical reactions are caused depending on the thermal stress.

The graphical representations of Figs. 102a, b and c as well as the listings of Table 72 illustrate how much the heat-release rate is influenced by the external radiation.

While the material-related heat-release rate of wood and rigid PUR foam increase in a direction to the maximum by increasing the test temperature, e.g. by external radiation, the boundary value, i.e. the net calorific value, is not reached [352].

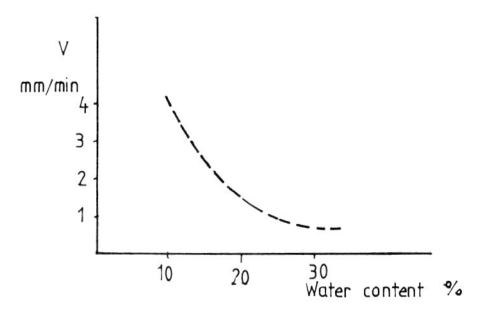

Figure 101a Influence of the humidity on the burning rate of wood

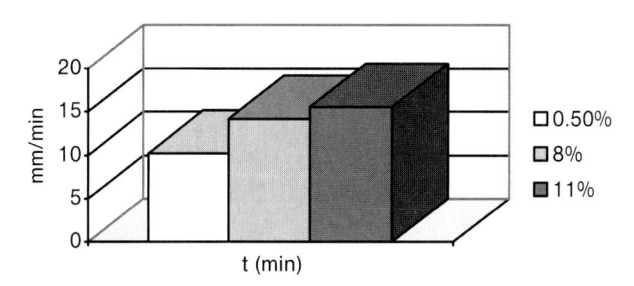

Figure 101b Influence of the humidity on the burning rate of chipboard

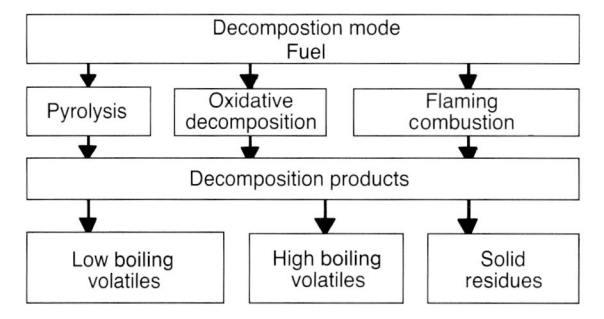

Figure 101c Thermal decomposition

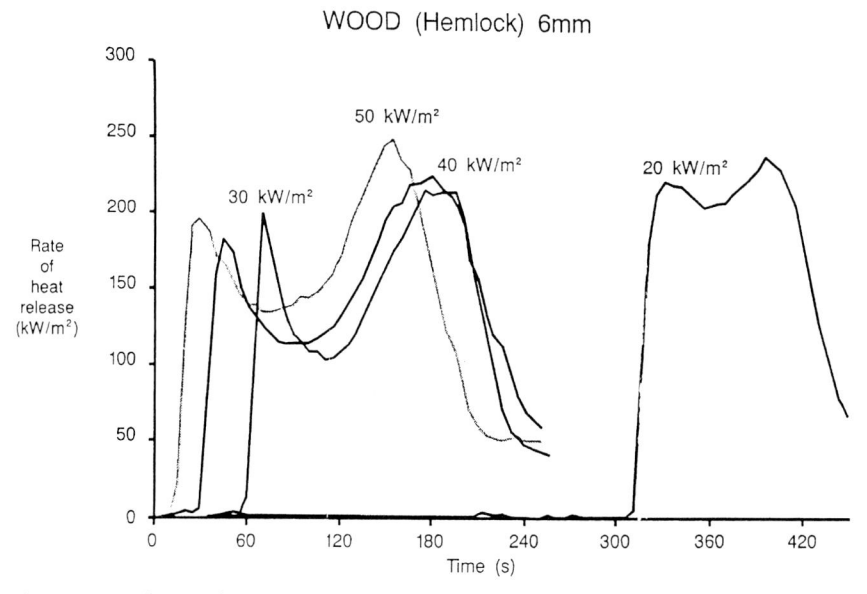

Figure 102a Influence of the temperature on the heat-release rate of wood

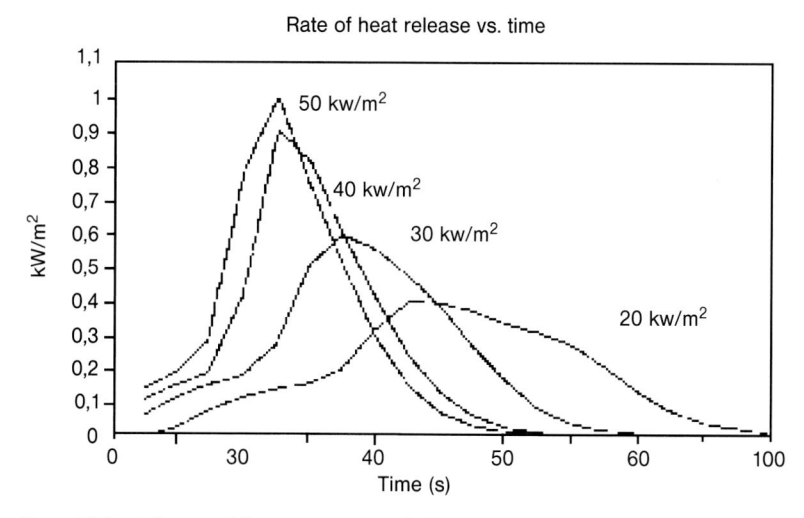

Figure 102b Influence of the temperature on the heat-release rate of flexible PUR foam

Table 70 Influence of the density to the heat release rate of Polyurethanes

CMHR (kg/m³)	RHR total (KJ/cm²)	RHR 3 min – value (KJ/cm²)
35,36	3,27	2500
25,12	2,260	2000
22,4	2,04	1800

The data records of Table 71 demonstrate the influence of the density on the heat potential of rigid PUR foam. The volume-based heat potential values of the reference product wood reach a significantly more unfavorable depreciation niveau.

Basic studies on mass-, volume- as well as area-related temporal process of the heat-release rate have been carried out also from Tewarson [353] in accordance with Table 71. Tewarson varied the heat flux level. The horizontally arranged test piece has a diameter of 100 mm and a thickness of 20 mm. The combustion with flames was initiated through an ignition source acting simultaneously. The data records of Figs. 103 and 104 give a view of the material-related heat potentials measured with various materials. The heat-release rate determined with the aid of the O_2 consumption and the release records of CO and CO_2. Tewarson illustrated by means of the characteristics that the heat release is not an elementary material related size, but it depends strongly on the fire behavior. He differentiates as shown by the listing of Table 71 and the graphical representation of Figs. 103, between the potential and the current value, which fit together according to Fig. 104 from the convection and the radiation components (see also Table 72).

Table 71 Influence of the recipe to the heat release rate of polyurethanes

Material	RHR(KJ) 3 min	CO_2/CO 3 min	RHR(KJ) 5 min	CO_2/CO 5 min	RHR(KJ) 10 min	CO_2/CO 10 min	tmax (s)
PUR-NFR 20kg/m³ (7.6g)	12351	165	13440	143	15282	142	3660
PUR-FR 22kg/m³ (8.1g)	10369	33	11648	26,5	12486	24,9	4760
PUR-HR 37kg/m³ (13.7g)	19441	193	20908	136	22542	130	5280
PUR rigid 30kg/m³ (11.1g)	6495	19	6955	23	6452	20	1540
PIR 29kg/m³ (11.2g)	1035	29	1483	22	2518	23	442

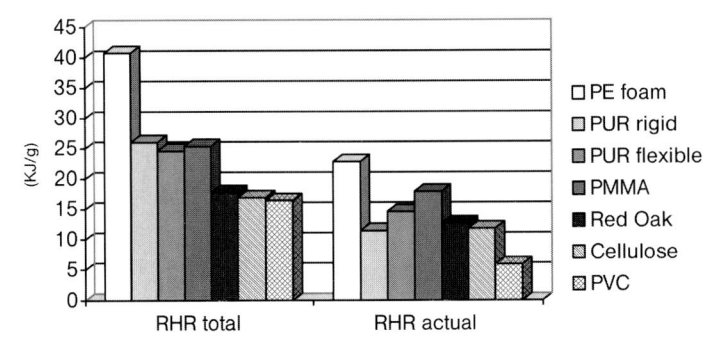

Figure 103 Influence of the burning behavior on the scale of heat-release rate

5.3 Polyurethanes

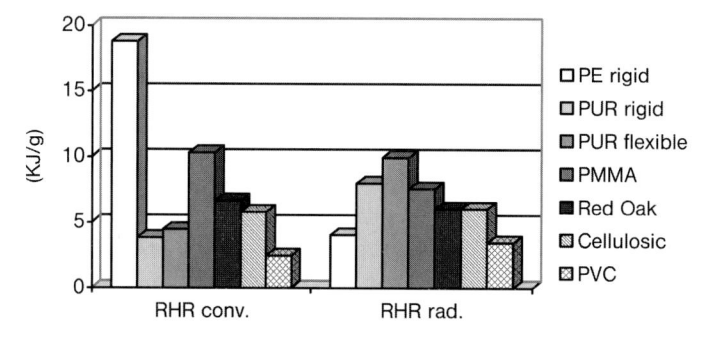

Figure 104 The current heat-release rate in the Tewarson apparatus

The graphical representations underline this statement from Tewarson that information about the heat release is supposed not to be always and exclusively oriented on the net calorific value. These numerical values clarify, furthermore, that the radiation component predominates as a rule. The graphical representations of Fig. 104 clarify that during the current release only a part of the potential values will become effective. Herrington et al. report in [354] on studies with the OSU chamber. The investigations of Herrington with the OSU chamber on the temporal course of the heat release clarify, e.g. also the strong influence of the recipe variation with the example of polyurethane foams in accordance with Table 73. The test pieces of dimensions 100mm × 150 mm × 25 mm were thermally stressed in a horizontally position in the stream of air for the duration of 10 min at an intensity level of 10 kW/m². Herrington et al. verify as shown by the records of Fig. 104 and Table 73 that the

Table 72 Data of the heat potential of various plastics

Material	RHR total (KJ/g)	RHR actual (KJ/g)	RHR convective (KJ/g)	RHR radiative (KJ/g)
PVC	16,44	5,89	2,43	3,46
Cellulose	16,88	11,80	5,78	6,02
Red oak	17,78	12,69	6,65	6,04
PMMA	25,40	17,89	10,33	7,56
PUR flexible	24,64	14,61	4,46	9,95
PUR rigid	26,02	11,48	3,85	7,99
PE rigid	40,84	22,83	18,79	4,04

Table 73 Influence of the recipe to the heat release rate in the OSU test

Foam type	HRA 3 min.	RHR – Max.	Time to Max RHR (s)
CAL 117	1872	1372	99
CAL 117 HR	2520	1660	115
Mel.mod.HR	200	124	57
CMHR	517	276	46

5 Material-specific Fire-Performance Characteristics of PUR

heat-release rate of flexible PUR foam will be changed significantly by recipe variation. For the judgement of the measured values divergencies of the density must be duly noted. In the present situation the cold-cure quality was of a considerably higher density.

The records of Fig. 105 also show that already after a test duration of 3 min the predominant part of the heat potential is released. These records also provide the impression that there is a significant strong influence of the recipe variation. These deviations due to the formula variation are presumably mainly attributed to the different densities (see Fig. 106). As soon as these results are based on homogeneous densities to be standardized, e.g. to 30 kg/m^3, a direct proportionality is assumed for the polyurethane foam qualities insert in approximately comparable values as Table 74 documents very vividly. The extremely clean combustion behavior of the cold-cure quality is to be put down presumably to their marked melting behavior that guarantees a perfect pool-fire situation.

Parrish et al. [355] describe the effectiveness of fire retardants as, e.g. the addition of melamine onto the heat release of cold-cure qualities. In the OSU chamber test according to ASTM-E 906 test pieces in outer dimensions of 10" × 4" × 2" of flexible polyurethane foam manufactured with 20 parts of melamine are exposed to the thermal stress of 17.5 kW/m^2. The listing of Table 74 clarifies the influence of the raw density.

The different raw density values were manufactured by variation of the water parts in the formula (2.4 – 3.5 – 4.0 parts).

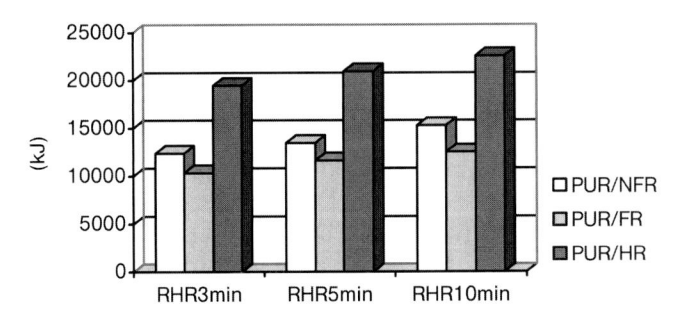

Figure 105 Influence of the test duration and the density to the heat release in the OSU chamber

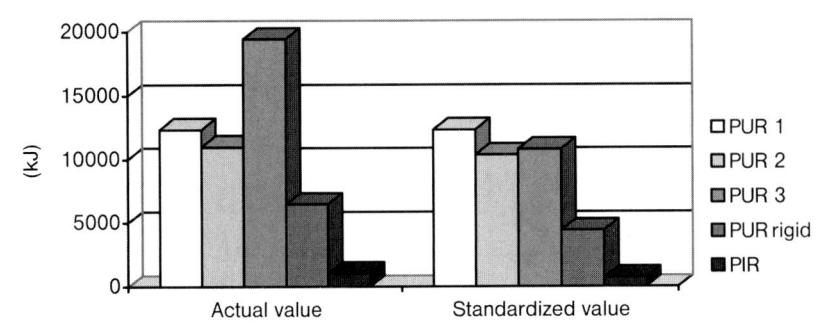

Figure 106 Influence of the test duration and the density to the heat release in the OSU chamber

Table 74 Influence of the fire retardant additive to the heat – release rate

Flexible foam grade	Melamine 1)	CM-additive 1)	Resistance against	3 – min 3) Peak RHR (kW/m²)	medium 4) RHR (kW/m²)
PUR-NFR	0	0	Failing	297	105
HR	0	3	source 2	299	112
CMHR	15	5	source 5	283	95
CAL	40	5	CAL 133	181	42

Details concerning the fire – retardant additive
1) Proven resistance

2) Heat release rate – peak after 3 min.
Medium heat – release rate

These numerical values clarify that the two foam lots of lower raw density already within the 3-min test duration burned completely. The 3-min value will be changed through the addition of melamine. Without melamine addition the heat-release rate increases, as illustrated by Table 74 significantly with increasing raw density. The laboratory test, the OSU chamber test, demonstrated that the recipe variation of the highly filled HR system (6 pcf) lead in comparison to HR and FR qualities under the irradiance level of 25 and 30 kW/m² (17 and 19.5 kW/m²) to a significantly smaller heat-release rate.

In [356] the influence of the formula on the heat-release rate of four different PUR grades is described from cone-calorimeter studies at the University of Gent. In accordance with IS 5660 horizontally arranged test samples with outer dimensions of 100 mm × 100 mm × 50 mm, will be thermally stressed with an intensity level of 25 kW/m². The listing of Table 75 clarifies the influence of fire-retardant additives to the heat-release rate.

Table 75 Evaluation criteria of the smoke potential.

Smoke density

Criteria	Definition	Dimension
T	Transmission	% . % min
D	Optical density	1/m
Ds	specif. optical density	
k	Coefficient of extinction	m²/s
SAE,Ao	Area of extinction	m²/kg
V	Smoke volume	m³

Smoke toxicity

Criteria	Definition	Units
td	The exposure time that leads to death	min
ti	Time to incapacitation	min
CO-Hb	Carboxyhemoglobin content	%
Mortality	Death rate	%
LT50	The exposure time that lead to 50% mortality	min
RD50	The smoke concentration that leads to a 50% reduction of breathing	g/m³
EC50	The smoke concentration that leads to a 50% reduction of the reaction	g/m³
LC50	The smoke concentration that leads to a 50% lethality	g/m³
L(ct)50	The product from concentration and exposure time that leads to a 50% lethality	g/m³

Table 76 Influence of the flame retardant additives on the heat release rate (17,5 kW/m²)

Foam Type	Density (pcf)	HRA 3 min.	Max HRH	Time to max HRH (s)
CAL 117	1,34	1304	1169	64
CAL 117	1,84	1814	1233	94
CAL 117	2,88	2497	1714	139
HR-FR	2,36	2158	1516	99
HR-FR	2,76	2587	1681	120
HR-FR	3,13	2816	1782	125
Melam-mod.HR	3,01	199	112	52
Melam.mod.HR	2,96	201	136	63
CMHR	3,0	548	272	47
CMHR	4,7	486	277	45
CMHR	5,3	517	280	45

The fire performance of flexible PUR foam systems analog to the Furniture Calorimeter and/or NT 032 in the test mock up (seat cushion 650 mm × 550 mm × 100 mm, back cushion 480 mm × 550 mm × 100 mm) by using crib 5 as the ignition source according to BS 5852-1990, resulted in comparison to the cone calorimeter as illustrated by the records (PHRH:NFR-825 and HR – 725 kW) in different values as well as a divergent evaluation. Stone et al. [357] try by means of comprehensive fire-performance studies with flexible PUR foams to illustrate the possible correlation of laboratories experiments, e.g. LOI test and torch test as well as OSU chamber with a realistic large-scale test like the Michigan Roll Up test. Stone [358] reported by means of numerous investigations on the influence of the fire retardance to the burning behavior of polyether and polyester qualities. The influence of an external radiator was studied with an air gap of 0.5" and 1.0" between radiator and test piece. The gas-fired radiator produced a surface temperature of 370 to 440 °C. The thermal stress of the foam surface conducted was 9.2 to 14.7 kW/m². The external dimensions of the test piece were 4" × 6" × 1.5". The ignition of the different foam surfaces was observed within 19 s to 142 s. No ignition occurred with the highly filled system.

Smoke formation

The fire-gas formation was influenced similarly to all fire-performance characteristics by material- and environmental-related factors (see the main characterizing assessment criteria of Table 76). In dependence of the enduse conditions the material will contribute to the fire-gas formation corresponding to the fire development. In relation to the net calorific value of the material specific potential is partly released in accordance with the thermal stress/loading. A partial set arises in relation to that at the complete combustion necessary air requirement in dependence of the test temperature as well as the ventilation conditions. The listing of Table 77 provides a view of the material-related data of some furniture materials.

The relevance of the fire scenario to be examined and evaluated is to be considered at the selection of the decomposition model sufficiently. These investigations show

Table 77 Material-related smoke potency data.

Material	Density (kg/m³)	Calorific value (kJ/kg)	Calorific value (kJ/m³)	Smoke (Nm³/m³)
Wood	450	17.5	7650	2385
Particle board	800	16.5	13200	4160
Mineral fiber board	250 x)			
EPS	015–0,30	38,9	585–1165	148–297
PUR-flexible	030	27	800	
Latex	060	44	2200	

x) Binder

very vividly that the experimental parameters such as, e.g. temperature, heating rate and air supply both influence the smoke release seriously as do also the composition of the stream of effluents and the CO_2/CO relationship. The importance of material-related variables as, e.g. external form of the fire load (granular materials, slabs and boards or cellular constructions) as well as order and orientation, is discussed. Hillenbrand et al. [359] and Bußman et al. [111] have demonstrated the importance of the CO_2/CO ratio for the characterisation of effluents. Within the framework of ISO TC 92 SC3 Toxic Hazard in Fire key data for the characterisation of effluents of the main fire scenarios were agreed and demonstrated in the TR 9122. Figure 48e offers a view of the key data in packed form for the three main fire scenarios. In the smoldering fire field, as well as in the developing stage of a fire one distinguishes in each case between three different stages. During the fully developed fire differentiation is done with regard to the present ventilation only. The experiences of the preventive fire protection as well as the results of the practice-relevant full-scale fire trials showed that the dominating position of the CO_2/CO ratio is the characteristic parameter for the effluents. Characterisation became underlined by Japanese investigations. Basic Japanese investigations illustrated that highest priority is the CO_2/CO ratio both with regard to the smoke density and the acute toxicity of the stream of effluents. Figures 107 and 108 show essential findings of Morikawa et al. [360, 361] with high CO_2/CO ratios.

For very well ventilated fires, the stream of effluents as a rule can be classified as noncritical both with regard to smoke density and acute toxicity. Decreasing CO_2/CO

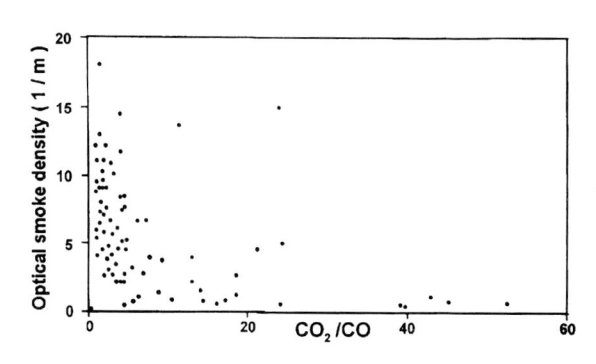

Figure 107 Influence of the CO_2/CO ratio to the characteristics of the smoke density

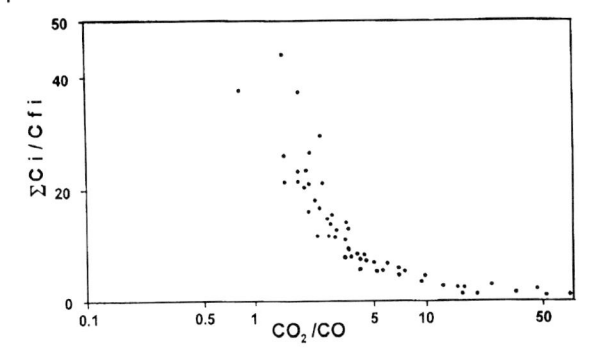

Figure 108 Influence of the CO_2/CO ratio to the characteristics of the smoke toxicity

ratio reduced visibility as well as crucial smoke concentrations. Smoke particles are formed under smoldering as well as flaming conditions. Particles with a large diameter are produced under flaming conditions. Extensive investigations on the influence of different test parameters on the number and size of the fire-gas particles are reported 1988 in [362].

The scale of the thermal loading, the level of the external intensity of radiation determine both the kind of decomposition and the rate of smoke production. The decomposition rate and the extent of the surface area already burning have to be considered, to carry out a risk-related estimation of the expected smoke volume. The graphical representations of Fig. 108 illustrate that a significant reduction of the released smoke volume occurs at the moment of ignition.

The philosophy of smoke-density measurement

A comprehensive overview of the smoke-testing evaluation philosophy is given by Steinert [363]. The objective of any smoke-density test method is a risk-relevant evaluation of the reduced visibility in the case of a fire. The various test methods based on three fundamental test principles.

a) Measurement of the current weight of the smoke particles with the so-called gravimetric method.
b) Determination of discoloration of defined filter cartridges through the stream of effluents. The assessment is carried out by means of a calibrated filter system and
c) Photoelectric measurements of the luminous absorption through the stream of effluents

An evaluation by means of the filtered fire-gas particles as well as the filter discoloration does not correspond any longer to the state of the art. The logic of the above statement is understandable if one considers how strongly steam can reduce visibility. The currently practised procedures use the physical regularities for the evaluation of visibility in clear media. The mathematical connection according to Bougners and Lambert's law clarifies the information of Fig. 109. The extinction coefficient k is the faintness of light through smoke particles. k is independent of the intensity of the light.

$$k = 1/L \ln(I_o/I) = 1/L \ln(100/T)$$

$$I_t = I_0 * 10^{-K*c*L}$$

$$T\% = \frac{I_t}{I_0} * 100 = 100 * 10^{-K*c*L}$$

$$D = \log \frac{I_0}{I_t} = K * c * L$$

I = Initial transmitted flux
I = Transmitted light flux
K = Proportional factor
c = Concentration
L = Length of the optical path
T = Transmission
D = Optical density

Figure 109 Smoke-density determination

The reduction of visibility through smoke today is mainly characterized through the optical density D. Both characteristics correlate as follows:

$$k = \ln 10\, (D/L) = 2.303\, D/L$$

The extinction coefficient k is a common measure for the evaluation of the smoke density. k is defined as the effective cross-sectional area of the smoke particles released by unit mass. From the engineering point of view the optical density D can be accepted as directly proportional to the number of smoke particles. The widely used SEA value, the specific extinction area, is in no way a measure for the smoke volume, but a conversion factor only, if there is a 100% transfer conversion. The SEA value can be used for the characterisation of the fire-gas potential. The following formula clarifies the connection of the SEA value and the optical density D.

$$Sm = SEA = KV/mp = Dv/lmp\ (m^2/kg)\ 2.303$$

The extinction coefficient is multiplied accordingly by the volume flow of the flue gases and divided by the temporal mass loss. For the decomposition model DIN 53436 the simplified mathematical relationship shows:

$$Sm = SEA = 2.303\, D1/LmV\ (m^2/kg)$$
m = corresponds to the mass of smoke particles
V is the volume filled from the smoke
L is the light path and
D is the optical density

Sometimes a rating is carried out by means of the measured transmission numbers. It is overlooked that the percentage translucence does not show a linear depen-

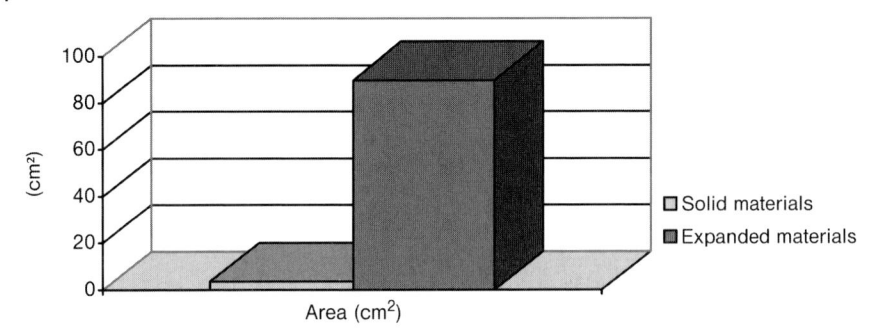

5 Material-specific Fire-Performance Characteristics of PUR

Figure 110 Smoke-density measurement in the XP2 chamber

dence on the fire-gas concentration. The data records of Fig. 110 show the material-related deviations in the XP2-chamber test method, where the surface spread of flame and the melting behavior influence the results possibly significantly [364]. The PIR quality produces, due to the increased carbonisation tendency, a substantial reduction of the smoke-density data.

The correlation of optical density D and visibility (range of vision) by means of measured values clarifies Fig. 111. In the log-log plot both characteristics are inversely proportional and the product of optical density D and visibility is constant. Basic investigations have been carried out in particular by Jin et al. [80, 365].

Rasbash recommends in [83] instead of the optical density D the characteristic ob (obscura) and the relevant derivations for the smoke volume and the mass- and/or area-related smoke-release rates. The smoke potential D/L (ob) correspond to where according to Rasbash L (1/m) is not as well defined as dB/m (ob). With that, D char-

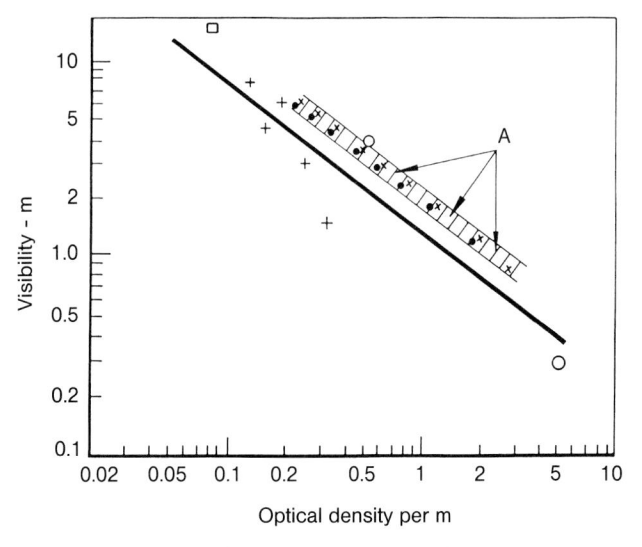

Figure 111 Correlation of optical density D and visibility S

acterizes the smoke quality, D/L the smoke quantity and Do the smoke potential. The relationship 0.1 ob (m^3/m^2) that means $(bel/m)/(m^3/m^2)$ is valid for the specific optical density Ds.

Smoke-density records
For the assessment of the various smoke-density data must be guaranteed that only the records of volume- or mass-related test pieces are taken, which were produced under comparable thermal loading. Provided that different test methods are used, it is to be guaranteed that the effluent concentrations were produced under comparable conditions. The graphical representations of Figs. 112a and b are to be interpreted in such a way that the dependence of the material-related records of the test temperature is variable. The respective peak values are to be determined as functional characteristics. The determined data sets of the dual-chamber box, an accumulative test procedure, underline the differing material behavior. Carbonizing cellulose-based materials produce their peak value under smoldering conditions. A significant reduction of the smoke release occurs after the ignition of the test piece takes place. Materials that burn after ignition without any additional external heat stress produce large amounts of smoke in relation to the increased mass-burning rate. Such products fall below the defaulted conditions peak values at the end of the test duration depending on the external radiation.

Figure 112 c illustrates that comparable transmission values are achieved if the same quantities of different material groups are tested under smoldering conditions. Latex rubber, animal hair and coconut fiber produced in the tube furnace in accordance with DIN 53436/37 comparable values. Significant differences were found only at a decomposition temperature of approx. 200°C. The divergencies are explained by the different decomposition behavior of the three products. No significant difference is found at a reference body temperature of about 300°C [366, 367].

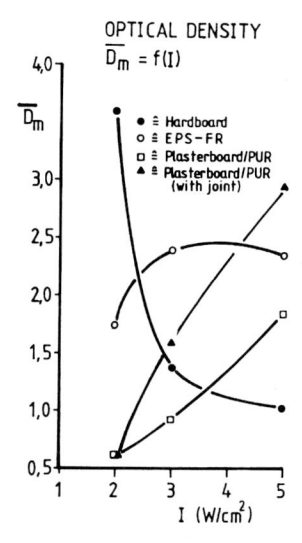

Figure 112a Influence of the irradiance level

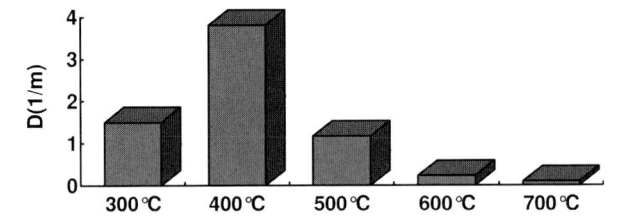

Figure 112b Smoke-density ratings depending on the test temperature

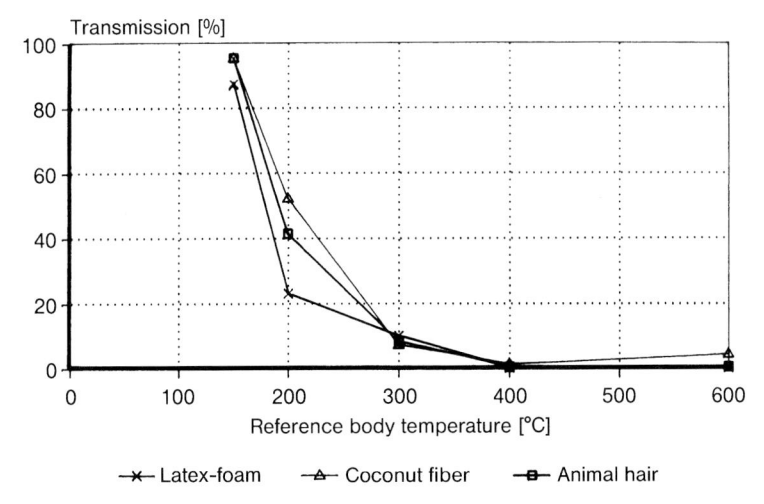

Figure 112c Transmission values depending on the test temperature

The research studies of Grayson et al. [368] showed to what extent the visibility-reducing properties of the thermal decomposition products of polyurethane foams can be changed by formula variations. The graphics of Fig. 113a illustrate for the decomposition products produced at a thermal heat stress of 600 °C to 850 °C that the variation of the formula in comparison to the temperature effect is to be regarded as negligible. A marked effect of the recipe variation is recognizable in the case of low temperature levels. The graphical representations of Fig. 113b give a view of the temporal course of the smoke-density characteristics of two CMHR qualities tested in the NBS Chamber. There seems to be no influence of the recipe.

The selected criteria for the assessment of the smoke release with regard to a possible reduction of visibility due to thermal decomposition products do not consider, as a rule, the most important parameters of the fire-gas concentration. The records of the optical density D determined by means of the DIN equipment depending on the nominal concentration confirm this interpretation as illustrated by Fig. 19. A considerably larger volume of the flexible polyurethane foam must be involved in the fire to produce comparative to latex foam an equivalent smoke release and/or an equivalent visibility. As an additional evaluation criterion, e.g. the mass optical density (MOD) was used, i.e. values based on the tested mass of material. Chien and Seader [369]

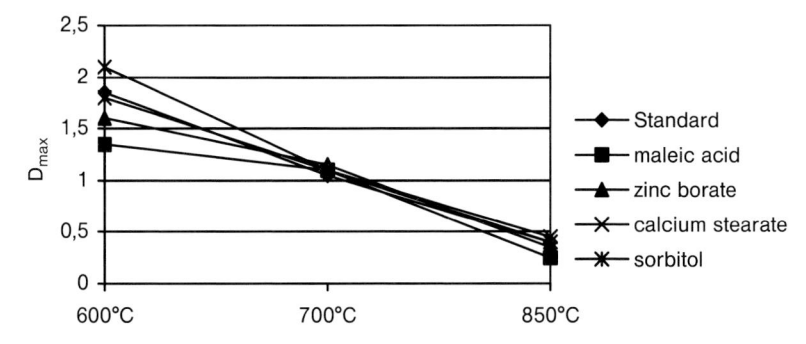

Figure 113a Influence of the formula on the records of the optical density D

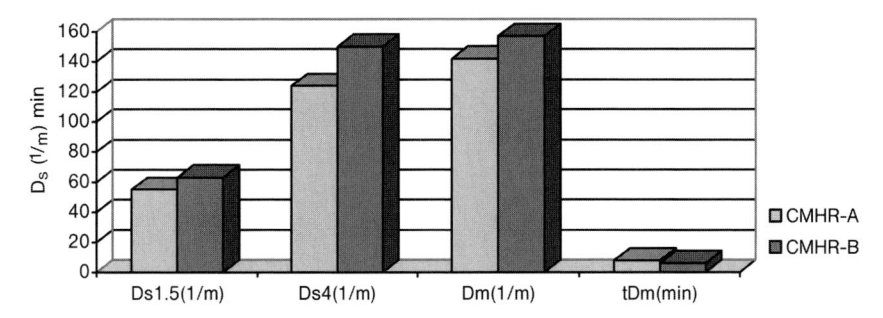

Figure 113b Smoke-density measurements of CMHR qualities

demonstrate in accordance with Fig. 114 by means of a modified test chamber the strong influence of the intensity of radiation in the field of 25 to 75 kW/m² to the mass-related records of the mass optical density D (MOD). Material-related basic data for numeric approaches under enduse conditions are achieved, if the records are based on the tested volumes. Also, such material-related mass-based potential values are, as a rule, suitable only to achieve classifications for a specific, simulated fire scenario. The data records of Fig. 115 are to be interpreted in such a way that the variation of external radiation produces only negligible modifications of the smoke-density data under the forced flaming conditions of the cone-calorimeter device. These deviations were probably caused by the increased decomposition rate due to the increased intensity. How much these smoke-density data produced under flaming conditions with the cone calorimeter deviates from the possible fire-gas danger under actual fire conditions, illustrate the smoke-density data determined in the ISO Smoke Box (see Fig. 112a).

The characteristics of Fig. 116 show very vividly that the material-related smoke-density potential can be varied by variation of the test temperature. The data clarify furthermore that the data level depends crucially on the concentration. The data set also illustrate that the material-related volume-based records of fire endanger than those of EPS and PUR rigid foams, which provide a two-fold mass-related level only. The data set of Fig. 116 show that the smoke-density data changes significantly

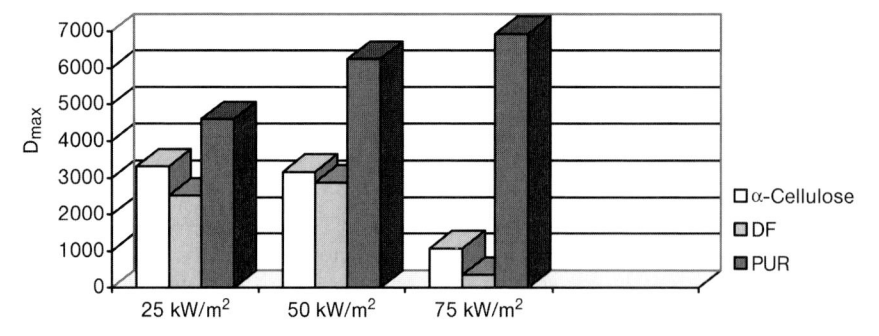

Figure 114 Influence of the radiation intensity to the optical density D

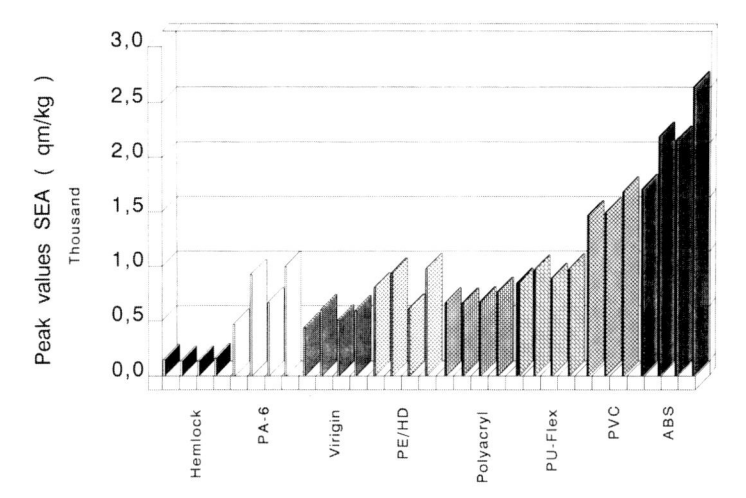

Figure 115 Influence of the radiation intensity on the optical density D in the cone calorimeter

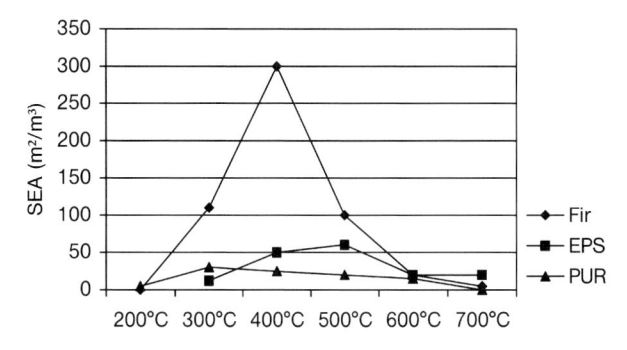

Figure 116 Influence of the testing temperature on the records of the optical density D

depending on the test temperature. The records of Fig. 117 illustrate a similar behavior for the fire-gas toxicity. Both characteristics pass a maximum depending on the test temperature.

The listing of Table 78 demonstrates that differing densities influence the data set levels far more strongly than, e.g. the variation of the recipe, based on different polyol qualities. The use of melamine amplifies the carbonisation tendency and therefore leads in particular to high intensities of radiation in comparison to ester qualities to more favorable values.

Table 78 Influence of the recipe to the smoke potency data.

Foam quality	Ds 1,5 (1/m)	Ds 4,0 (1/m)	D m (1/m)	t −D m (min)
CMHR −A	55	124	142	7,8
CMHR −B	63	150	157	6,0

By means of the Blow Torch Test (see Fig. 118) the carbonisation tendency of the different foam qualities/grades is examined. The exposure time varies from 15 s to 2 min. While in the case of standard products flame penetration and consumption occurred, filled grades behaved quite differently. Penetration of the samples occurred within 15 to 30 s, however, it did not result in complete combustion. The test procedure obviously characterizes in particular the tendency to carbonisation.

The listing of Tables 79 and 80 illustrate that a high filling content (Al_2O_3) or the addition of melamine, change the smoke release significantly. For three density attitudes (4.1–3.6–6.1 pcf) the smoke-density characteristics $Ds_{90}"$, $Ds_{240}"$ and Ds_{max} were determined depending on the radiation intensity. The influence of flame-retardant additives was determined for a CMHR quality based on a 60/40 and 50/50 polyol mixture.

The smoke release was also part of a series of investigations. The fire-gas release of two different PUR systems based on PHD polyols with specific fire retardants was assessed by the NBS Smoke Density procedure according to ASTM -E 662. The records of Table 80 demonstrate the effectiveness of the formula variation.

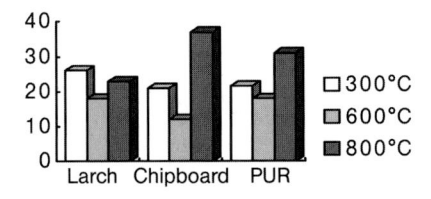

Figure 117 Influence of the test temperature on the toxic potency

Table 79 Influence of the recipe to the smoke density data

Smoke density	Ds90			
Intensity (KW/m^2)	25	17	10	1
HR	281	294	330	316
EST	191	91	84	78
CMHR	114	52	45	46
Melamine	102	248	263	187
	Ds240"			
HR	322	391	340	398
EST	248	158	179	126
CMHR	267	193	144	93
Melamine	251	314	279	291
	Dsmax			
HR	331	397	370	429
EST	260	158	186	128
CMHR	268	229	179	101
Melamine	270	356	300	428

Table 80 Influence of the recipe variation to the smoke density data

Polyol	I=25	I=25	I=25	I=17	I=17	I=17	I=10	I=10	I=10	I=1	I=1	I=1
FR	90"	240"	max	90"	240"	max	90"	240"	max	90"	240"	max
60/40	128	161	163	78	125	127	99	150	157	164	203	203
50/50	115	149	150	95	132	132	87	130	133	165	207	209

Stone et al. [370] report on comparative smoke-density measurements with differing foam qualities. They examined conventional polyether foam, polyester, CMHR and, e.g. with Al_2O_3-filled PUR systems. Prager reported on smoke-density records gained with rigid PUR and PIR foams. A similar view is given by Brochhagen et al. The listing of Table 81 provides an overview of these with different national fire-reaction test methods determined material-related data. Table 82 illustrates the fire-performance classification of polyurethanes according to Dutch and Swiss requirements [371, 372].

The listing of Table 83 provides an overview of the influence of the formula variation to the smoke-release rate. The results of these studies have to be interpreted as follows:

- Due to the possible large variety of polyurethane grades, generalisation should be done with great care.
- The variation of the density lead to significant deviations of the record level.
- CMHR qualities lead, as a rule, to significantly increased smoke-density data.
- The polyol component does not seemingly have any influence on the value.
- The melamine quality, unlike the polyester grade, provides a more intensified burning behavior at higher radiation levels.

Table 81 Smoke density data

Test method	PUR	PIR	PCD
F.P.Test BS 476 pt6 max.obscuration (%)	50	10–40	–
XP2- BS 51111 max.obscuration	90	30–70	–
NEN 3883 Vlamoverslag D	$200*/10^2$	40–50 as wood	< 40 as wood
XP2 – Schweiz max.absorption (%)	~ 100	60–80	40–55
NT 0 Brown'sche Box	class 1 poorer than wood	class 2 – 1 as wood	class 3–2 –
ASTM E 84 rating	~ 450	40 – 200	–
NBS Kammer flaming D m	180–190	145–150	~ 70
smoldering D m	110–120	80 – 90	~ 30
DIN 53436/37 Transmission (%)	300–600 °C 60	300–600 °C 95	

Table 82 Classification of PUR according to national requirements

Material	Vlamuitbreiding	Vlamoverslag
PIR	I	I
PCD	I	I

Material	Fire – reaction rating	Smoke test
PUR rigid foam	II	1–2
PIR rigid foam	V	3
PUR Elastomer	V	–

- The measurement of a lower level of radiation can lead to a significant misjudgement.
- The determined characteristics of smoke density and the CO concentration values are to be interpreted in such a way that the cone-calorimeter results do not allow any forecast about the smoke release during the actual combustion process. The correlation of these data records is obviously too small.

The smoke characteristics of the smoke potential determined with the DIN equipment illustrate according to Fig. 118 that estimates as have been undertaken, e.g. by Ames, according to Fig. 21, for the combustion of 2 kg of material in a room of 100 m³ [361], need not only the relevance to an actual fire situation but also presuppose detailed knowledge of the functionality. These characteristics confirm the estimates of Woolley and Ames [362], which assumed, e.g. for the burning of 5 kg EPS and PUR rigid foam samples within a fire room of 1000 m³ a visibility reduction of

Table 83 Smoke volume as material-related data

Material	Smoke volume (m³)	Visibility (m)
EPS	1452	1
PUR	503	2

1 m (EPS – 1452 m³ smoke volume) and/or 2 m (PUR – 502 m³ smoke volume). The conversion of such material-related records within a framework of a danger estimation shows that the range of visibility can be assigned to the relevant smoke-concentration data. Figure 112c shows very impressively that material-related transmission values are not only suitable in themselves as base values for risk-relevant danger estimates. The dominant influence of fire-performance parameters are also shown by risk-oriented full-scale field trials depending on the variations of material-related potency values. The simulation of the Maysfield Leisure Park Fire at the Fire Research Station of Boreham Wood hardened this statement with regard to flexible polyurethane foams. Flexible polyurethane mattresses produced in the developing stage of a fire an extremely small smoke release [7]. This regular burning process due to the temperature and ventilation circumstances developed into an optimal after-burner situation.

The scale of the burning rate and the surface involved in the fire occurrence is decisive for the amount of smoke released in the case of a fire. The listing of Table 84 and those of Table 78 provide a view of the fire-gas amounts to be considered in theory determined by means of the calorific value formula.

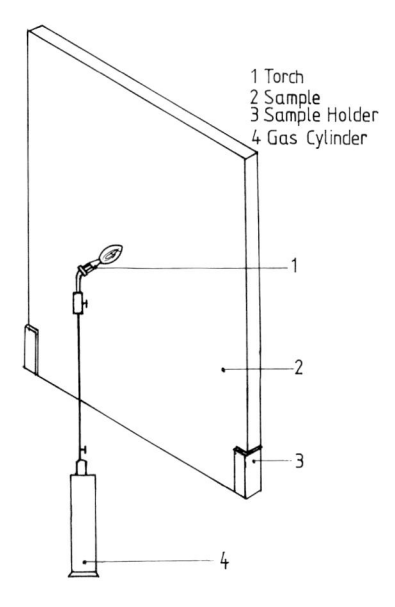

Figure 118 Blow Torch Test

Table 84 Calculated smoke values

Material	Density (kg/m³)	Calorific value (MJ/kg)	Calorific value (MJ/m³)	Smoke volume (m³/kg)	Smoke volume (m³/m³)
Wood	800	16,5	13 200	5,2	4160
Particle board	450	17,0	7650	5,3	2385
EPS	15–30	38,9	585–1165	9,9	148,5–297
PUR	50	26,1	795	7,2	360

The listings clarify the dominating influence of the material volume being involved in the fire occurrence. As the following list shows, the danger is increased by these material-related potential values crucially from the scale of the possible contribution to the surface spread of flame. In dependence of the enduse conditions, the material compound and the spatial factors including the ventilation conditions, preheating effects play in particular an important role. The records of Table 85 demonstrate that the extent of the participation on the fire can be decisively changed by preheating effects. The rise of the room temperature from 10 to 20 °C produces a fivefold increase of the surface spread of flame of the tested material toluol. Being aware that there is a linear dependence of the mass-burning rate to the fire duration and on the other hand for the space of time a second potency, attention is to be paid to relevant testing methods by the estimation of the possible contribution to the fire-gas generation.

Table 85 Influence of the temperature to the burning rate

Burning behavior			
Material surface temperature		Flame spread	Burning rate
Toluol	10 °C	10,2 m/min	2,31 kg/m² min
	20 °C	50,4 m/min	

In connection with a smoldering process in the upholstered chair field a fast burning rate can be caused at the moment of flaming ignition. This very specific phenomenon initiated press reports, e.g. in Great Britain, about exploding chairs [375]. In comparative investigations with the Mini Michigan-Roll-Up Mattress Test it was verified by means of the listed results of Table 86 that the mass-burning rate will be changed considerably by the composite with different outer coverings unlike the PUR formula variations.

Test philosophy for the assessment of the toxic potency of effluents

The dominating role of the side effect "acute toxicity" was documented for the first time in 1929 in connection with the Cleveland Clinic Fire [88]. The fire gases released during the combustion of X-ray films stood in causal connection with the high percentage of fire victims. Meanwhile, a line of different approaches was developed for the evaluation and classification of the acute toxic potency of effluents. First and widespread implementation of approaches based on the chemical nature as well

Table 86 Burning behavior in the Michigan Roll Up Test

Modified HR-Systems	Burning time t (min/s) Test I=15 kW/m²	Burning time t (min/s) Test I=118 kW/m²)
HR 281-27B	3:00	7:15
HR 281-28C	2:05	10*)
HR 281-41D	1:50	10*)
HR 281-42A	2:20	7:30
HR 281-42B	2:35	7:30
HR 281-42D	2:30	10*)
CMHR-DensiteII	7:00	–
Neopren LS200	6:00	8:00

*) Extinguishing of the test samples

as on analytically determined concentration data of the materials to be assessed. Furthermore, risk-related implementation approaches are based on animal-experimental studies. According to the view of toxicologists only these were suitable to register possible additive, synergistic or antagonistic effects of individual flue-gas components and to allow their assessment for security reasons. The findings from different fire disasters showed that effluents act in the form of poison and irritants. Provided that the elementary composition of the combustible is known, the volume of the dry combustion products can be determined on the basis of the calorific value formula [128, 376]. The amount of effluents can be determined for the complete combustion with the aid of the calorific value formula (V_r) and the air supply (L_{min}) with the equations:

$V_r = 0.89 \times H_u/4190 + 165$ (N m³/kg) for solid combustibles
$V_r = 1.11 \times H_u/4190$ for oil
$L_{min} = 0.85 \times H_u/4190 + 0.5$ (N m³/kg) for solid combustibles
$L_{min} = 0.85 \times H_u/4190 + 2.0$ (N m³/kg) for oil

The scale of the possible fire-gas formation in the case of a fire will depend basically on the different fire parameters. With the aid of the chemical structure of the fire load the flue-gas amount released maximally can be determined by means of the, in theory, necessary amount of air to the complete combustion by calculation.

The nominal oxygen supply results from the relationship

$L_{min} = O_2 \text{ min}/0.21$ (N m³/kg)

The experts of SC3 of TC 92 Toxic Hazard in Fire determined for the characterisation of effluents a line of characteristics and their key values. Figure 119 clarifies the key data stipulated in TR 9122. These general conditions of the experts of SC3 have a pilot function for all other relevant standardizing committees [87, 377]. The SC3 of TC 92 "Toxic Hazard in Fire" where toxicologists and experts of the preventive fire protection, from science, economy and administration participate, also established

Characterization of fire effluents

Fire risk situation	Fire temp. [°C]	Room temp. [°C]	Fire effluents		
			O_2 [%]	T [°C]	CO_2/CO
Smouldering fire					
self-sustained	500	RT	21	>20	<50
Oxidative pyrolysis	<500	>RT	21	>100	<30
Nonoxidative pyrolysis	>200	>50	0	>100	<10
Development stage					
Ignition phase	>700	<100	21	>200	<100
Combustion	>700	>500	21	>200	<50
Flashover	>700	>500	<10	>700	<10
Fully developed fire					
Good ventilation	>700	>700	>10	>700	<50
Poor ventilation	>700	>700	<15	>700	<20

Figure 119 Fire-gas characterisation

the basis for all other fields as there are electromechanical appliances, furniture and furnishings and transportation. The relevant standardisation activities at IEC, CEN and/or CENELEC must be basically oriented on the framework of SC3 [374–376]. Meanwhile these were modified due to newer findings in particular by means of a detailed analysis of experiences of the defensive fire protection as well as by relevant large-scale test results [111, 378–380].

Cornish et al. [381, 382] examined the possible influence of the so-called static and dynamic test procedures on the toxic potency of effluents by means of animal-experimental investigations. Advantages and disadvantages of the different test parameters were demonstrated as well as the importance of the decisive evaluation criteria. At the selection of the decomposition model the relevance to the fire scenario is to be considered sufficiently. The relevance of the combustion model to the actual fire case must be assessed by means of a series of criteria:

- the temperature and/or radiation-time-profile of the stream of effluents
- the CO_2/CO relationship of the smoke layer
- the concentration–time profile of the main fire-gas components
- the hold-up time of the stream of effluents in the temperature zones
- the repeatability of the fire course
- the usability of the procedure for analytical and bioassay testings as well as
- the usability/handling of the test report.

The graphical representations of Fig. 120 illustrate that the decomposition process can occur in a static and/or dynamic way. It must only be guaranteed that the concentration of the effluents have a comparable level. This can be achieved, as demonstrated by Fig. 120, through a regular burning process as, e.g. in the case of the DIN 53436 system or through gathering the different streams of effluents [383–388].

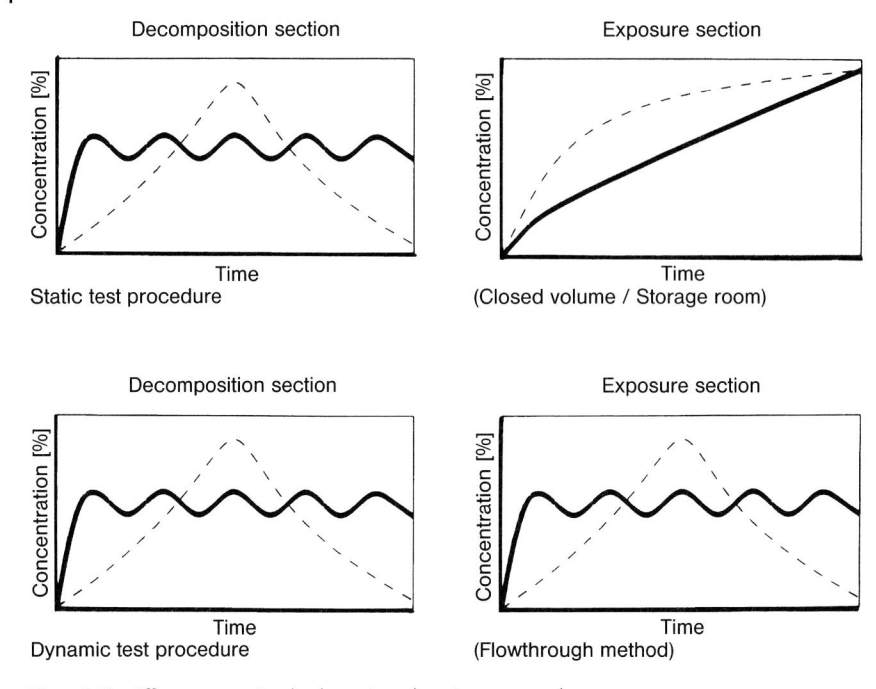

Figure 120 Effluent generation by dynamic and static test procedures

The investigations of Alerie [389] and Jouany [219] showed very instructively that experimental parameters like temperature, heating rate and air supply are able to influence significantly both the release of effluents as well as the composition of the fire-gas streams. The CO_2/CO relationship of the stream of effluents is changed correspondingly. Morikawa shows among other things in [361] with the example of the box- and crib-model that the ventilation is a dominating parameter during the effluent formation. Saito [223] explains also by means of the CO_2/CO ratio the importance of the ventilation. Rasbash et al. [83, 390, 391] explain that the fire controlled by the ventilation leads to a stronger fire-gas formation than does the fire-load controlled one. Baker et al. [392] explain the necessity of the selection of a risk-related combustion device using the example of Teflon as model substance. The comparison of the records of the toxic potential that were determined with different decomposition models like Potts Pot, U-Pitt and USF, illustrate (see Fig. 121) that a deviation up to 400 times of the ACL value (Acute Lethal Concentration) for Teflon are observed in comparison to wood. The data records of Fig. 121 clarify these extreme differences of the material-related characteristics of the toxic potential depending on the test criteria of the different procedures, that a 400-fold critical value for PTFE is reached in the Potts Pot protocol, if the LC_{50} data of Douglas Fir is defined as basis 1, a 100-fold value is produced in the U-Pitt Method and an equivalent level is demonstrated by the USF Method.

The extremely high acute toxic potential of the thermal decomposition products of fluorpolymers was linked to an PTFE oligomer with a particle diameter of 0.04 μm

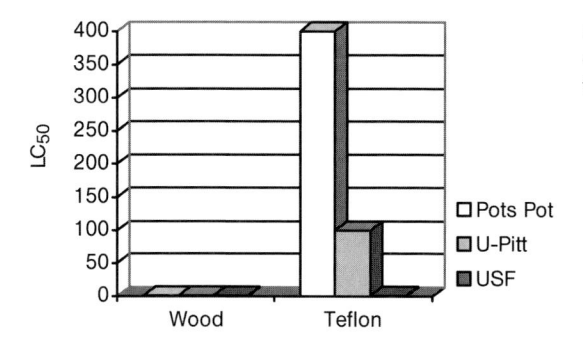

Figure 121 Influence of the decomposition protocol on the toxic potency records

with the aid of a detailed trace analysis. The evidence was produced with the aid of IR and NMR spectroscopy. This specific fire-gas component was neutralized by the use of filters as well as by water parts in the decomposition products.

ALC_{50} 30-min value was determined with 20 mg/m³ for the effluents, while for the particles a value of 0.2 to 0.4 mg/m³ was reported. The mass-related ALC_{50} value determined by the NBS Chamber was 0.02, i.e. 20 mg/m³. The perfluorpolymer aerosol reached a value of 0.0003, i.e. 0.3 mg/m³ indicated. Babrauskas attempts in [393] to identify advantages and disadvantages of static and dynamic combustion systems. The assessment of the characteristics of the toxic potential is based on the N-Gas-Computation Model [394] developed at NBS. Against the currently most used combustion system, the cone-calorimeter test method, which is favored by testing experts in the most international standardizing committees, e.g.,

- ISO TC 92 -SC1 Fire Reaction Testing (WG5, WG8)
- ISO TC 92-SC4 Fire Safety Engineering
- CEN TC 127 Fire Reaction Testing
- CEN TC 207 Furniture
- IEC TC 89 WG 7 Toxicity

give all essential facts of the know-how for a possible effluent danger in the fire case. Since the possible fire-gas danger must mainly be concentrated on the impact of the CO component, this question is of fundamental importance. Babrauskas et al. [394] recognized the obvious weakness of this method, which he developed. He propagates a computational correction of the CO concentration of the stream of effluents by means of the chemical nature of the material to be evaluated. He assumes that the concentration of other essential fire-gas components such as, e.g. HCN, HCl or SO_2 is not influenced in the same scale of the fire course as the CO. A detailed study of the relevant expert literature shows that such an assumption can not be maintained. Figures 14a and 14b clarify among other things the influence of the testing temperature on the formation of HCN. The investigation of Morikawa [395] illustrated that, e.g. also the concentration of aldehydes and acrolein can be modified significantly both from the test temperature and also the air supply. If one takes the influence of the reduction of the visibility as a further indication of a possible smoke danger, the objections increase then against the implementation approach from Babrauskas. In analog mode to the material-related toxic potential all relevant investigations show that the visibility-reducing effect of a stream of effluents also depends on the fire course considerably. This effect was documented in the relevant principle basic papers of the testing experts of the SC3 and SC4 of TC 92 as well as CEN TC 207 and IEC TC 89. This knowledge remains, however, completely unconsidered during the determination of the testing and evaluation criteria. The influence of the testing temperature seems to be due to the flaming combustion principally negligible (see Fig. 115). This result is inconsistent with all the other relevant studies that prove that with increasing temperature a rise of the smoke-potential values up to peak values of approx. 500 °C to 700 °C [367, 396]. Therefore the standardized test and evaluation criteria of the cone calorimeter are opposite to any relevant experiences. The listing of Fig. 122a clarifies by means of the Oestman results that the cone-calorimeter method of test, as it is standardized, is unsuitable for the general characterisation of the stream of effluents. Recent research work in Australia with a risk-related modified equipment offers a solution to these problems (see Fig. 122b) [397]. The thermal decomposition of the products to be evaluated occurs in the different combustion devices with mass-, volume- or area-related test pieces in the stream of air with constant temperatures and/or constant temperature increase. The concentration of the essential smoke components is determined continuously in the stream of effluents or in the collector. While in the tube furnace in accordance with DIN 53436 per unit of time the same mass-, volume- or area-related test piece is being subjected to thermal stress, to guarantee an approximately constant stream of effluents. A continuously changing stream of smoke must be expected in the remaining procedures, where the entire test piece is exposed to heat. A direct reference to the individual criteria is not possible without effort. The repeatedly propagated advantage of a continuous weight-loss measurement is not to be classified as useful by any means in case of realistic consideration as a rule. Those in Figs. 123a and b opposed smoke-density data, which are based, on the one hand, on the weight loss and, on the other hand,

Material	Density [kg/m^3]	Smoke Potential medium value [ob m^3/g]	[ob m^3/m^2]	CO_2/CO medium value
fiber board	250	0,4	(20 mm thick) 17,5	250
wood panel	450	0,4	31,5	130
particle board	670	0,4	90,0	250
PUR	32	3,8	18,5	4
EPS	18	5,5	20,5	24

Figure 122a Fire-gas evaluation based on the Östman cone-calorimeter records?

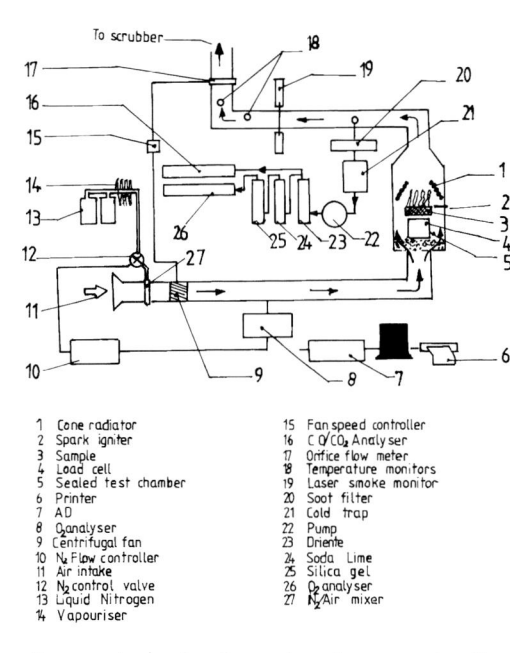

Figure 122b Modified Australian Cone Calorimeter

1 Cone radiator	15 Fan speed controller
2 Spark igniter	16 CO/CO$_2$ Analyser
3 Sample	17 Orifice flow meter
4 Load cell	18 Temperature monitors
5 Sealed test chamber	19 Laser smoke monitor
6 Printer	20 Soot filter
7 AD	21 Cold trap
8 O$_2$ analyser	22 Pump
9 Centrifugal fan	23 Driente
10 N$_2$ Flow controller	24 Soda Lime
11 Air intake	25 Silica gel
12 N$_2$ control valve	26 O$_2$ analyser
13 Liquid Nitrogen	27 N$_2$/Air mixer
14 Vapouriser	

Cone calorimeter for reduced oxygen testing

on the tested volume, illustrate the programmed faulty interpretations, which seem to be compulsory in the early stage of a fire.

The composition of the effluents depends primary of physical and chemical phenomena. Temperature level, exposure time and availability of oxygen are the dominating parameters. The irradiance level can influence both the nature and the rate of the decomposition process. By estimating the release rate the area of burning and the burning rate had to be taken into account sufficiently. The numerical values of Fig. 122a and Table 83 explain the influence of the oxygen availability on the smoke composition using the example of the smoke-density measurement. A comparison of the material-related characteristics illustrate that the smoke-density data pass through a maximum depending on the test temperature. Through the selection of mass-, volume- or area-like test pieces the material-related value is displaced correspondingly. A classification by means of mass-based material-related data only is not risk relevant. The determined test parameters will correspond to the factors of the fire-risk situation to be evaluated in exceptions only. Material-related effluents and oxygen-deficiency will became effective in remote areas if a local flash over occurs and the smoke layer moves forward without essential dilution (see also Fig. 12).

Three groups of persons are to be considered

- the group that stays in the room of origin, e.g. careless persons and arsonists
- the group that stays in neighboring fields and
- the group of fire-fighting people.

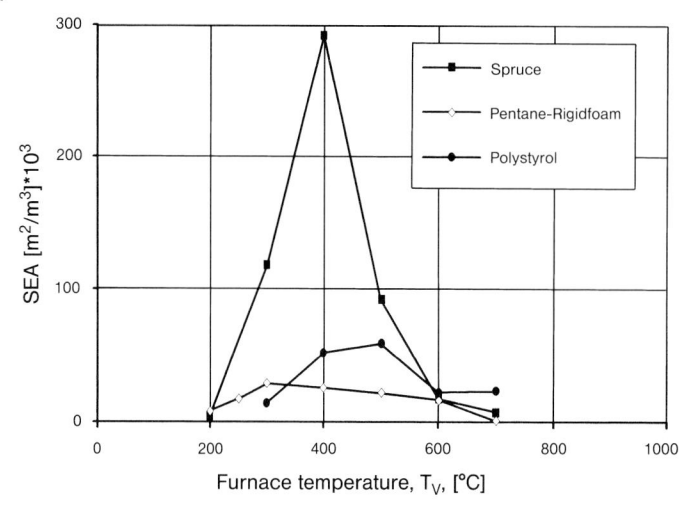

Figure 123a Influence of the test-piece dimensions on the characteristics of the optical density D

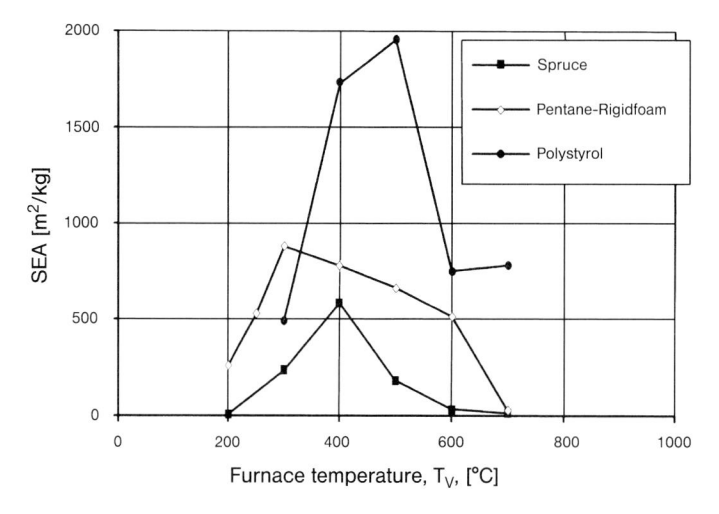

Figure 123b Influence of the test-piece dimensions on the characteristics of the optical density D

Essential aspects for a positive behavior of the people in the fire case are, as the evaluation of the different findings show, e.g.

- Local knowledge
- The familiarity with the emergency exits increases [178, 398–400] the escape probability. Escape groups with leaders who know their way around are up to 1.5 times faster
- Sufficient visibility

- Sufficient visibility offers [80, 402] favorable assumptions for promising escape operations. From [176] men go, e.g., rather more readily through smoke than women.
- Excellent physical condition
- Good physical condition allows escape operations within a short space of time. With old and sick people who need possibly the aid of other persons a corresponding delay of the escape operations must be taken into account.
- Avoidance of panic

Panic happens [177] when the hope for subjects to escape fades away. The demand for a delayed fire course is raised repeatedly in connection to fire disasters to be guaranteed that more time is available for effective escape and extinguishing measures. Neither oxygen defiancy nor raised temperatures should hinder or made escape and rescue activities in borderline impossible by damage of respiration organs. The listing of Table 87 gives an overview of usual, partly standardized procedures that are offered by law as fixed problem solutions and also partly applied.

Table 87 Test methods for evaluating the toxic potency of effluents

Method	Thermal stress	Test sample (mm)	Sample volume (cm3)	Duration (min)
ATS 1000	25 kW/m² with and without flame	75x76 × 25 for 2mm	144	6
			11,6	20
NES 713	Burner	450 × 450 d = 2 mm	101	20
				6
U-PITT	100–1000 °C 20 °C/min	ca.1 g	–	30
DIN 53436	200–900 °C	300 × 20 × 10	60 (ca.5 g)	30,60
GOST 12.1	200–800 °C	40 × 40 × 8 (1–10 g)	12,8	20
NF-T51	200–900 °C	270 × 5 × 2	2,7	10–15
Cone Calorimeter	10–100 kW/m2	100 × 100 × 50 d = 2 mm	500 20	60

A line of different evaluation criteria have been defined next to the CO-Hb content while trying to solve the problem by animal experimental investigations [76, 89, 198, 402–404]. These characteristics can be determined both for pure gases, gas mixtures and effluent mixture. For animals without haemoglobin, as, e.g. insects, CO acts like N as a pure, suffocating gas. As assumption for a realistic evaluation of the acute flue-gas toxicity, detailed knowledge is regarded as vital about the respective fire development and the cause of death of the fire victims. Pathological and toxicological investigations are necessary for this. Biochemical and chemical analysis as, e.g. gas tests on the pulmonary tissue, which can support together with medical investigations possibly autopsy results. During the appraisal of lungs normal airborne particles that can become effective possibly as a base for toxic components have to be taken into account. The toxicological investigations can be carried out, e.g. on urine and blood samples as well as on tissue samples of liver, bile spleen and/or kidneys.

The analytical determination of the concentration of the essential fire-gas components is indispensable for all three different implementation approaches

- analytical model
- biological model and
- computation model

These material characteristics are necessary for the analytical implementation approach in order to allow the rating of products to be tested anyway. With the biological implementation approach these characteristics are, as required by Einbrodt [405], the assumption of an evaluation of the animal-experimental results from an toxicological viewpoint. He assumed a solution based on bioassay testings a row of evaluation criteria have been defined beneath the CO-Hb values [88, 90, 93, 406]. The third implementation approach, which lead to an essential reduction of the necessary bioassay, needs these analytically determined material characteristics for the mathematical determination of characteristic records such as, for example, the value LC_{50}. The degree of poisonousness of the effluents is unambiguously a function of the concentration, the exposure time as well as the possible preinjuries of the victims, e.g. through heart illnesses. Next to the specific poisonousness of the fire-gas components the smoke distribution and the absorption must also be considered among other things. The degree of the absorption is influenced, e.g. also from the ventilation of the lung and with that the breath volume. The degree of poisonousness is thus influenced also by the breath volume and the different rate of breathing. For the noxious fire-gas components one differentiates among other things between suffocating gases (asphyxiants), irritants, inhalation toxica and inorganic and metalor-

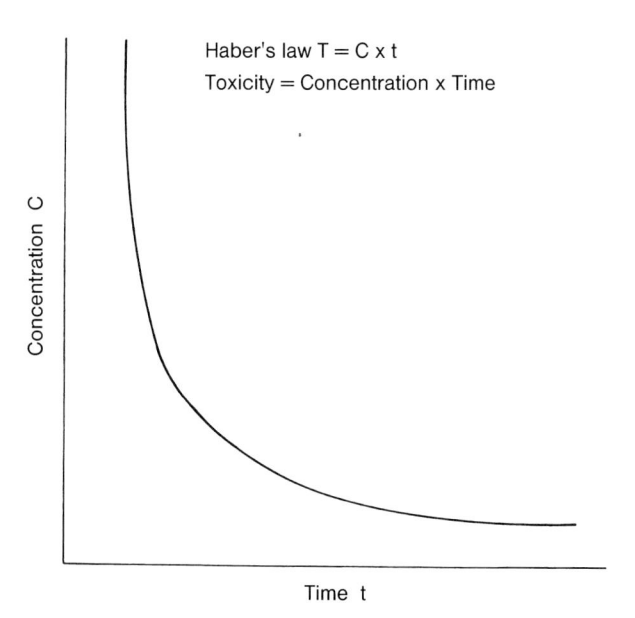

Haber's law $T = C \times t$
Toxicity = Concentration × Time

Figure 124a Haber's Rule

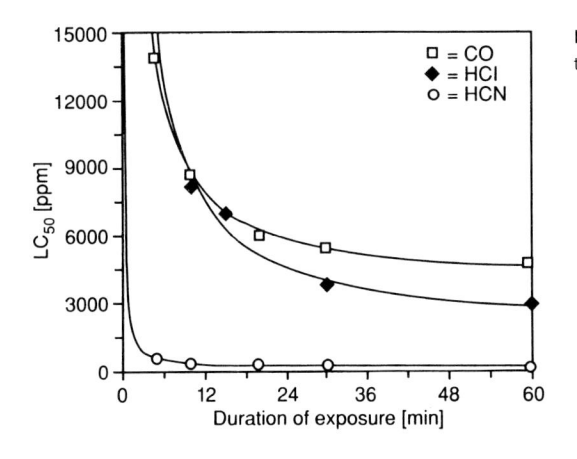

Figure 124b Acute inhalation toxicity

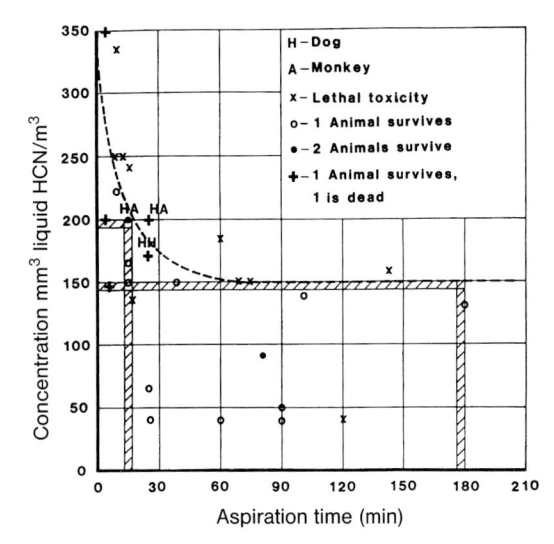

Figure 124c Concentration–time relationship

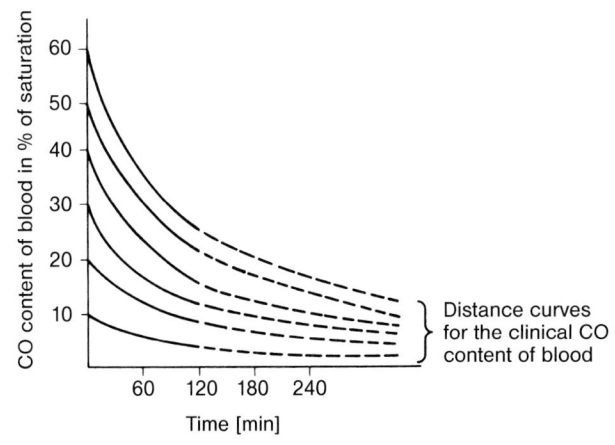

Figure 124d Removal of the CO-Hb concentration

ganic components. Such subdivisions, mainly due to the action of specific components to individual organs such as, e.g. eyes, upper breathing ways in the windpipe area and larynx or lungs are carried out. To asphyxiants belong simple suffocating gases as nitrogen and helium that separate the oxygen mechanically as well as suffocating gases, performing chemically as CO and HCN, which prevent the replacement of the oxygen in the blood and in the tissue. Also, irritants are often assigned as asphyxiants according to Flury and Zernik [407] due to the suffocating effect of the pulmonary edemas caused by them. For the irritants one differentiates as a rule between lung-stimulating components such as chlorine and phosgene, blistering and corrosive such as, for example, sulfur and arsenic combinations and tears and sneeze gases like formaldehyde. As a rule, many effects are associated with each other and depend decisively on the respective concentration. The connection between concentration and exposure time of toxic components can be demonstrated most simply by Haber's Rule in accordance with Fig. 124a, which defines that the product concentration c and exposure time t for the attaining of a specific effect $ct =$ constant. Figure 124b clarifies this connection for some of the essential fire-gas components.

Flury and Zernik [407], Packham and Hartzell [408] as well as Hilado and Huttlinger [409] pointed out possible deviations. While Flury and Zernik confirm experimentally the relationship $ct =$ const. for irritants, they refer to deviations of the suffocating gases. They indicate a modified relationship for hydrocyanic acid $(c-a)t =$ const (see Fig. 124c). They also report that the information of a "ct value" without simultaneous announcement of the concentration is nearly worthless. The effect may possibly enter only after an infinitely long period of time. Short exposure can lead to health problems if high pollutant concentrations are effecting. The refrigerator fire described in [410] can be stated here as an example. Saito [236] describes relevant considerations for the evaluation of the acute smoke toxicity, by means of analytically determined concentration data of essential flue-gas components.

Suffocating fire-gas components
The effect mechanisms of CO poisoning were cleared up according to Petry [411] already in 1857. CO acts as an asphyxiant, because there is a 250-fold greater affinity to the red blood dye than oxygen and can therefore regarded as a blood poison. If CO acts on haemoglobin

$O_2Hb + CO = COHb\ O_2$ from it then after the equation for the reaction. For the affinity constant factor

$$K = \frac{COO_2Hb}{O_2COHb}$$

a value is indicated in the literature of 200 to 300. According to Petry only about 15 % of the carbon monoxide of the pulmonary blood available in the breath air are included after investigations by Großkopf. The CO absorption takes place faster with smaller and younger individuals with lively metabolism. Children are endangered by the CO action particularly, because of their relatively large breath volume. This circumstance has been taken into account in the determination of the permissible

Table 88 Effect of CO intoxication

CO-Hb	Symptoms
10%	No noticable effects
20%	Slight headaches
30%	Headaches, irritable light nausea
50–60%	Headaches, collapse after strong work
60–70%	Coma, weak respiration
80–90%	Death in less than an hour

boundary values for the fire-gas toxicity of A2-classified (DIN 4102) materials with a 35% level according to Einbrodt [412]. Typical for CO-hypoxia (oxygen deficiency in the blood) is the fact that through the CO-effect unlike pure oxygen deficiency, no raised respiration excitement is caused. The following symptoms (see Table 88), as described by Petry [411], Kimmerle [88], Müller [413] and others (TR 9122) have to be distinguished for acute CO-poisoning.

Forbes et al. [414] discuss the problematic nature of the CO takeup by people in the acute fire case. The submitted results reveal very clearly that the CO uptake can be significantly influenced by the activities of the subjects (resting position, light work and heavy work). Levin et al. show in [415] that there is a synergistic effect for the two fire-gas components CO and CO_2. Children and sick people are damaged already [412] at considerably smaller CO-Hb values. According to Purser [416] the Stewart Formula provides a mathematical determination of the CO-Hb content. Detailed information about the experimental technical determination on blood samples are given by Fretwurst et al. in [417]. Petry refers to experiences from fire disasters that illustrate that the action of extremely high CO concentrations immediately produce fatality through a sudden saturation of the pulmonary blood. One speaks of stupefaction with subsequent death and rigor mortis. Petry, in addition, gives a very full view of possible late results through acute CO intoxication. Rescue illness and death can perhaps appear after longer times without symptoms. Such late results can also lead to permanent illnesses with corresponding permanent damage. The main result of the CO poisoning happens in the brain. The more rapidly the CO deficiency is eliminated, the more likely it is to avoid the possibility of permanent damage or fatality. Einhorn and Grunnet [418] reported on such late results through CO intoxication. Petajan et al. confirmed in [419] that the fire-gas component CO is able to cause influences and damage to the brain system. Also in [420] specific side effects were reported on due to CO intoxication. Daunderer [421] warns before such damages and Zirkria [422, 423] verified how strong the CO content can be removed in particular in the first 30 to 60 min after the inhalation through punctual processing. The graphical representation of Fig. 124d shows this very clearly.

In addition to CO the second performing chemically asphyxiant gas HCN is put repeatedly as a possible cause of death for fire fatalities [365, 421, 424]. HCN reach with the ferrous breath ferment and blocks the intracellular oxygen transport. According to Kimmerle [88] the action of HCN causes in human beings the following reactions in accordance with Table 89. The crucial, fatal concentration in the blood is, according to Kimmerle, 5 µg/ml.

Table 89 Effects of HCN-Exposition.

HCN (ppm)	Symptoms
10	MAK value
18–36	Slight symptoms (headache) after several hours
45–54	Tolerated for 1/2 to 1 h without difficulty
100	Death after 1 h
110–135	Fatal after 1/2 to 1 h, or dangerous to life
135	Fatal after 30 min
180	Fatal afteer 10 min
280	Immediately fatal

Reliable statements on the importance of the HCN component are very difficult since, as the clinical investigations of Harland et al. [425, 426] demonstrated, high CN values were verified in the blood of fire victims as a rule in connection with very high CO-Hb values only. Figures 124 clarify the connection of the exposure duration and the concentration. The investigation results of Flury and Zernik demonstrate this connection by means of bioassay test results with different species. Bonsel explains in [427], by means of a current fire case, the risk of a massive HCN dose. Kimmerle discusses very fully in [88], under observation of the relevant literature, the toxic effects of further fire-gas components as, e.g. of NO_2, SO_2, NH_3, H_3S as well as formaldehyde and acrolein ($CH_2 = CH\text{-}CHO$) and the findings of experiments with pure gases that have been reported in various papers. Sakurai explains in [428] extensive studies with pure gases. The test animals were exposed for a duration of 10 or 20 min with pure gases and defined mixtures. The damage scale was evaluated by means of the respective damage to their behavior. The results demonstrate an additive-effect mechanism. Hartzell also came to analogous conclusions and Levin also reported on such studies [429, 430]. The exposure duration was varied between 1 and 60 min. Einbrodt discusses in [431] all aspects of the fire-gas toxicity. Next to the essential components such as CO, CO_2, HCN, NO_x and SO_2 also polycyclic hydrocarbons as well as dioxins furan are taken into account. Higgins et al. report in [432] on the acute inhalation toxicity of pure gases and mixtures. Gray discusses in [433] the acute inhalation toxicity of nitrogen oxides. The LC_{50} value is indicated for NO_2 with 174 ppm. The exposure duration was 30 min. The damage potential of NO and NO_2 was described by Oda in [434]. Skidmore and Sewell [435] report about the evaluation of nitrogen oxides from heated polymers. Plesser [436] reports on animal-experimental investigations with NO and NO_2. The results can be interpreted in such a way that NO shows a slightly higher acute toxicity value. In [437] Birky describes investigations with pure gases, e.g. acrolein. Morikawa [395] reports on extensive investigations where this component A LC_{50} value of ca. 30 ppm was determined in bioassay test runs with an exposure duration of 10 min. Saito explains in [235] further analogous Japanese studies. A much-discussed problematic nature is also the possible injury effects of specific fire-gas components.

The question of the so-called supertoxicants was caused last but not least due to animal-experimental results [43, 438]. In the stream of decomposition products of tetrafluorethylene components of extremely high acute toxicity were determined depending on the combustion systems [439, 440]. Comparative investigation results were interpreted in such a way that with such specific combustion devices unrealistic recirculation processes must be reckoned on. Byron explains in [441] the possible danger by effluents in the case of a fire from a medical viewpoint. Warnitz describes in [442] bioassay test results with the decomposition products of fabrics coated with tetrafluorine ethylene films. Crucial concentration levels were observed with temperature levels above 359 °C. The test animals were exposed up to 4 h with the decomposition products produced at 50 °C to 750 °C. Oil and butter were thermally attacked as reference products. Caunter tests with thermal decomposition products of foods are reported in [199] and [443]. A second case with rigid polyurethane foam was unjustly cited, because these unfavorable results were caused by the decomposition products of the fire-retardant additive only. This special gas component showed the extremely high acute toxicity only for itself. The toxic potential of the combustion products of the foam material did not produce, however, any analogous LC_{50} values. Kimmerle describes this TMPP (trimethylpropane phosphate-bicyclic phosphate ester) problem in [444]. This extremely highly toxicity of this specific component (see Fig. 125) does not play a dominating role for the toxic potential of the total mixture of effluents due to the low concentration. Einhorn discusses in [445] this special problem.

Munn evaluates the TMPP problematic also from a medical viewpoint. In comparison with nicotine he explains the importance of the current concentration for the acute inhalation toxicity. The assessment of the TMPP effects have to be, according to Munn [446, 447], put into perspective. Einhorn reports in [193] about analogous reactions of exposed test animals due to the exposure with the decomposition products of electro-isolating liquids. Norpoth et al. discuss in [448] the problematic nature of alkylating combinations as a component of fire gases. The fire-gas mixtures of the examined material, as reference products were set in cellulose and paper, were produced with the tube furnace according to DIN 53436 and evaluated from the toxicological viewpoint. Schilfknecht treats in [449] possible intoxications with active substances. The importance of the concentration is presented in this case. It is demonstrated that a danger is possible first through the interplay of exposure duration and concentration. The admissibility of rigid PUR and PIR foams because of their acute fire-gas toxicity was and is to be accepted according to the state of the art. Rigid PUR foams like Baytherm ® meet the relevant requirements of the Seeberufsgenossenschaft. In connection with 1-mm steel facing this compound complies with class B1

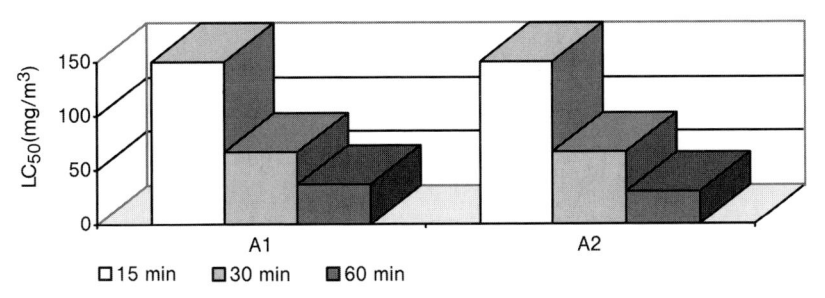

Figure 125 Toxic-potential records of TMPP

in accordance with DIN 4102 (PAIII – 2.95, 2.96 and 2.276) and is in accordance with the experts judgement on the basis of bioassay investigations. The requirements concerning the thermal decomposition products were passed [450].

Evaluation of the fire-gas toxicity by means of analytically determined data.

In the former fire-performance guidelines of the Fire Insurances of Switzerland (Guidance for Police Regulation 1969) as shown by the listing of Table 90 a solution by means of the chemical nature of the materials to be targeted.

Table 90 Swiss requirements concerning effluents

Class	Beneath CO, which have to be expected in each fire, other components are released
1	Poisons like HCN, NO_x, phosgene and others
2	Strong irritants like strong acids, formaldehyde etc.
3	Weak irritants, weak acids and weak bases etc.
4	No poisons and irritants, others

The rating due to the chemical nature, e.g. in the case of doubt proved as insufficient, an on-going investigation was to be carried out in the form of a fire-gas analysis. Fire gases have to be produced through smoldering of 0.5 g substance on a steel sheet (1 mm thick, 120 mm diameter) heated by means of a bunsen burner to approx. 450 °C. This implementation approach became standardized in SI 755 and SI 92. Ausobsky [451] carried out extensive investigations and strove for a risk-relevant classification of combustion gases. Approximately 120 mg of substance were burned on a heated glow-spiral (900 °C) with a sufficient oxygen offer as also under reduced O_2 content. The fire-gas component was assessed in relation to the CO content of the effluents of fir wood. Sumi and Tsuchiya [452] produced their combustion products at 800 °C. The exposure time was 2.5 min. The evaluation was carried out by means of the toxicity factor. The authors assumed an additive effect of the individual components. LC_{50} characteristics of the pure gases that lead within 30 min to the death of 50% of the test animals were selected as base values. Einhorn et al. gave in [453] a comprehensive view of a detailed analysis of effluents determined by the Fire Research Center (FRC) Utah. The effectiveness of the computerized system enables comprehensive investigations of fire-gas mixtures and the identification of the most varied components. The effectiveness of the system was demonstrated by fire-gas mixtures of PUR foam produced by 300 °C to 1000 °C. Hileman et al. report in [454] detail about the possibilities of the fire-gas analysis of PUR decomposition products by means of gas chromatography and mass spectrometry. Kallonen explains in [455] toxic polancy tests and in [456] the effectiveness of the FTIR method for the continuous estimation of fire-gas components, e.g. HCN. Within the framework of ICUP/III full-scale field trials [56] the BIS equipment according to Fig. 126 formed the basis for material-related analytical fire-gas investigations. The substance to be evaluated was thermally stressed in the air stream in the temperature range of 400 to 1000 °C.

Within the framework of the standardisation activities of DIN modifications were carried out. Merz et al. [457] explain this modified decomposition device in detail that became, as the VCI apparatus, the basis in various studies for the evaluation of the

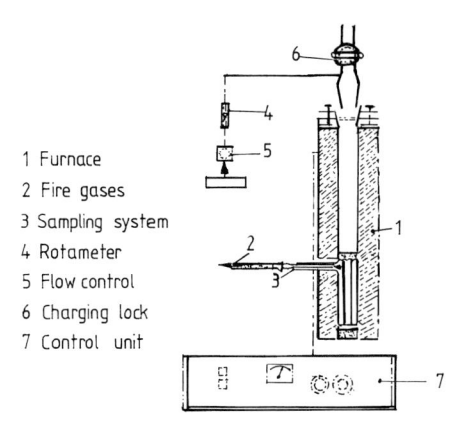

Figure 126 BIS-decomposition apparatus

1 Furnace
2 Fire gases
3 Sampling system
4 Rotameter
5 Flow control
6 Charging lock
7 Control unit

smoke toxicity [458]. Analogous investigations were carried out with a two-stage model developed at Ciba-Geigy. Kübler et al. explained in [98] the testing background. In the first stage a thermal decomposition occurs at 550 to 600 °C. The produced gas mixture is burned in the second stage of this protocol at 200 to 600 °C and/or 900 °C. The air supply flow was 3 l/min. The concentration of the main components like CO, HCN or NO_x was determined and registered as a mass-based material-related record. Pohl discusses the disadvantages of the BIS/VCI apparatus in detail in [459]. The investigation results of Sistovaris [460] with wool in accordance with Fig. 127 illustrate that the fire-gas release of different fire scenarios can be simulated by variation of the mass of the test samples. The results of Sistovaris demonstrate that in the case of the high test temperature of 950 °C also relatively high CO_2/CO values can be reached by minimizing the test-piece mass. Only if the test piece mass is small enough is sufficient air/oxygen available during the sudden thermal loading.

In the field of railway vehicles the combustion system according to NF-T51, (see Fig. 77c) was agreed as a central method, e.g. in France, Spain and Portugal. This model has a comparable construction to the tube furnace DIN 53436. The essential longer circular tube furnace permits a sudden thermal loading in the stream of air of the entire test piece. Approximately 1 g of substance is thermally stressed at 600 to 800 °C for 20 min in a stream of air of 120 l/h. in the equipment according to Fig. 77c. The stream of effluents is collected and analysed in the approx. 40-m³ large measuring room. The fire-gas toxicity is evaluated by means of the relevant animal-

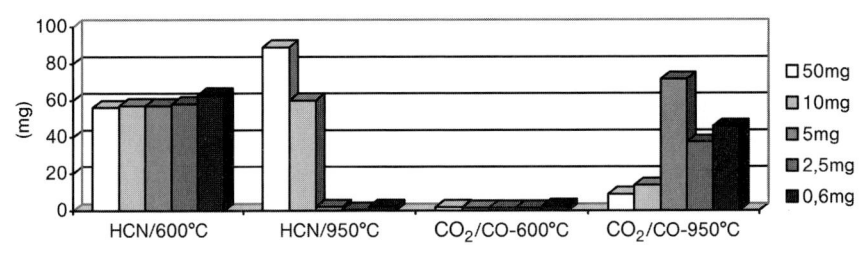

Figure 127 HCN content and CO_2/CO ratio of flue gases given off by wool

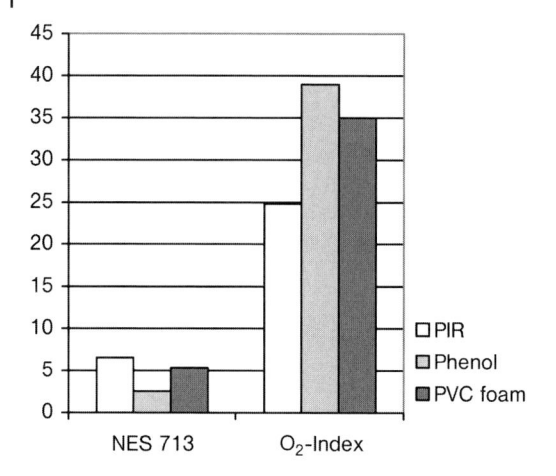

Figure 128 Fire-reaction properties

Chart showing NES 713 and O₂-Index values for PIR, Phenol, and PVC foam.

experimental results determined at an exposure duration of 15 min. An additive effect of the singular components is assumed. In the naval field, guidance is given by the Naval Engineering Standard (NES) 713 [461]. The substance to be evaluated is ignited in accordance with Fig. 74c by means of a bunsen burner and/or brought to the burning process. Provided that no ignition occurs, this specific combustion process is forced by the continuous application of the bunsen burner. With regard to the chemical nature of the material, 1 g of the substance is thermally attacked for differing times. The determined concentration data are projected on $100 \, g/m^2$. The representatives of Fig. 128 provide an overview of NES 713 material-related characteristics of PIR, PF and PVC foam in comparison to their O_2 index data. These records demonstrate that there is no relationship.

The responsible IMO committee agreed temporary regulations on the basis of the FAA requirements [462]. All these implementation approaches do not reach the targeted protection destination. On the one hand, these characteristics of the toxic potential are not sufficient to carry out a classification of the possible fire-gas danger and, on the other hand, the selected combustion systems do not correspond to the criteria that were defined by the experts of the SC3 of TC 92 "Toxic Hazards in Fire" [87]. The French "Nitrogen-Chlorine Law" can be stated here as a classical example. As already explained, not only was the dominating fire-gas component CO excluded by the classification system, but also the parts of the two selected crucial fire-gas components HCN and HCl, provided that they are yielded by natural products in the fire case.

Implementation approach based on animal-experimental investigations

Toxicologists [87–89, 463, 465, 469] agree that all the combined effects of all fire-gas components can be qualitatively determined and evaluated by means of biological testings only. Bioassay testings enable fire gases to be assessed with regard to their narcotic and irritative potential, and they are also able to detect fire-gas components of unexpectedly high toxicity. Animal tests were conducted to determine the relative toxicity for fire effluents For various polyurethane grades and materials of reference as a function of the test conditions including test temperature, specimen size and weight and ventilation. The test animals were exposed to effluents produced to simulate the smoke formation in the case of a fire. The listing of Table 91a provides information about the most essential evaluation criteria of the acute toxicity. The test methods selected by the experts of SC3 of TC92 for simulating the formation of effluents for the most important fire scenarios according to Fig. 120 have been listed in Table 91b.

Table 91a Criteria for Assessment of Fire Models

Relevance to real Fires

Validity to smoke, hazard assessment

Flexibility to sample composition and configuration

Experience and documentation

Procedural criteria
   Repeatability and reproducibility
   Safety in use and operation
   Relevance of data interpretation

Table 91b Evaluation criteria of the smoke potential

Relevance to real fires

Validity of the smoke formation
Decomposition temperature
Ventilation conditions
Dilution effects
Effluent composition
Concerntration of components
Reproducibility and repeatability
Relevance of interpretation

A line of essential criteria was agreed in the WG1 Fire Model that must be met by a combustion system within the framework of the long-standing standardisation of SC3 from TC 92 "Toxic Hazard in Fire". Assumptions for the narrower selection are:

- extensive experiments with bioassay testing in different test laboratories
- sufficient documentation of the test protocol and
- feasability and satisfactory reproducibility

With this it is guaranteed that the released stream of effluents, both with regard to the composition and also to the concentration of the main components, corresponds to the stream of effluents of the fire scenario to be simulated.

The modified US-Radiant procedure (NIBS-Test), one of the selected prototypes of the SC3 of TC 92, as well as the cone-calorimeter test were included with respect to their expected future importance. The criteria elaborated in the SC3 concerning the acceptance of a combustion system were not a prerequisite for the selection of the prototypes. It is not to be expected that all criteria of the essential fire scenarios can be simulated by a single test. The selection of every combustion system presupposes the evidence of the relevance of the test method. The acceptance of the determined material-related potential values depends on this, it is exactly verified that the combustion device is sufficient suitable for simulating the effluent production of the fire scenario surely, which have to be certificated. Essential criteria to be considered in this case are:

- the CO_2/CO ratio of the stream of effluents
- the thermal circumstances of the smoldering and/or combustion process
- the dwell time of the combustion products in the flame action field as well as
- possible recirculation effects including the afterburner effects combined with that.

Figures 13 and 121 provide a view of those for the characterization of the stream of effluents necessary characteristics. The newest findings of relevant investigations of the fire departments as well as the outcome of realistic full-scale field trials were considered at the determination. The determinations in the TR 9122 [87] are modified during the pending revision correspondingly. The predominant animal model is the white rat (e.g. Wistar III and Sprague Dowley). The similarity of the mechanism of the CO poisoning in the arrangement to the human being supported the choice of the rat in particular. Some methods also use white mice for financial reasons. Selected investigations were carried out also with other species [219, 468, 469].

The listings of Tables 92 and 93 provide detailed information about the various test protocols that are used worldwide to simulate the smoke formation of different fire scenarios for bioassay investigations.

The listings illustrate that the different test protocols use not only differing combustion systems. Significant differences also result concerning the kind and duration of the exposure. The exposure duration varies up to 4 h, although 30 min were agreed by the majority as the common exposure time. Next to the head-nose expo-

Table 92a Selected Fire Models by SC3 of TC 92

"Box" Furnaces
  NBS – Cup Furnace
  U PITT Box Furnace

Tube Furnace Model
  DIN 53436 Tube Furnace

Radiant Heat Fire Models
  US Radiant Furnace (modified NIBS)
  Cone Calorimeter
  Japanese Cone Furnaces
    BRI (Building Research Institute)
    Cone Furnace
    RIPT (Research Institute for Polymers
    And Textiles) Cone Furnace

Japanese Ministry of Construction Fire Model

Table 92b Test parameters of the selected decomposition models

Decomposition mode

Method	Furnace	Thermal stress	Ventilation	Quantity of materialö	Air Flow
DIN 53436	Tube 40 mm	200-600 81000)°C	100, 300 l/H	5 g, same G / V	Dynamic
Potts Pot	Crucible	200-1000 °C 50 below/above SI		Up to 8 g, 13 mg/l	Static
U-Pitt	Tube-furnace	200–800 °C 20 °C/min	9 +11 l/min	Variable	Dynamic
US Radiant/ Nibs	Radiator	20 – 80 KW/m² +flame		Variable	Static
JIS A 1230	Radiation+flame	BS 476 pt.6		Same area	Static
JIS – RTB	0–50 KW/m²	0–50 KW/m²		Same area	Dynamic
Cone Calorimeter	0–75 KW/m²	0–75 KW/m²		Same area	Dynamic

Exposure mode

Method	Duration	Exposure chamber	Test animal	Toxic results
DIN 53436	30, 60 min	Through flow	Rats, rabbits	Tk, CO-Hb, LC50
Potts Pot	30 min variable	160 l, 200 l	Rats	td, CO-Hb, LC50ti,
U-Pitt	30 min variable	200 l	Mice	ti,td,LC50,Lt50, CO-Hb
US Radiant/ Nibs	30 min variable	160, 200 l	Rats	ti, td,CO-Hb LC50,LCt50
JIS A 1230	6, 15 min	126 l	Rats, mice	ti, td, EC50
JIS – RTB	variable	125 l	Rats, mice	ti, td, EC50
Cone Calorimeter	30 min variable			Gas analysis

Table 92c Variations of the Test Method DIN 53436

Author	Furnace	Thermal Stress	Ventilation	Animal model	Exposure time	Toxic.Result
Effenberger	Tube,40 mm	200–600 °C	5 l/min	Rat, a)	30, 100 min	CO-Hb, LC50,c)
Einbrodt	Tube,40 mm	200–600 °C	300 l/h	Rat, a)	30, 60 min	CO-Hb, c)
Kimmerle	Tube,40 mm	200–600 °C	100u.300 l/h	Rat, a) b)	30 min	CO-Hb, Tk, LC50
Klimisch	Tube,40 mm	200–600 °C	100 l/h	Rat, b)	30 min	CO-Hb, LC50
Norpoth	Tube,40 mm	300–500 °C	300 l/h	Rat, a)	30 min	c)
Pauluhn	Tube,40 mm	200–600 °C	100u.300 l/h	Rat, b)	30 min	CO-Hb, LC50
Pohl	Tube,40 mm	250–600 °C	100u.200 l/h	Rat b)	30 min	CO-Hb, LC50
Oettel	Tube,40 mm	200–600 °C	100 l/h	Rat, a) b)	30 min	CO-Hb, Tk, c)
Hoffmann	Tube,40 mm	200–600 °C	100 l/h	Rat, b)	30 min	CO-Hb, LC50, c)
Herpol	Tube,40 mm	200–800 °C	200 l/h	Rat, b)	30 min	CO-Hb, LC50, td
Reploh	Tube,30 mm	300–500 °c	300 l/h	Rat, a)	30 min	CO-Hb, c)
Jouany	Tube,40 mm	200–800 °C	100 l/h	Rat,Rabbit b)	30 min	CO-Hb, LC50
Purser	Tube,40 mm	300–900 °C	1 l/min	Rat, b)	30 m in	CO-HB, LC50

a) whole body   c) mortality
b) head-nose

Table 92d Further decomposition models

Decomposition Model

Method	Furnace	Thermal stress	Air system	Ventilation	Quantity of material
FAA	Tube furnace	625 °C	Static + recirculation	4 l/min	0,75 g
NBS Smoke Chamber	Radiation+pilot flame	20-75 KW/m^2	Static	0,5 m^3	Same area
Gost 12.21	Crucible	18–60 KW/m^2	Static	100–200 l	Same area (840 × 40)
USF	Tube furnace	40 °C/min 200–800 °C	Static, dynamic		Max.1 g
Skornik	Tube furnace		Static		Max.1 g
JIS Box	Burner	Crib fire	Static		Same area
SRI	Radiator	20–80 KW/m^2	Static		Same area

Exposure Mode

Method	Duration	Test animal	Exposure	Toxic result
FAA	30 min	Rats, mice	Whole body	ti, td, CO-Hb
NBS Smoke chamber	30 min	Rat	Whole body	ti, td, CO-Hb
Gost 12.21	30 min	Mice, rats	Whole body	td, CO-Hb, LC50
USF	30 min, variable	Mice, rats	Whole body	ti, td,
Skornik	30 variable	Rats	Whole body	ti, td, LC50
JIS Box	30 variable	Rats	Whole body	ti, td
SRI	15, 30 min	Rats	Whole body	ti, td, mortality

Table 93 Further decomposition models

Methode	Thermal stress	Test sample (mm)	Sample volume (cm^3)	Duration (min)
ATS 1000	25 KW/m^2 with and without flame	75 × 76 × 25 for 2 mm	144 11,6	6 20
NES 713	Burner	450x450 d=2 mm	101	20 6
NF-T51	200–900 °C	270 × 5 × 2	2,7	10–15
Cone Calorimeter	10–100 kW/m2	100 × 100 × 50 d=2 mm	500 20	60

sure mode also the whole-body exposure mode is standardized. In the exception it results also in the forced respiration of the test animals. The essential parameters are to be considered during the selection of the decomposition model. A key question of the PRC 5-million dollar program was the acute toxic potency of the thermal decomposition products of polyurethanes. The DIN tubular system was also offered for solving this problem, but was rejected by the responsible expert group. Nearly all investigations were run according to the Potts Pot and the U-Pitt protocols [91, 92,

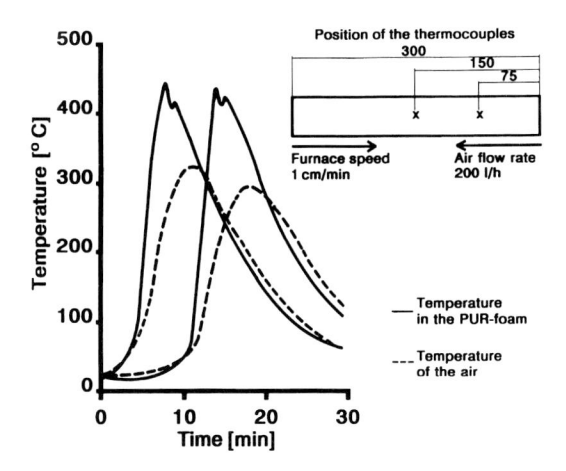

Figure 129 Temporal course of the smoldering temperature of flexible PUR foam

184, 185, 470]. The LC_{50} value was the most frequently used criteria for the evaluation of the toxic potential of the thermal decomposition products from the toxicological point of view. The common exposure time was agreed as 30 min. Only one special study was run within the PRC program [14] with the DIN tube system. Nearly all experts of the SC3 agreed that the DIN tube system according to Fig. 50 was the one model that nearly complied with most of the necessary criteria for simulating the most important fire scenarios as defined by Fig. 121. Last but not least, the evidence for the simulation of the self-sustained smoldering fire of upholstered furniture was raised by the British experts in connection with the British Furniture Regulation. Through skillful selection of the various test conditions a self-sustained smoldering process of flexible PUR foam can be simulated. The graphical representations of Fig. 129 show the temporal course of the smoldering temperature and the Figs. 130a and b provide an impression of the carbonisation process of flexible polyurethane foam in the DIN tube furnace. The stream of effluents produced in the DIN equipment is used as a rule, for a head-nose exposure mode (see Fig. 130c) [88, 219, 471, 472].

Test methods, similar to the cone-calorimeter test, like NIBS protocol or the RTB protocol are limited due to their standardized test conditions to the flaming combustion process. Although the decomposition conditions lead to a well-reproducible stream of effluents, the mandatory ignition with simultaneous radiation stress result, both for carbonizing as well as melting materials, in considerable deviations of the effluent release. The time to incapacitation t_i is one often used criterion for evaluating the toxic potency of fire effluents. The percentage of killed animals is another indicator of the toxic potential of effluents.

Hilado reported in [476] on investigations with thermal decomposition products of flexible and rigid polyurethane foams produced according to the USF protocol. The graphical representations of Figs. 131a and b provide information about the toxicological data t_i and t_d. The data records of Fig. 131c illustrate to what extent the behav-

Figure 130a Charring process of flexible PUR foam sample

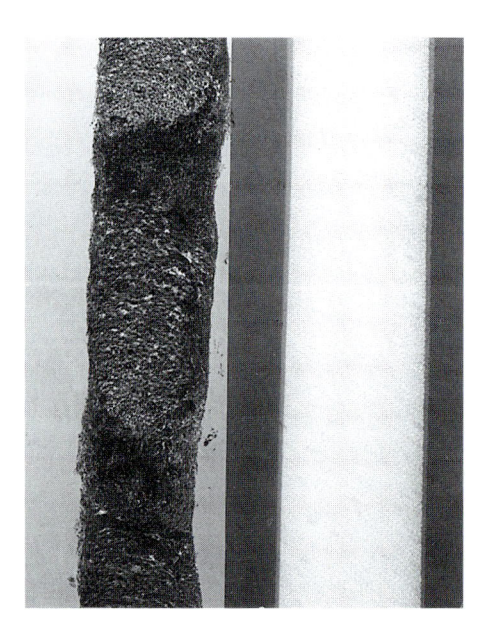

Figure 130b Charred flexible PUR foam sample 1007

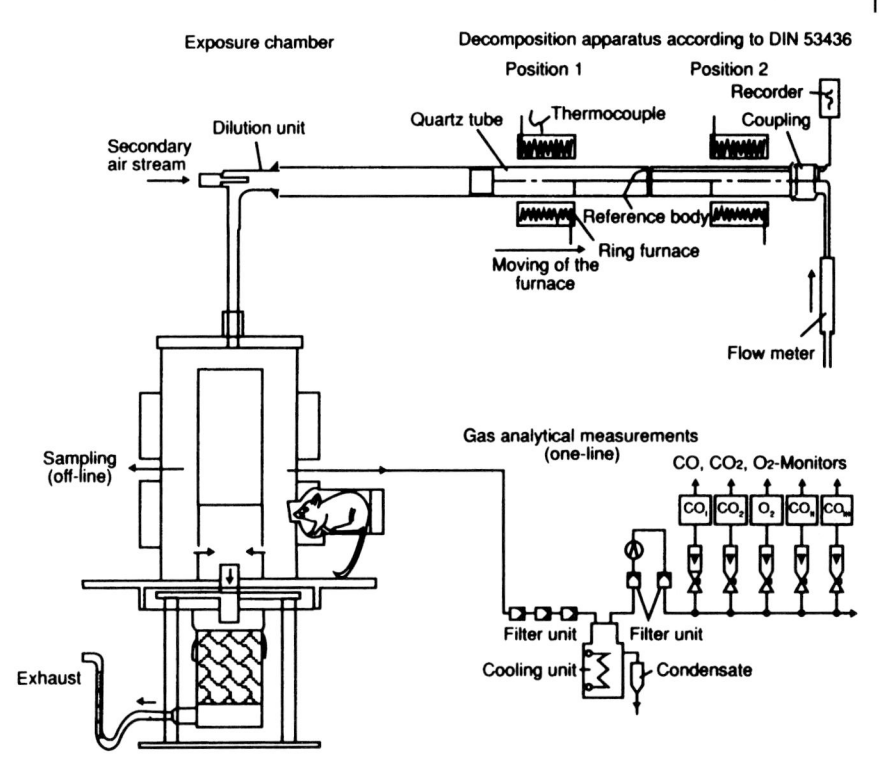

Figure 130c Decompostion model DIN 53436 furnace

ior of the exposed test animals will be influenced by the exposure with the thermal decomposition products of flexible and rigid polyurethane foams. One goal of these studies was to find out if the aging process of these flame-retarded grades will influence the reaction behavior. The data set show that the decomposition products given off by the aged flexible PUR foam under the condition of the USF test cause somewhat extended reaction times. The graphical illustration of Fig. 131c illustrates the influence of the exposure time on the response of the test animals exposed to the thermal decomposition products of flexible and rigid PUR foam. The decomposition

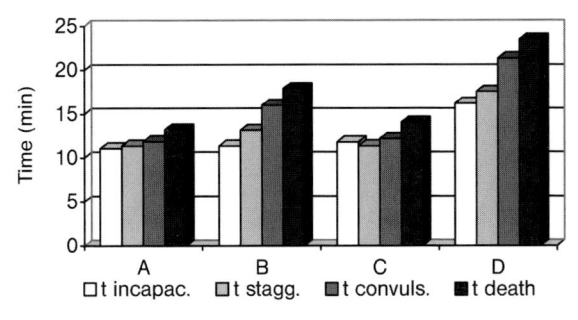

Figure 131a Influence of the effluents on the reaction behavior of the test animals

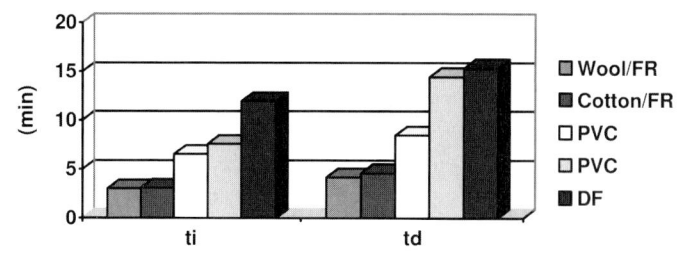

5 Material-specific Fire-Performance Characteristics of PUR

Figure 131b Toxicological characteristics t_i and t_d

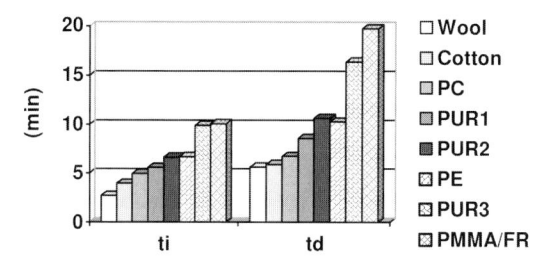

Figure 131c Influence of the effluents on the reaction behavior of the test animals

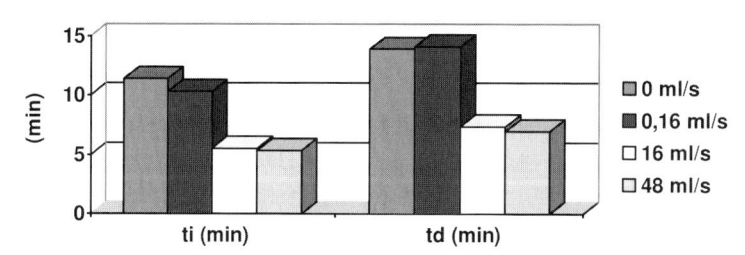

Figure 131d Influence of the ventilation on the characteristics of the toxic potential

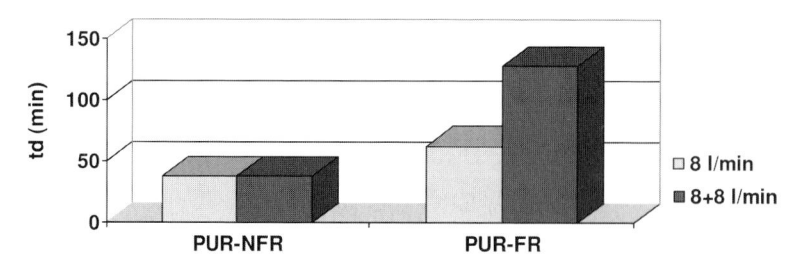

Figure 131e Influence of the ventilation on the characteristics of the toxic potential

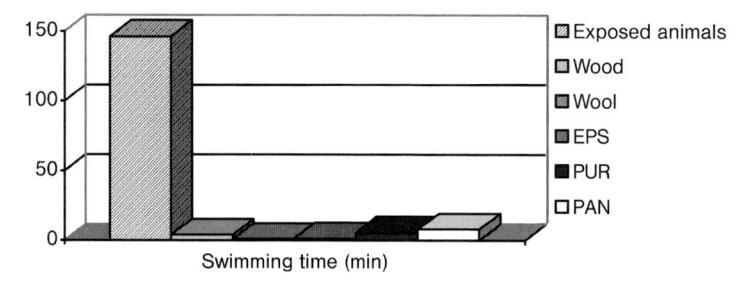

Figure 131f Influence of the effluents to the swimming behavior of the test animals

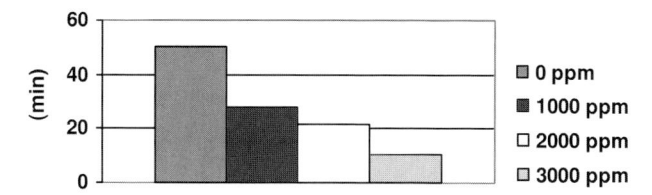

Figure 131g Influence of the CO intoxication on the swimming behavior

products released by the flexible polyurethane foam initiated, similar to the aging process, a somewhat earlier reaction.

The decomposition process is sped up by intensified ventilation, as illustrated by Fig. 131d and e. The reaction times of the test animals became reduced correspondingly as proved by the investigations with decomposition products of the flexible PUR foam.

Specific investigations were carried out among other things without additional air supply as well as with a limited supply from 1 l/min. Kaplan et al. [477] reports, e.g. on the influence of the ventilation. At the additional ventilation of 8 + 8 l/min he achieved reaction times 122-38 (FR quality) and 38 (NFR quality) in comparison with 63 (FR quality) and 38 (NFR quality) at 8 l/min. This criterion is determined in a line of procedures, e.g. the USF method and is used for the comparative evaluation. The records of Figs. 131d and e deliver information about the influence of the ventilation on the concentration of thermal decomposition products and therefore t_d. These characteristics of the material-related potency show that the modified ventilation conditions basically reflect dilution effects.

The intensified ventilation leads obviously to a considerably increased burning rate. In the case of the quality with the fire-retardant additive higher fire-gas concentrations cause considerably reduced reaction times. The characteristics of Fig. 131e demonstrate, using the example of PUR foams, the importance of this criterion for the U-Pitt procedure. The data records show that the formula as well as the aging process lead to a somewhat increased time. Effenberger [478] reports on the swimming behavior of the exposed test animals. The exposure with the decomposition products of various materials led (see Fig. 131f) to a significant reduction of the ability to swim. The staying power to swim was very low in comparison to the nonexposed animals. Kimmerle [88] confirmed the usability of the swim test conducted by

Effenberger. Figure 131g provides test data of bioassay tests conducted with the fire gases of spruce wood.

Variation of the test parameters can significantly change the material-related characteristics. The key parameters for assessing the toxic potency of thermal decomposition products are the test temperature and the exposure time. The time to a significant reduction of the test animal activity, e.g. t_i incapacitation occurs, up to the time to death t_d, are quite common criteria of different test protocols. The percentage of killed animals is another indicator for the toxic potential of effluents. Kimmerle [88] reported in Salt Lake City on research work with various polyurethane grades in the DIN tube system. Kimmerle summarized beneath the toxic potential of PUR effluents all the basic problems of bioassay testing with various components of fire gases. Results are given in Table 94. The critical test temperature range, Tc, in which mortality of the exposed animals can be expected, was one of the most important evaluation parameters determined for the decomposition products produced with the tube system according to DIN 53436.

Figure 132a provides information about the influence of the test temperature on the fire-gas concentration of various materials. The decomposition products of flexible PUR foams produced, with increasing test temperatures, at a higher temperature level a decreasing death rate. This effect was shown by earlier tests of Kimmerle (see Figs. 132b and c) and confirmed by tests conducted by Sand and Hofmann [479, 480]. A contrary effect was observed for rigid PUR foams. In the case of the products A, B and C, a maximum of the fire-gas concentration seems to be obviously already passed through within the testing temperature range. With temperatures above approx. 500 °C, the concentration of the decomposition products is already reduced significantly. In the case of product D the maximum seems to lie above 500 °C to

Table 94a Criterion Tc of the toxic potency

Material	Lowest temperature (°C) indicating mortality
Rigid polyurethane foam	600 and > 600
Rigid polyisocyanurate foam	500, 600 and > 600
Semi-rigid polyurethane foam	500, 600 and > 600
ABS	400
Polycarbonate	600
Spruce – Wood	350
Cork	300
Flexible polyurethane (polyether) foam	300 and 400
Flexible polyurethane (polyester) foam	450
High Resilience foam	500
Coconut fibre	250
Rubberized hair	300
Latex foam	> 200
Flocks	300
Wool	300
Cotton	250

Table 94b Influence of the test temperature to the toxic potential

Material	300 °C mortality (%)	300 °C CO-Hb (%)	400 °C Mortality (%)	400 °C CO-Hb (%)	500 °C Mortality (%)	500 °C CO-Hb (%)
Cork	0	17	28	69	100	86
Rubber	0	<15	100	78	100	75
Lamb wool	17	<15	100	19	100	27
Pine	25	32	100	86	100	85
Leather	100	67	100	69	100	63
PVC	83	25	92	45	100	70
PA6	75	–	0	<10	100	30
EPS	0	<15	0	<15	100	72
PE	0	<15	100	79	100	85

600 °C. While in the case of the products A and B test samples of the same weight (5 g) were examined, the tests of products C and D were conducted with samples of the same volume (300 mm × 15 mm × 10 mm) at an air throughput of 100 + 100 l/h. The products were examined in detail.

Among other things, Kimmerle et al. [93] report on the Tc and $D(LC_{50})$ values, the critical temperature and the critical dilution of the stream of effluents, in which mortality of the exposed animals can be expected. Tables 95a, b and c provide information concerning the Tc and $D(LC_{50})$ records of different PUR materials, their chemical nature and reference products. The data records of Fig. 133 show very vividly using the example of six different PUR foams that a variation of the CO-Hb content in the blood of the test animals can be caused by the change of the formula due to a modified stream of effluents. Fleischmann [481] investigated in his thesis among other things the influence of the recipe of PUR foams at the smoke release in the case of thermal loading in the DIN tube furnace. The determined CO-Hb value illustrates in accordance with Fig. 133, that next to CO further fire-gas components become effective very rapidly. The thermal loading of the test pieces with the same volume (600 mm × 20 mm × 15 mm) was stopped at 400 °C. The nominal fire-gas concentration varied due to the different densities (27 to 80 kg/m^3) between 16.2 to 48 g/m^3. The data set of Fig. 132c show that the decomposition products of paintings induce an increased mortality of the exposed animals with increased numbers of paint layers. The TUF (Time of Useful Function) data records are obviously reduced at a higher, 4-fold painting rate, the reason for this must be a reduced decomposition rate.

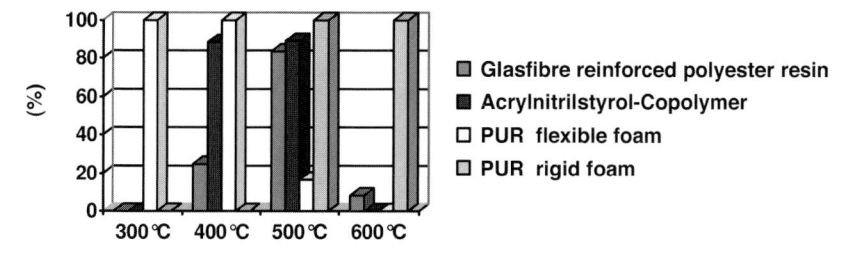

Figure 132a Influence of the test temperature on the mortality data

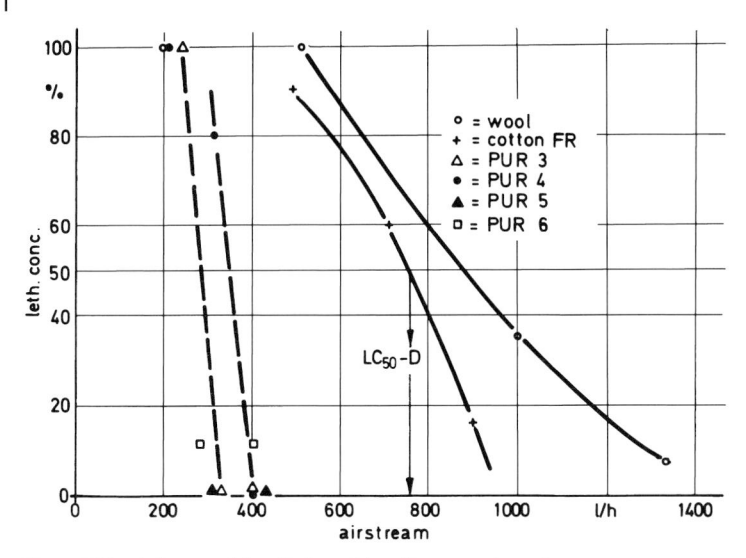

Figure 132b Influence of the dilution of the effluents to the toxic potency

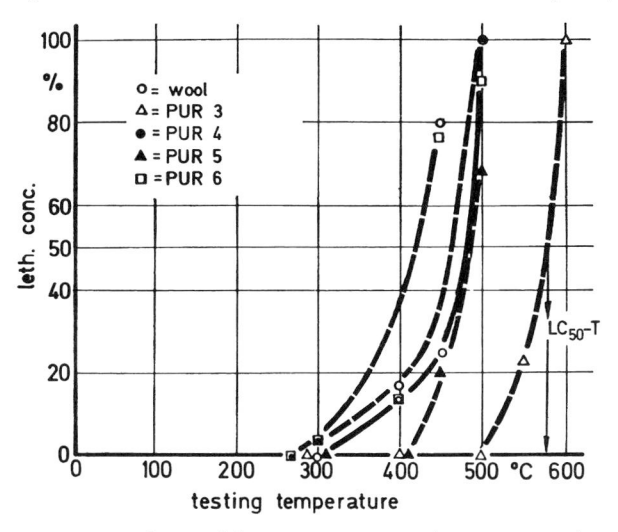

Figure 132c Influence of the test temperature on the toxic-potency data

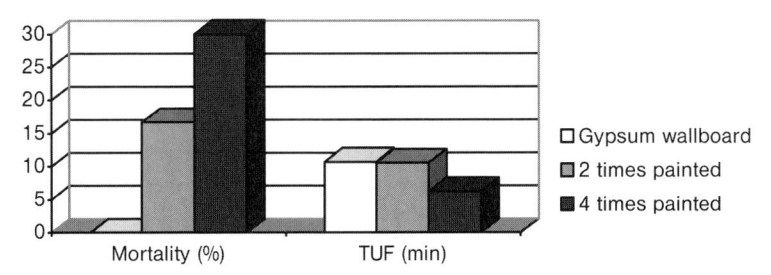

Figure 132d Toxicological results of paint layers

Table 95a Bioassay test results with effluents of polyurethanes and wood

Test Conditions				Toxicological Results[1]		
T (°C)	L1 (l/h)	L2 (l/h)	t2 (min)	Fir wood	PUR-FR1	PUR-FR2
250	100	100	30	–	–	–
300	100	100	30	0/14	–	–
350	100	100	30	19/20	–	–
400	100	100	30	12/14	0/14	–
450	100	100	30	–	20/20	–
500	100	100	30	10/20	20/20	–
300	100	100	30	–	–	0/20
400	100	200	30	–	5/20	–
		300	30	–	20/20	17/20
		400	30	–	1/20	7/20
		900	30	–	0/20	0/20
500		100	30	16/20	–	–
		200	30	7/15	–	–
		300	30	0/15	–	–
		400	30	0/15	–	–
		900	30	–	19/20	16/20
		1200	30	–	9/20	0/20
			30	–	0/20	–
600	100	100	30	–	20/20	20/20
		900	30	–	12/20	1/20
		1200	30	–	1/20	0/20

Sample weight = 0.12 g/cm

[1] first number = number of dead animals; second number = number of animals exposed

5 Material-specific Fire-Performance Characteristics of PUR

Table 95b Toxicological results with Combustion Products of Isocyanate based foams and reference products

Test conditions			Tested materials									
Tv (°C)	L1 l/h	L2 l/h	Spruce wood Co-Hb %	tox. result	Cork Co-Hb %	tox. result	PIR 1 Co-Hb %	tox. result	PIR 2 Co-Hb %	tox. result	PCD Co-Hb %	tox. result
300	199	100	5,2	0/20/20	5,0	0/20/20	0	0/20/20		0/20/20	0	0/0/20
350	100	100	49,98	0/20/20	36,4	0/20/20		–		–		
400	100	100	59,2	14/20/20	47,7	7/20/20	12,1	0/20/20	27,2	0/20/20	40,3	0/20/20
450	100	100						–		–		20/20/20
500	100	100					*)	20/20/20	*)	20/20/20	*)	
600	100	100	69,9	19/20/20	63,2	14/20/20	*)	20/20/20	*)	20/20/20	*)	20/20/20
600	100	200	69,9	19/20/20	63,2	14/20/20	*)	20/20/20	*)	20/20/20		20/20/20
600	100	300	54,3	3/20/20	56,5	4/20/20						20/20/20
600	100	400	51,8	0720/20	43,2	0720/20	*)	20/20/20	*)		45,4	7/20/20
600	120	600					48,5	4/20/20	48,9	10/20/20	30,1	0/20/20
600	100	600					44,9	6/20/20	45,0	4/20/20		
600	100	800						–		–		
600	100	900										
600	100	1000					36,0	0/20/20	46,8	1/290/20		
600	100	1200						–		–		

t = 30 min – sample weight = 0,12 g/cm
1) first number = number of dead animals
   second number = number of animals with symptoms
   third number = number of exposed animals
*) death occurred during exposure

5.3 Polyurethanes

Table 95c Toxicological test results

Test conditions			Tested materials										
Tv (%)	L1 (l/h)	L2 (l/h)	Wool CO-Hb %	result[1]	Cotton FR CO-Hb %	result[1]	PUR 3 CO-Hb %	result[1]	PUR 4 CO-Hb %	result[1]	PUR 5 CO-Hb %	result[1]	PUR 6 CO-Hb %	result[1]
250	100	100		1)	50,2	0/20/20	0		45,2	0/20/20			1,5	0/20/20
300	100	100	0	0/20/20	56,1	4/20/20		0/20/20	46,2	1/20/20	25,3	0/20/20	49,0	1/20/20
350	100	100			*)	20/20/20								
400	100	100	4,7	2/20/20	*)	20/20/20	0	0/20/20	60,6	3/20/20	43,3	0/20/20	51,8	2/20/20
450	100	100	10,2	5/20/20					82,8	16/20/20	58,4	4/20/20	58,2	15/20/20
500	100	100		20/20/20			28,1	0/20/20	63,8	20/20/20	58,6	14/20/20	69,2	18/20/20
550	100	100					60,2	5/20/20						
600	100	100		20/20/20	*)	20/20/20	67,6	20/20/20		20/20/20				
600	100	100		20/20/20	*)	20/20/20	67,6	20/20/20						
600	100	200					47,6	0/20/20	63,8	16/20/20	44,4	0/20/20	46,7	2/20/20
600	120	300					46,0	0/20/20	50,7		40,1	0/20/20	42,5	2/20/20
600	100	400		20/20/20	*)	18/20/20			52,8	0/20/20				
600	100	600			54,7	12/20/20								
600	100	800			50,2	3/20/20								
600	100	900	17,6	7/20/29										
600	100	1200	21,7	1/20/20										

t = 30 min – sample weight = 0,12 g/cm
1) first number = number of dead animals
second number = number of animals with symptoms
third number = number of exposed animals
*) death occurred during exposure

Table 95d Toxicological test results

Tested materials	T_c (°C)	$T(LC_{50})$ (°C)	DC_{600} (l/h)	$D_{600}(LC_{50})$ (l/h)	$D_{600}(LC_{50})$ (g/m³)
Spruce	350–400	375	300–400	250	29
Cork	350–400	475	300–400	250	29
Wool	300–400	475	>1300	900	8
Cotton	250–275	275	300–400	300	24
Cotton FR	250–300	325	>900	750	9,6
PUR1 600	400–450	425	<1300	1100	6,6
PUR1 500			1200–1300	950	7,5
PUR1 400			500–1000	250	29
PIR 1	400–500	425	>900	700	10
PIR 2	400–500	425	<1100	700	10
PCD	400–450	425	400–450	350	20,6
Wool	300–400	475	<1300	900	8
PUR 3	500–550	575	200–300	250	29
PUR 4	250–300	425	300–400	350	20,6
PUR 5	400–450	490	200–300	300	24
PUR 6	250–300	425	>400	300	24

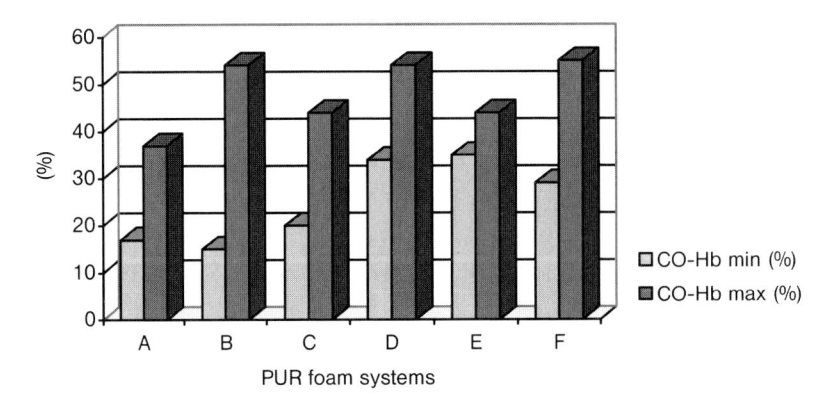

Figure 133 Influence of the recipe to the CO-Hb values

The data set of Tables 95 to 97 provide information concerning the test conditions and the toxicological results of extensive studies with the effluents of various polyurethane materials listed in Tables 96 and 97b. The toxic potency criterion LC_{50} was the goal of various investigations.

A comparison of the toxicological results demonstrate that the extent of the variation of the test conditions may alter the toxic potency of the produced effluents. The comparison test conducted with wood and cork underline this statement. The $T(LC_{50})$ and $D_{600}(LC_{50})$ data demonstrate that the relative toxicity of the decomposition products of rigid polyisocyanate based foams seems to be less than those of the reference products under the conditions of this test, whereas $D_{600}(LC_{50})$ values showed that with samples of equal weight the foam products produce effluents of a

Table 96 Tested materials

	Rigid foam – closed cell
PUR 1	Polyurethane foam based on polyfunctional polyols and polymeric Diphenylmethane-diisocyanate (MDI) with a reactive flame retardant
PUR 2	Polyurethane foam based on polymeric MDI
PIR 1	Polyisocyanate foam based on polymeric MDI with non reactive flame retardants
PIR 2	Polyisocyanate foam based on polymeric MDI with non reactive flame retardants
	Rigid foams – open cell
PCD	Polycarbodiimide foam based on polymeric MDI
	Semirigid and flexible foams
PUR 3	Semirigid polyurethane foam based on bi- and trifunctional and polymeric MDI
PUR 4	Flexible polyurethane foam based on bi- and trifunctional polyols and Tolylendiiso-cyanate (TDI)
PUR 5	Flexible polyurethane foam based on bi- and trifunctional polyols and TDI
PUR 6	Flexible polyurethane foam based on bi- and trifunctional polyols and TDI

1) The behavior or symptoms of exposed animals were noted
2) Lethality as a function of test parameters was determined
3) The carboxyhemoglobin (CO-Hb) content in the blood of exposed animals was measured

greater toxicity. The toxic potency criterion LC_{50} was the goal of various investigations. Prager et al. [482] report on bioassay tests conducted with the thermal decomposition products of polyurethanes. The effluent formation was produced by the DIN tube system 53436 using samples of the same volume (300 mm × 15 mm × 2 mm) in an air stream of $L1 = 100$ l/h. Special test series were conducted at the same reference body temperature of 600 °C and an air stream of $L1 = 300$ l/h. These tests were conducted under comparable conditions. The listings of Tables 97a and b illustrate that the selected test conditions will influence the test results, e.g. the LC_{50} data, the CO-Hb values and the analytically determined concentration data of the most important effluent components. These records demonstrate the importance of the time factor for the relevance of the CO-Hb and the LC_{50} values.

Tables 98a and b provide information about the influence of the test temperature and the mode of decomposition to the LC_{50} value. The data records of Table 98a, determined with the DIN tube system, give an overview of the toxic potency of the effluents of the reference product wood. Comparing the LC_{50} records for an exposure period of 30 min with the relevant data of reference products, substantial divergencies have to be stated, especially for volume-based data. These records, determined with quite differing decomposition models (see Table 99), demonstrate that polyurethanes produce effluents that do not characterize an unusually marked acute toxic potential. In the majority of these investigations, the exposure duration was 30 min, as agreed by the experts of SC3. The fire gases produced at a reference body temperature of approx. 500 °C reaches the most unfavorable LC_{50} data.

The graphical representations of Fig. 134a illustrate that PUR/PCD (polycarbodiimid foam) decomposition products, produced in the DIN tube system DIN 53436, contain a reduced acute toxicity in comparison to the fire gases of wood. The thermal decomposition products of fir wood caused a significantly shorter exposure duration

Table 97a PUR – relative toxicity – measured and calculated values a) b)

Material	T_v (°C)	L_1 (l/h)	L_2 (l/h)	ΔG (%)	CO_2/CO	CO (ppm)	HCN (ppm)	NO_x (ppm)	CO-Hb (%)	LC_{50} (g/m³)
PUR – Coating systems										
System A	600	100	≤300	≤46	8	≤3000	117	58	68	115
System A	600	300	100	46	4	2500	42	23	54	29
System B	600	100	≤600	≤58	7	≤4000	≤261	8	72	57
System B	600	300	≤400	≤57	65	≤1000	≤19	117	33	93 c)
System C	600	100	≤600	≤96	27	≤3000	≤244	≤30	73	29
PUR – foams										
System D	600	100	≤400	100	10	≤3000	≤137	≤37	64	28
System D	600	300	100	99	17	930	≤22	≤6	34	16 c)
System Dd)	600	100	≤400	100	9	≤2300	≤68	≤18	74	19
System E	600	100	≤600	≤90	5	≤7000	≤360	28	62	12
System E	600	300	700	90	2	≤7050	300	2	60	9
PUR – solids										
System F	600	100	100	36	44	1100	67	80	49	158
System F	600	300	≤400	≤38	27	≤5500	≤167	≤40	87	53
System G	600	100	≤800	≤31	2,5	≤3500	≤256	≤3	59	43
System G	600	300	≤600	32	2	≤3200	≤206	≤4	71	41

a) Data recorded are not linked to the calculated LC_{50} values
b) Most unfavorable results have been used for calculation and/or listing
c) Mortality (%)
d) Exposure time: 1 h

5.3 Polyurethanes

Table 97b Materials tested

Polyurethane – based coating systems

System A	MDI based
System B	MDI based
System C	Baytec® reactive, based on polyols and MDI polyurethane – based foams
System D	Flexible foam, based on polyether polyols and TDI
System E	Rigid foam, based on polyether and PMDI polyisocyanate-based solid systems
System F	Cast system, based on epoxy and PMDI, fillers, silica foam
System G	Based on epoxy and PMDI, fillers, silica foam

Table 98a Toxic potency data of the decomposition products of wood

Material	T_V (°C)	LC_{50} (g/m³)	References
Spruce	600	29	Kimmerle
	600	30	Klimisch
	500	30	Klimisch
	400	20	Klimisch
	500	36	Effenberger
	495	30	Müller
Pine	500	37,5	Herpol
	475	21,2	Herpol
	400	19,6	Herpol
Fir	600	25	Sand
	500	15	Sand
	400	20	Sand
	500	22,7	Thyssen
	400	27,9	Jouany
	850	29,2	Jouany
Mahagony	400	30,4	Jouany
	850	34,3	Jouany

Average: 26,9 ± 6,35

Table 98b Toxic potency data

Material		T_v (°C)	LC_{50} (g/m³)	LC_{50} (cm³/m³)
PUR	Rigid foam (49 kg/m³)	600	≥ 6,6	≥ 165 a)
	Flexible foam (30 (kg/m³)	600	≤ 19	≤ 666 a)
Wood	Fir (500 kg/m³)	600	≤ 28	≤ 54 a)
	Pine (500 kg/m³)	600	≤ 27	≤ 54 a)
	Beech (500 kg/m³)	600	≤ 24	≤ 48 a)
		500	≤ 18	≤ 36 a)
Wool		600	≤ 7	

a) lower densities lead to larger LC50 values

Table 98c Influence of the temperature onto the LC$_{50}$ values – DIN 53436 (g/m^3)

Material	400 °C	400–500 °C	500–600 °C	600 °C
Wood	20	10–20	28–30	20–30
Cotton	–	18	20	–
Cotton FR	–	18	9,6	–
Wool	–	–	6	–
PVC	–	–	17	–
PUR flex.	28	18	16	13
PUR rig.	50	20	6–7	6–7
PES fiber			<18	
PA	–	–	18	–
UF	–	–	6	–

to result in the death of 50% of the exposed animals. Figure 134b illustrates the range of LC_{50} data of four different PUR systems. In the sequence coating, enamel, flexible and rigid PUR foam, more critical LC_{50} data were determined. At the peak values, i.e. the data set of decreased acute toxic potency, significant deviations are to be seen. Earlier test series, conducted under comparable test conditions produced zero mortality, if the CO_2/CO ratio of the stream of effluents reached values above 100. Effenberger [450] confirmed these results. The T(LC_{50}) data, the concentration for a 50% mortality, approx. 7.5 g/m^3 for rigid polyurethane foam was also confirmed by Yammamoto [475]. Jouany [467] report on the influence of the direction of the primary air stream on the toxic components of the thermal decomposition products of PUR foams and the reference materials wood and PVC. The results are illustrated in Fig. 134c. The effluents of the rigid PUR foam reached in the cocurrent procedure an increased toxic potency level. The material of reference reached a significantly lower level, but by assessing the records it must be taken into account that the PUR effluents were produced at a somewhat higher temperature, 500 °C instead of 400 °C.

These records demonstrate the strong influence of the temperature level. The graphical representations of Fig. 135a and Table 98 demonstrate that the volume-based, material-related data of the flexible polyurethane grades reach much more favorable toxic potency values than the reference products. The influence of the testing temperatures was determined also with different streams of effluents produced by means of the Gost Apparatus [230] shown by the graphical representations of Fig. 117. The mortality records obviously pass a marked maximum depending on the testing temperature. The listing of Table 99 shows the records gained with various wood products. In the majority of these investigations the exposure duration was also 30 min, as agreed by the experts of SC3 the decomposition conditions are changed by fillers or fire-retardant additives, significant deviations of the potential values result as illustrated by Fig. 135c. The example with cotton FR demonstrates that the use of fire-retardant additives can perhaps substantially increase the material-related toxic potency (see Fig. 135d).

Hartzell [14] report on the characteristics of the toxic potential of various PUR grades, EPS and a PUR-Asbestos compound. The effluents were produced at SWRI within the PRC program under smoldering and flaming conditions of the Potts Pot

Table 99a Comparison of LC$_{50}$ data of different methods.

Material	DIN 53436	Potts Pot	NIBS	U-Pitt
Wood	18	–	60–100	64
PUR flex.	16	26	–	11,2
PUR rig.	7	11	.-	10,5
Wool	7	7	37	5
Cotton	27	27	–	–
Cotton FR	8,5	–	–	–
PA	8	100 / 49	–	–
PVC	17	25 / 35	33.4	9,3
PA6.6	–	100 / 49	37	8,8

Table 99b Influence of the intensity onto the toxic potential

Intensity	L(ct)$_{50}$ (g/m³min)	Ale (g/m³min)
25 KW/m²	576	489
50 KW/m²	752	804

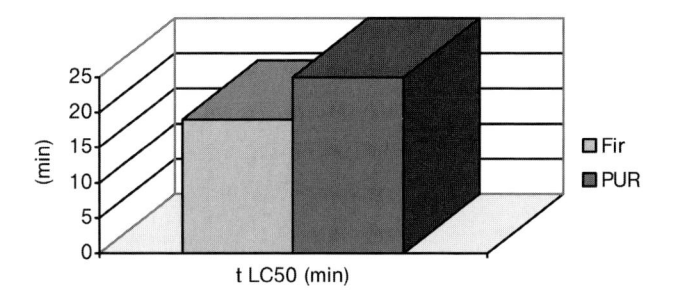

Figure 134a tLC50 values of PUR effluents

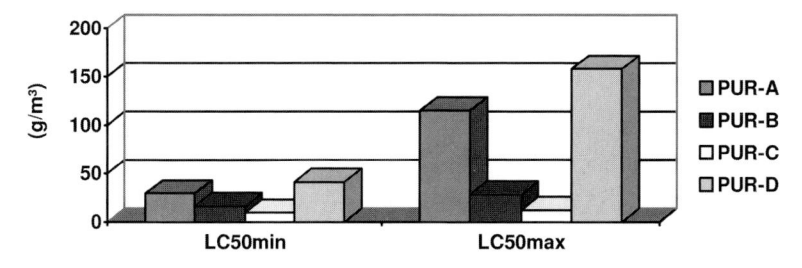

Figure 134b Influence of the recipe on the toxic potency – Range of deviation

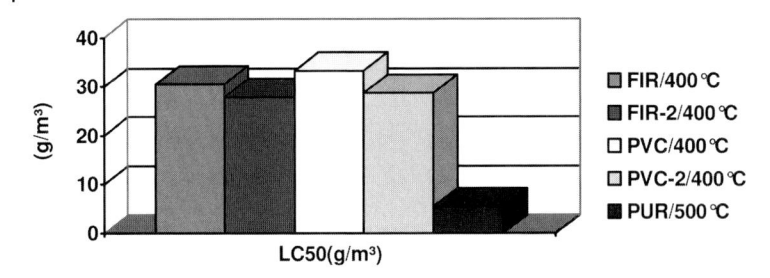

5 Material-specific Fire-Performance Characteristics of PUR

Figure 134c Influence of the co current conditions in the DIN Tube System on the toxic potency data

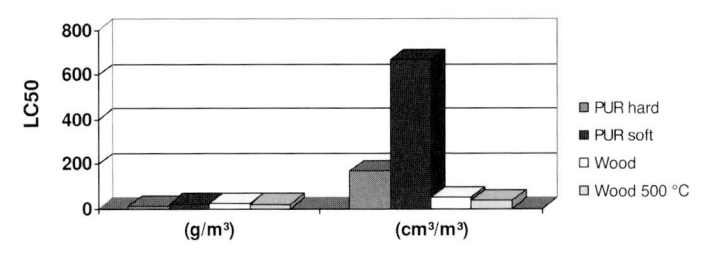

Figure 135a Toxic-potency data as a function of the dimension

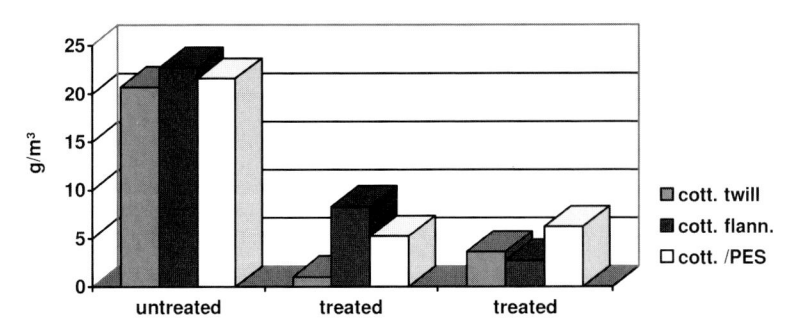

Figure 135b Influence of the fire-retardant additive on the toxic potential $LC50$ (g/m^3)

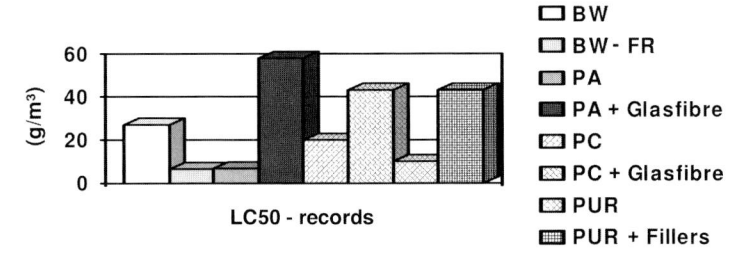

Figure 135c Influence of fillers and fire-retardant additives on the toxic potential $LC50$ (g/m^3)

Procedure. The test results illustrated by Figs. 136a and b and Table 100 show that the variation of the test conditions seems to play no important role for the EPS and the flexible PUR grades. With the rigid PUR foams, flaming conditions obviously produced effluents of a significantly higher acute toxicity. Contrary to this, the PUR-Asbestos compound produced somewhat more unfavorable results under smoldering conditions. Table 100 provides detailed information concerning the measured deviations. In the Fire Research Center (FRC) of Utah at the University of Salt Lake City, an investigation has been conducted with the modified NBS Chamber [483]. The graphical representations of Fig. 136c show the results of decomposition products produced under smoldering and flaming conditions with various PRC – PUR grades. Most of the tested PUR grades produced the same level of toxic potency. The material-related data records of the toxic potency $L(ct)_{50}$ (g min/m^3) seems to be a good basis for evaluating the effect of different exposure times, generally a key parameter of the U-Pitt method. The graphical representations of Fig. 136d, clarify that the characteristics $L(ct)_{50}$ of the effluents of PIR grades determined by Alexeeff [190] under different heat stress according to the NIBS specification reach an equivalent record level similar to ALE values. The increase of the irradiance level from 25 to 50 kW/m^2 caused a reduction of the toxic potency.

Alerie [484] and Alarie et al. [189] demonstrated the possible significant deviations by means of mass- and volume-related samples of flexible polyurethane foams. The danger of the possible faulty interpretation was illustrated by means of material-related records. The graphical representations of Fig. 137a and the listed records of Tables 101a and b show, in a very informative way that the differences determined with samples of the same weight did not occur in the case of the same volume. In the case of the evaluation of the material data attention is to be paid to ensure that there are comparable exposure times. Assuming that in the case of real fires primarily comparable volumes and/or areas are involved in the fire occurrence, intensified observation must be given to these material-related characteristics.

Table 100 Toxic potency data of the effluents of Polyurethanes produced under smoldering and flaming conditions

Material	NF	F
GM 25 HR	36,9 ±6,0	> 37,5
GM 27 HR / FR	30,5 ±7,4	33,1
GM 35 PUR rig. Frigen	37,7	12,1 ±1,5
GM 37 PUR rig. CO$_2$	36,7	10,9 ±1,5
GM 39 PUR/Asbestos	10,9 ±1,6	16,6
GM 51 EPS	> 40	33,8 ±3,1

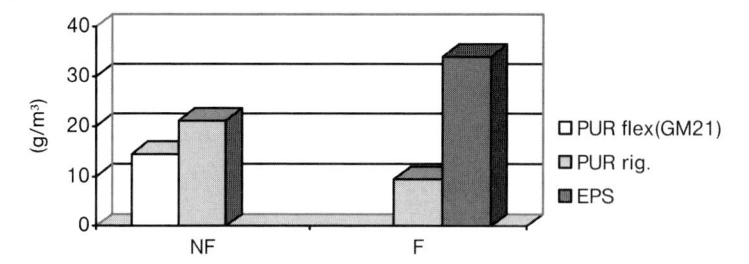

Figure 136a SWRI toxic-potency data – PRC Program

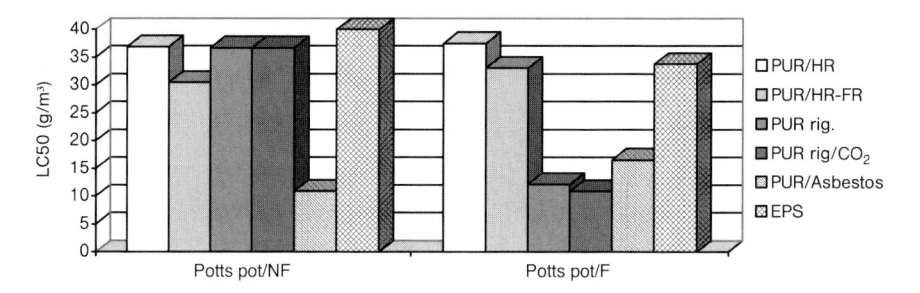

Figure 136b SWRI toxic-potency data – PRC-Program

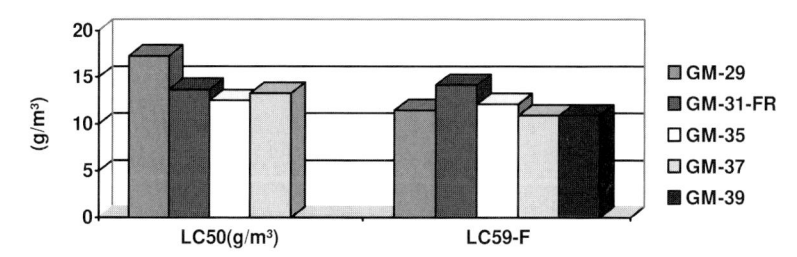

Figure 136c PRC toxic potency data

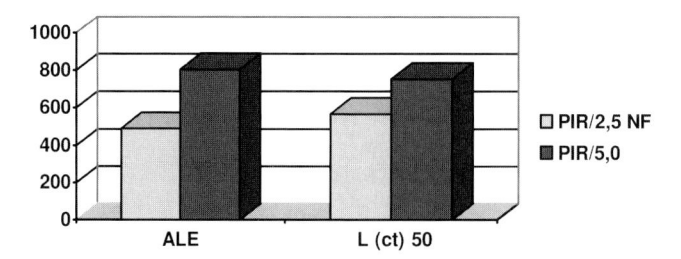

Figure 136d Toxic-potency data – NIBS test procedure

5.3 Polyurethanes

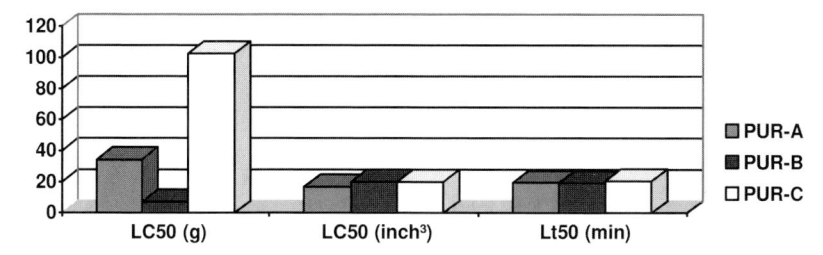

Figure 137a Influence of the test parameters on the various evaluation criteria

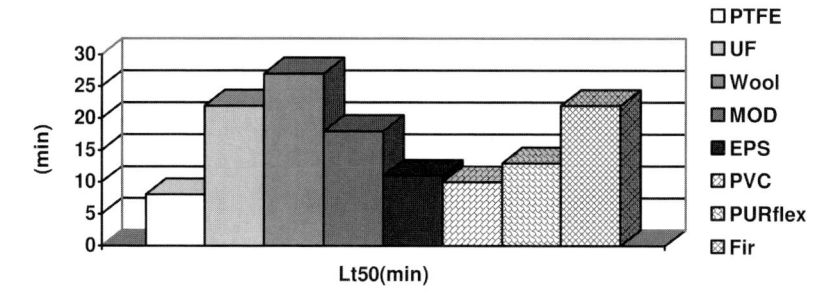

Figure 137b Data records of the toxic potency $Lt50$

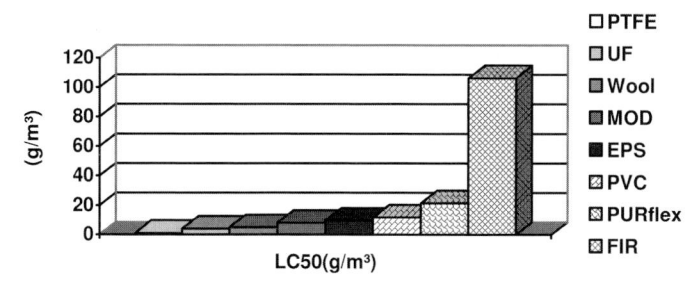

Figure 137c Data records of the toxic potency $LC50$

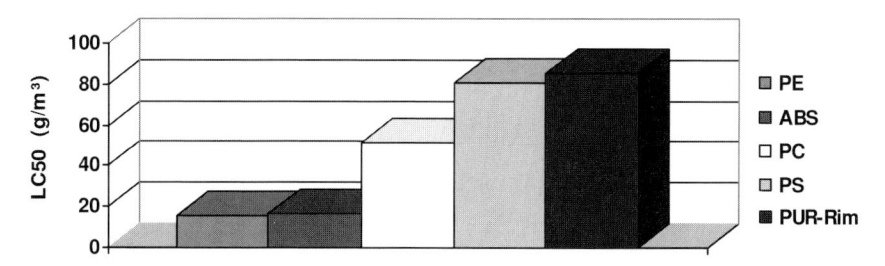

Figure 137d USF – Protocol – $LC50$ data (g/m^3)

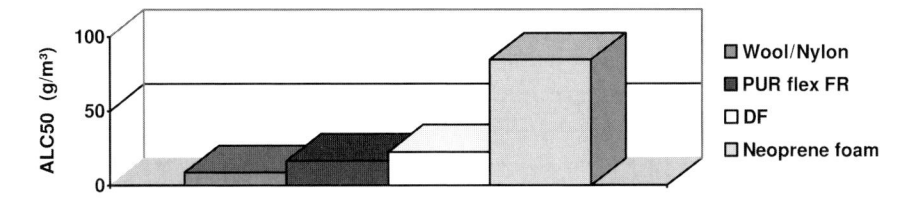

Figure 137e USF – Protocol – ALC50 data (g/m^3)

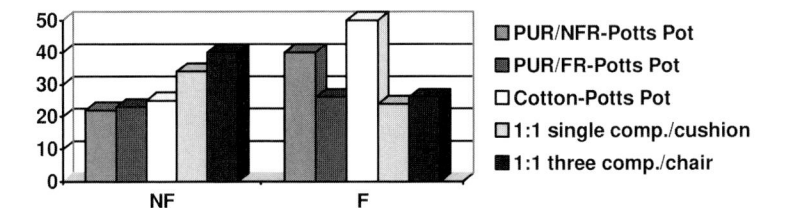

Figure 137f NIBS Test Protocol – LC50 data (g/m^3)

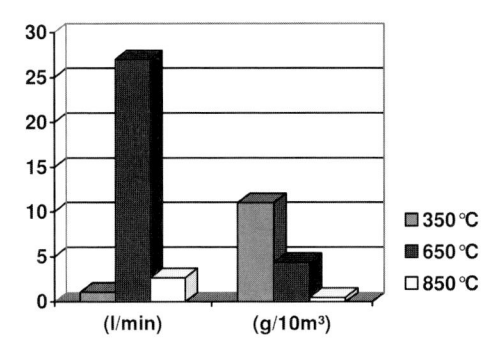

Figure 137g Purser evaluation system

The graphical representations of Figs. 137b and c show after [484] the strong deviations of the corresponding Lt_{50} values. The material-related LC_{50} values of the U-Pitt method are by no means based on a standardized exposure duration of 30 min. These characteristics must be assigned to obviously different exposure times. Table 101b illustrates the relatively good correlation of the LC_{50} and the ALH values.

Hilado [476, 485] report on the material-related toxic-potency data ALC_{50} and LC_{50}. The graphical representations of Figs. 137d and e demonstrate that the thermal decomposition products of polyurethanes, produced according to the USF method, cause obviously toxic-potency data comparable to those of the reference products wood and wool/nylon flooring material. The ALC_{50} data, produced by Hilado [476], are to be interpreted in such a way that with the exception of neoprene foam the fire-retardant grade of the flexible PUR foam produce thermal decomposition products under the thermal stress according to the USF protocol with comparable ALC_{50} data as wool/nylon carpet and wood (DF), (see Fig. 137d and e).

Table 101a Toxic potency data – U Pitt model

Material	Code	LC$_{50}$ (g)	Lt$_{50}$ (min)	L(ct)$_{50}$ (gmin/l)
PUR flex	GM21	12,9	13	49,6
PUR flex	GM21 FR	10,4	18	28,9
PUR-HR	GM25	8,3	19	21,8
PUR-HR/FR	GM27	14,4	15	34,7
PUR rig.	GM29	10,4	28	18,6
PUR/FR rig.	GM31	8,2	23	17,8
PUR rig.Frigen	GM35	7,5	17	22,1
PUR rig.CO$_2$	GM37	8,0	15	26,7
PIR	GM41	6,4	18	17,8
PIR / PUR	GM43	6,1	16	19,1
EPS	GM47	5,8	11	26,4
EPS / FR	GM49	10,0	9	55,5
PF	GM57	6,3	20	15,8
Wool	–	3,0	27	5,6
UF	–	2,5	22	5,7
DF	–	63,8	22	145,0

Table 101b Toxic potency data in comparison of LC50 data and ALH records

Material	LC50 (g)	ALH	Material	LC50 (g)	ALH
GM 21	13,0	7	GM 41	6.4	22
GM 23	10,4	9	GM 43	6,1	22
Gm 25	8,3	3	GM47	5,8	17
GM 27	14,4	20	GM 49	10,0	10
GM 29	10,4	11	GM 57	6,7	106
GM 31	8,3	16	DF	63,8	200
GM 35	7,5	33	UF	2,5	83
GM 37	8,0	41,3	PTFE	0,64	100
			Wool	3,0	–

Babrauskas et al. [486] compared the toxic potency of the thermal decomposition products of rigid PUR, PVC and DF produced with different combustion systems and under real scale conditions (see Fig. 138a and Tables 102a and b). Being aware that the CO concentration of the effluents will be significantly influenced by the combustion parameters, a correction of the CO concentration is proposed on behalf of the chemical nature of the tested materials. Babrauskas et al. [487] assesses the toxic potency of the thermal decomposition products of PUR, EPS and MW that have been produced in the ISO room-corner test and in the DIN tubular furnace at a reference body temperature of 500 °C and 700 °C with regard to the FED concept. The tested mineral-fiber product deliver, according to the calculation in the ISO room scenario, a 13- to 20-fold positive result. For the PUR quality he also calculated with the DIN tube system a 7- to 8-fold positive classification. The assessment of these records is acceptable only if the transformation to a realistic fire scenario is guaranteed.

Table 102a Large Scale Validation

Method	DF	PUR rigid	PVC
Potts Pot	41–51	10–13	18–22
NIST	100–200	20–30	20–30
Real scale	>70	30–40	35–45
Potts Pot CO – corrected	21–24	9–12	16–19
NIST – CO corrected	21–23	14–19	13–17

Table 102b Toxic potency data determined by FED

Method	PUR	EPS	MW
ISO room	>136	>205	10
DIN 500	62	2	8
DIN 700	58	4	8

Table 103 Criteria LC_{50} and LCt_{50} of the toxic potency of effluents

Material and test protocol	LC_{50} (g / m³)	LCt_{50} (g min / m³)
PUR flex. – DIN $_{600}$	11,0	330
PUR rig. – DIN $_{600}$	2,1 calculated	63 calculated
PUR rig. – 600	29,0	870 calculated
PUR flex. – 600	6,6	198
PUR flex. – 500	7,5	225
PUR flex. – 400	22,5	960
TPU-FR – DIN $_{600}$	26,0	995
FPU – LS 1090	0,03 calculated	

The newest findings with the SC3 Toxic Hazard in Fire, which allow a computational determination of the LC_{50}'s ratings, will increase the acceptance of this evaluation criterion with high probability. Tables 103 and 104 provide an overview of some results of Purser [488], and demonstrate the necessity to take into account the relevance of the L(ct)50 data. There seems to be a lack of assessment of the calculated test data of the defined fire scenarios. Purser reports on the influence of the test period on the temperature range of the stressed surface. Purser et al. [488] compare the toxic-potency data of thermal decomposition products gained by bioassay testing as well as by calculation based on the FED concept. Purser calculated extremely unfavorable LC_{50} data for polyurethanes based on large-scale investigations. The records in Table 104 illustrate that calculated data are far less favorable than the experimental data. None of the data listed in Table 104 confirm the calculation of Purser. The overwhelming majority of these worldwide, with all the differing test protocols determined bioassay test results demonstrate the necessity to back calculated records at least by one biological test of confirmation, if legal demands are necessary. These investigations demonstrate that calculated values have to be handled very carefully. The records of Table 106 provides information concerning LCt_{50} and LC_{50} data of

5.3 Polyurethanes

thermal decomposition products produced under smoldering, early flaming and postflashover conditions of rigid and flexible polyurethane foams.

Table 104 Bioassay Test Results with decomposition products of polyurethanes in comparison to calculated records of Purser

PUR Rigid Foam

Test Method	1	2	3	4	5–6	7–8	9-10	11-15	16-20	21-30	31-50	> 50
1 DIN Tube					1x	2x	5x	2x	2x	3x	2x	5x
2 Potts Pot						3x	3x	14x		2x	9x	
3 U-Pitt							2x	3x				
4 US / NIBS									1x	1x		
5 USF									1x	1x		
5 Others									1x			
6 Calculation Purser		2×	1×									
LC$_{50}$ (g/m^3)	1	2	3	4	5–6	7–8	9-10	11-15	16-20	21-30	31-50	> 50

PUR Flexible Foam

Test Method	1	2	3	4	5–6	7–8	9-10	11-15	16-20	21-30	31-50	> 50
1 DIN Tube									2×	4×	10×	
2 Potts Pot									1×		1×	3×
3 U-Pitt									1×	2×	2×	
4 US / NIBS									1×			
5 USF												
5 Others												
Purser						1×					1×	1×
LC$_{50}$ calculated (g/m^3)	1	2	3	4	5–6	7–8	9-10	11-15	16-20	21-30	31-50	> 50

References:
Kimmerle G.Prager F.H. J.of Combust. Tox. 7(1980)
Kimmerle G.Pauluhn J.Prager F.H. Kautschuk+Gummi Kunststoffe 45(1992)Nr,2 p.141-148
Prager F.H.Kimmerle G.Maertens T.Mann M.Pauluhn J. Fire and Materials 18(1994) p.107
Boudene C.Jouany J.M. Report to III 4.April 1978
Jouany J.M.Truhout R.Boudene C. Arch.Mol.Prod. 38(1977) Nr.9 p.751-772
Jouany J.M. PRC Conference Tuscon 19
Prager F.H. J.of Fire Sci. 6(1988) p.3-24
Kallonen R.von Wright A.Tikkonen L. Kaustia K. J.of Fire Sci. 3(May/June 1985) p.15
Herpol C.Fire and Materials 1(1976) p.25-35
Herpol C.Vandevelde P. Fire Safety J.4(1981/82) p.271-280
Alarie Y.Anderson R. Toxicol.and Appl.Pharmacol. 51(1979) p.341-362
Alarie Y.Anderson R.Toxicol.and Appl.Pharmacol. 57(1981) p.181-188
Hilado C.J.La Bossiere M. J.of Combust.Tox. 38Aug.1976)
Paabo M.Levin B.C. Fire and Materals 11(1987) p.1-29
Babrauskas V.Harris R.H.Braun E.Levin B.C. Paabo M. Gann r.G. J.of Fire Sci 9(1991)
Purser D.A Fardell D.J.Rowley J.Vollam S.Bridgemwan B Ness E.M. Interflam 1995

Table 105 Characteristics of the toxic potential according to the RTB protocol

Material	EC_{50}^* t min (g min/m³)	EC_{50}^* t max (g min/m³)
Wood (Lauan)	120	530
Nylon	50	130
Wool	27	28
PUR flex.	84	200
PAN	22	25

Table 106 Comparison of the corrosive potential of the thermal decomposition products

Material	(pH)	(µS / cm)	CNET (%)
PVC Standard	2,08	3350	14,2
PVC FR	2,51	1500	8,8
PTFE	3,49	172	10,4

The potential value for wood, plywood and rigid polyurethane foam determined with the Gost method show the chair dependence on the test temperature very vividly (see Fig. 268). The decomposition products produced at temperatures of 400, 500 and 850 °C, demonstrate that in the temperature range of 400 to 800 °C a peak value of the fire-gas toxicity is passed. Alexeeff et al. [489] refer to the good agreement of the crucial $L(ct)_{50}$ values with ALE (approximately lethal exposure) characteristics. The data records of Fig. 136d clarify, using the example of a PIR foam, the good agreement of the values and demonstrate that the burning behavior obviously leads at the higher intensity levels to a smaller toxic potential of the decomposition products.

Testing philosophy of the smoke corrosivity
The corrosive damage caused by gases in the case of a fire was one main topic of fire science at the beginning of the 1970s in Europe. The succession damage to be expected in the case of a fire became, in particular, the central point of great interest in connection with the participation of PVC [489]. Condensation of the fire gases, especially at cool metal faces, caused consequential fire damages. Application-oriented full-scale field trials illustrated that the problem does not represent a particularly tractable problem. Provision and damage limitation were the emphasis of extensive experimental studies. These relevant investigation programs showed that in particular in the field of the precision mechanics as well as with electrical and electronic contacts, attention is to be paid to possible corrosion damages. The entire spectrum of the corrosion problematic nature was fully treated on the occasion of the conference of Stockholm (Skyd 69) in 1969. The corrosion problematic nature of the fire gases also determined the subject in [490] and [491]. Theoretical aspects and cable fires formed the emphasis of these studies. Nowlen [490] verifies that the concentration of fire gases is the dominating parameter. In the fire case the effected room are smoke filled, as demonstrated by relevant full-scale field trials. The damage

scale is influenced from the solid particles in the smoke. At increasing pool diameter larger smoke particles are released. Babrauskas et al. discuss in [491] the entire problematic nature of the smoke corrosivity. He refers to the relevant American standardizing work and compares the corrosivity meter based on the cone calorimeter with the CNET-protocol proposed as IS standard. The corrosimeter according to ASTM-E activities uses the cone-shaped radiator of the cone calorimeter procedure (see Fig. 38) as the combustion device. The arising decomposition products are guided into the exposure chamber. The scale of damage is evaluated and registered. Sandman et al. report in [492] about experiments where stress-crack formation of steel samples has been selected as evaluation criteria. Humphries et al. describe in [493] the development of acid components due to the decomposition process at 500 °C in an air stream of 600 cm^3/min. By means of the pH measurement acid and alkali parts of the effluents are determined. Jin [178] also refers to the serious influence of irritants to the visibility. The emphasis of the possible consequential fire damages is, however, in the insurance technical field unambiguous. A measure for the damage limit is the immediate reduction of humidity as a necessary result of the extinguishing of a fire. The fat neutralisation of the deposit immediately after the fire occurrence proceeds as a most important, very efficient damage-reducing measure. Wet and dry procedures were developed for damage abatement. Corrosion protection and possible danger potentials in the acute fire case must be discussed from case to case again and decided upon [494].

6
Use and Interpretation of PUR-Test Results Determined under Enduse Conditions

The target of the predominant majority of the different national fire-reaction test methods is the proof of the resistance of materials against defined ignition sources. This evidence is required for end-application-oriented composites for exceptional cases only. The material examination field strives for the perfection of the test procedure as well as the improved reproducibility in the stage of standardization, as a rule, in particular. It is not examined and evaluated to what extent the modification of the test parameter endanger possibly the attaining of the primary targets. Here the question is raised whether supplementary statements concerning the relevance of the measured values to the targeted protection destination are required. This is only so that parties responsible for the orders are in the position to state to what extent the appropriate procedure is suitable to stipulate risk-relevant requirement profiles.

6.1
Relevance of Combustion Systems

The relevance of laboratory testing was the subject of numerous studies. Lawson [495] interprets fully the data records determined in the round-robin tests. Waterman [496] discusses the correlation of laboratory and large-scale tests for insulating materials. Hanlon clarifies in [497], using the example of the fire disaster of the Isle of Man, the deficiency of laboratory testing. Schuhmann [55] focuses on the relevance of laboratory test methods to the furniture field. By means of cushion experiments it is illustrated that the scenarios smoldering fire and flaming combustion of upholstery furniture are to be examined and evaluated separately. Among others McGuire et al. [497] also clarify the limited informative value of laboratory test results by means of relevant American test methods. McGuire explains among other things in [497] the relevance of the Radiant Panel Test according to ASTM-E 162. Extensive investigations showed, among others things, that Al-foil coverings, in connection with rigid PUR boards, lead to misleading, favorable ratings, which are not obtainable in the full-scale field trial. Results that correspond to those of the full-scale field trial could be achieved by blackening of the covering foil in the laboratory test. Martin et al. [498, 499] give an analogous evaluation for the Australian test methods. The

Polyurethane and Fire. Franz H. Prager and Helmut Rosteck
Copyright © 2006 WILEY-VCH Verlag GmbH & Co. KGaA, Weinheim
ISBN: 3-527-30805-9

Early Fire Hazard Test, prescribed for building materials by law in accordance to AS 1530 was standardized in agreement with numerous application full-scale field trials [234, 500]. The limited informative capability of ratings determined in laboratory tests, caused a corresponding textual restriction in the American standardizing texts that stipulates a safeguard by relevant full-scale field trials. For fields of raised fire risks specific test methods were agreed as well. Each of the individual fire-technological test methods simulates a specific, real-fire situation. Basic findings have to guarantee that the various determined characteristics allow significant conclusions that promise that the simulation of the essential fire scenarios is possible. After a careful study of the general conditions for the different fire scenarios of SC3 of ISO TC92 the three fire sources are to be considered for the evaluation of possible fire-gas dangers in the case of a fire (see Table 107).

Table 107 Definition of relevant Fire Scenarios

Fire scenario	Definition
1	Due to thermal reaction self-sustained smoldering process
2	Combustion with flames – pool fire conditions
3	Flash over conditions and oxidative decomposition under smoke-layer attack

Especially in the case of a fire involving flaming combustion the spatial circumstances determine next to material-related variables the heat and fire-gas release. Practice-oriented investigations clarify, to what extent findings from laboratory research may be generalized [501]. A risk-relevant evaluation can be carried out only by considering the functional dependence. Data records of (1-dot) single-point test methods are meaningful as a rule only for the product check in the production and supplying fields (see Table 108).

Table 108 Primary and secondary ignition sources

Cigarette
Overheated wire / glowing particles
Match and candle flames
Burning paper
Burning waste-paper basket, clothes
Burning item (TV– upholstery item – Christmas tree)

6.1.1
Relevance of the Procedures for the Risk of Ignition

The furnace test (see Fig. 41a) is used worldwide at a test temperature of 750 °C to determine the incombustibility. The most recent findings have shown that the verified classification "noncombustible" is by no means always the accepted privilege by the authorities of the preventive fire protection. During the inspection of noncombustible classified rockwool, all eight checked qualities passed the requirements of the furnace test, but failed partially, in the chimney test, the A2 and the B1 requirements concerning the contribution to the surface spread of flame. The results of these orienting investigations [322, 502] show that the furnace test seems not to suffice alone to provide a safety expertise with regard to the ranking "noncombustible". The safety declaration of the expert commission [1], "Under real fire conditions these products do not burn or to a negligible extent (only small quantities of organic components). Moreover, they do not contribute to the production of smoke and toxic gases." is accordingly under scientific aspects, in such a general form, inadmissible. Risk-oriented testing methods pursue two different destination routes.

"On the one hand", there are material-related characteristics, which do not allow a realistic classification for themselves. These mainly functional characteristics are qualified as basis values for mathematical modelling. For different fire-risk situations the relevant variable parameters must be considered.

On the other hand, there are single-point measuring procedures that simulate a very specific risk situation. The results are significant for this situation only. A transfer of the results is usually not possible. The conversion of the findings for other scenarios requires detailed knowledge that is not available as a rule.

Table 109a Simulation of smoke release in different fire scenarios.

Test methods	Thermal stress	Scenario 1	Scenario 2	Scenario 3
ASTM-E84	Burner	no	(yes)	no
DB Chimney	Burner	no	(yes)	no
Radiant Panel	10–1 KW/m^2	no	(yes)	no
XP2 Chamber	Burner	no	(yes)	no
NBS Chamber	25 KW/m^2	(yes)	(yes)	(yes)
NBS modificated	20–50 KW/m^2	(yes)	(yes)	(yes)
ISO Smoke Box	20–50 KW/m^2	yes	yes	yes
Gost 12.1	20–80 KW/m^2	(yes)	(yes)	(yes)
DIN 53436	200–900 °C	yes	yes	yes
NF-T51	200–900 °C	yes	yes	yes
IEC TC 89	Flames	no	(yes)	no
NES 7123	Burner	no	no	no
Potts Pot	200–1000 °C	yes	yes	yes
Box Japan	Flames	no	(yes)	no
BRI Japan	10–50 KW/m^2	no	(yes)	(no)
RTB Japan	10–50 KW/m^2	(no)	(no)	(no)
SRI	10–80 KW/m^2	(no)	(no)	(no)
U-Pitt 2	10–70KW/m^2	(no)	(no)	(no)
Cone Calorimeter	10–100 KW/m^2	(no)	(no)	(no)
NIBS/ US Radiant	10 – 80 KW/m^2	(no)	(no)	(no)
U-Pitt	200–1000 °C	no	no	no
NES 713	Burner	no	no	no

Szenario 1 – self – sustained smoldering
Szenario 2 – developing fire with reduced oxygen availability
Szenario 3 – fully developed fire with flash over conditions

6 Use and Interpretation of PUR-Test Results Determined under Enduse Conditions

The different national fire-statistics data from the United States of America, the United Kingdom and Japan, show very vividly that the majority of dwelling fires start in the building content. The representations of Figs. 24a to 37f illustrate which piece of furniture is first ignited by which ignitions source. The different data records show very clearly that smokers materials are identified as the dominating ignition source, if the number of fire victims is defined as the decisive evaluation criteria. Table 109a provides a view of primary and secondary ignition sources of different intensities of the daily life, which were characterized, e.g. in the Technical Report Categories of ignition sources of TC 61 [503].

Table 109b Assessment of fire performance standards

Codification

Code	Criteria
1	ignitability
2	spread
3	heat release
4	burning droplets
5	smoke density
6	smoke toxicity
7	smoke corrosivity

X	classification of standard
0	no standard classification
XX	standard classification is realistic for the simulated scenario only
+	realistic classification – mass-, volume- or area – based potency data
++	realistic classification might be possible by variation of test parameters
–	no realistic classification with mass-,volume- or area – based potency data possible
A	thermal attack
A1	furnace
A2	flame
A3	radiation
A4	flame + radiation

Method	A	1	2	3	4	5	6	7
AS 1530	A4	x, xx	x, xx,++	x, xx,++		x, xx,++	0, -	0, -
AS Zig crib.	A3	x, xx	x, xx	0, –	0, -	0, -	0, -	0, -
ASTM E84,NFPA 255, UL 723	A2	x, xx	x, xx	x, xx	0, -	x, xx,-	0, -	0, -
-ASTM.E648, DIN 410 pt.142, Ö B3800	A4	x. xx	x, xx	0, -	0, -	x, xx	0, -	0, -
ASTM-D2843,DIN 4102 pt.1,Ö3800	A2	0, -	0, -	0, -	0, -	x, xx,-	0, -	0, -
ASTM-D2863,NT-013, O2-Ind.	A2	x, xx	x, xx	0, -	0, -	x, xx,-	0, -	0, -
ASTM-E136,DIN 4102 pt.1 BS476 pt.4, ISO 1184	A1	x, xx	0, -	x, xx	0, -	0, -	0, -	0, -

6.1 Relevance of Combustion Systems

Method	A	1	2	3	4	5	6	7
ASTM-E162, E-1354, D-3675	A4	x, xx	x, xx	x, xx	0, -	x, xx,-	0, -	0, -
ASTM-E662,FAR 25.853, ATS 1000	A4	0, -	0, -	0, -	0, -	x, xx,++	x. xx,++	0, -
ASTM-E906, ATS	A4	x, xx	x, xx	x, xx,++	0, -	0, -	0, -	0, -
ASTM-Zig.,UFAC,BS 5852, BS 5852 pt.1 ISO 8191 pt.1, DIN/EN 1021-1	A3	x, xx	x, xx	0, -	0, -	0, -	0, -	0,+
Box	A2	0, -	0, -	0, -	0, -	0, -	x, xx,-	0, -
BS 5852 pt.2„ISO 8191-2,DIN/EN 1021-2	A2	x, xx	x, xx	0, -	0, -	0, -	0, -	0, -
BS476 pt.6	A4	0, -	0 , -	x, xx	0, -	x, xx,-	0, -	0, -
BS476 pt.7,NEN 3883	A4	x, xx	x, xx,++	0, -	0, -	0, -	0, -	0, -
CAL TB 133	A2	x, xx	x, xx	0, -	0, -	0, -	0, -	0, -
CAL TB121	A2	x, xx	x, xx	x, xx,-	0, -	0, -	x, xx,-	0, -
CSE RF-2/75A	A2	x, xx	x, xx	0, -	0, -	0, -	0, -	0, -
CSE RF-3/77, ISO DP 5658	A4	x, xx	x, xx,++	0, -	0, -	0, -	0, -	0, -
DIN / EN 13823 (SBI)	A2	x, xx	x, xx	x, xx,-	x, xx	x, xx,-	x, xx,-	0, -
DIN 4102 B1 pt.15	A2	x, xx	x, xx	0, -	x, xx,-	x, xx	O, -	O, -
DIN 4102 B2,ÖNorm B3800, CSE RF-1/75A	A2	x, xx	x, xx	0, -	x, xx,-	0, -	0, -	0, -
DIN 4102 Tl.7	A2	x, xx	x, xx,	0, -	0, -	0, -	0, -	0, -
DIN 5510	A2	x, xx	x, xx	0, -	0, -	x, xx,-	0, -	0, -
DIN/VDE 0472 Tl.804, 804c								
DIN/VDE 0472 Tl813c								
DIN7VDE 0266 Tl 85								
FAR 25.853 burner	A2	x, xx	x, xx	x, xx,-	0, -	0, -	0, -	0, -
FAR 25.853..Nr.191 M.5903	A2	x, xx	x, xx	0, -	0, -	0, -	0, -	0, -
GOST 12.01	A1	0, -	0, –	0, -	0, -	x, xx,++	x, xx,++	0, -
-Herpol Fire reaction	A4	x, xx	x, xx	x, xx	x, xx	x, xx,-	x, xx,-	0, -
IEC 695	A2	x, xx	x, xx	0, -	0, -	0, -	0, -	0, -
IEC Glow Wire	A1	x, xx	xx	0, -	0, -	0, -	0, -	0, -
IEC Glühstab	A1	x, xx	xx	0, -	0, -	0, -	0, -	0, -
IMO–	A4	x, xx	x, xx,++	x, xx,-	0, -	x, xx,-	0, -	0, -
IMO Spread FP..	A4	x, xx	x, xx	x, xx,-	0, -	x, xx,-	0, -	0, -
IS 5960, ASTM-E..,NT-..CEN	A4	x, xx	x, xx	0, -	0, -	x, xx,++	x, xx,++	x,xx
IS 9705, ASTM-E..,NT.. CEN	A2	x, xx	x, xx	x, xx	0, -	x, xx,-	x, xx,-	0, -
JIS A 1230	A4	x, xx	x, xx	x, xx	0, -	x, xx,-	x, xx,-	0, -
NEN3883	A4	0, -	0, -	x, xx	0, -	x, xx,-	0, -	9
NES 713	A2	x, xx	0, -	0, -	0, -	0, -	x, xx,-	0, -
NF-T51	A1	0, -	0, -	0, -	0, -	x,xx,++	x, xx,++	0, -
NF-X70	A1	0, -	0, -	0, -	0, -	x, xx,++	0, -	0, -
NT-002	A2	x, xx	x, xx	0, -	0, -	0, -	0, -	0, -
NT-004	A2	x, xx	0, -	x, xx,++	0, -	x, xx,-	0, -	0, -
NT-006	A2	x, xx	x, xx	0, -	0, -	0, -	0, -	0, -
NT-007	A2	x, xx	x, xx	0, -	0, -	x, xx,-	0, -	0, -
ÖNorm B3800	A2	x, xx	x, xx	0, -	0, -	0, -	0, -	0, -
ÖNorm B3825	A2	x, xx	x, xx	0, -	0, -	0, -	0, -	0, -
PSA Nr.5,6,7	A2	x, xx	x, xx	0, -	0, -	0, -	0, -	0, -
Roland CEN	A4	x, xx	x, xx	x, xx,-		x, xx,-	0, -	0, -
RTB	A4	0	0, -	0, -	0, -	0, -	x, xx,++	0, -
SBI DIN EN 13823	A2	x, xx	x,xx	x, xx, -	x, xx,++	xx, -	0, -	0, -
UIC.564-2.,DIN 54341..EN 45504	A2	x, xx	x, xx	x, xx, -	0, -	0, -	0, -	0, -
UL 7461 (hot wire)	A1	x, xx	x, xx	0, -	0, -	0, -	0, -	0, -
UL 94	A2	x, xx	x, xx	0, -	x, xx,-	0, -	0, -	0, -

Igniting risk by ignition sources of low intensities

The cigarette test is certainly the perfect example for the simulation of real-fire risks. The selection of the ignition source corresponds in the cigarette test fully and completely to the igniting risk. The technical test parameters do not, however, by any means fully and completely fix on the factors of the final application.

The important factors for the smoldering fire risk including the probability of occurrence, as there are location of the ignition source, area-, corner- and joint configuration as well as coverings, which are capable of glowing such as, e.g. cushions and or bedding, remained to a large extent unconsidered. Only especially crucial cushion combinations are registered by the current standardized procedures as a rule. The smokers utensils are to be classified as the dominating ignition sources, if the number of fire victims is used as a basis for comparison. All fire statistics show this worldwide. The graphical representations of Figs. 24a to 37f underline this statement and confirm the necessity of a risk-relevant simulation. The definition in the various test protocols, as illustrated by Figs. 68a, 69a or 69e, allow, in principle, risk-oriented evaluations. Figures 138a and b clarify the ignition potential of a glowing

Figure 138a Smoldering behavior of a cigarette

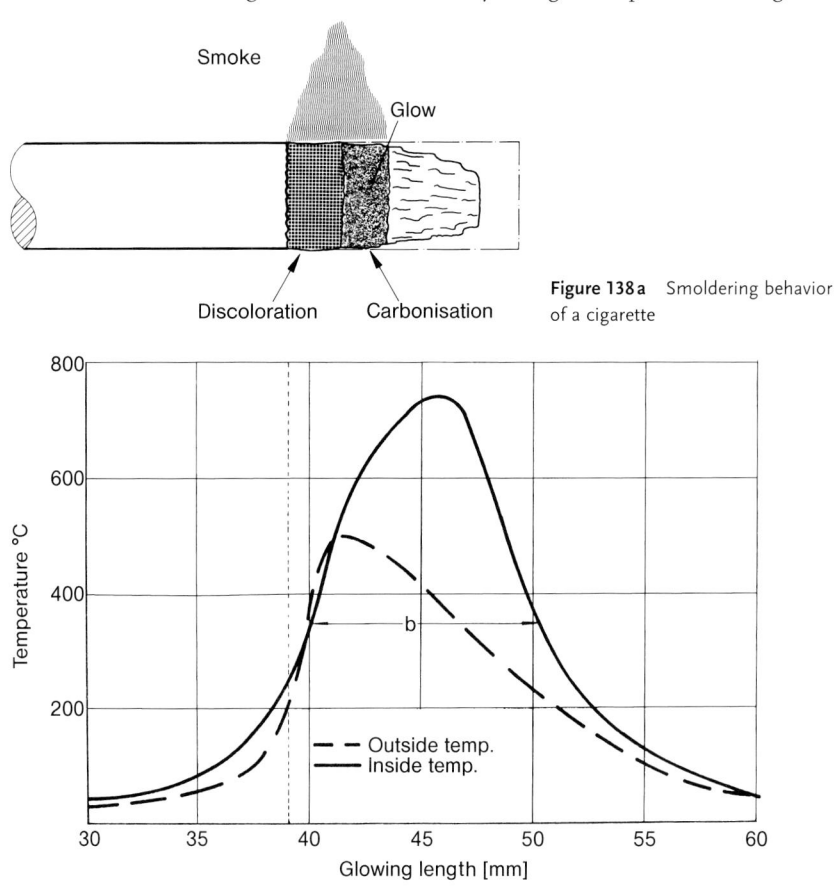

Figure 138b Smoldering behavior of a cigarette

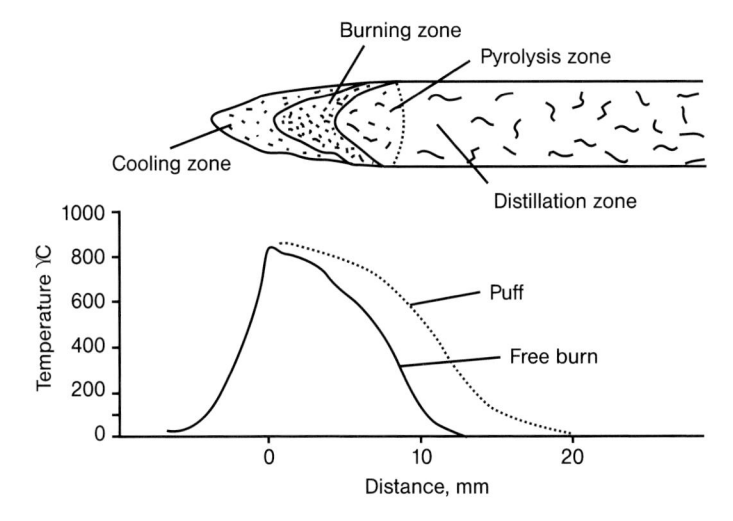

Figure 138 c Thermal potential of a cigarette

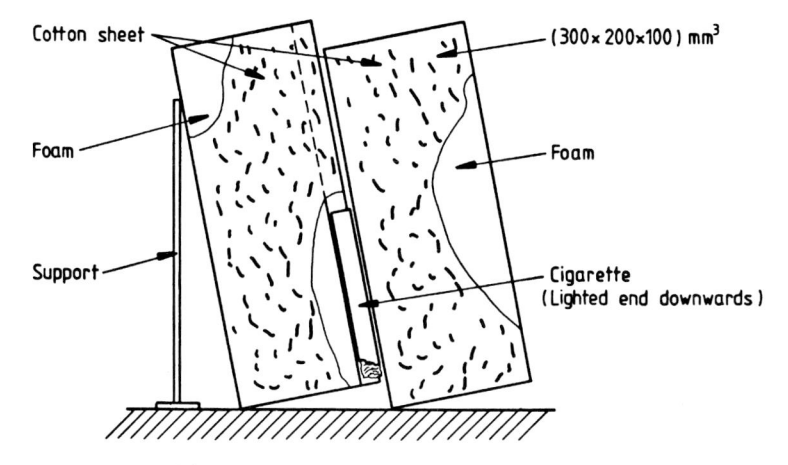

Figure 138 d Modified cigarette test

cigarette and the course of the temperature–time profile of this ignition source in the upholstery furniture field. Figures 138c and d demonstrate also the temperature–time profile caused by a glowing cigarette in an upholstery combination.

These illustrations clarify the temporal process in the glowing cigarette. Temperatures up to 800 °C were reached in the glowing zone. Summerfield et al. [504] determine, with the aid of thermocouples, the temperature–time process in the smoldering cigarette. Baker [392] discusses the temperature–time profile of a glowing cigarette. The possible heat release and the CO_2/CO ratio of the stream of effluents are discussed in detail. Extensive investigations in large-scale, measure 1:1, have demonstrated that the chemical nature of the filling, the textile structure of the covering as well as the location of the ignition source and radiation losses conditional through that can influence the developing stage of the smoldering fire. For the improvement of the reproducibility of the cigarette test an electrically heated glow-wire spiral (see Fig. 71a) was 1985 elaborated at the CSIRO [499, 505] as a draft standard. The glowing spiral acting with a constant heat-release rate does not consider the heat loss by melting and carbonisation processes. In reality this heat loss is possibly just sufficient that no smoldering process will be initiated. The reproducibility of the test methods is altogether proved by such measures without doubt. Before the variation of the test parameters it should be guaranteed that the practical factors were taken into account. If the protection effect of outer coverings is randomly eliminated by unrealistic parameters, then this test procedures should not be used for regulation. Such modifications of the cigarette test are certainly of great benefit for basic research work. For orders they are usable only if they allow the simulation of a real-fire scenario. They must be based on the real danger scenarios as far as possible and must not degenerate. If round-robin tests are carried out in order to verify the utility of the targeted procedure within the framework of the standardization activities, a list of fundamental limiting conditions are to be considered, because round-robins are usually based on limited test numbers only. The validity of conclusions should be checked by means of the relevant expert literature. When, like the cigarette experiment at the BSI round-robin shows that the cotton covering of the glowing cigarette leads to more favorable values, then a such conclusion should not be generalized without further proof. Such conclusions are treated all the more cautiously when, as in the current case, the worldwide present findings came to a contrary conclusion. Reproducibility should not be the primary goal of standards in the field of security. The primary protection destination must be covered primarily. It must be expected of relevant classifications/ratings that they lead to a significant reduction of the danger potential. A higher probability, to cover the danger potential risk-relevant, could be achieved, e.g. by variation of the test parameters. For example, from the beginning more unfavorable limiting conditions can be simulated. The current/present general conditions of BS 5852 as well as the corresponding ISO/CEN standards simulate circumstances where combinations fail whose resistance is from experience to be designated very low. The schematic representation of Fig. 138d illustrates a modified calibration setup that is able to simulate extreme crucial limiting conditions of this dangerous situation.

The testing and evaluation criteria should be adapted during the routine revision of such standards to the possible modified limiting conditions. Of course, the pure continuation of such standards guarantees no increased safety. A typical example of the continuation of a specification robbed of their protection destinations to a large extent is the O_2-index test method. Material testers prefer this test procedure in particular because of the excellent reproducibility. The validation of the determined data is no basis for the interpretation of the results. As long as the American space industry worked with an oxygen-rich atmosphere, the procedure developed for that supplied results with high testimony value. These special work-place conditions were changed after the Challenger fire disaster, but this modification of the test parameters was not taken into account by all sides of the standardizing committees. The candle-like burning of test pieces in the oxygen-rich atmosphere determines gradings that need now a very careful interpretation and are wrongly estimated by the nonexperts and deliver an illusory safety. The statements of Damant clarify the problematic nature of reproducibility. He reports on the error rate during the UFAC-cigarette test classification of upholstered furniture. These data records show very clearly that the inspection of the self-certification of the industry discloses significant deviations on the basis of the UFAC guidelines. The possible influence through formula variation is demonstrated in this case in the same way, as the dominating influence of the cushion construction that underline present findings in the mattress field. 50% of the sisal mattresses failed in combination with a cotton cover. The graphical representations of Fig. 139a illustrate the influence of the mattress construction. The combinations with PES coverings proved altogether as resistant [506–508].

The value of this statement must be reviewed, since the number of the required examinations is not sufficient to make a mathematically perfect, safeguarded statement. Figure 139b illustrates the deviations during the inspection of the UFAC classification for upholstery combinations with different PUR grades that certainly are not to be assigned to laboratory outliners but rather to the relative insecurity of ciga-

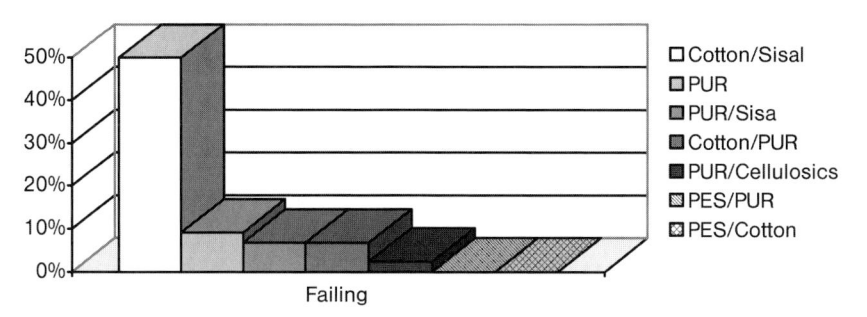

Figure 139a Failing of classified mattresses according to the UFAC guidelines

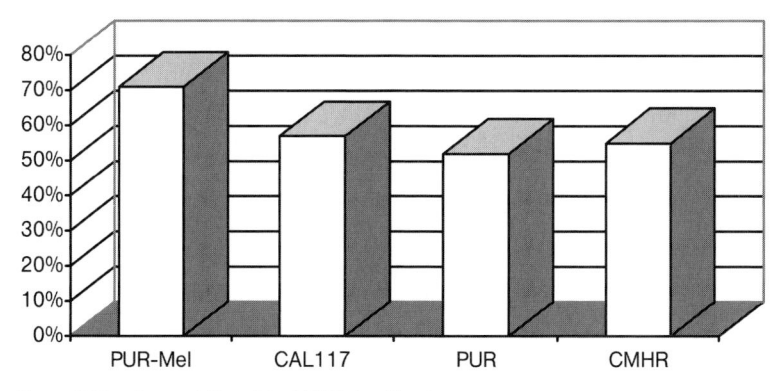

Figure 139b Repeatability of the UFAC classification

rette tests. When Damant et al. [507] report that about 30% of the UFAC-rated upholstery combinations fail in the cigarette test, then this shows the necessity to prescribe possible realistic deviations in the standards information.

The investigation results of the Bureau of Home Furnishing demonstrate that the probability of the failure is changed by recipe variation. Damant [220, 508] shows by means of numerous investigations that not only the nature of the outer covering but also the location of the ignition source have a considerable influence on the classification. The location of the flat surface means a considerably smaller ignition risk in relation to the wall position, for the slit position or for the location in the field of the arms. For the examined coverings cellulosic-containing products depending on the weight area unit contained a considerably more serious ignition risk than combinations with thermoplastic covers or leathers (100%). Upholstery combinations with thermoplastics at 87.7%, 100% cellulosics at 32.6% and leathers at 95%. These records demonstrate how much the upholstery construction that is the nature of the covering, the interliner and the formulation variation of the upholstery filling, influence the resistance against glowing cigarettes. The protection effect of the interliner in the case of a fire depends decisively on the smoldering ability. The fire retardancy of the interliner on the cellulosic basis can obviously not generally suppress the smoldering tendency of upholstery combinations.

The question of the reproducibility was the subject of various research studies. Partly, it was attempted, similar to the Australian draft standard to replace the glowing-cigarette ignition source by a glowing-spiral (see Fig. 71a). With this the protection effect of outer coverings was circumvented. Melting coverings, which prove resistant against glowing cigarettes failed under the impact of this modified ignition source. Day et al. [510] report on the simulation of a glowing cigarette by a 100-W heating spiral. The differences in relation to the normal cigarette test are shown. In dependence of the intensity of radiation the starting time for the self-sustained smoldering process could be reduced from 19–28 min to 3–4 min. Increased density of the cushion material to be examined led to decreased ignition times. The characteristics of Fig. 67d clarify, using the example of the accumulation of ignitions, to what extent the different cushion construction is able to influence the probability of the ignition in the cigarette test according to TB 117.

Figures 91b and c demonstrate that in the case of resistant combinations only an insignificant contribution to the fire occurrence is given. There, where heat is withdrawn from the glowing process of the cigarette sufficiently, e.g. through melting effects of the covering, it results in the limitation of the smoldering process. Provided that in the current fire case there is an analogous behavior, no fire-gas danger is to be expected. The different laboratory test procedures for the assessment of the fire risk through glowing tobacco products (ISO and CEN procedures as well as national standards) are based relevant to risk on a glowing cigarette as the ignition source. The various national and international standards do not use due to a limited BSI round robin an additional cotton cover of the glowing cigarette although it was well known that the smoldering process might be intensified. These BSI experiences stand in blatant contradiction to all published test results [509–511]. While at the basis test (see Fig. 67a) possible variants of the ignition source location were considered, this

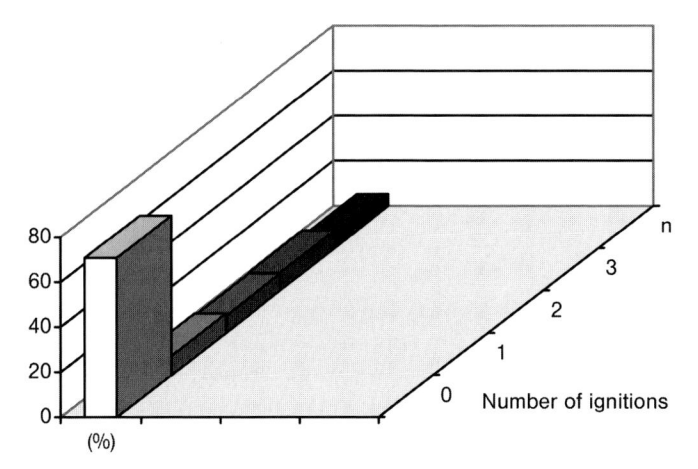

Figure 140a Reproducibility of cigarette test results

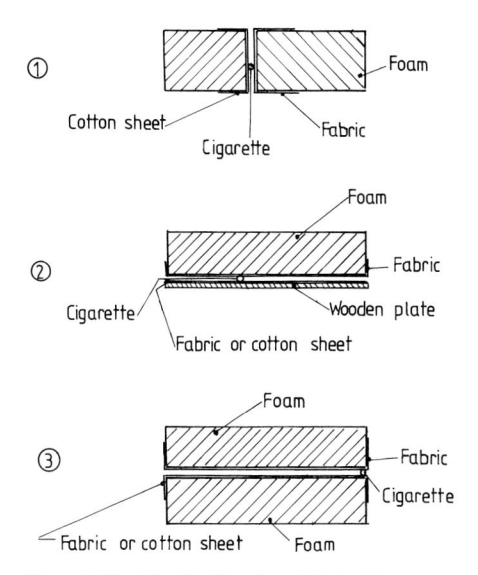

Figure 140b Modified location of the ignition source cigarette 1068

effect was mainly neglected at the standardisation of the current test methods for upholstery furniture. The mock up according to Fig. 69a does not consider either the aggravation/intensification possible in the current fire case through modified test configurations according to Fig. 140b nor crucial coverings, e.g. bedding, cushions and others tending to smoldering fires. The damage-scale classification criterion defined in the different lab procedures must be designated as risk relevant, provided that the smoldering tendency through an unrealistic modification of the ignition source (2 and more cigarettes, electrical heater spiral) and with that the temperature–time profile of the glowing cigarette is not modified. The fringe effects in the schematic illustration of Fig. 140b show that in the real-fire case different coverings of the glowing cigarette have to be expected. The location of the test cigarette in the slit will, due to the raised accumulation of heat, result in considerably intensified feedback effects.

Overheated wire and glowing particles

A practice-oriented imitation of the ignition source overheated wiring and filament is represented by the glow-wire test (see Fig. 77a) [503]. The temperature of the glow wire can be varied in accordance with the orders. This test methods is the main item of the preventive fire protection in national and international guidelines for the fire safety of electromechanical products and is supposed to guarantee that electromechanical products do not become, in the failure case, an ignition source for their environment. Figure 141a gives a view of possible causes of fires in the electromechanical sector. The listings of Fig. 141a clarify, as illustrated by the numerical values of Figs. 55, 58, 60 and 74, to what extent the failure of electromechanical products cause the initiation of fires. Due to glowing contacts, e.g. in bad connections, as illustrated by Fig. 141b, localized heating and in borderline cases the ignition of the

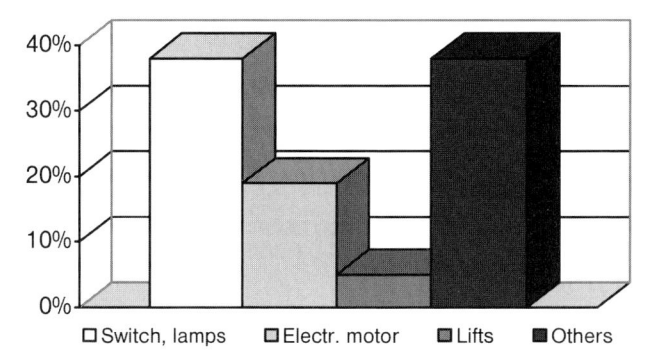

Figure 141a Failure of electrotechnical products as primary ignition sources

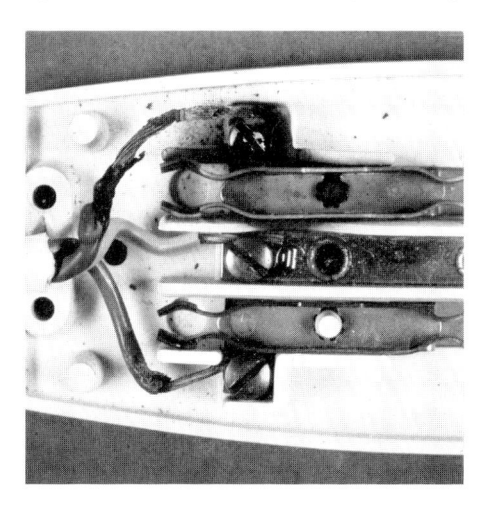

Figure 141b Failure of electrotechnical products causing fire

decomposition products takes place. Not very often do impurities form the basis for tracking processes that finally lead to the local ignition. The failure of washing machines and refrigerators are examples.

Ettling [512] clarifies that the smooth heat transition can be disturbed, e.g. by bad contacts. The formation of glowing contacts leads possibly to a raised ignition risk. The problematic nature of electromechanical ignition sources was also the subject of the studies in [513–515]. The danger of spontaneous ignition of button batteries and their role as a secondary ignition source is well known. Pohl [516] explains for the transportation field that in vehicle fires, second to the fuel area, the electronics are to be classified as a main responsible fire cause. The importance of welding beads as an ignition source was recently quoted in evidence through the Düsseldorf Airport Fire [1]. Practice-relevant testing and evaluating criteria have up to now not been standardized. While only a locally and, within a given time, limited decomposition process is being caused by torch beads on open surfaces as a rule, this ignition source can produce, due to feedback effects in a narrow slit ignition and consumption by burning. Figures 92a and b illustrate this state of affairs. The igniting risk combined with hot works (welding processes) are discussed in [517]. The investigations illustrate that the ignition temperature is to be addressed not as a product, but rather a system attribute. Preheating effects, heating rate and accumulation of heat can perform, due to spatial factors, a decisive influence. By means of extensive investigations, Jach [338] showed that welding beads, depending on environmental conditions, cause a glowing fire and consumption by burning even in the case of melting products like polystyrene foam. To what extent the glowing contact is also suitable to imitate the risk of ignition by blow-torch work and welding processes will depend presumably on whether the preheating effects can be trialled in a risk-oriented way. From experience, environmental factors play a decisive role.

Match and candle flames

Also, open flames are to be considered as practical ignition sources, which are based on a negligent or playful contact dealing with match or candle flames, with torch flames, e.g. blow or welding torches. Burning matches attained in particular a special meaning in the case of negligent and playful arson through children playing. The effectiveness of the flame-ignition source depends decisively in which form the flame front becomes effective. The schematic representation of Fig. 142a shows the possible influence while acting on an upright wall.

A list of national and international standards stresses the risk-oriented imitation of the match flame, e.g. DIN 4102, Ö Norm B 3800, CSE RF- 1/75A, SNV 198897 and BS5852. Also during the match test altogether realistic imitations of the match flame are used. Seekamp and Roeske [518] describe, among other things, extensive studies of the characterisation of match flames. The small-burner procedure (see Fig. 49a) was standardized based on this study. The exposure time of the propane flame was stipulated with 15 s. The procedure was standardized also by ISO TC38 and is part of the standardisation work of CEN TC 127. Troitzsch [519], pointed out that the match test standardized in DIN 4102 does not signal the resistance a priori against match flames. This building-material test method limits the flame height, however, within

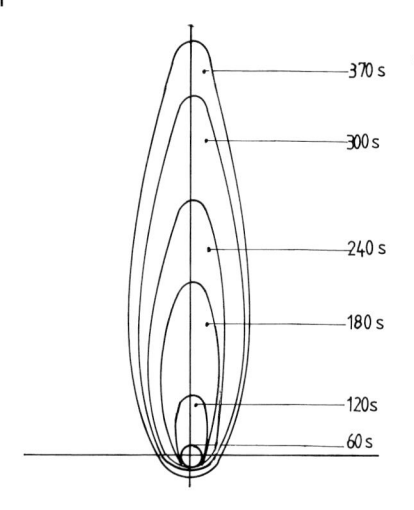

Figure 142a Temperature profile due to the acting of a flame at a vertical wall

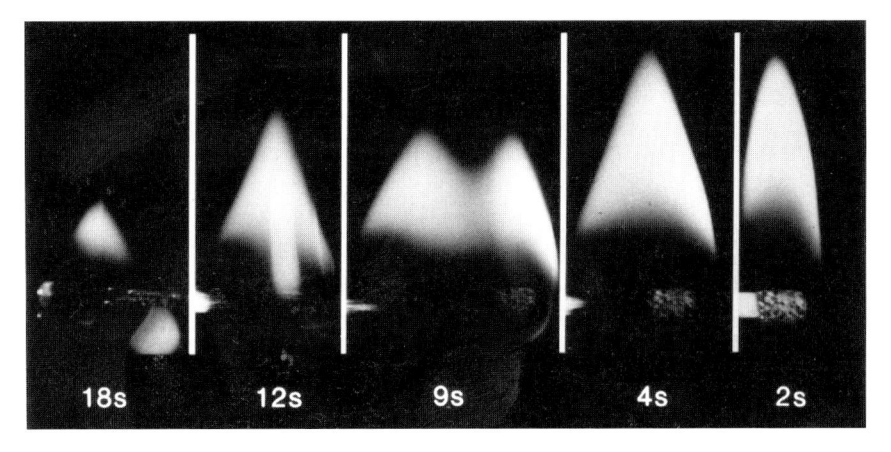

Figure 142b Match-flame impingement

a defined period of time of 20 s (including the flame impingement of 15 s) only but not the flame spread itself. The surface spread of flame after 20 s is not limited. During the determination of the classification criteria boundary values become determined due to reasons of the procedure specification and the reproducibility as for example the afterburn time of dripping material that are not absolutely relevant for the targeted resistance against burning matches. The orders on individual components are not covered by experiences from real fires and risk-relevant fire tests. Figure 142b demonstrates the effectiveness of this ignition source. The shifting flame was one of the reasons to limit the exposure time of the burner flame in the relevant CEN standard to 15 s and not to take over the 20 s of the butane-gas burner of BS 5852 standardized through BSI and ISO TC 136 [19, 520]. Within the framework of the relevant standardization activities of CEN TC 207 the experts agreed after intensive studies of the relevant literature that the practice-oriented simulation of a match flame has to be a 15-s flame impingement. Mehkeri and Dhawan [521] discuss the temporal process of a burning match in detail. The maximum temperature is indicated as 838 ±31 °C.

The possibility of the negligent arson through discarded burning matches was the subject of extensive investigations at the Carpet Research Institute in Aachen [340]. In the case of an edge impingement, as also practised in the small-burner test procedure, intensified preheating effects must be assumed. Procedures, such as for example the FAR vertical test according to Fig. 67d, provide results that must not be transferred to arbitrary fire-risk situations. The classification criteria are not universal and can be applied only where their validity is covered. Only the composite determines the fire risk of the upholstery unit. Filling materials that are tested by themselves and that are classified as resistant according to Fig. 142a and/or b fail possibly as a composite. Due to the "pot effect" it can result in a complete combustion similar to the vandalism conditions according to Fig. 69f. After [503, 522] ignition sources are mostly small or medium-sized, such as, e.g. matches up to burning wastepaper baskets. The importance of cellulosic materials for the ignition hazard is reported among other things in [510, 523]. Paul features in [19] the usual ignition sources of low intensity. One of his conclusions is the confirmation of the determination in the UK Furniture Regulation concerning the simulation of the match flame. Butane and propane are proposed as fuel gases. Paul gives in [524] a comprehensive comment for the match-flame simulation. Babrauskas and Krasny [523] provide a view of ignition sources in the furniture field including the usual national simulations. Paul [525] demonstrates this also through formula variation, e.g. of the upholstery filling material flexible PUR foam, the resistance against ignition sources of different intensities can be influenced. Analogous evidence was produced by Szabat [343]. The possible influence through formula variation is also reported in [526]. The influence of the formula variation on the risk of ignition of flexible upholstery PUR foams was the subject of numerous studies, about which Haas et al. [341] and others like Marchant [527] as well as Andersson [528] reported. Haas et al. examined the influence of, e.g. humidity, additives and fire-retardant additives like melamine or TCAP on the resistance of flexible PUR foams against ignition sources of low intensity. Conventional polyether foam and cold-cure qualities were used as experimental materials. The open-cell structure of the foam grades proved a most effective actuating variable. The listing of Table 110 clarifies that also in the case of a variation of the ignition-source intensity the principle safeguard of the protective aim remains guaranteed. The classification criteria with the destination of a spreading out of the classification are obviously not oriented as a rule on the protection destination. These conclusions can be made by means of the classification system of vertical test procedures like FAR 25 (see Fig. 67c) as well as the small-burner test procedure according to DIN 4102, Ö-standard 3800 or CSE 1/75A.

The experiences showed that also in the case of procedures with ignition sources of increased intensity a risk-oriented classification by no means always is guaranteed. The requirements in NW for buildings of special kind and use, e.g. for the seating in theaters and movie theaters, to use only B1 (difficult to ignite) coverings, disregard-

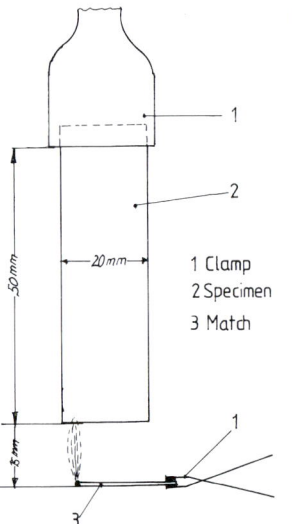

Figure 143 a Evidence of the resistance against ignition sources of low intensity

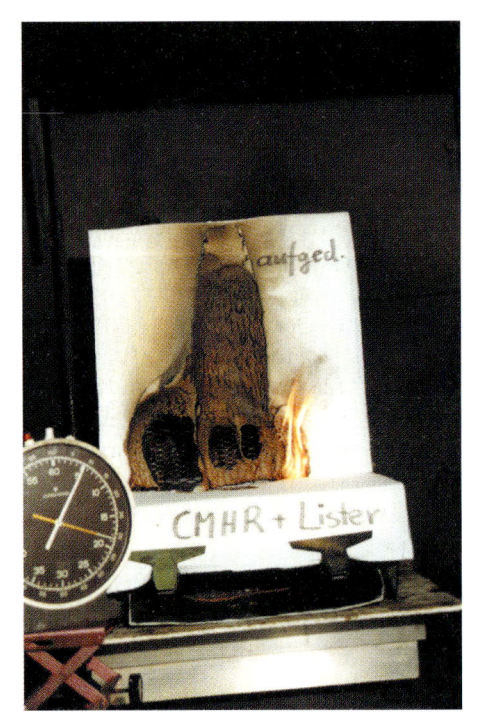

1 Clamp
2 Specimen
3 Match

Figure 143 b Evidence of the resistance against ignition sources of high intensity

Table 110 Standardized ignition source paper

Ignition source	Mass of paper
UIC-cushion	100 g
California TB 121	98 g
Boston paper basket	98 g
Australian	98 g
ISO TC 136	20 g
ISO TC 136	100 g
CEN TC 256	20 g
CEN TC 256	100 g

ing the experiences from the upholstered furniture field that had shown that only the relevant classification of the upholstery composite guarantees a sufficient safety. The chimney test is an excellent example to demonstrate that scientifically defined test parameters might possibly anticipate a misinterpretation of the danger estimate in the case of a fire. The disaster of the Düsseldorf Airport Fire illustrate this discrepancy [1].

Risk of ignition with ignition sources of high intensities

The listing of Tables 110 and 111 provides a view of the various national simulations from ignition sources of increased intensity. Paper is employed as the dominating secondary ignition source of raised intensity in accordance with DIN 5510. The listing of Table 110 clarifies in which diverse manner paper is employed for the simulation of negligent and playful arson. Paul et al. give in [530], as illustrated by Table 111, a view of a list of different ignition sources, simulated by gas flames and burning cribs.

Ramsay and Dowling [259] explain test details for ignition sources of higher intensities in Australia. Kirkby and Schmitz [30] discuss the possible influence of various factors that lead to a more stable generation of flames. The degree of mixing decides the stationary circumstances of the diffusion flames. The detailed information of the Figs. 144a to d show very vividly that the combustion behavior of the ignition source paper can be changed considerably by variation of the ignition source parameters such as, e.g. paper mass, paper type, location or pillow construction. The schematic representation of Fig. 144b shows the construction of a measuring system for the definition of the heat release during the combustion of paper pillows [531]. Paul [532] examines the influence of the mass of paper to the ignition intensity of the paper pillows. Time of burning, temperature development and flame height were the decisive evaluation criteria. The data of Fig. 144d provide an impression about the achieved results.

Differing information concerning the ignition-source intensity are related as a rule to deviations in the burning rate. Due to extensive investigations, wooden cribs were standardized as a paper substitute [532]. The improved reproducibility of the ignition source, required by the legislation, was the aim of these studies.

The effectiveness of the flame-ignition source depends perhaps decisively on the flame-front formation. The present fire-performance requirements of the Fire

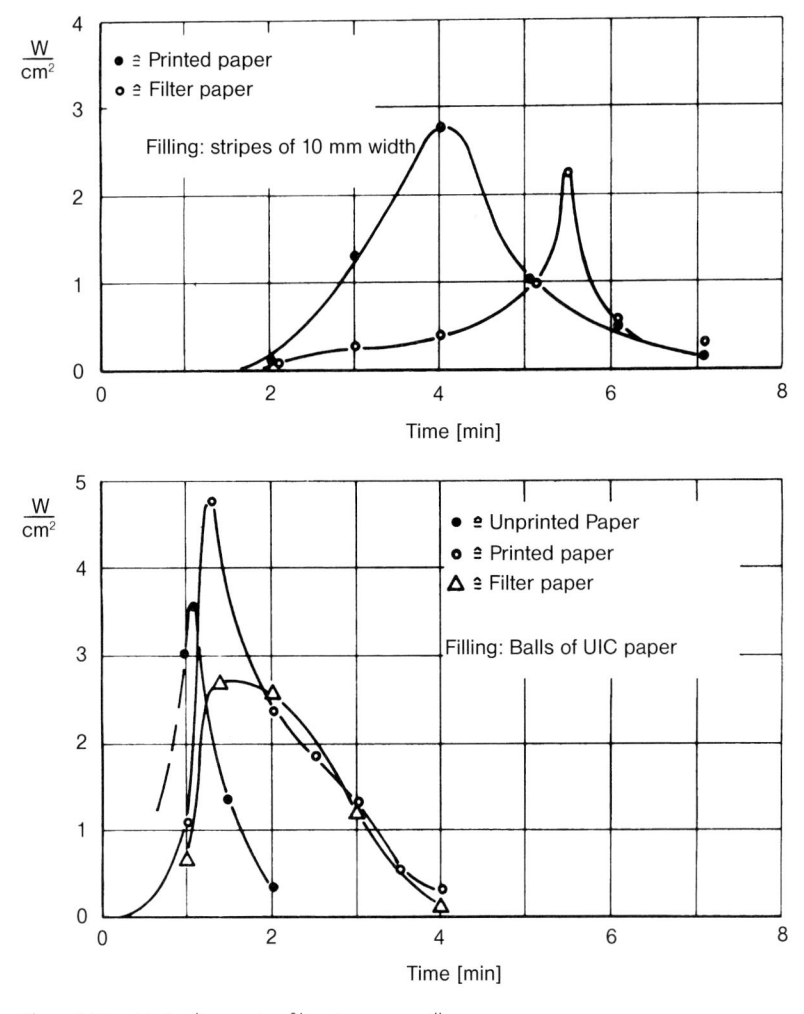

Figure 144 a Heat-release rate of burning paper pillows

Table 111a Characterisation of ignition sources

Ignition sources	Burning time (s)	Power (KJ)
Electric arc	1	0,4
Electric arc	5	15
50 g paper (shredded)	36	840*)
20 g paper	152	340
100 g paper	210	1689
400 g torf	330	9500

*) 17–24 KW/m^2

6.1 Relevance of Combustion Systems

Table 111b Characterisation of ignition sources

Ignition source	Flame Nr. 1	Flame Nr. 2	Crib Nr. 5	Crib Nr. 7
Burning time (s)	20	215	203	392
Heat release (KJ)	21,7	143	285	2100
Max.intensity (kW/m^2)	15	18	18	49

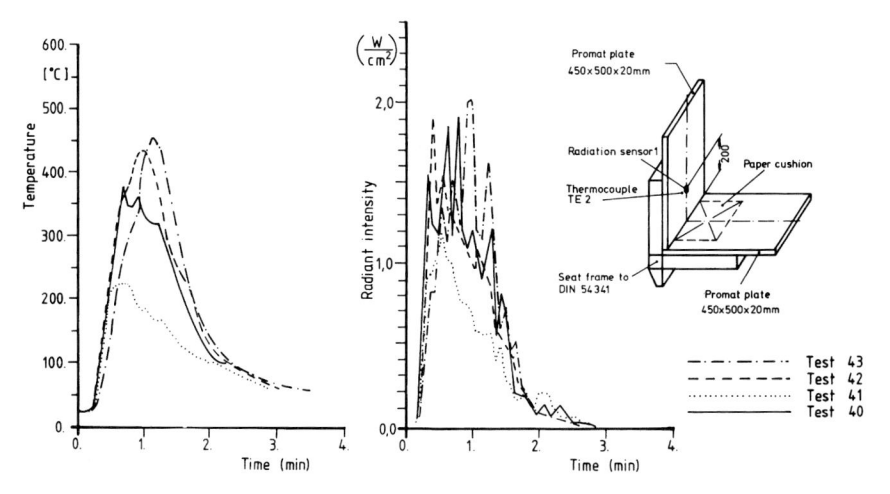

Figure 144b Calibration setup for the heat-release determination

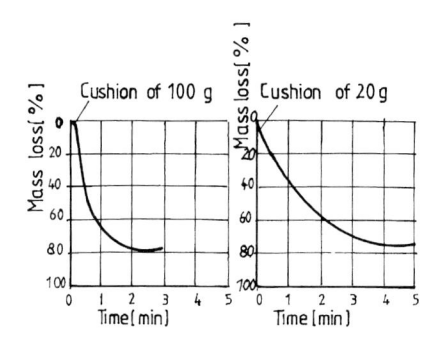

Figure 144c Influence of the mass-burning rate

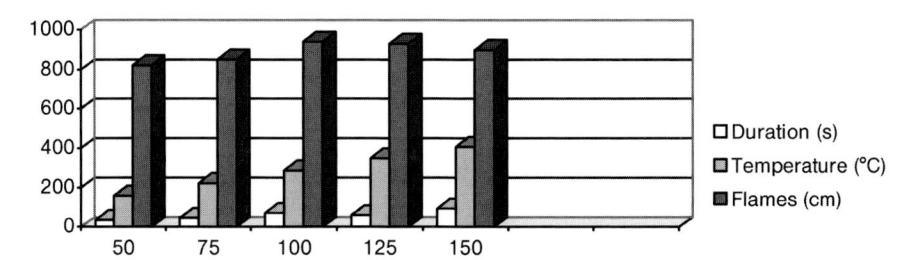

Figure 144d Influence of the paper mass on the characteristic parameters of the ignition source

Department of Boston concerning the ignition risks of upholstery furniture are explained in [533]. Next to the impact of burning paper a limited contribution to the fire spread in accordance with ASTM-E162 (see Fig. 42c) is required in analogous to the order of the New York Port Authority [534]. If the resistance against an acting flame is simulated, than the so-determined test data makes conclusions possible in principle only on the simulated fire situation. Deviations in the ignition-source intensity, the heat stress duration as well as the test configuration, that is for example horizontal or vertical orientation of the test sample, e.g. wall or corner position, will significantly modify the ventilation or the possible preheating effect as a rule the result. Igniting flames of greater intensity are then justified where corresponding ignition sources are to be reckoned on in the real-fire case. Dealing with lighter flames, soldering flames and torch flames are examples as well as the failure of electrotechnical products. To scrutinize crucially in each case the classification criteria that are not based as a rule on the danger potential of the initiated fire scenario. If electrical devices and plants are constructed and practised according to regulations, they are then to be regarded as fire-proof. Deficiencies as, e.g. wrong fixing of the life units, missing stress relief, ineffective protectors (temperature controller, guardian) nonholding of the tracking distances and wrong ventilation conditions can lead for the failure and due to glowing contacts, overheated wiring or short-circuits to the ignition and to the combustion. Possible igniting risks in the failure case of electromechanical products are discussed [535]. The ignition hazard initiated by television receivers, possibly short-circuits,is discussed in [536, 537]. Ignition risks in the cable field, including possible damage caused by the fire are reviewed in [538, 539]. A practice-relevant igniting risk due to the failure of electromechanical products is demonstrated in [540] in detail. The ignition hazard through electrical appliances or secondary ignition sources are closely examined in some studies [541]. The data records of Fig. 145a reflect the results of Finnish studies, and clarify the great igniting potential of burning electromechanical items that become effective in the case of failing. The effectiveness of such ignition sources is determined from their combustion behavior. The danger of an intensified flame spread increases in the developing stage of the fire with increasing size of the ignition source. The data records of Fig. 145b provide a view of the combustion behavior of cushions, which become effective as secondary ignition sources. The intensity of these secondary ignition sources is possibly significantly modified by different filling materials as well as by variation of the outer coverings. The cushion filled with latex produced, in relation to PUR, feathers and PES fibers, a considerably higher peak value of the heat-release rate. At the interpretation of the igniting potentials of the different cushion it is to be considered that due to uneven outer coverings a direct comparison of the single values is not possible. As an additional igniting-potential about 0.112 kg covering and 0.123 kg paper (6 clews) had to be taken into account.

In [542] it is reported on the efforts at NIST to replace the paper-pillow ignition source by a burner. To get an improved reproducibility of the results a torch was developed at NIST for the simulation of the ignition source in an analogous way to the regulation CAL-TB 133 mattress test in accordance with Fig. 68b. Figure 145c provides information about constructive details in the scheme. Ohlemiller and Villa

6.1 Relevance of Combustion Systems | 287

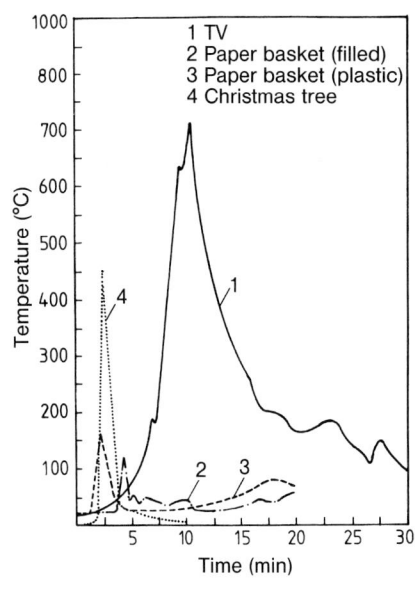

Figure 145 a Ignition-source intensity-influence of the burning duration and the fire load

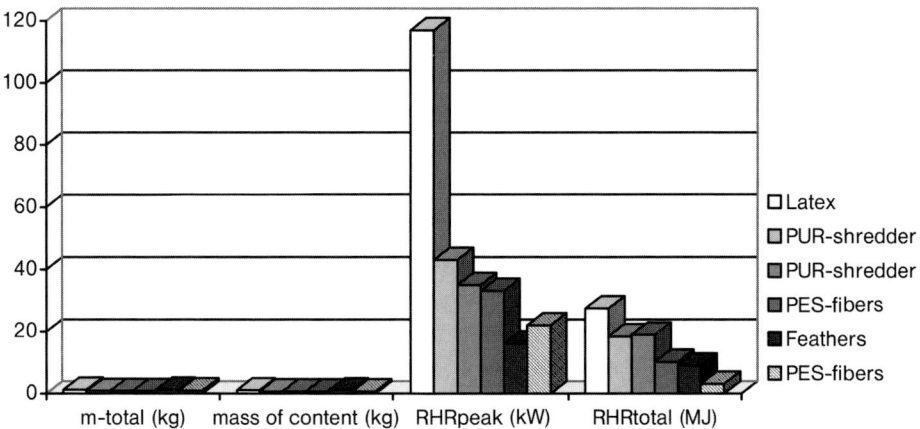

1 TV
2 Paper basket (filled)
3 Paper basket (plastic)
4 Christmas tree

Figure 145 b Ignition potential of secondary ignition sources

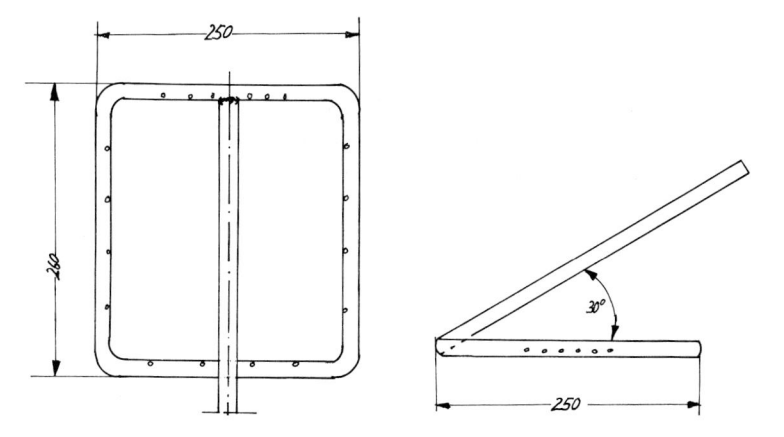

☐ Latex
☐ PUR-shredder
■ PUR-shredder
■ PES-fibers
■ Feathers
☐ PES-fibers

Figure 145 c NIST-burner for the simulation of the ignition source according to TB 133

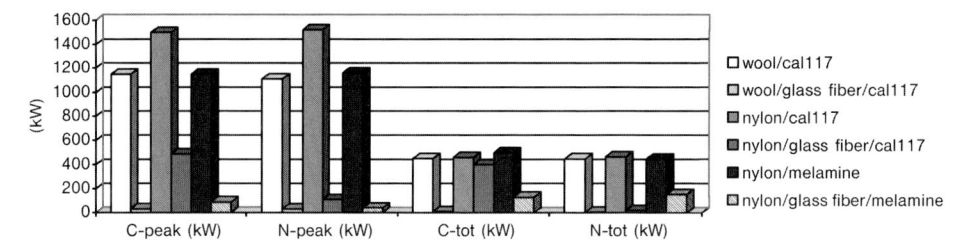

Figure 145d Comparative investigations with the NIST burner and CAL TB 133

[543] report on experiences with the NIST burner as a substitute for the wastepaper basket, the ignition source in the CAL-TB 133 mattress test.

As a rule, 2/3 of the tested piece of furniture unit was also destroyed. Comparative investigations with different cushion combinations have shown (see Fig. 145d) that a mainly good agreement has to be assumed. This torch was taken into account in modified form as an alternative also for the ignition-source intensities of 20 and 100 g paper planned by the European Commission [9]. According to Babrauskas [544] no torch is necessary, since the ignition of the upholstered furniture was altogether possible by means of a burning paper.

The results of the III research studies at RAPRA illustrated that both the design of the upholstered chairs as well as the position of the ignition source are able to influence the resistance of the upholstered combinations. While during the studies of the CBUF program, which was carried out for the EG, acting ignition sources of raised intensities (see Table 111), a list of upholstery combinations produced a limited contribution to the flame spread only, caused the ignition source "3/2 balled paper pages of Daily Mirror" of the III research program depending on the chosen design and the location of the ignition source in accordance to Fig. 146 ignition and combustion in all examined combinations [100].

This investigation showed a good agreement of the results in the main. The studies carried out with different upholstery combinations showed very vividly that the

Figure 146 Arrangement of the ignition source (20 g newspaper) at the ICUP/III research program

use of interlayers like woven glass fabrics reduced the heat-release and the fire-gas production significantly. With thermoplastic coverings both the peak values and the total values of the heat release were reduced to a great extent. Within the framework of the CBUF research program for the commission the effectiveness of ignition sources of high intensity was studied next to the standardized ignition sources cigarette and match to guarantee the combustion of the stressed upholstered item (see Table 112) [545].

In the case of the pool-fire configuration that exists, e.g. as well in the floor and in the roofing field, a strong decay of the intensity already near the flame action has to be assumed. Through ventilation effects the scale of the thermal stress can change by routing of the flame front substantially. A rapid fire spread in the ventilation direction can be one of the results. Large-area furniture units are assessed nationally by building-material-specific orders. Within the framework of the European harmonization a homogeneous testing and evaluation system is striven for. The different national testing and evaluation criteria resist, as the already carried out comparative investigations in the 1960s at the SC1 of ISO TC92 have shown, all harmonization attempts. The evaluation carried out by Emmons (see Fig. 82) underline the above statements. The graphical representations of Fig. 147a and b illustrate the scale of the thermal loading with the ring burner of the chimney-test procedure of DIN 4102. Figure 147a shows the profile of the intensity of radiation about the test piece length, while Fig. 47a provides a view of the surface-temperature profile measured on sheet metal samples by means of thermocouples.

Studies carried out with different rigid PUR foam qualities demonstrated also the influence of the test configuration according to Fig. 148a. The strong influence of the test configuration on the test results is shown by the graphical representations of Fig. 148b. These experimental results clarify, however, too that the intensifying test configuration, the simulation of a duct scenario, leads to a considerably intensified flame spread. In relation to an end-application-oriented room-corner configuration signifi-

Table 112 Comparison of ignition sources – CBUF Program.

Ignition source	Max RHR (KW)	Fire duration (s)	HRH total (KJ)
20 g Paper scrap	2,4	ca.125	301
20 g Paper ball	2,7	ca.132	356
100 g Paper scrap	14	ca.120	1688
100 g Paper ball	12	ca.142	1720
crib 7	10	ca.182	1820
CAL TB 133	20	80	1600
NIST Burner	20	80	1600
Burner	1,7	90	153
Burner	5,8	90	522
Burner	20	300	6000
Burner	30	180	5400
Burner	40	90	3800
CBUF – Burner	30	120	3600

Figure 147a Irradiance profile of the chimney burner

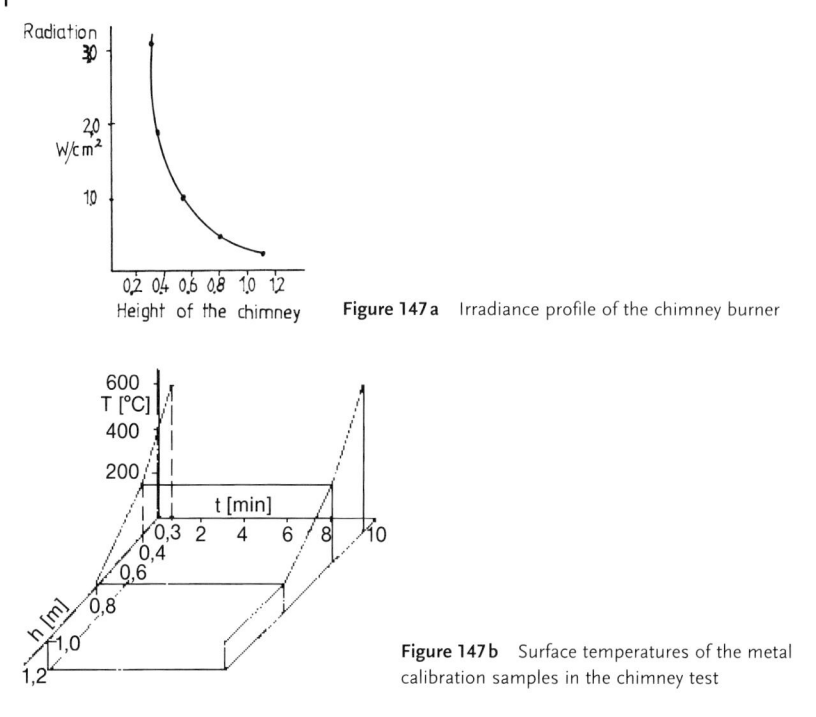

Figure 147b Surface temperatures of the metal calibration samples in the chimney test

cantly less favorable results were achieved in these investigations. Figure 148c illustrates the sharpness of the classification criteria of this chimney-shaped test method. In the case of a chimney height adapted to the normal room height, a much larger margin of the undestroyed length of the tested samples would have been available. The PIR foam test samples tested in the long-chimney configuration demonstrate the limited contribution to the surface spread of flames very emphatically.

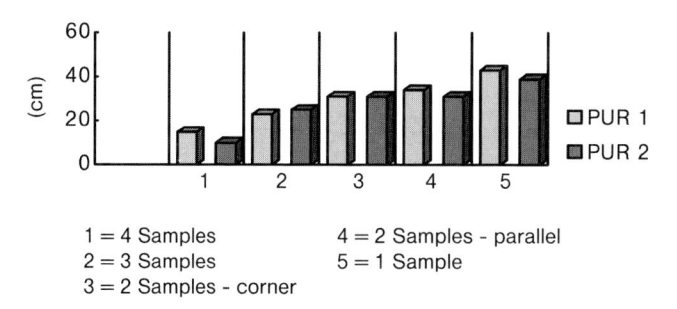

1 = 4 Samples
2 = 3 Samples
3 = 2 Samples - corner
4 = 2 Samples - parallel
5 = 1 Sample

Figure 148a Influence of the test configuration on the test results

6.1 Relevance of Combustion Systems | 291

Figure 148b Influence of the height of the chimney on the test results

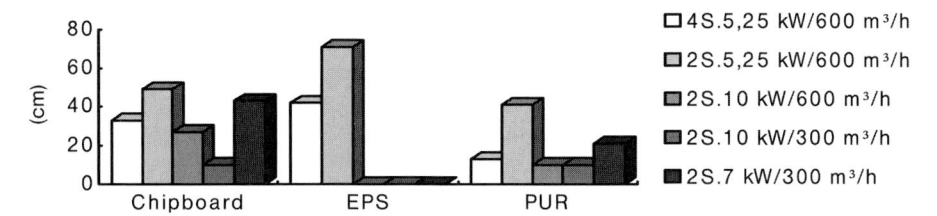

Figure 148c Influence of the variation of the test parameters on the classification

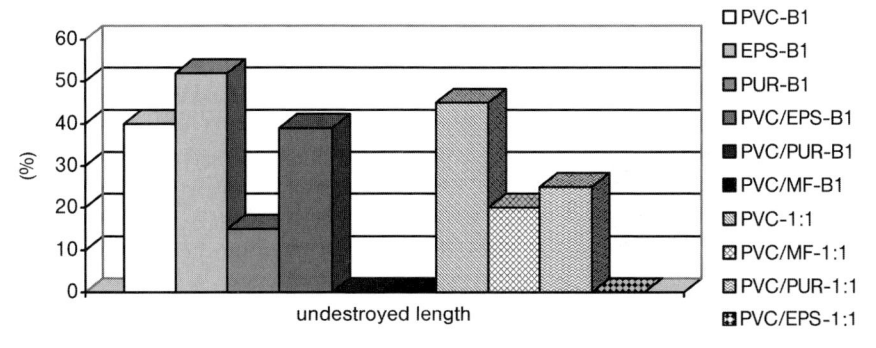

Figure 148d Extent of the surface spread in the chimney test and in 1:1 the facade test

In the test methods now pending for the vote as they are featured in Table 27, material-related characteristics are obviously further mixed with the data of the combined test methods. The SBI (Single Burning Item) Test and the Radiant Panel Test are promising end-application-oriented evaluations of the ignition hazard. With regard to the fire-related characteristics all findings of the SC3 of TC 92 Toxic Hazard in Fire have now been taken into account.

Burning dripping and smoke release are coupled also in future at single-point test methods, which as the state of the art are currently completely unsuitable to allow risk-relevant grading of these qualities. With the example of rigid EPS foam it can very vividly be illustrated (see Fig. 97b) that the defaulted limiting conditions of the chimney-test procedure can lead to significant deviations in the case of different ignition source intensities as well as modified ventilation conditions. The results of these studies illustrate very clearly that already small changes of the check criteria, e.g. the ignition-source intensity or the test configuration modify the chimney-test classification significantly. Both the surface spread of flame and the flue-gas formation as well as the burning dripping or falling off are influenced possibly decisively. Through the arrangement of the test pieces, e.g. hanging free, which is justified without doubt for theater curtains, faulty interpretations both with regard to the possible fire speed and the danger of burning dripping might be programmed as demonstrated by the current sad experiences of the Düsseldorf airport disaster. EPS foam boards classified class B1 (difficult to ignite) in the chimney-test procedure failed in the chimney due to the Al-lamination. The findings of the Düsseldorf airport disaster have also shown very clearly that possibly the horizontally fixing, up to now recognized by the preventive fire protection as the more harmless enduse condition in the case of a fire, can

Figure 149a Extent of damage of the aged PC horizontal glazing in the 1:1 facade testing

presumably prove crucial. In this case, the laminated EPS boards surprisingly failed even in the examination in the Radiant Panel Test according to DIN 4102 for flooring materials [1]. The certainly not always risk-oriented fire-reaction classification of the chimney-test procedure according to DIN 4102 was also dealt with in the large-scale test in the fire-test house of the Hoechst AG company [546] with heat-insulated PVC facade linings. The graphical representations of Fig. 148d show very clearly the deviations of the ratings in the standardized chimney test from the behavior in realistic full-scale field trials. In these investigations a wooden crib of 50 kg was used as the ignition source.

While the EPS composite achieved the most favorable result in the laboratory test, the same composite failed as the only one of the tested systems in the realistic 1:1 facade test. These full-scale field-trial results, which demonstrated that in the chimney test misleading classifications are possibly produced, were basically confirmed also by the findings of the Düsseldorf Airport Disaster. The experiences illustrated with regard to the example EPS rigid foam boards that the results gained in full-scale field trials guarantee without doubt a higher safety in the case of a fire. Deviating behavior of the chimney-test classification to the realistic 1:1 experiment among other things have been achieved as well with PC glazing elements. Even aged PC glazing elements, which failed in the chimney-test procedure, passed the relevant requirements concerning the surface spread of flame for B1-classified building materials in the facade test (see Figs. 149a and b), and also while acting a strong thermal stress in a room fire [72, 547].

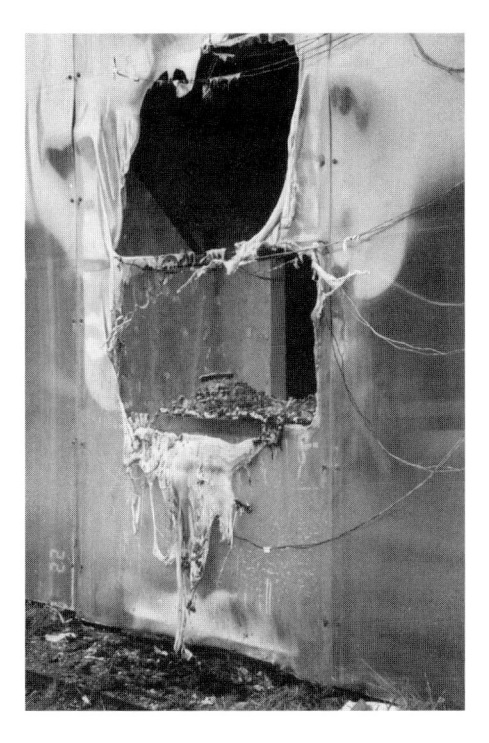

Figure 149b Extent of damage at the PC vertical glazing

Such experiences in the United States of America initiate that the positive classification in laboratory test methods has to be confirmed in a risk-relevant large-scale test [14]. Within the framework of the harmonization of the European Fire Reaction Test Methods, the test tandem cone calorimeter and room-corner test became propagated as the basis for a risk-relevant classification, especially by the members of the fire-science field [223, 237, 240], although the material-related characteristics achieved no correlation with that for fire scenarios defined in the Robust Solution by the expert group appointed by the Commission according to Table 4. The standard project, room-corner experiment, was initiated on the occasion of the ISO TC 92 session 1979 in Sydney to provide, analogous to standardizing activities of ISO TC 61 Plastics, an appropriate, material-related bench scale test of acceptable size [547]. These testing activities took as a basis the problematic nature of the ignition hazard under the impact of ignition sources of different especially raised intensity. Williamson proposed the burning wastepaper basket as a realistic ignition source for such kinds of test procedures [548]. For the safeguard in the industrial field, burning wooden cribs were designated as ignition sources [549]. With the wish to produce basic data for mathematical modelling, the testing and evaluation criteria were modified in the course of the research work. A risk-relevant differentiation of the fire scenarios, as done by the SC3 of ISO TC 92 Toxic Hazards in Fire was not the intention of the SC1 Fire Reaction Testing. The primary goal of the present ISO standard IS 9705 Room-corner test Procedure is exclusively the assessment of the flashover situation. The practice-proximity of the test tandem cone calorimeter and room-corner test is obviously limited to very specific fire scenarios. Hirschler [550] challenges therefore for the benefit of the cone-calorimeter test the practice-proximity of the Flooring Radiant Test, which was developed on the basis of extensive realistic large-scale tests at NBS [68, 182] and confirmed, e.g. in Austria by further large-scale tests [551], because the cone-calorimeter test results correlate with those of the room-corner test and is, according to his view, therefore much more practice-relevant. Hirschler bases the smoke-density evaluation on the large-scale tests with flooring materials in Gent [552] and Dortmund [553]. Hirschler neglects that these studies provide only informative values of fully developed fires but not for the developing stage of a fire. The targeted key test method within the framework of the European Harmonisation of the fire-reaction testing, the Single Burning Item (SBI) Test, is due to the experiences from a lot of practice-oriented large-scale tests with a room-corner test configuration, more suitable than the previous national procedures to guarantee risk-relevant classifications for vertical applications.

Table 110b gives a view of the intensity of different standardized ignition sources in relation to the heat-release rate of upholstered chairs. The graphical representations of Table 110b point also to a realistic simulation of an upholstery chair fire through the ISO sand-bed torch. Since in a real-fire case a strongly varied heat stress by the burning upholstery chair must be assumed, a generalization of the severely localized flame attack by means of the sand-bed torch is extremely doubtful. The relevance of these general conditions must be verified for every individual case. The graphical representation of Fig. 150b illustrates for the different ignition-source simulations, e.g. ring burner of the chimney test, wood-wool basket, the ISO sand-bed

6.1 Relevance of Combustion Systems | 295

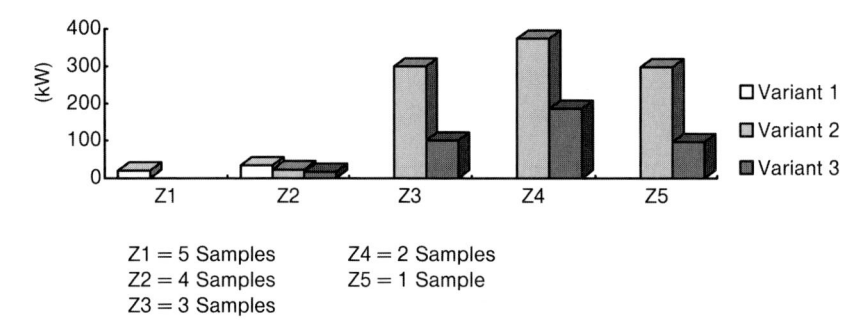

Figure 150a Ignition sources of different high intensity level; Z1 = chimney, Z2 = wood basket (300 s, 450 s, 600 s), Z3 = ISO Brenner 10 min., Z4 = upholstered chair (6/12 min.), Z5 = upholstered chair (12/28 min.)

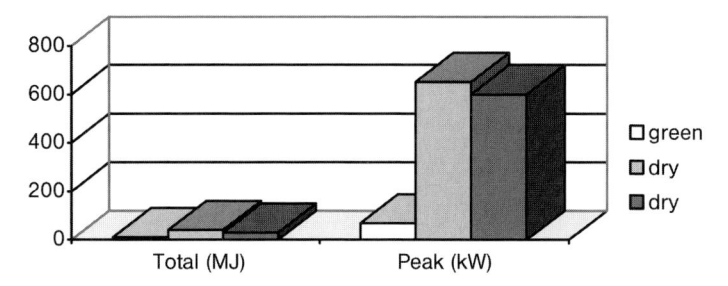

Figure 150b Christmas tree as ignition source

burner and upholstery units burning the possible influence of the time variant on the respective level of the intensity of radiation.

As practical ignition sources of raised intensity the Christmas tree as well as televisions can be named as examples. The ignition potential of the Christmas tree will have, e.g. a different effect depending on the moisture content. A rapid burning and an extremely high thermal stress must be reckoned if ignition occurred after a longer drying stage. The graphical representations of Table 110b and b as well as the listing of the Table 113a illustrate how the effectiveness of this ignition source can be changed.

The devastating effects of the Christmas tree ignition source were demonstrated by some spectacular fire disasters [554, 555]. The fire disasters of Quebeq and Anvers demonstrated among other things the danger of dried Christmas trees as secondary ignition sources. In Quebeq the combustion by playful ignition (just for fun) of firwood decoration caused the fire disaster with several dead visitors. The rapid burning of two Christmas trees caused in the hotel Switel of Anvers the death of in total 12 persons, although the fire remained locally and temporarily limited. Nine of the twelve persons succumbed to their fire injuries. van Hees et al. [556] report on this fire disaster based on the technical simulation at the Gent University. The ignition and burning behavior of the furniture item, e.g. the upholstery chair, decides the effectiveness of this secondary ignition source. If there is a longer self-sustained

Table 113a Fire safety characteristics of burning christmas trees

Ignition source Christmas tree	RHR (KW)	Mass–burning rate (g/s)	Heat–release potential (MJ/kg)
Artificial tree untreated	483	17,6	18,4
Norwegian fir			
small – dried	765	49,2	15,6
Large – dried	1590	111,1	19,3
Large – humid	670	100,7	6,5
Silver fir			
Small – dried	1117	65,9	16,9
Medium – dried	1817	104,3	17,4
Large – dried	2883	102,7	17,7
Large – humid	540	88,2	5,7
Large – dried decorated	3381	230,4	19,4
Large – treated	2736	179,5	15,2

Table 113b Ignition source intensity

Desription	Standard	Gas Flow Total mass	Power [kW]	Duration of burning [s]	Heat of combustion [kWh]
Small burner 20 mm	A ÖNorm B3800 CH SIA 183/2 D DIN 4102	0,025l/min propane	0,039	15	0,163 10^{-3}
Small burner 20 mm	I CSE RF 2/75A	0,025l/min propane	0,039	30	0,325 10^{-3}
Small burner 40 mm	I CSE RF 1/75A	0,062l/min propane	0,097	12	0,323 10^{-3}
Schlytertest	A ÖNorm B 3800	1,08 l/min propane	1,69	900	0,42
Brandschacht	D DIN 4102	35 l/min CH_4	20,9	600	3,48
Woodwool basket	D DIN 4102	600g woodwool	34,5	300	2,88
			23,0	450	2,88
			17,3	600	2,88
Ignitability 50 KW/m²	ISO TC 92, ISO 5657	–	1,6 x)	max.900	max.0,4
Cone Calorimeter 100 KW/m²	ISO TC 92 ISO 5660	–	5 x)	max.3600	max.5
Room Corner	ISO TC 92 ISO 9705	62,4l/min propane 192,7l/min propane	100 300	600 600	16,7 50
Wood crib 10 kg	TGL 10685/12	10 kg wood	267 133	600 1200	44,4 44,4
Wood crib 40 kg	ISOPA Leipzig	40 kg wood	533 356 267	1200 1800 2400	178 178 178
Wood crib 9 × 30 kg	ISOPA Leipzig	270 kg wood (30 kg/m²)	1200 900 720	3600 4800 6000	1200 1200 1200
Square gas burner	CBUF	19,5l/min propane	30	120	1

smoldering process, a rapid combustion of the preheated upholstery piece of furniture takes place, after the incidence of the ignition. Hallman et al. [557] discuss the various influencing parameters, intensity of radiation, surface structure, spectral absorption distribution influence in the case of direct flame impingement on the ignition t_i. The time to ignition t_i stays practically the same, above an intensity level of 75 kW/m^2. Hicks [558] and Kashiwagi [559] explain the theoretical connections between the ignition hazard and the external intensity of radiation. The importance of the different influence parameters such as, e.g. temperature distribution and absorption behavior as well as the concentration of the fire-gas mixture are shown and reviewed. For radiation as an ignition source it can be assumed that already burning items occupy the ignition-source function through their radiation. The pyrolysis parameters dominate the decomposition course of solid fuel. Reference [16] provides information on how far the different experimental parameters influence the characteristic t_i, the time to ignition. The possible influence of the necessary minimum ignition energy and the oxygen availability were reported in [20] and [14]. Berlin [560] discusses the thermal load depending on the flame length. The thermal stress is indicated with 1 kW/m^2 for small flames, with 20 kW/m^2 for about 0.3-m high flames and 40 kW/m^2 for 1.6-m high flames. Hansen and Asibulkin [561] discuss in this connection the temperature levels produced by the flame impact on a vertical wall. They carried out basic investigations with PMMA samples (see Fig. 151) and determined the degree of damage depending on the time of burning.

In the cone-calorimeter test according to Fig. 21 it is intended to assess the ignition hazard by means of the exposure time of the external, variable radiation. The time to ignition t_i, after the heat-release rate, is in this case a relevant characteristic. The graphical representation (see Fig. 152a) clarifies the strong influence of the external intensity of radiation. Increasing external intensity of radiation leads generally to a reduction of this characteristic. During the estimate of the danger potential by means of mathematical calculation the characteristic t_i plays a key role. The graphical representations of Fig. 152b illustrates that the value can be influenced significantly by a purposeful selection of the components. The safety level can be already changed impressively by constructive measures. The practice relevance of this procedure is postulated due to its good agreement with the room-corner-wall test results [562]. Simmes et al. [563] report

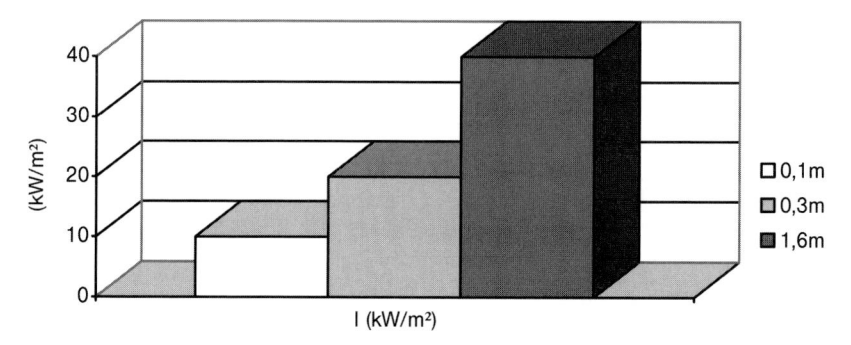

Figure 151 Influence of the flame height on the thermal stress of the impingement

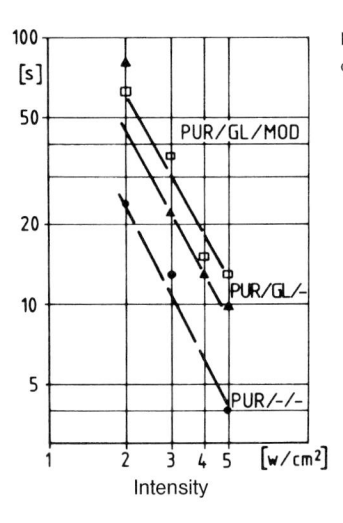

Figure 152a Influence of the external radiation on the ignition time

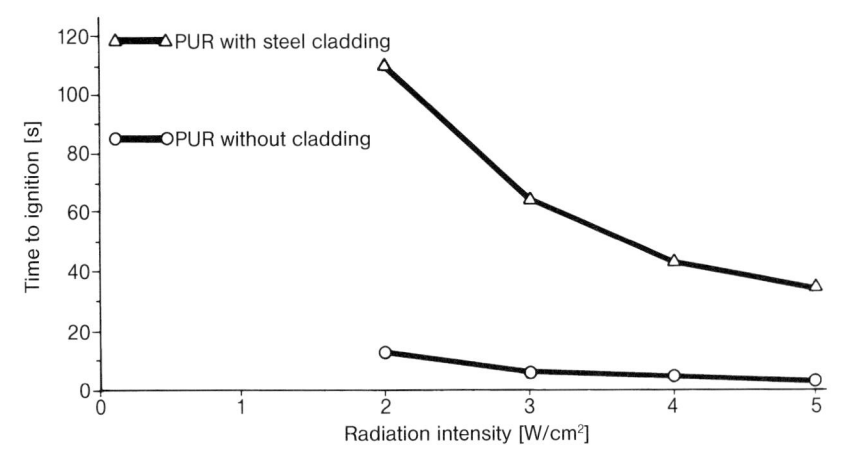

Figure 152b Time to ignition – Influence of the radiation intensity

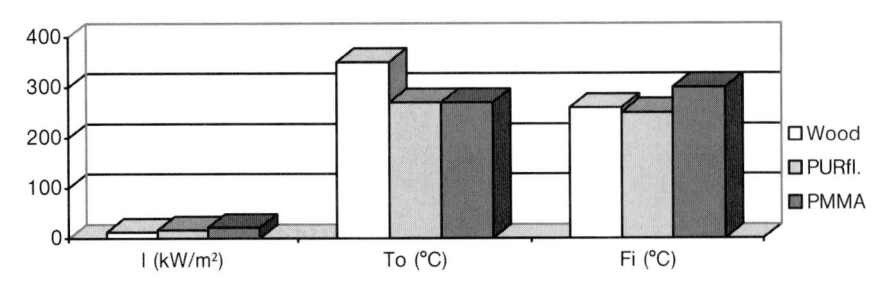

Figure 152c Risk of ignition at the impact of thermal stress

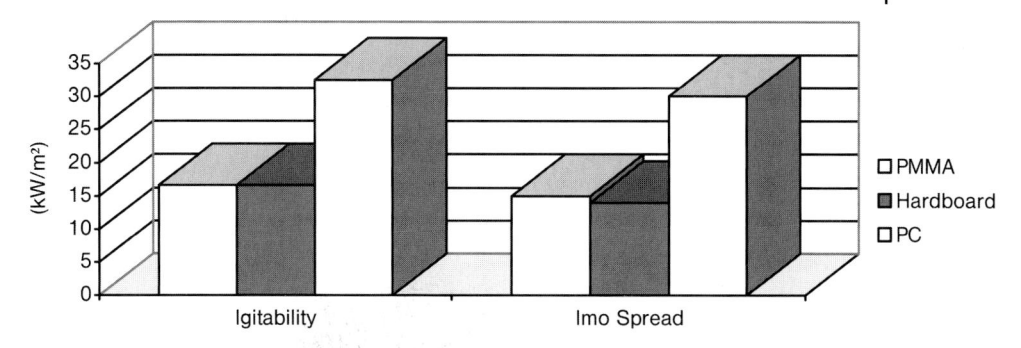

Figure 153 Influence of the test method on the records of the critical flux

on the influence of the humidity on the igniting risk in the case of external radiation. The higher humidity increased significantly the time to ignition t_i.

The data records of Fig. 152c, which illustrate the risk of ignition of various products by means of relevant surface temperatures as well as corresponding external radiation, differ considerably from the flash-ignition temperatures of the Setchkin apparatus [327]. The comparison of these data sets clarifies the necessity with material-oriented characteristics also to definition determinations.

To = Surface temperature °C)
FI = Flash-ignition temperature (°C)

Information is given in [16] to what extent different experimental parameters are in a position to influence the characteristic value t_i, the time to ignition. The data records of Fig. 153 show that the value is determined decisively from the external intensity of radiation. Provided that different procedures, e.g. the ISO Ignitability Test and the ISO Spread of Flame Test correlate, identical classifications are achieved.

Babrauskas [564] studied the influence of the external radiation on the ignition hazard of upholstered furniture. The data records of Fig. 154 [565] prescribe that a

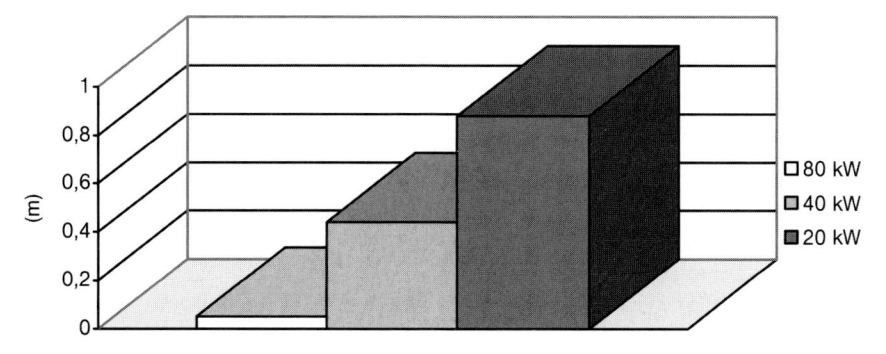

Figure 154 Safety distance in relation to the intensity of the burning item

rise of the intensity of radiation from 20 to 80 kW/m² considerably intensifies the ignition hazard. The safety distance of the burning chair to the neighboring item is determined mainly by the heat release of the already burning item. To reduce the ignition hazard, a considerably higher safety distance is necessary (>0.05 m – 20 kW, >0.44 m – 40 kW, >0.88 m – 80 kW). For 20 kW the average length of time to ignition is approx. 30.6 s, for 40 kW this is decreased to 14.5 s.

The risk of ignition in the transportation field is among other things fully discussed in [566]. A survey of the ignition causes of the essential fire disasters in tunnel plants is given with regard to statistical documents [567]. About 2/3 of these fires were caused by technical failures. The igniting risk problem due to the impact of external radiation was the subject of numerous studies, both in the lab, then also in the 1:1 measure. Drysdale et al. [568] demonstrate that the time to ignition t_i varies depending on the external radiation, while on the other hand the ignition temperature remains constant. Roberts [569] clarifies among other things how the product from thermal conductivity, density and specific heat, the so-called Kpc factor, is able to influence the ignition and combustion behavior. Papa and Proops [570] report on the possible influence of material-specific parameters as, e.g. recipe, fire-retardant additives or carbonisation and melting effects. O'Neill [571] reports on the ignitability risk of different products due to the external intensity of radiation and the heat-release potential acting during the combustion. Within the framework of the CBUF program these characteristics were determined also in dependence of the upholstery combination. The data records of Table 113b provide a view of the possible scale of the modification of ignition sources.

Reference [566] provides information about essential fire disasters in the tunnel plants based on fire-statistics data. The marked influence of the ventilation on t_i is discussed in detailed in [572]. According to [573] and [574] larger test pieces lead to somewhat smaller t_i values that may be connected presumably to the somewhat higher concentration of the fire-gas mixture.

The effect of fire-retardant additives after [575, 576] with ignition sources of low intensity is positive on the resistance against ignition sources. This influence is, as a rule, negligible with fully developed fires and with correspondingly high intensities of radiation in the cone-calorimeter test. Babrauskas and Parker [565] discuss to what extent the ignition hazard of upholstery combinations can be assessed by means of the cone-calorimeter test. Realistic investigations demonstrate that the risk of ignition of upholstery items can be significantly changed also by constructive measures [577]. Neither the individual components, such as, e.g. postulated by the English Furniture Regulation, nor the aimed relevant grid system can generally reduce the ignition hazard. This interpretation is confirmed by various investigations [578]. Room fires as ignition sources were regarded as a realistic basis for the examination of flooring materials in escape routes. The investigations of the University of Gent [552] have shown that during the selection of such ignition sources the aims of protection including all relevant test parameters have to be defined to guarantee a risk-relevant interpretation. The needed increase of the heat release of the burner to 600 kW in order to achieve a corridor flashover, illustrates that the flooring plays a subordinate role. This markedly increased the value achieved in the corridor flashover but they are obviously not suitable to carry out risk-relevant evaluations. A strong incipient fire leads in the ceiling field to temperatures of approximately 400 °C. In the flash-over situation a stream of hot gases over 800 °C has to be dealt with. The assignment to the individual stages of the fire is discussed fully. The 10-min value was mentioned among other things as the critical impingement time period. Since high intensity levels have to be expected always in the case of a flame impingement, material-specific differences of t_i are basically negligible in the direct ignition source action field. Figure 155a illustrates the possible effectiveness of secondary ignition sources using the example of a jute curtain fire.

Possible dangers through the ignition source of burning Christmas trees were investigated at NFPA [578]. Essential experimental parameters were tree species,

Figure 155a Combustion of a jute curtain

Figure 155b Simulation of a room fire

humidity and decoration in particular. Flame height, maximum heat-release rate, maximum burning rate, fire duration and the time at maximum temperature were used as evaluation criteria. The data records of Table 113a illustrate very vividly that the moisture content must be classified as a dominating parameter. The possibly escaping flames out of the door in the case of a room fire are illustrated by Fig. 155b. This picture resulted in the combustion of a fire load of 30 kg/m^2. If the burning fire load comes into play as a secondary ignition source for the neighboring facades or adjoining rooms, flashover was reported. If the British Fire Statistics Data were used as a basis, then this statement is valid for about 10% of dwelling fires only as demonstrated by Fig. 62. According to Babrauskas [579] this statement is valid for about 60% of the dwelling fires in the United States of America. Babrauskas reports that most of the fatalities in dwelling fires are caused by a flashover situation. The graphic presentations of Fig. 33a and b demonstrate that such general statements may not be correct at all. Recent fire statistics data concerning the average extent of smoke and fire damage in US high-rise occupancies shows quite a different picture. Differentiations might be necessary for towns, countryside, etc. In [580] Babrauskas reports on large-scale Japanese tests, with fully furnished rooms, to illustrate the fire-risk situation of houses endangered by earthquakes. The fire load conducted was 35 kg/m^2. Within the 60-min fire duration no flashover with fire spread to the neighboring fields in the stressed flats occurred amazingly in all three experiments.

The hot gas stream emerging from the window possibly causes surface spread of flame in the facade field during the flashover situation. The ignition in the fire section situated above is caused as a rule by the radiation component of the hot gas stream (see Fig. 156). Curtains and wall panelling played, as shown by relevant fire disasters, a decisive role [106, 581]. If the hot gas stream of the burning room emerges, e.g. in the door field, then the burning behavior of the wall and ceiling linings as well as the flooring materials in the bordering escape routes then become possibly the dominating parameter.

The problematic nature of the sudden fire spread due to the flash ignition of the fire-gas mixture was the subject of various research studies [582, 583]. This effect is discussed in detail by Waterman [584] by means of extensive large-scale test results. The evidence of a sudden ignition of the fire-gas mixture was produced by using positioned paper strips as indicators. According to Morikawa [114] the result of large-scale tests, which included two floors, demonstrate that the flashover situation led to a strong rise of the acute toxic potential of the stream of effluents. The period of time up to the flashover is basically dominated by the fire intensity, the spatial factors as well as the ventilation [54]. During the flashover situation extremely high intensities of radiation as well as a strong decrease of the O_2 content in the stream of effluents has to be reckoned on. The investigations, described in [585], illustrate also that thin deposits, even with a somewhat more favorable fire-reaction evaluation, can initiate a fireball situation, provided that they come with a large surface to the decomposing stage. The correctness of this experimental interpretation is confirmed by the King Cross fire disaster, whose course is explained in [5] and [586] by the findings of technical experiences. In full-scale field trials with textile wall coverings flashover was indicated by the effluent temperatures of ca. 600°C. The fire-gas temperatures were

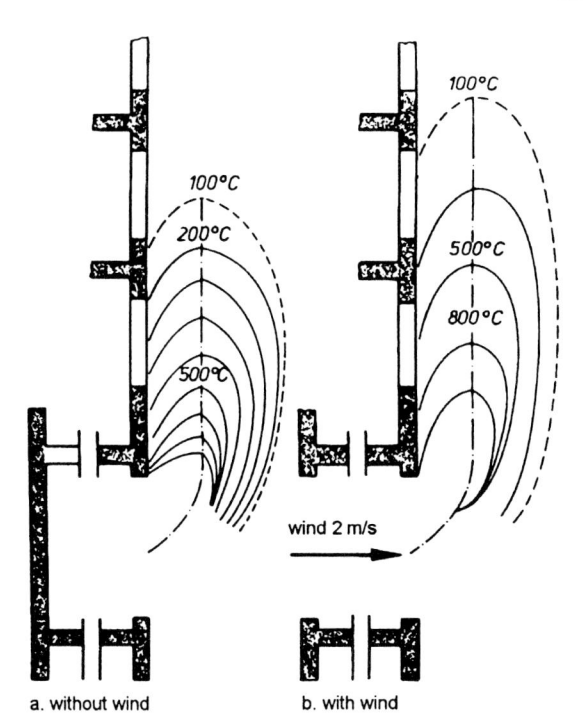

Figure 156 Temperature profile of the hot gas stream escaping in the window area

a. without wind
b. with wind
wind 2 m/s

produced by means of a defined sand-bed burner (40 kW–5 min and 100 kW–10 min) analogous to the room-corner test. Emmons [107] attempts to analyse the course of the Beverly Hill Club Fire by means of the possible flashover situations. The reason for the extremely fast fire spreading is, according to his understanding, primarily the ignition of the new decomposition products produced by the impact of the smoke layer. Quintiere [588] confirmed that the flashover situation is not caused through unburned flue-gas parts, but through the decomposition products of ceiling tiles. Quintiere [589] and Friedmann [582] underline that the flashover situation is characterized by a fast flame spread in the flue-gas mixture. This statement stands in contradiction to the determination of Sundström et al. who report in [179] on full-scale field trials with upholstery furniture items. This Swedish and Norwegian joint program demonstrated, according to Sundström, that a flashover situation can happen without combustible wall and ceiling linings. These different findings could be interpreted in the way that Sundström et al. defined the flashover occurrence in a somewhat different way, which is in line with the ISO room test configuration but not with a real room-fire situation. Lawson et al. [496] discuss the risk of a flashover situation with regard to large-scale fire tests of upholstery furniture. Denyes and Quintiere [183] report on the necessary fire-load size to initiate a flashover situation in a room-corridor configuration. According to Briggs et al. [279], room fires of extreme heat-release rate can also cause flashover conditions in an adjacent corridor. The relevance of the extremely high fire load for the common fire risk in residential areas is not shown by this. An additional heat insulation of the flooring material plays, after [546], no decisive role. Relevant 1:1 experiments caused only minimum deviations. Budnik et al. [591] verified in practical full-scale field trials that wall and ceiling linings are able to influence the origin of a flashover situation next to the furniture. The large-scale tests with passenger coaches produced a relatively good agreement. The combustion behavior of the interior content, including the wall and ceiling lining, determines decisively the period of time until the flashover occurs. Ödeen [592] produced in large-scale tests with passenger coaches the flashover within 5 min, a time length that was achieved also in Germany [80] as well as in Japan [593]. In the case of a fire the flashover must be reckoned on within 5 to 10 min. Morikawa indicates this length of time as >5 min. The stream of effluents reached, as illustrated by Fig. 157a, within these experiments temperatures of approx. 900 °C. According to Morikawa the fire load in the living area conducts, as a rule, 28 kg/m^2 [114, 424]. The fire load in railway vehicles is, as a rule, about 50–80 kg/m^2 [566].

Fiala and Dussa [108] achieved similar results in the case of a large-scale test with an airbus cell. The impingement by means of a kerosene torch for 1.45 min led to the flashover situation with an extreme flue-gas temperature rise and an oxygen decay. The maximum temperature was reached within 2 min and 15 s. Herrington [354] tried to explain the fire-gas problematic in aeroplane fires using the example of flexible PUR foams. The stream of effluents were simulated by means of the OSU-Chamber. A correlation to current fires was not found. The addition of fire-retardant additives caused, in the laboratory tests, high CO and HCN values as well as extremely high smoke-density levels. Fang [594] explains the results of room fires from the viewpoint of the reproducibility. A good agreement was achieved for both the temperature evolution and also the intensity of radiation, in the opinion of the authors. The deviations were under 6%. The temporal process of the fire-room temperature can be desig-

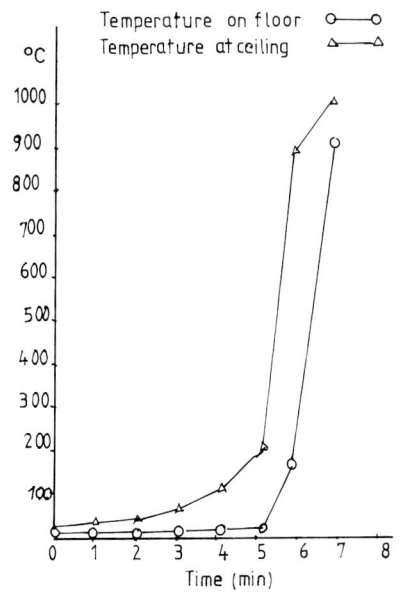

Figure 157a Fire-gas temperature evolution in the flashover situation

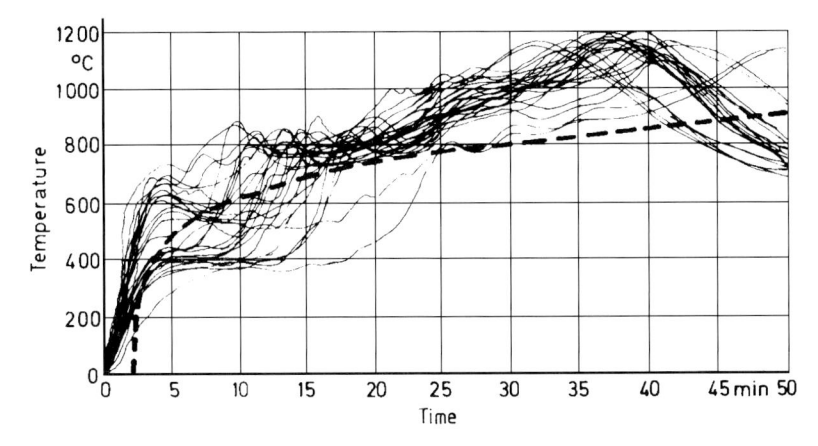

Figure 157 b Fire-gas formation at the ETK combustion of wooden cribs

nated as relatively uniform. Significant deviations can not be excluded in the case of such experiments. The realisation of two experiments only does not suffice obviously to carry out a dependable interpretation of the results. Full-scale field trials are due to the complexity of the actuating variables, as a rule used as demonstration objects. In the case of the interpretation of the results a corresponding bandwidth must be taken into account provided that deviations, not possible for the essential parameters, were limited. The question of the reproducibility of full-scale field trials simulating real-fire cases was the subject of different studies. After Tran [595] a good agreement of OSU Chamber and cone calorimeter results with the fully developed fire can be confirmed concerning the heat-release rate. The Delft experiments for the evaluation of the trapezoidal roofing elements showed a relatively good agreement of the combustion behavior of the defined fire load, wooden cribs, as illustrated by Fig. 157b.

The temporal course of the fire-gas temperature at the simulated fire of a small storage room showed a relatively good agreement also as a result of the defined edge parameters. In spite of exactly defined test parameters differences of the fire course could not be avoided during the developing stage of the fire. Malhotra [596] and Benjamin [597] showed very vividly, using the example of flooring material, that in the case of correct selection of the test parameters a sufficient agreement of results is attainable from the laboratory and large-scale experiments. A fully developed fire is the starting point for mathematical modelling. Theoretical mathematical aspects of the ventilation-controlled fire are discussed by Heskestedt [46] and Harmaty [232, 233]. The verified relatively good correlation of cone-calorimeter and room-corner test results clarify the difficulty to initiate the fire-gas formation by means of the cone calorimeter in a developing stage of a fire. The standardized cone-calorimeter test method simulates up to now only a well-ventilated pool-fire situation of a fully developed fire.

6.1.2
Relevance of the Procedures for the Assessment of the Side Effects of a Fire

Paul and King [598] report on large-scale tests with upholstered furniture units. Topics of the examinations were, among others, the influence of the test parameters; size of ignition source, exposure time as well as cushion construction/upholstery combination including the recipe variation. Wiliamson [599] discusses possible performance-influencing factors in the pre- and postflashover situations. Ventilation and possible preheating effects proved to be particularly significant performance-influencing factors. The key question was among others defined by Babrauskas [564] "Will the second item ignite?" The temporary trend, to define the fire-performance orders with the aid of mathematical modeling, disregarded the present state of knowledge about the obviously still existing deficiencies for the scenario of the incipient fire. Also, Beard [600] questioned the reliability of computerized modelling for the characterisation of the temporal course of fires. The entire fire situation for the different fire scenarios is still too complex for universal numerical approaches for the forecast of possible fire processes. Friedmann [39, 40] reports due to the unclear compatibility of the extremely complex influence parameters in the stage of the incipient fire on existing reservations against the prediction of the course of a real fire. Kawagoe and Hasemi [601] explained the progress of the Japanese fire science in this connection. The academic implementation approaches are, in their opinion, still afflicted with so much uncertainty that currently a generalisation is not possible. The complexity of the fire problem and the uncertainty about the effectiveness of different parameters is why the numerical representations are very much limited. The assumptions made include:

- the fire-gas mixture relates as a perfect gas
- the molecular weight of all components of the fire-gas mixture are identical
- the values of the specific heat and the different coefficients of all components stay constant
- the chemical reactions in the test phase are negligible
- the boundary layer is laminar and
- the mass-loss runs according to an Arrhenius formula and depends on the surface temperature only

They are in the acute fire case not verifiable. The geometrical order of the shared components is very variable and within a given time stationary. Emmons [12, 113] explains the entire situation of the computation models and recommends a cooperation of the teams responsible for the relevant models. In his opinion true solutions are possible only for partial aspects of a fire. Emmons clarifies that due to the innumerable variants all possibilities are not computable for the time being. During the characterization of the fire three different scenarios had to be considered: smoldering fire due to exotherm reactions and combustion in particular under generation of flames with a) reduced and b) sufficient oxygen offer.

In order to be able to take into account the safety attempts in the sense of the second essential requirement of the planned furniture directive, the general conditions of the second basic demand of the preventive fire protection "the spreading of fire and smoke is to be precluded" must be taken into account sufficiently. The scale of the possible danger in the case of an expanding fire comes in particular from the fire occurrence including the defaulted spatial circumstances. Provided that there is sufficient oxygen the spatial combustion occurs according to the law of free burning and is marked therefore from the free surface of the fire load. The combustion runs according to the formula

$m'' = K_x F$,
where
m'' = burning rate in (kg/m²)
F = free surface of the fire load (m²) and
K_x = an index that depends on the nature of the fire load as well as on the temperature evolution in the fire room.

Provided that the available air for the burning process does not suffice, the fire development is controlled by the ventilation conditions. Due to extensive investigations with wooden cribs, Thomas [28] developed a relationship on the basis of simplifying assumptions.

$m'' = K_x \times A_v\, H^{1/2}$
In this case
m'' = mass-burning rate (kg/min)
A_v = the ventilation area (m²)
H = height of ventilation area
K_x = a characteristic, whose value for a window area smaller than $1/2$ of the wall area is 5.5 to 6.0 (kg/min.m$^{1/2}$)

Seeger [35] points out that the communicability is not guaranteed on arbitrary fire cases. The complexity of the fire problem and the uncertainly about the effectiveness of the different parameters is after [601] the cause that the pure mathematical modelling is very much limited. Kashiwagi's [574] explanations can be interpreted in such a way that the usual computation models require, to a large extent, considerable simplifications so that the results of such numerical approaches are still afflicted with significant uncertainties for the incipient fire. They are accordingly still for the time being unsuitable for layer determination. Thomas [602] discusses model ideas for the preflashover stage. Bukowski [603] explained possibilities to carry out such computational predictions by means of basic characteristics. He also considers in this case feedback effects, spatial conditions and ignition-source criteria. Levin [604] discusses model presentations for the predictions and Lee [605] explains model ideas of the possible contribution to the spreading of fire in the furniture and furnishing fields. The dominating influence of the upholstery furniture compound is illustrated. Babrauskas and Parker [542] attempt the computational forecast on the course of furniture fires by means of material-related characteristics of the cone-calorimeter test. Jianin [606] pursued an analogous aim by means of the results from practical Swedish full-scale field trials and relevant computation models. Due to the functional connections such mathematical modelling delivers results that do not allow generalisation for the course of a developing fire. Friedmann refers in [607] to the possible fire-gas danger due to a rapid fire development. The developing stage of a fire is convicted in his case very fast in the fully developed fire, which then meets the assumptions for the numerical approaches. Kawagoe and Hasemi [601] discuss the different partial aspects, such as temperature–time profile, fire spread and smoke-layer formation. Favorable assumptions for numeric approaches are given in his opinion only for partial aspects where due to a flashover a fully developed adopt/prepare and spreading of heat and smoke orients on the connections of the ventilation-controlled fire. During the real fire different influences will possibly dominate others, for example, ventilation or preheating. Woolley and Fardell [608] explain the trend of future computation models to the estimating of possible fire-gas dangers. Jolly and Saito [609] demonstrate the influence of surface spread of flames on the fire-gas generation by means of the results of relevant full-scale field trials. According to [38] the burning rate of the fire load depends decisively on whether there is a fire load or ventilation-controlled fire. Wakamutsu [610] explains the numerical model ideas for the forecast of fire-gas movements that does orient itself on physical regulations. He clarifies that in the current fire case due to unknown limiting conditions significant deviations can appear. In [611] it is demonstrated that for a universal numerical computer program standardized rooms including the furnishings and their arrangement as well as further simplifications are conditional. Numerical approaches for the predictions of the fire-gas spreading depending on the spatial factors, expect after [612, 613] a promising chance of success. The fire-gas movement remains accordingly computable as long as there are no outside influencing occurs. Waterman and Domijanaitis [614] explains the theoretical connections of the smoke movement. Wickström et al. [615] report on the problematic nature of smoke control in hospitals. The smoke-layer formation and the smoke spreading in bedding fires were essential aspects of these studies. Questions of the fire-gas distribution in the corridor, including possible temperature and pressure gradients, are treated in [616]. The determination of the temperature gradients in the corridor field was also part of the studies in [601]. Quintiere and Denbraven [617] illustrate a computation model to the characterisation and forecast of the movement of the smoke layer. The effectiveness of the computation model became tested in a model-room corridor layout in the measure 1:7. In the endfield of corridors a considerable turbulence of the stream of effluents must be taken into account. Quintiere et al. [618] discuss very fully the regularities of the smoke movements. Christian [619] explains the principle basic questions of the fire-gas movements in staircases. Hinkley also clarifies in detail the problematic nature of the smoke spreading at FRS [620]. Relevant model approaches were demonstrated among others in various studies [618, 621, 622] and computational programs are described for the determination of the smoke-layer movement in tunnels. According to Sato [623] the fire gases of a developing fire do not represent any immediate danger in the tunnel due to the dilution effects. These conclusions are based on the results of risk-relevant full-scale field trials with passenger coaches. Within the framework of a detailed search of fire investigations it showed that the fire-gas formation mainly leads to a significant interference of the escape possibilities in accordance with Fig. 80. Basic characteristics were gained, for example, in the extensive investigations of the operation "Schoool Burning" [624]. The practice-relevant large-scale tests in a multistorey building in Paris demonstrated the extremely fast fire-gas spreading. Fackler [625] reports that the smoke of the large-scale test spread within 12 min from the room of origin in the second floor up to the 14th floor. A smoldering fire caused by a cigarette spread the smoke in 25 of 42 floors. At the UL investigations the exit signs were not to be seen within 5–12 min. The danger of uncontrolled fire-gas spreading is shown by means of current fire cases. Fire disasters such as, for example, Kings Cross Fire [5], Dublin Leisure Park Fire [7], Saint Laurent du Pont [626], Star Dust Fire [634], Beverly Hills Club Fire [107, 173], Cleveland Fire [88], Manchester Woolworth Fire [351], Glasgow Fire [627], Canberra Fire [628], New York Plaza Fire [583], New York BOAC-Terminal Fire [249], Old Nürnberg Fire [629], Hotel Fire Zaragoza [633], Turin Movie Theater Fire [629], DuPont Plaza Fire [8], Ferienzentrum Holm [628], Varig Orly-Air Plane Fire [630], Hotel Augarten Fire [548], Sao Paulo Fire [631], Bradford Stadium Fire [632], etc., illustrated the extremely complex flue-gas problem. Easton [636] explains this problem using the example of the 1942 Coconut Grove Night Club Fire where 492 persons were killed. CO and aldehydes are named as the essential fire-gas components. It becomes clear that not the material-related potential but rather the contribution to the fire spread dominated the danger potential. Bryan [637] reports on the human behavior in the MGM Hilton Hotel Fire. Carmack [638] explains the relevant literature about human behavior in fire and Bryan [639] discusses all relevant performance-influencing factors. He refers, among other things, to the experiences of UL and Fackler [625] and warns before the possible, fast spreading of smoke. The possible fire-gas danger due to a mattress fire in a hospital is discussed in [640]. On behalf of theoretical considerations it is clarified that the length of time of 10 min between two control walks is too long. Teaque [641] discusses the possible fire-gas danger in fires in school buildings. Vandalism and playful arson are named in this case as the possible main cause. The effect of the findings from fire disasters illustrated that in the fire case not always the shortest, but the most familiar way is searched presumably [113]. Wood [176] discusses the behavior of the effected group of persons. Bryan [595] discusses diverse experiences of the fire degretament on site. Detailed information also supply the relevant court records [106], where extremely inconsistent statements are made by the effected subjects. The influence of the fire gases on the escape behavior was examined in detail by Phillips [177]. He illustrates that the escape behavior can depend considerably on the degree of knowledge and the physical factors, e.g. also from the ownership circumstances. Men are rather more ready to go through the smoke. Men considered primarily extinguishing activities, while women strive for warning activities as a rule first. Berlin et al. [642] discuss the evacuation problem in the current fire case. Mathematical model ideas are developed in order to examine possible influences of the essential parameters such as movement, reaction behavior and stress situation. Under observation of the essential parameters the planning for two buildings was tested. Stahl [643] reports on analogous studies. Supported by an extensive literature research the author considers the course of the fire and the movement of the effected persons. It is to be assumed that there is not always a rational behavior. Panic leads, according to Abe [398], to irrational acts. According to Bryan [637] panic lets forget all ties as an increased feeling of the fear. Only a fast reaction helps to overcome the fear. Friedmann discusses in [644] the escape chances. Due to the possible diverse influence parameters a general computational problem solution is not possible. The length of time available for rescue operations can be, according to the present fire scenario, possibly very short. Jin reported in [80] and [178] on studies where the behavior of people in smoke was observed. Reduced visibility, irritant gases as well as knowledge of the neighboring areas proved in this case to be important parameters. As boundary values of visibility relating ships $D = 0.5/m$ and $D = 0.05/m$ are indicated. Mepperly and Sewell [645] discuss fire-gas dangers due to CO and HCN intoxication. Possible effects of nitrogen oxides and isocyanate components are shown. In [646] the problematic nature of the escape behavior of passengers is examined in the case of a fire at sea. The authors are engaged to explain the behavior of passengers during the Scandinavian Star Fire with the aid of the present experiences at fire disasters ashore. Shorter [647] clarifies, e.g. that through measures of the preventive fire protection, e.g. an effective compartmentation that uncontrolled spreading of the fire as well as the resulting fire gases can be markedly suppressed. Maguire [648] and Philipps [177] and Julga [649] refer to the possibility through measures of preventive fire protection, e.g. education, to create a more favorable initial position with simple living units extinguishing operations will have, as a rule, priority. The listing of Table 107 provides a view of the essential evaluation criteria of the different national certification tests concerning the contribution to the fire spreading. Foreseen from some specific exceptions, where, e.g. also the flame-spread rate is to be considered as a decisive criterion, the limited scale of the fire spreading forms the base quantity for the classification. The self-sustained smoldering process is a material-related quality that depends on the air. The main criteria of the various national tests concerning the spread of flame are listed in Table 108. The schematic representation of Fig. 158 illustrates a smoldering fire caused by a glowing cigarette.

Carter [650] describes studies on possible influences of the smoldering behavior of cushion combinations through formula variations. The smoldering fires were initi-

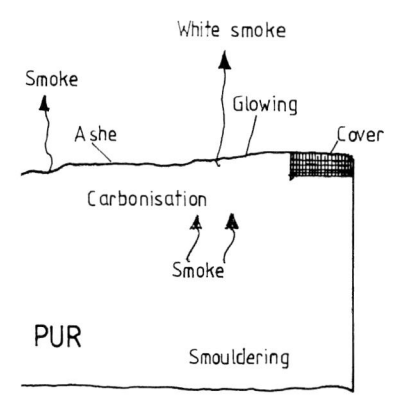

Figure 158 Smoldering fire

ated with glowing textile stripes with a weight per area unit from approx. $590\,g/m^2$ and smoldering temperatures of 250–400 °C (see Fig. 158). Provided it results in the smoldering process, effluents are released whose CO/CO_2 ratio, as demonstrated vividly by Fig. 166, is extremely low. The glowing rate is dominated by the ventilation conditions [651]. A value of approx. 0.6 cm/min was determined. Melinek [652] clarifies in the strong, combined effect of the components "smoking" and "drinks" to the ignition risks and the damage scale combined with that. Drysdale [653] discusses the spreading of smoldering fires. He explains the physical connections by means of the models of Williams. A raised smoldering fire risk exists after [654] in the case of slack fillings. Wiendl [66] discusses the igniting risk of smoldering fires and the essential parameters leading to the smoldering fire spread of dispersal bulk material. The phenolic resin as binding agents is the critical component of mineral-fiber mats. Jahn [655] clarifies the propensity of the spontaneous combustion through accumulation of heat from sawing dust. The spontaneous combustion was demonstrated within the framework of the FTC regulation. Such risks exist in particular in the case of cellulosic materials, dusts and open-cellular combustible substances. After initiation of the fire, spreading occurs without additional external application of heat [95, 76]. Ohlemiller [656] describes smoldering experiments with cutting dust. The smoldering velocity is approx. 4×10^3 cm/s. The determined smoldering temperature was for three standardized products at 330–450 or 620 °C. The spreading of the smoldering fire of flexible PUR foams was the subject of various studies [657–659]. The smoldering fire zones of about 350–500 °C contains 2–3 cm and expands, depending on the ventilation, at 10 cm/min. The released amount of heat is 200–300 cal/g. The data records of Fig. 19 give a view of the determined material-related data (Salt Lake 74). The representations of Fig. 19 show very vividly that a raised ventilation of the spongy material tending to smolder lead to an increased smoldering velocity. The surface spread of flame is determined only in exceptions in the usual laboratory procedures as the classification criterion. Almost altogether the limited contribution to the surface spread is condition of the consisting. The upholstery combination with the easy cotton fabric showed in the modified ISO/IMO Surface Spread of Flame Test according to Fig. 159a in relation to the pure flexible PUR foam as well as the PES-PUR combination, a significantly faster combustion. Differences in the damage scale form as a rule the basis of the respective rating. The data records of Figs. 159b and c clarify the influence of the cushion compound on the scale of the surface spread of flame and the burning rate combined with that. The essential faster surface

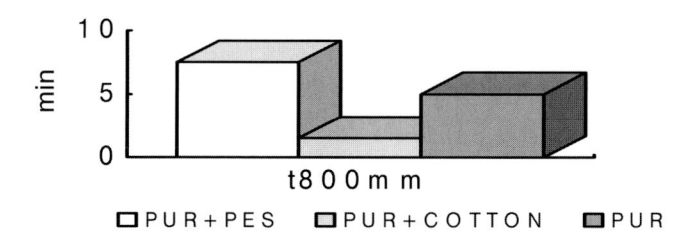

Figure 159a Influence of the upholstery combination on the surface spread of flames in the IMO test

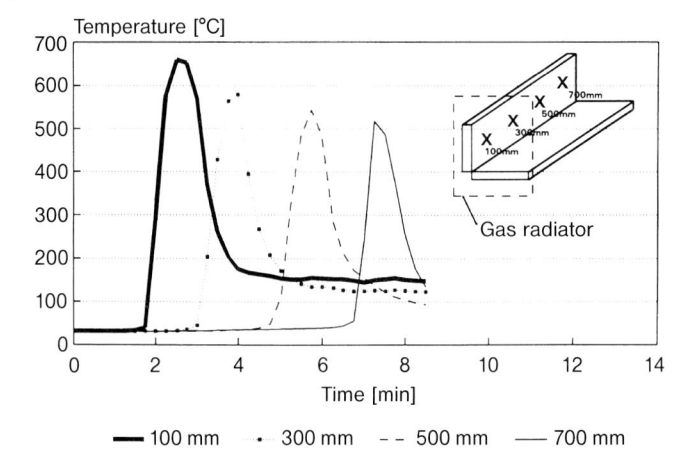

Figure 159b Modified IMO test for proofing the spreading in the upholstery combination

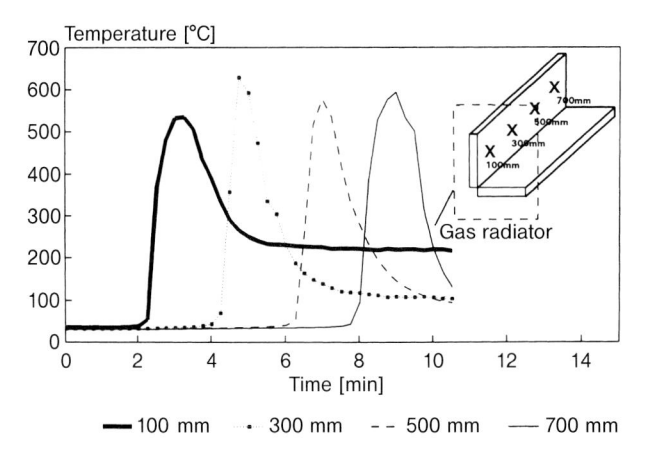

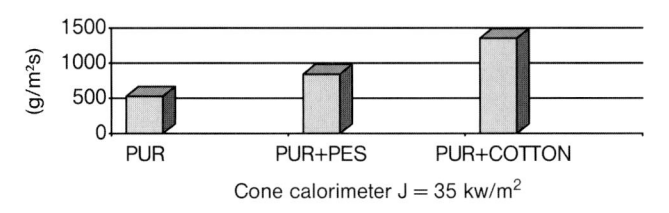

Figure 159c

Figure 159d Influence of the upholstery combination on the burning behavior in the cone-calorimeter test

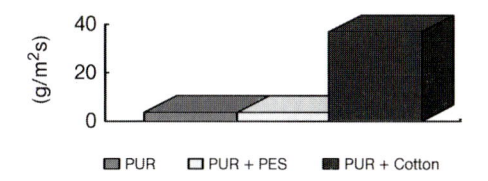

Figure 159e Influence of the surface spread of flame on the mass-burning rate

spread of fire of the tested cotton composite caused larger areas of the upholstery compound became part of the fire. A significantly increased spread occurs under preheating conditions in combination with the burning behavior of the composite in the modified IMO-Spread procedure [276]. An equivalent differentiation of the tested combinations with cone-calorimeter results is not possible.

The graphical representations of Fig. 159d demonstrate that those differences can not be realized by the pool-fire model of the cone-calorimeter procedure. The data records of Fig. 159e underline this statement. The fire spreading across the surface occurs in this test configuration in a considerably shorter space of time. The present results confirm that both the outer covering as well as the interliners and the fillings are able to modify the contribution to the surface spread of flames. Figure 159e also provides information about the intensified combustion activities per unit of time and the fire-gas release combined with this.

The graphical representations reveal that the differences of the heat release verified in the cone calorimeter are negligible in comparison to the fast burning rate of the seat unit caused by the increased spread of flame due to the cotton cover. As long as the fire remains limited on an individual subject, such as for example a chair or a lounge according to Figs. 160a and b and the second piece of furniture is not located next to that, it has to be assumed that the radiation component will not attain the dominant position.

The data records of Fig. 161 illustrate the influence on the upholstery combination of the fire spreading across the surface. The investigations, within the scope of the PRC (Product Research Committee) project, demonstrate that the mass-burning rate can be crucially changed by constructive measures. These studies were carried out in

Figure 160a Fully developed fire of a chair unit

Figure 160b Pool fire of a settee

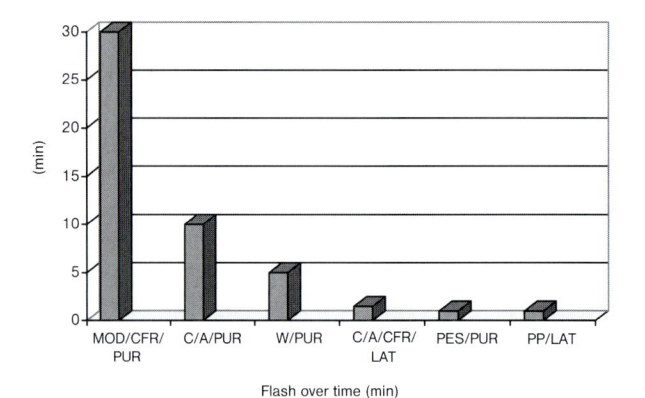

Figure 161 Influence of the upholstery combination on the mass-burning rate

the fire-test house of RAPRA (Rubber and Plastics Research Association) (see Fig. 162a). Figure 162b illustrates schematically the selected 2-chair arrangement of the investigations.

Practice-relevant investigations with upholstery units have illustrated that ignition sources of larger intensity, such as for example paper, will lead to the combustion of the ignited item, depending on the arrangement of the ignition source and the design of the item. Babrauskas and Parker [660] come to analogous conclusions. Specific torches, such as, e.g. the NIST-Burner (see Fig. 145c), are according to their

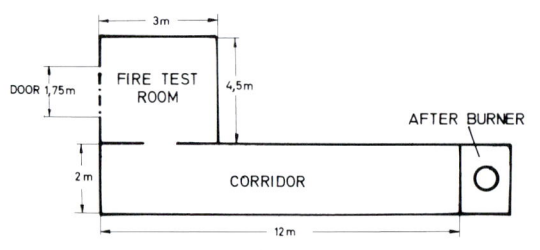

Figure 162a Fire test house of RAPRA

6.1 Relevance of Combustion Systems | 315

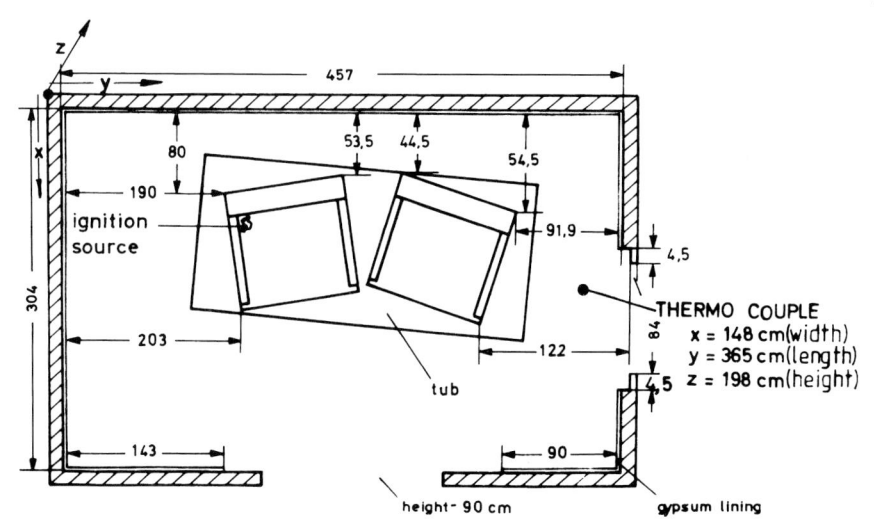

Figure 162 b 2-Chair – Arrangement

meaning not necessary, since all usual examined combinations could be brought to combustion using the secondary ignition source paper. The flashover on further furniture items depends on the intensity of the flames of the item already burning, the spatial distance as well as the feedback effects and the ventilation conditions. The data records of Fig. 163a reveal that the fully developed fire of the ignited item develop possibly very fast. The specific characteristics of the upholstery fillings themselves determine the degree of the fire spread by upholstery combinations with melting coverings. The investigation program initiated and financed by the International Institute of Isocyanates (III) disclose that the flashover to the next chair depends in particular on the fire course and with that on the size of the flames and the intensity of the radiation. The intensified burning behavior of the second chair unit was

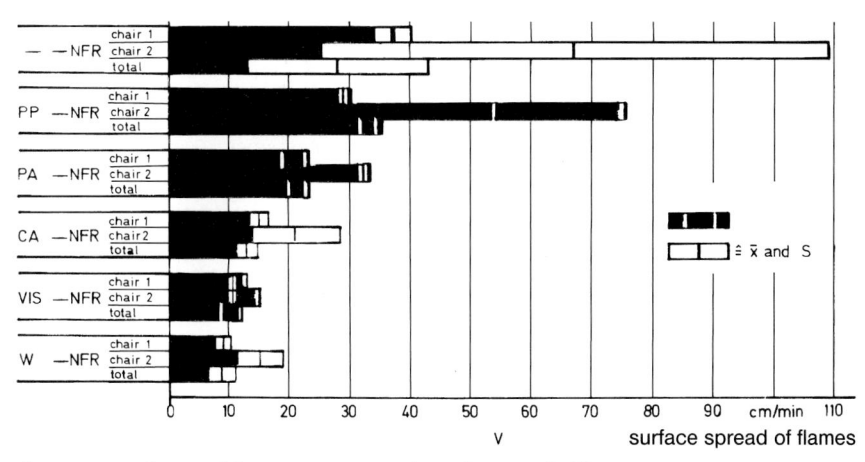

Figure 163 a Influence of the outer covering to the surface spread of flame

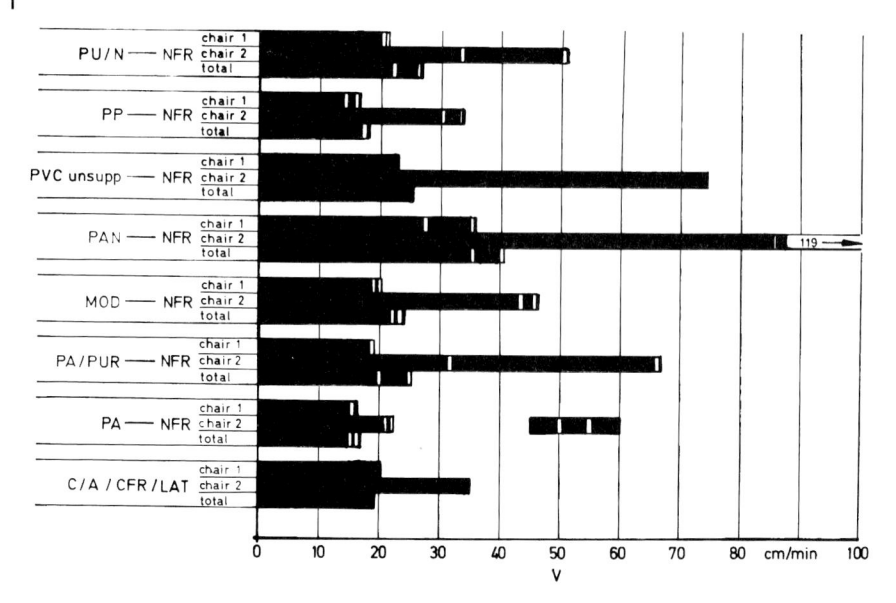

Figure 163 b Influence on the fire spread

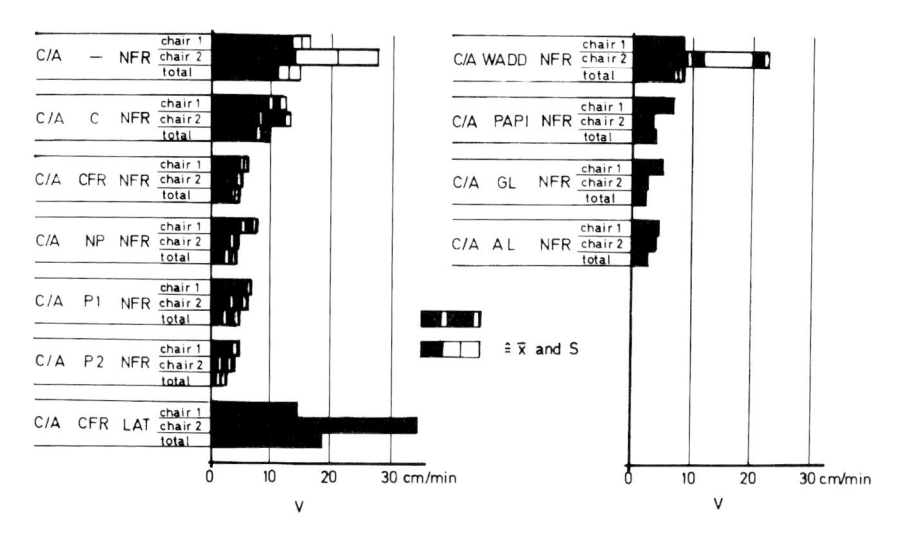

Figure 163 c Influence on the fire spread

effected by the preheating of the chair already burning. The flame size during the burning behavior depends on the pool size of the item already burning and from the upholstery construction.

The investigation results of this research project disclose very clearly that the fire spread across the surface is significantly changed through the material selection and through design variations. The results of Figs. 163b, c and d demonstrate that the

6.1 Relevance of Combustion Systems

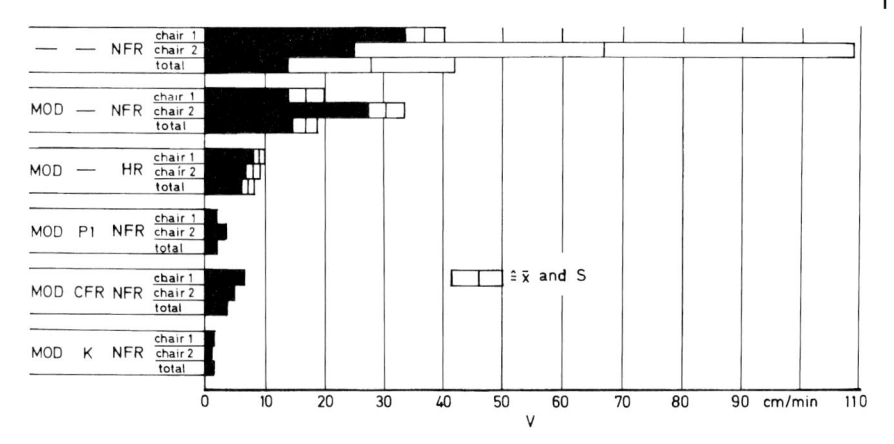

Figure 163 d Influence on the fire spread

chosen experiment conditions, including the spatial circumstances, depend on the recipe variation. In the sequence standard polyether, polyether FR and high resilience/cold-cure quality a reduced surface spread of flame is seen. The graphical representations of Figs. 163b and c underline, however, the strong influence of the outer covering. Coverings that are easily ignitable and that enable the direct impingement of the upholstery filling by a marked peeling effect, obviously lead to a faster fire spread. This behavior could be verified for covers with thermoplastic behavior such as, e.g. PA, PAN, PES, PVC or MOD fibers. The burning rate could be considerably reduced by the use of suitable interlayers. When the stage of the fully developed fire according to Figs. 163a and/or b is reached with the first-ignited item, the fire course as a rule depends on whether the second item will ignite or not. The graphical representations of Figs. 163c and d give a view of the results of these risk-relevant full-scale field trials with different upholstery combinations. These data records show very vividly that through variations in the upholstery construction, both the flashover onto the second chair unit as well as the destruction by burning can be modified significantly. The graphical representations of Figs. 163c and d demonstrate the significant influence of the outer coverings and the interliners. Figure 163d illustrates the investigation results with different settee constructions that also demonstrate that the burning rate will be significantly modified by the variation of the upholstery combination. The comparison of the two different upholstery filling materials PUR and latex foam show that the latex-combination produces, even with interliner protection, a faster combustion behavior than unprotected PUR combinations. These graphical representations underline the strong influence of the composite. The extremely positive qualities of the modacrylic fibers could be demonstrated at least in connection with effective interlayers. The graphical representations of Fig. 163c show very vividly that both the burning rate of the first ignited unit and the flashover to the second item are changed clearly. The excellent behavior of upholstery combinations with outer coverings based on of modacrylic fibers and effective interliners such as cotton FR, cotton with Al-foil lamination, glass fabric or carbonizing man-made fibers was also found in other studies [661].

The fire disaster of Kings Cross also brought new findings about influence parameters on the fire occurrence [5, 662]. This fire disaster demonstrates that already combustible coatings or paints can contribute to a flashover situation with a corresponding sudden mass of smoke release in the ceiling and wall field. In connection with this the possible fire contribution of paint and paintings was reported in [663]. Realistic full-scale trials carried out by the Atomic Energy Authority of Harwell as well as investigations at the HSE's and Flammability Laboratory Buxhem revealed the dominating influence of the so-called trench effect. If a sufficiently large ignition source becomes effective in a trough, like a moving staircase field, an explosive flame spread occurs then in the trough in accordance with Fig. 164. The effect was verified by large-scale tests [664].

Practical large-scale tests with seat bowls confirmed the importance of the seat distance in the area of fire origin to the danger of fire spreading in the developing stage of a fire. Also ignition sources of raised intensity lead only to a limited flame spread of the tested materials as shown by Fig. 165. The radiation component dominates basically during the advanced stage of a fire without ventilation. If there is a significant ventilation effect, than a direction change of the flames occurs and with that an intensified influence of the convection part of the heat stress. Roberts and Clough [665] verify among other things that with increasing ventilation a considerably intensified spreading in the direction of the ventilation occurs. The basic studies of ISO TC 92 SC1 Fire Reaction Tests already clarified that the orientation of the test samples is of special importance to the evaluation of the surface spread of flame. Figure 120 illustrates the orientation recommended for the basic examination. In this procedure the material-related contribution to the surface spread of flame under the thermal stress of external radiation is determined. As in the computation models it is assumed that radiation is the dominating performance influencing parameter. This assumption is, from experience, valid for the horizontal spread of large pools, e.g. tank fires [666]. Only, if a sufficient influence of the convection part is given, will a significant spreading of the fire occur with small pool diameters. The results of relevant full-scale field trials with flooring materials confirm this statement. The graphical representations of Fig. 166 illustrate the strong radial decay of the intensity using the example of a wooden crib as ignition source.

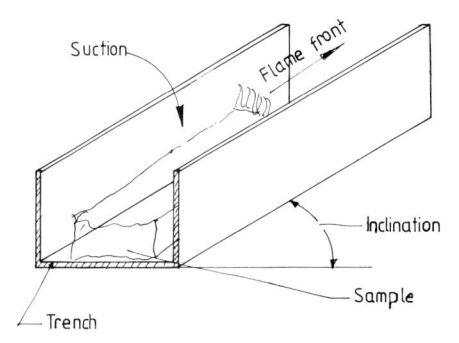

Figure 164 Trench effect

Figure 165 Fire spread due to stadium seats

Although unfavorable t_i values exist for the B2-classified rigid PUR foam, no spreading in the TNO investigations with a 4 × 4 m² trapezoidal steel sheet construction occurs in the neighboring roof section (see Fig. 167). These full-scale field trials confirmed the corresponding evaluations in the Flooring Radiant Panel Test. Using the example of this flooring examination in accordance with Fig. 41c it can be illustrated that orientation plays possibly very much a dominating role in the case of fire spreading. While at wall and ceiling tiles as a rule the hot gas stream of the already burning items cause preheating effects, contrary to that a cooling effect of the incom-

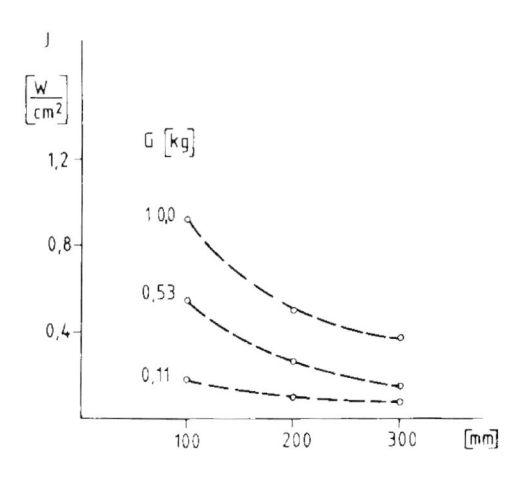

Figure 166 Wooden crib fire – Radial decrease of the intensity of radiation

ing fresh air in a direction towards the source of the fire for the pool-fire situation of flooring materials must be reckoned on. The possible preheating effect through the intensity of radiation of the developing stage of the fire can be neglected. The conclusion of the nonexisting correlation of the cone-calorimeter test and the radiant panel test results given by Hirschler [550] that the cone-calorimeter test only is representing and producing practice-related records, seems to be entirely proven. While the large-scale room-corridor test studies at NBS and the Austrian large-scale trials [548] confirm the practice-relationship of the radiant panel test, it is quite clear that the room-corner procedure in the standardized conditions will not be able to produce meaningful results for this specific flooring fire scenario. The differentiated test and classification criteria of the harmonized European test protocol [DIN EN 13823] are a further indication for this statement. How much, e.g. the intensity of radiation from flames becoming effective can be influenced by ventilation effects is shown in Fig. 168a very vividly. The graphical representations illustrate the possible influence of the ventilation on the fire spreading across the surface. Already a wind rate of 10 km/h causes a remarkable increase of the radiation impact through inclination of the flame front. By routing of the flame front a continuous concentration of ignitions in direction of the ventilation due to the higher intensity of radiation occurs. The radiation stress due to the burning activities of a wood-wool basket for simulation of the impact of flying brands is increased due to wind influence. Figure 46 shows schematically the calibration setup for the measurements.

A list of model ideas were developed for the stage of a fully developed fire. The extensive studies of Thomas [28, 45, 669], Harmathy [233, 670] and Friedmann [39, 40, 582, 671] for the pre- and postflashover fields clarify the complexity of the thermodynamic connections. Computational ideas were sought by considering the essential parameters. Simplifications and assumptions allow the stipulating of model procedures. The instabilities in the developing stage of a fire are the reason that up to now no universal regularities have been verified [601, 672]. An analogous preheating occurs in the case of upright flame spread due to the thermally conditional buoyancy forces of the fire gases. The convection part determines the scale of

Figure 167 TNO roof experiments – Flame-spread behavior

the surface spread of flame. Kashiwagi [51] reported the applications of mathematical modelling will only be possible through a list of simplifications due to the high non-linearity and complexity of fire gases. The simulation of some of few typical ignition sources is proposed or a practical arrangement for the attainment of the targeted raised safety against negligent and/or intended ignition. They are supposed to initiate the essential safety levels so that the financial expenditure is justified. The following ignition-source levels provide themselves in this case as problem solutions.

- resistance against glowing tobacco products and burning matches
- resistance against glowing wirings, glowing particles and glowing contacts
- resistance against burning paper and burning items.

The acting of primary and secondary ignition sources, e.g. glowing contact, a burning match or a burning wastepaper basket leads basically to a locally and time-limited pyrolysis and ignition of the combustible material brought into contact with the ignition source. The ignition risks depend, e.g. on the density, the surface finish, the humidity, the nature and the exposure time of the ignition source, the ventilation or the oxygen content of the environment. Integrated electrical and electronic devices or, e.g. overheated wiring and glowing contacts in electric blankets can also cause ignition in the upholstered furniture and bed field.

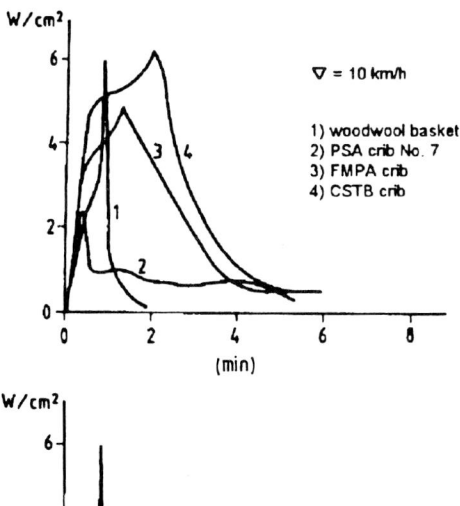

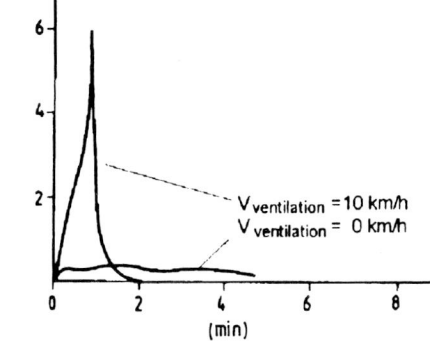

Figure 168a Influence of the ventilation on the radiation section of flames

Burning dripping

The evidence of the characteristic burning dripping occurs mainly within the framework of general fire-safety testing methods for building products and materials. In the building field this special side effect is used, for example in France, as an evaluation criterion. In the Federal Republic of Germany this characteristic is determined indeed in addition in the material testing methods, but it does not have, however, any direct influence on the classification. Due to the requirements of the building authorities this side effect of a fire achieves, however, the status of an expulsion criterion [329, 673, 674]. In [43] it is stated that some material can produce burning dripping or burning parts and add through that to fire spread and possibly impede the rescue of people and animals. The use of such products is inadmissible for specific applications. The question to what extent burning dripping or burning falling material can become effective as a secondary ignition source in the case of a fire and endangers rescue and extinguishing measures, depend on constructive and environmental conditional factors (see Fig. 169). Purposeful large-scale tests showed that this criterion causes only a limited contribution to the fire spread. The falling off of burning parts and burning dripping does not include any additional danger to rescue and extinguishing measures provided that this occurs only in the action field of the primary ignition source [72]. The demand of the expert commission concerning the Düsseldorf airport fire, to classify the application-oriented composite, must from the viewpoint of the preventive fire protection be designated as risk-relevant. Through the determination in the guidelines for combustible materials in the building construction [674] the insulating materials have to be tested by themselves and not

Figure 169 Burning outflow in the case of trapezoidal roofing construction

receive in the recognised laboratory test methods, e.g. the chimney test, the relevant classification of the composite.

In roofing applications the classification "burning dripping" next to the dominant material- related insulant testing also in the end-application-oriented construction, at a section of the real construction, the impact of the wood-wool basket fire is determined, the classifications are carried out in facade field, e.g. for glazings exclusively due to the observations during the pure material examination in the lab. The contradictions in the conversion/ implementation of the fire-performance requirements of the building codes (LBOs) through

- pure material-related building material examination
- composite examination in the laboratory standard with exit stipulations like "insulants must be tested by themselves only" as well as
- through realistic investigations with component sections as for example the evidence of the resistance against flying brands is manifest.

Analysing the evaluation criteria for the rating burning dripping, it is illustrated that the ignition risk by the limited quantities of material is in relation to the respective ignition-source intensity as a rule of minor importance only. Flow of burning material or dripping of burning parts that is adding to flames of the running off or coming off units can lead to a primary danger only, where their intensity and does the intensity of the primary ignition source. Provided that the resistance against defined larger ignition sources has been proven, the burning/falling off of dripping material of small intensity can not represent a primary risk for the environment. Only where materials of smaller resistance govern the environment, e.g. stored goods, furniture or building products, a possible danger must be feared. This basic philosophy is valid basically for all applications, e.g. building area, the electromechanical field or the transport section. In particular, in the electromechanical field it must be guaranteed that the electromechanical product in the failure case will also not become a secondary ignition source for the environment. The evaluation of the burning dripping is coupled with national-specific fire-reaction test methods as a rule. As a rule there is a connection with procedures such as proof of the resistance against ignition sources of small and medium intensity, such as for example the small-burner test and/or the FAR vertical bunsen-burner test [250]. The test pieces arranged mostly upright become in general for example in the FAA perpendicular test or in the UL Subject 94 from below a thermal impact by a bunsen-burner flame for a defined test duration. The evaluation/rating is done due to the burning dripping, possibly falling off or dripping including their afterburn time. The horizontally or vertically arranged test sample becomes, in this case, by means of a bunsen burner an edge or area flame impingement in accordance with the schematic representation in Fig. 80a. Intensity and exposure time differ possibly considerably as demonstrated by the B1 and B2 classification according to DIN 4102. Using the example of the DB requirements for materials of the interior fittings and furnishings and of upholstery seatings it was clarified that the requirements concerning the burning dripping according to Table 114 is not in agreement with the real risk in the case of a fire. In Table 114 it is attempted, furthermore, to put the extent of the burning dripping in relation to the intensity of the relevant ignition source. Using the comparison of energy balances it becomes understandable that with the evidence of the resistance against the relevant ignition sources a possible danger through the burning dripping material is to be denied.

Table 114 Ignition risk by burning dripping (ca.1g) in relation to relevant ignition sources.

Material	Intensity (KWh)
1 g PE	$12{,}92 * 10^{-3}$
1 g ABS	$10{,}0 * 10^{-3}$
1 g PC	$8{,}61 * 10^{-3}$
1 g PMMA	$7{,}22 * 10^{-3}$
Burner flames	
UL-94 20 mm – flame	$0{,}175 * 10^{-3}$
UL 94-125 mm – flame	$0{,}8 * 10^{-3}$
FAR pt.25 Vertical Flame Test	$2{,}06 * 10^{-3}$
DB Burner-DIN 5510	$39 * 10^{-3}$
UIC-Paper cushion-DIN 5510	$380 * 10^{-3}$
FAR pt.25 – Kerosene Burner	$3300 * 10^{-3}$

Comparable differences result from the upholstered furniture composite examination in the match test in relation to the UIC pillow test and the kerosene-torch procedure. The setup of these test methods is mainly combined with the condition to achieve only a temporally and locally limited contribution to the fire spreading. In the Epiradiateur Test the material will be thermally stressed with a fixed intensity of 35 kW/m². A generalization onto other fire-risk situations is not allowed as a rule. The small-burner procedure example, which is introduced for the B2 rating, moderately ignitable, among other things is in Germany and Austria obligatory for the rating of class B2, illustrates by any means very vividly that such real definitions do not guarantee a priori a risk-relevant combustion system. The rating "burning dripping", limited to 15 s flame impingement and 5 s observation time, demonstrates the free existence of the targeted protection destination, which cannot be reduced to a time length of 20 s. The limitation of the impingement time can be justified without doubt, but not, however, the extremely short observation time. Figures 169 and 170 demonstrate that a real assessment of the burning dripping process can not be given within a time of 20 s, the impingement time of 15 s included. A detailed analysis of the quality criteria of the B1-classified materials, considering a valuation scale in accordance with Table 114, shows that this determination includes little information in relation to the ignition-source intensity with regard to the preventive fire protection. The igniting capability of the burning dripping material is, furthermore, evaluated by means of a receptor such as, e.g. tissue paper [DIN 4102]. This side effect of a fire becomes also in the case of testing methods to raised ignition-source intensity such as, e.g. chimney test, UIC paper-pillow test and the kerosene burner test registered and evaluated. In France, Spain, Portugal and Belgium this side effect of a fire is assessed also by means of a specific testing method in accordance with Fig. 54d

Figure 170 Burning behavior of a PMMA lamp covering

under the action of radiated heat. Figure 170 demonstrates very clearly that the possible contribution to the surface spread of flame influences the scale of the burning dripping decisive. At the PMMA lamp covering the stress with a burning UIC paper pillow according to Fig. 170 initiated ignition and destruction by burning and a continuous burning dripping. Research studies demonstrated in a very informative way, that specific coatings and paints led with EPS ceiling tiles to a rapid surface spread of flames combined with strong burning dripping [675]. Figure 97b illustrates the extent of the expected burning rain. Provided that in the case of room fires a surface spread of flame about the facade taken place to the roofing material, a burning outflow can result as shown by Fig. 169.

Heat release

The heat release is a continuously changing characteristic depending on the fire occurrence. The heat-release rate is a measure of the assessment of the fire development. The released amount of heat determines the temporal process of the fire occurrence (see Fig. 8a) decisively depending on the feedback effects increased by the environmental conditions. The heat potential is not decisive, but rather the time-controlled heat-release rate. Products with marked carbonisation tendency will behave, as a rule, differently from products that are oriented on the melting behavior and therefore on the pool-fire conditions. The reaction of the individual parameters is different. The heat-release rate of materials in the case of a fire was always an essential evaluation criterion of the fire-performance examination. The net calorific value as a measure of the complete combustion was the check parameter. Meaningful characteristics were determined as a rule through fire-gas temperature measurements within the framework of the relevant fire-reaction test methods. With these procedures such as, e.g. the Chimney Test, the Epiradiateur as well as the IMO Spread of Flame Test the heat-release rate depends decisive of the contribution to the surface spread of flame. With the Brown's Box, the fire-propagation test as well as the Vlam Overslag Test, which results altogether in the ignition of the total entire test surface subjected to thermal stress not the contribution to the surface spread of flame but rather the mass-burning rate, will determine the scale of the temporal process of heat release. The so-determined characteristics can be taken as the potential value of the agreed scenarios. Figure 171 provides a view of some determined boundary values of fire-gas temperatures.

In order to be able to carry out a better appraisal of this characteristic in dependence of the fire occurrence, specific testing methods, such as for example the Tewarson equipment (see Fig. 9), the OSU Chamber (see Fig. 73b) or the cone calorimeter, shown in Fig. 21, were developed. These procedures are obviously suitable to supply material-related records depending on the external radiation. After studies lasting for years at the SC1 of ISO TC 92 the cone calorimeter was selected as the basis for standardisation [676]. The use of an ignition source guarantees the burning behavior with flames under pool-fire conditions depending on the external radiation. The general conditions of these procedures, which supply material-related potential characteristics, cause the burning with flames and result in corresponding data records. The simulation of a smoldering fire is not possible under the standardized test conditions. The heat potential as a rule becomes only partly released. Basic investigations have been conducted by Tewarson [353] and Herrington [354] (see Figs. 103–106 as well as Tables 71 and 72).

Modak et al. [36] report on the influence of the pool size on the heat-release rate under the action of external radiation. The disc-shaped PMMA test pieces were arranged horizontally in the OSU Chamber and subjected to a thermal stress of 10 kW/m^2. The data records of Fig. 172a illustrate that with increasing pool diameter the mass-burning rate and thus the heat-release rate increase significantly. Friedmann [52] reported also on this investigation program at the FMCR. The graphical representations of Fig. 172b reveal that due to the intensified feedback effects an intensified mass-burning rate occurs with an increasing pool diameter.

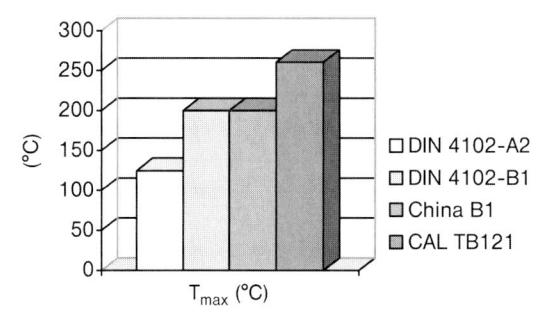

Figure 171 Permissible maximum fire-gas temperatures in laboratory testing

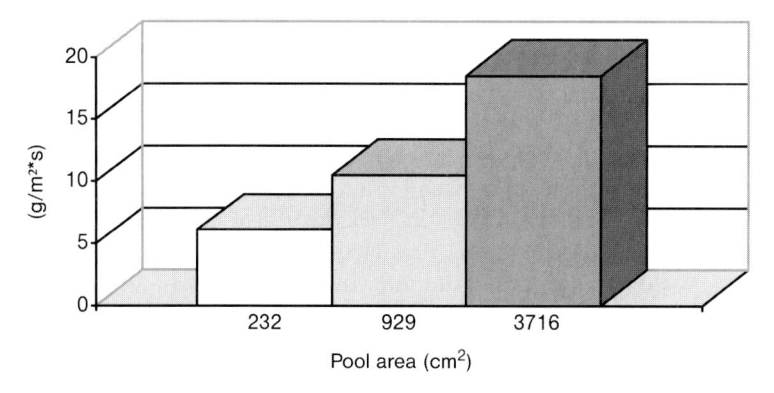

Figure 172a Influence of the pool size to the mass-burning rate of PMMA

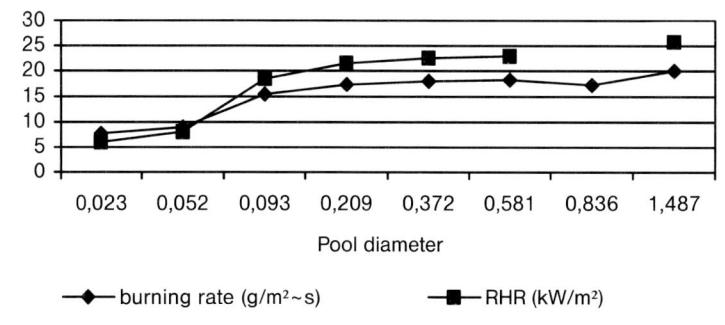

Figure 172b Influence of the pool diameter on the heat-release rate of PMMA

While increasing intensity produced insignificant changes of the heat-release rate only in those products with carbonisation tendencies, it can be increased in the case of melting materials with marked pool characteristics significantly. The cone calorimeter simulates a pool-fire situation under variable preheating. At a constant pool diameter the burning behavior into the material determines exclusively the scale of the fire-gas liberation. The mass-burning rate is, in particular, dependent on the nature of the tested material to be directly proportional to the fire duration. The burning rate is influenced decisively by the carbonisation tendency of the material. If a fire propagation across the surface is assumed in the current fire case than the mathematical relationship $V_R = Kv^2ut^3$ with

V_R = burning rate
K = constant
v = spread across the surface
u = spread into the interior
t = fire duration

shows that the mass that shared the burning-process is increased by the third order of magnitude and is able to influence the fire-gas release correspondingly.

From the viewpoint of the Combustion Behavior of Upholstered Furniture (CBUF) consortia, the action community of essential European testing institutions, the practical relevance of the cone-calorimeter combustion system became targeted within the framework of the research program to the characterisation of the fire performance of upholstery furniture. The investigation program was planned and carried out in order to elaborate key data for the planned European Furniture Directive. The representation of Table 115 provides a view of the structure of the CBUF program whose main item is the cone-calorimeter examination. The proximity to reality is shown by the proven correlation to the Furniture Calorimeter (see Fig. 68c) and the room-corner fire scenario. The characteristics of Fig. 173 give a view of the results of a round-robin test gained with five different upholstery furniture combinations. These numerical values indicate/illustrate that the upholstery combination significantly influences both the value and the temporal process of the heat-release rate.

Within the framework of the CBUF program the influence of ventilation as well as the design on the mass-burning rate were, among others, in detail examined. The data records of Fig. 174 illustrate the possible influence of the design on the characteristics of the heat release. In the case of identical construction additional arm rests lead to a strong rise of the peak values. The modified design led obviously to a changed burning behavior of the upholstered chair unit. The CBUF consortium reports concerned the reproducibility to former round-robin tests with padded chairs and settees. The data records of the heat release achieve, for the padded chairs, an

Table 115 Structure of CBUF Program

Test phase	Test specimen
Phase 1	Component Material
	Cone Calorimeter
Phase 2	Composite
Composite Model	Cone calorimeter
Phase 3	Full-sized Furniture
Furniture Model	Furniture Calorimeter
Phase 4	Full-sized Furniture
Room-fire Model	Room-fire Scenario

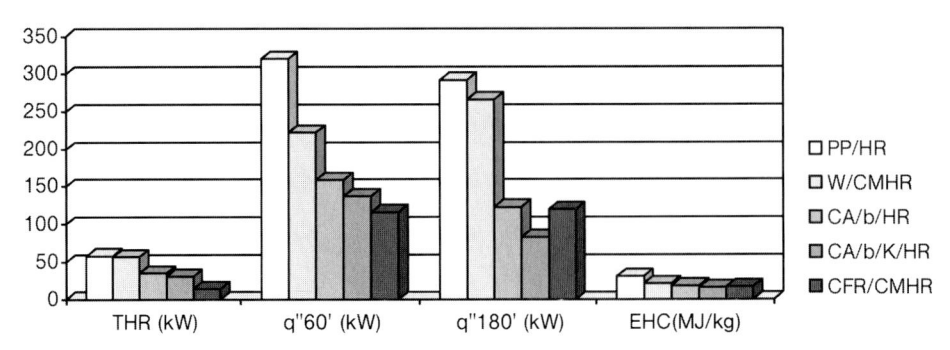

Figure 173 Influence of the upholstery combination on the heat release

appealing agreement. The influence of the ventilation was examined with four different upholstery combinations in the Furniture Calorimeter. The values determined in the ISO room-corner cabinet in accordance with Fig. 175 were measured with a completely opened door, and with 1/4, with 1/8 and/or 1/16 opened door. Comparing these numerical values with those of the different upholstery combinations in accordance with Fig. 176, then the possibility to influence the value seems to be stronger through variation of the ventilation. The data records determined with different ventilation conditions demonstrate very vividly that the material-related heat-release rate is modified more strongly by ventilation than through different cushion constructions.

Further extensive studies of the industry confirmed the possibly strong influence of the upholstery construction on the heat-release rate during the rate of burning. Smith [677] reports on such investigations with numerous upholstered furniture combinations (see Tables 116a and b). The modification of the "Furniture Regulation" was the starting point of these studies in Great Britain. The influence of essential parameters that are constructive elements was examined, the recipe variation of the upholstery filling as well as the importance of individual test parameters. The realisation of fire tests with fully furnished rooms should provide information to what extent methods like the cone calorimeter are suitable to carry out risk-relevant

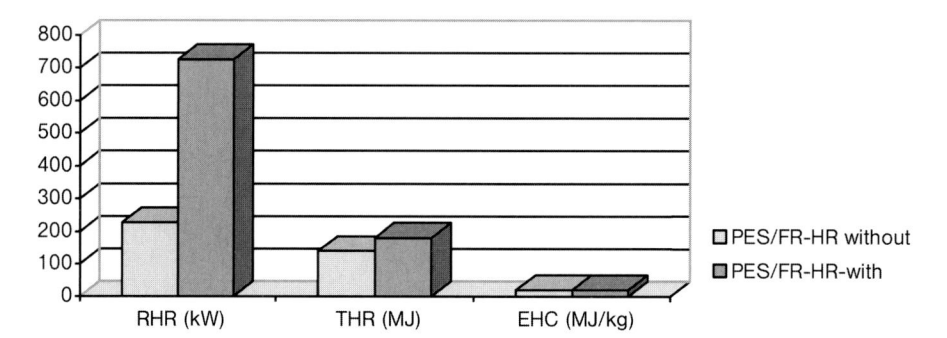

Figure 174 Influence of the design on the heat-release rate

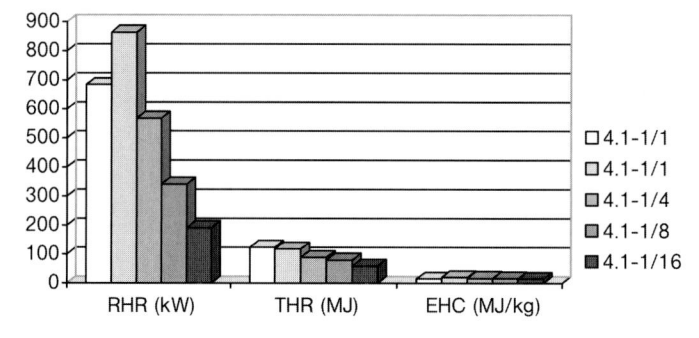

Figure 175 Influence of the ventilation on the heat release

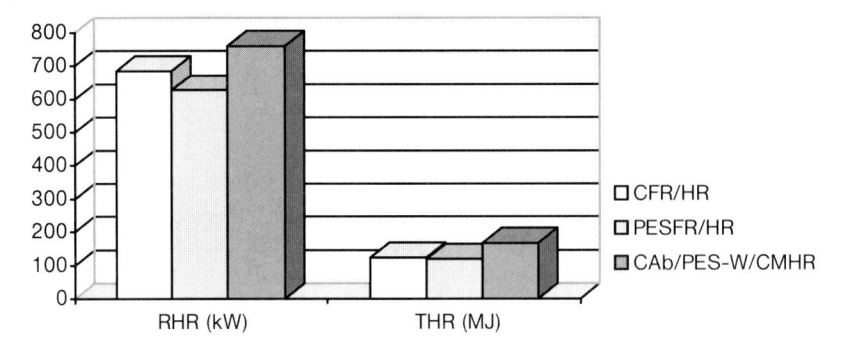

Figure 176 Influence of the cushion combination on the heat release

screening examinations. The listings of Table 116a demonstrate the strong influence of the cushion component's outer coverings, interliner and upholstery filling on the temporal course of the heat-release rate. The reference cover PP determines a significantly intensified combustion of the entire combination. The treated PAC fabric leads visibly to a small intensified heat release in the first stage of a fire. This effect is presumably caused by the fast combustion of the PAC pile. After this first stage a significant reduction is to be observed. The treated cotton covering brings a clear improvement in comparison to wool. The scale of the protection effect was raised considerably through fitting of a Kevlar interliner.

Table 116a Influence of the upholstery combination onto the records of the heat potential

Material	q 60 (KW/m^2)	q 180 (KW/m^2)	q total (KW/m^2)
PP/PUR-NFR	320	292	58
CAb/PUR-NFR	159	123	35
W/CMHR	222	266	57
C-FR/CMHR	116	118	40
CAb/K/CMHR	137	83	31

1-PP, 432 g/m^2, PUR-NFR, 21 kg/m^3
2-Polyacrylic/cotton backcoated, 540 g/m^2, PUR-NFR, 21 kg/m^3
3-Wool, 432 g/m^2, CMHR, 30 kg/m^3
4-Cotton FR, 422 g/m^2
5-Polyacrylic/cotton backcoated, 540 g/m^2, Interliner Kevlar 65 g/m^2, CMHR 30 kg/m^3

Table 116b Course of the rate of heat release data – Influence of the upholstery combination

Upholstery	5 min	10 min	15 min	30 min
Acryl / PUR	117	230	244	–
Acryl / MW	20	175	230	244
Acryl / CMHR	14	91	200	266
Cotton / CMHR	0,34	3,6	20	230

Table 116c Deviation of the heat release of upholstered furniture.

Outer covering	CMHR (35 kg/m³)	CMHR (50 kg/m³)	PUR (Standard)
Acryl	490	94	661
Cotton	145	674	630
Viscose	415	395	350
PP	482	402	430

NBSIR 87–3560

The listing of Table 116b provides information about the characteristics of the material-related heat potential values of different combinations determined within a round-robin test with the cone-calorimeter procedure. In these studies that were carried out with regard to the planned European Furniture Directive in English testing institutions, the comparability of the mass-burning rate in fully furnished room tests at RAPRA (see Fig. 162b) with that of the Furniture Calorimeter shown by Fig. 68c at the WFRC Test facility, was the aim of the investigation. The data records of Fig. 177 show the possible influence through the selection of suitable cushion components. PP as an outer covering causes a markedly greater heat-release rate in comparison to C/A. The composites with the HR quality as upholstery filling resulted in the sequence wool, cotton FR and C/A backcoated a decrease of the heat release. The listing of Table 116b shows, furthermore, that there is an analogous behavior also in the temporal course of the release rate. Within the CBUF programs it was confirmed that a strong influence of the heat-release rate can be initiated by the design of the upholstered furniture as illustrated by the characteristics of Fig. 174. On the scale in which the design of the upholstered furniture changes the mass-burning rate, also the characteristics of the heat-release rate are modified. The possible strong influence through the upholstery fabric became, among other things, also verified unambiguously in the CBUF program (see Fig. 177). If the differing densities of the foam qualities and the deviations in the variations of the fabrics are considered for the evaluation of the results, then the strong influence by the cushion constructions to the heat-release rate is confirmed. The graphical representations of the results in Fig.

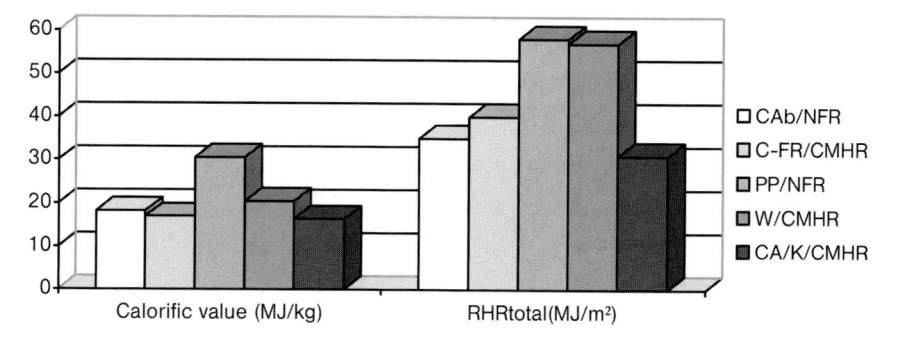

Figure 177 Influence of the chosen dimension on the HRH data set

177 reveal that a better differentiation is achieved if the results become subjected to comparable volume or area units.

These results illustrate that both the resistance of the outer covering as well as the density of the upholstery filling influence the heat-release rate determined in the cone-calorimeter test. In order to exclude faulty interpretations possible fringe effects are to be considered in the evaluation of the results. If a standardization of the density is carried out, then there will be no doubt that the CMHR quality produces an insignificantly smaller heat-release rate only in relation to the standard polyether quality. The effects determined with the pure foam grades according to Fig. 105 are confirmed. The graphical representations of Fig. 178 give a view of the comparability of the characteristics of the heat release determined in the room experiment (R: RHR) as well as in the Furniture Calorimeter (F:RHR). It appears that in the Furniture Calorimeter Test almost without exception material-related differences are determined, which do not occur in the room test. Only with rockwool upholstery combinations were the most favorable results in the furniture calorimeter achieved. While the mineral-wool combination in the room-test an approximately comparable value was produced, more favorable values up to 100% were determined in the Furniture Calorimeter Test with the rockwool combinations. The data records of Fig. 180 show the liberation rates of different upholstery combinations determined by BRMA/Europur in relevant full-scale field trials [321]. The graphical representations reveal very vividly that variation of the composite compounds changes the mass-burning rate significantly. The occurrence of the peak values of approx. 250 kW could be delayed approx. 10 min to 30 min by different cushion constructions.

Figures 159b to 160a illustrate, using the example of upholstery seat cushions, to what extent the surface spread of flame, which will be possibly decisively influenced by the fabric, dominates the fire-gas liberation. Risk-relevant full-scale field trials, e.g. at RAPRA [321, 677] have shown that the burning rate and the smoke temperature can be influenced decisively by both the nature of the fabric and the interlayer. The maximum fire-gas temperature of the upholstery combination with the standard polyether foam grades was varied by the fabric variation from 150°C to 700°C. This effluent temperature level could be reduced to 50°C to 200°C by the use of effective interlayers. Thermoplastic interliners such as, e.g. PE and PP fleece led to a corre-

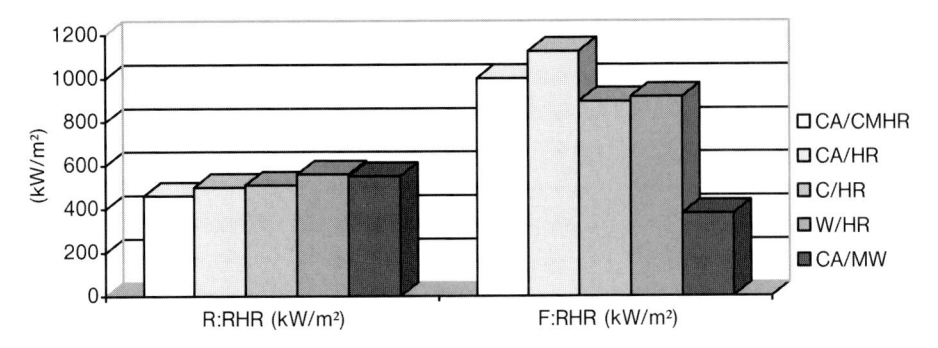

Figure 178 Influence of the test method on the heat-release rate in the large-scale test

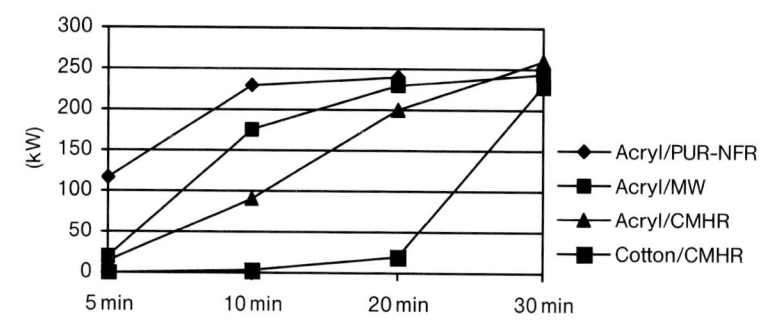

Figure 179 Influence of the upholstery combination on the heat-release rate in the full scale field trial

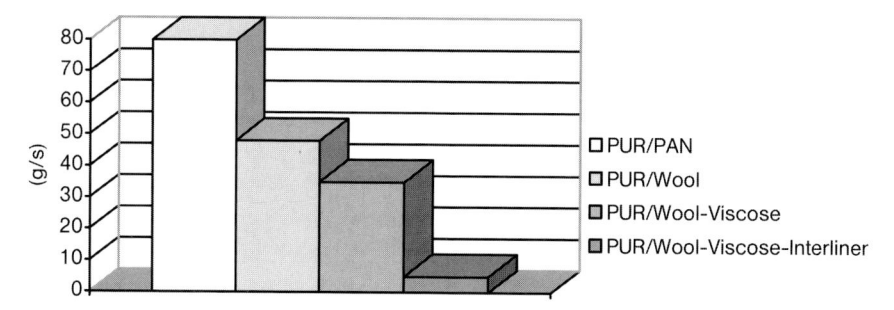

Figure 180 Influence of the upholstery combination on the mass-burning rate in the 1:1 experiment

sponding rise of the temperature–time relationship. The reducing effect of efficient interliners could be verified also in the case of the investigations with fully furnished room burns out in Moreton in the Marsh [102, 103].

The data records of Fig. 179 confirm the strong influence of the upholstery combination on the mass-burning rate. The results of these realistic large-scale tests carried out in Sweden in the measure 1:1 demonstrate the possible strong influence of the outer covering and especially the interrupting effect of effective interlayers that reduce the incipient fire impressively. The Swedish investigation confirms that the burning rate can be modified both through the choice of the fabric and through the composite combination, e.g. by the use of protective interliners.

The decomposition products of fast-burning upholstery units can, as reported in multiple ways, lead to a flashover depending on the spatial conditions. In the case of the valuation of such information the edge parameters must be considered carefully. Provided that, for example, this examination occurs in a room-corridor test house in accordance with ISO TC92, which was also the basis of the CBUF program, then the definition of the flashover situation must be sufficiently considered. The results must be classified in relation to the size and arrangement of the fire load, the size of the room and the ventilation and be evaluated correspondingly. These results must not be generalized without special evidence. The key question "Will the second item ignite" and the burning rate of the ignition occurres, depend decisively from the size of the ignition source, the arrangement, the design as well as the environmental conditions and must be considered as functional, variable characteristisc.

Smoke formation

As a summary of studies, lasting for years, general conditions were elaborated in the SC3 of TC 92 Toxic Hazard in Fire for the classification of thermal decomposition products. Figures 13 and 119 provide a view of the basic determination of the effluent characterisation. The different laboratory procedures for the effluent characterisation have to be examined in such a way by considering all those conditions and to find out, to which one of the fire scenarios recommended by the SC3 (see Fig. 119) they fit in. The graphical representations of the Figs. 20b–23a illustrate the present findings from relevant full-scale field trials with fully furnished rooms. The results clarify that the CO_2/CO ratio remains mainly below a value of 50 in the stream of effluents. The characteristics of Fig. 20a confirm this statement. These results are based on joint studies of the Harvard University and the Fire Department of Boston. The listing of Table 117 and 118 provides a view of the different combustion systems

Table 117 Relevance of the simulation of the smoke formation of various fire scenarios

Test methods	Thermal stress	Scenario 1	Scenario 2	Scenario 3
ASTM-E84	Burner	no	(yes)	no
DB Chimney	Burner	no	(yes)	no
Radiant Panel	10–1 KW/m^2	no	(yes)	No
SBI Test	Burner	no	(yes)	No
XP2 Chamber	Burner	no	(yes)	No
NBS Chamber	25 KW/m^2	(yes)	(yes)	(yes)
NBS modificated	20–50 KW/m^2	(yes)	(yes)	(yes)
ISO Smoke Box	20–50 KW/m^2	yes	yes	yes
Gost 12.1	20–80 KW/m^2	(yes)	(yes)	(yes)
DIN 53436	200–900 °C	yes	yes	yes
NF-T51	200–900 °C	yes	yes	yes
IEC TC 89	Flames	no	(yes)	no
NES 7123	Brenner	no	no	no
Potts Pot	200–1000 °C	yes	yes	yes
Box Japan	Flames	no	(yes)	no
BRI Japan	10–50 KW/m^2	no	(yes)	(no)
RTB Japan	10–50 KW/m^2	(no)	(no)	(no)
SRI	10–80 KW/m^2	(no)	(no)	(no)
U-Pitt 2	10–70 KW/m^2	(no)	(no)	(no)
Cone Calorimeter	10–100 KW/m^2	(no)	(no)	(no)
NIBS/ US Radiant	10–80 KW/m^2	(no)	(no)	(no)
U-Pitt	200–1000 °C	no	no	no
NES 713	Burner	no	no	no

Scenario 1 – self- sustained smoldering
Scenario 2 – developing fire with reduced oxygen availability
Scenario 3 – fully developed fire with flashover conditions

Table 118 influence of the sample mass and the size (Lab-equipment and Lab-Room) on the smoke density

Material	F (g)	NF (g)	Dm-Room F(19,3m³)	Dm-Room NF(19,3)	Dm-Lab F(0,51m³)	Dm-Lab NF(0,51m³)
Wood	759	209	13,6	211	15,5	213
Plastics	759	25	16,8	164	20,8	16,3
PVC	20	6	494	351	441	217

that come into play for smoke-density measurements, the assessment of the smoke toxicity as well as the determination of the smoke corrosivity. For these combustion systems it is to be judged to what extent they are directly or after variation of test parameters, valid to supply the necessary functional material-related potential values and with that to guarantee the cover of the targeted protection destinations.

The dependence of these material-related characteristics for the smoke potential of the respective fire occurrence become verified both for cumulative procedures such as, e.g. the NBS Chamber and the ISO Smoke Box and through-flow systems such as the DIN 53436 or U-Pitt Method. The schematic representations of Fig. 120 clarify that both procedures are able to supply comparable risk-relevant values assuming the concentration of the effluent mixture is the same. The mass-burning rate of the material is determined both through physical as well as geometrical and technical parameters. Orientation, thickness, roughness of the surface, preheating as well as material- and air-composition will change the burning behavior among others decisively. The listing of Fig. 181 gives a view of the energy levels in (cal/g) necessary for pyrolysis of different products.

The records of Figs. 116a and b show very vividly that, e.g. the smoke-density levels of the effluents change considerably depending on the testing temperature. The numerical values of Fig. 117 clarify an analogous behavior of the fire-gas toxicity. Both characteristics pass a maximum, seemingly depending on the testing temperature. Only a small part of the listed procedures allow an unlimited simulation of the fire-gas liberation.

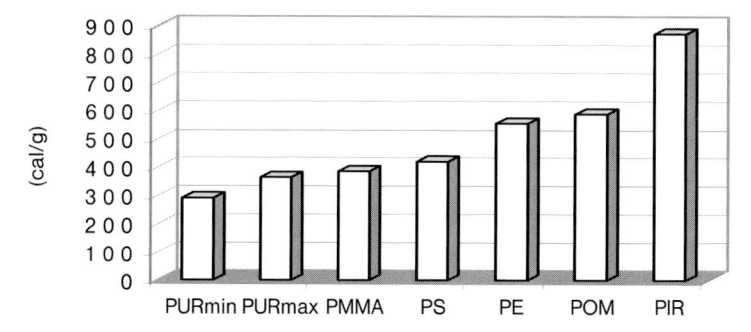

Figure 181 Energy levels necessary for the pyrolysis

the spreading of the self-sustained smoldering fire due to exothermic reactions.
the combustion with flames in the case of reduced oxygen offer and
the flashover situation including the large-area oxidative decomposition under the smoke-layer input.

Single point data procedures such as, e.g. smoke-density measurement in the XP2 Chamber, the evaluation of the fire-gas toxicity in accordance with NES 713 as well as the pH measurement according to DIN/VDE 0472 pt. 18 and the CNET procedure as proof of the smoke corrosivity are altogether unsuitable to supply functional material-related characteristics. The so-determined characteristics are coupled to the respective fire scenario of the test confirmation exclusively and can not be transferred without special evidence to other fire-risk situations. In the test method according to ASTM – D 2843 (see Fig. 47b) the test sample becomes affected by flames by means of a propane-gas burner. Figure 182 clarifies that at expanded plastic samples, 60 mm × 60 mm × 25 mm, and in comparison to solid materials, 30 mm × 30 mm × 4 mm, a significantly larger surface is thermally stressed. A direct comparison of such potential values must be carried out with the necessary care. The graphical representations of Fig. 182 demonstrate the balance using different sample sizes demanded by law.

In particular with decomposition devices where the entire test surface is suddenly thermally stressed, the general condition such as, e.g. mass-, volume- or area-uniform test pieces as well as ventilation conditions of the flashover situation must be assessed carefully including the extensive oxidative decomposition due to the thermal stress of the smoke layer. If the fire-gas components are determined in the through-flow system serious concentration differences must be reckoned on for samples with identical weight due to the variable thermally stressed surfaces. The graphical representations of Fig. 183 clarify using the example of the A2 classification of flooring materials the possible deviations of effluent concentration thermally stressed due to the test parameters.

The limitation to the pure combustion with flames limits, furthermore, the characteristics determined in the pure smoke-density test method in accordance with ASTM-D2843. This specific smoke-density test method, the XP 2 Chamber, which among other things is a component of the building-material examination in Germany, in Austria as well as in Switzerland, allows only the combustion with flames. Strong deviations of the values of the CO_2/CO ratio in the stream of effluents, strongly will considerably limit the data records. This fact is taken into account by the German authorities and additional examination in the DIN tube in accordance with DIN 53436/37 for the simulation of the fire-gas liberation under smoldering conditions for A2-classified materials has to be determined.

This essential aspect of the fire-gas liberation under smoldering fire conditions became duly considered in the predominant majority of the testing methods such as for example ASTM-E 662 (NBS Chamber, ATS 1000, ISO DIS, Gost 12.1, NF – T51 or DIN 53436/37). The field of the problem acute toxicity become regulated by single-point test methods, e.g. next to the bunsen-burner test NES 713 also by a 450 °C hot plate (Switzerland, Israel) and/or a 900 °C heated glowing spiral [517]. The NES 713 protocol requires that approx. 1 g of the material to be evaluated is ignited by means of a bunsen burner and brought to combustion and/or decomposed by a

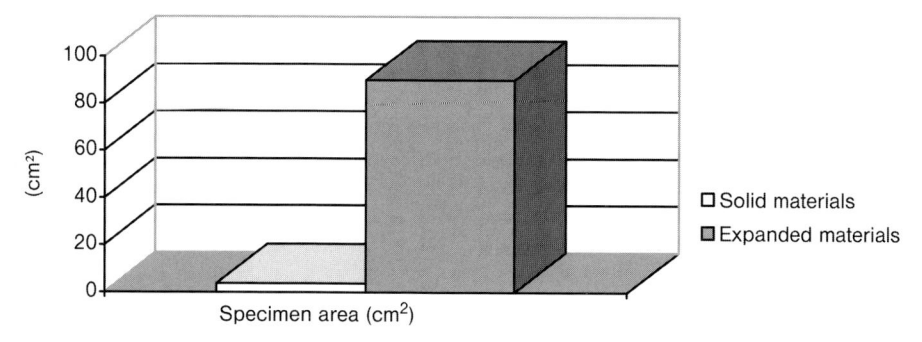

Figure 182 Test sample surface subjected at the determination of the smoke-density data

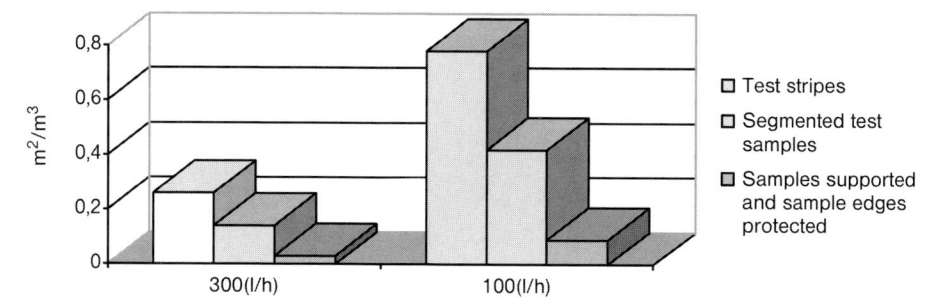

Figure 183 Influence of the primary air supply and the test piece size on the fire-gas concentration

continuous thermal stress. The determined fire-gas concentration values are projected onto a nominal load by 100 g/m³. In dependence on the carbonisation tendency of the tested material flue-gas concentrations are produced by completely different fire scenarios that exclude a comparability of the test results.

In the determination of the pH-value 1 g (DIN/VDE 0472 pt. 18 and/or 0.5 g (NF-C20-453) of the material to be evaluated in an stream of air of 10 l/h become decomposed at a temperature level of 750–800 °C. The thermal decomposition products are guided through wash bottles. The pH values and the conductivity were determined. Concerning the effluent corrosion situation a real distance is to be confirmed. Corrosion effects of fire gases of different fire scenarios are not so measurable. For the CNET procedure reality proximity is required from sides of the proponents. In the case of a closer examination of the combustion system it appears that, but on the one hand, indeed in the exposure part exact physical conditions, e.g. the air humidity, were determined and, on the other hand, the thermal decomposition products are produced under conditions that do not correspond to the general conditions of fire. Approximately 1 g is brought under addition of PE granular to the combustion process. The second group of the combustion systems in accordance with Table 109, set in as a rule a crucible for the generation of the thermal decomposition products. The

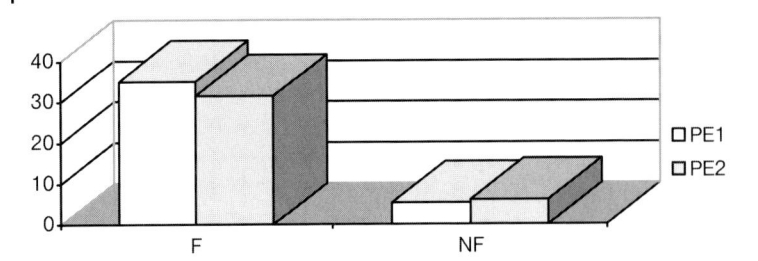

338 | 6 *Use and Interpretation of PUR-Test Results Determined under Enduse Conditions*

Figure 184 Influence of the test conditions on the characteristics of the toxic potential

most prominent representative of this class is the Potts Pot Method (see Fig. 43a). This procedure, whose standard was developed at the NIST (NBS), guarantees that thermal decomposition products are developed at 25 K below and above the self-ignition temperature. The stipulated test conditions allow the simulation of the flue-gas liberation under the smoldering and flaming conditions. The graphical representations of Fig. 184 clarify the instep broad one of the possible deviations. The determination of the band width of the thermal loading guarantees that smoldering fire conditions and flaming combustion are trialled realistically. The data records of Fig. 184 demonstrate very impressively that under smoldering fire conditions less favorable values are determined.

Detailed information and extensive testing-results contain in particular the studies of Babrauskas, Birky and Levin et al. and Cornelisson [185, 192, 678, 680]. Within the framework of numerous studies characteristics of the toxic potential such as LC_{50} and EC_{50} were determined for a large number of products such as, e.g. PUR, ABS, EPS and PA as well as wool, cotton and different sorts of wood as reference products. The combustion device employed in the CSR [681] corresponds to the disadvantages of the Potts Pot procedures. The thermal decomposition products are produced likewise at approx. 25 K below and above the self-ignition temperature. As experimental animals, however, differing from the Potts Pot procedure, white mice are used instead of white rats. The determined records in accordance with Fig. 185 demonstrate that the above final conclusions can turn themselves around depending on the examined products. With that, the limiting conditions must be very carefully checked

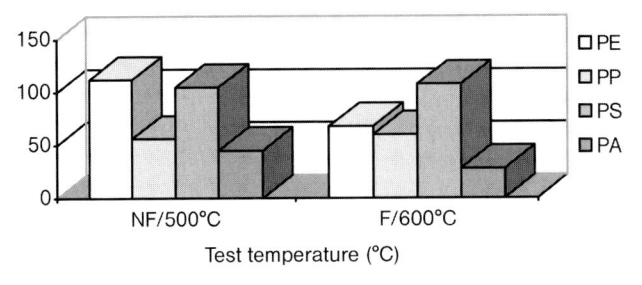

Figure 185 Influence of the test conditions in the crucible on the toxic potential of the decomposition products

also in the case of the arrangement of the results of similar combustion devices before a definitive judgement is yielded.

These data illustrate that the more unfavorable values of the polyethylene are gained in the case of flaming combustion. Skornik [195, 196], who produced the decomposition products in accordance with Fig. 99 in a stream of air of 100 l/min under recirculation conditions, employs an analogous combustion system. In the BIS apparatus, in accordance with Fig. 126, which was advanced as the VCI method [457], the decomposition products became released in the temperature range of 400–1000 °C. The investigation results of Sistovaris [458] in accordance with Fig. 127 for wool, clarify that the fire-gas liberation of different fire-risk situations can be simulated by variation of the test sample mass. The results of Sistovaris illustrate that in the case of a high test temperature of 950 °C also relatively high CO_2/CO value are possible by minimizing the test-sample mass. Only if the test sample mass is small enough is sufficient air-oxygen available during the sudden thermal loading to the test piece. The U-Pitt method, developed from Alarie at the University of Pittsburg, appertains also to this group. The data records of Fig. 186 clarify the weakness of this combustion system concerning the simulation of the thermal circumstances in the different fire scenarios. Grand [683] discusses the obvious weakness of this combustion system from the fire-protection technical viewpoint.

The constant heating rate of the U-Pitt method does not permit either the tuning of the fixed temperature level of a smoldering fire nor the simulation of the chronological/temporal temperature course of a natural fire. The data records of Fig. 187 show in a very informative way the influence of a variable heating rate on the characteristics of the toxic potential. Wood releases, unlike the high-pressure process polyethylene decomposition, products of raised acute toxicity with an increasing heating rate. The high-pressure process polyethylene releases decomposition products of smaller toxicity due to the better pool-fire conditions.

In [684] and [685] Alarie compares the effectiveness of the U-Pitt protocol with that of the Potts Pot procedure. The effects of the influencing parameters, heating rate as well as material-related characteristics such as, e.g. fillers content and exposure duration, are reviewed in detail [63, 686]. At the judgement of the material-related charac-

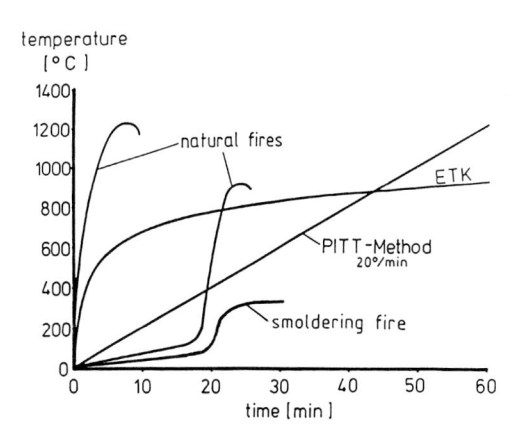

Figure 186 Thermal loading in the U-Pitt protocol

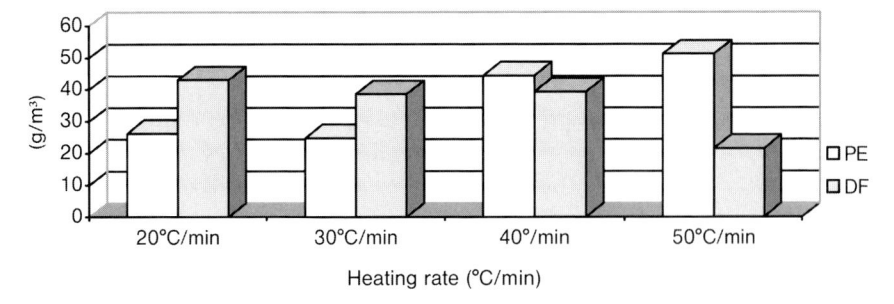

Figure 187 Influence of the heating rate on the characteristics of the toxic potential

teristics of the smoke potential is the possible influence of these marginal parameters to be considered as duly in the case of the evaluation of different combustion systems. The possible significant deviations were demonstrated by Alarie et al. [687] by means of volume- and mass-based test pieces of flexible PUR foams. The danger of the possibly faulty interpretation was shown by means of the relevant material characteristics. The graphical representations of Fig. 129 illustrate in a very informative way that the differences in the value of the toxic potential determined with mass-based samples did not occur during the examination of test pieces with equal volume. In the case of the evaluation of the material data attention is to be paid to the fact that there are comparable exposure times. Since primarily it has to be assumed that comparable volumes or surfaces are involved, these material-related characteristics must be observed exclusively under real-fire situations. Barrow [688] discusses the variation possibilities of the U-Pitt protocol for the determination of material-related characteristics of the toxic potential within the framework of his dissertation. Norris [689] reports in detail on experimental animal investigation results with the decomposition products of different sorts of wood. The present results show significant differences both with regard to the determined CO_2/CO relationship in the stream of effluents then also concerning the "c × t" products. The LC_{50} ratings, determined by Norris (see Fig. 188a), show very vividly that the U-Pitt method supplies under the same test conditions, different potential values, even in the case of identical sorts of wood. Norris found in the case of American red oak samples similar to southern pine three different values. The occurrence of several mortality maxima depending on the test temperature level might to be put down in particular to the weaknesses of this test system.

Constant test temperatures lead, contrary to the constant heating rate, to a not unimportant modification of the toxic potency values. The investigation results of Hilado with the USF method [690] are to be interpreted in such a way that combustion systems with a constant temperature increase, in the present case 40 K/min lead, as a rule, to a little more unfavorable characteristics of the toxic potential. The thermal decomposition products of PE, produced by a constant heating rate of 40 K/min, led to the determined influences first after a significantly longer exposure duration. The representative bioassay test results (see Fig. 188b) make it clear how vari-

6.1 Relevance of Combustion Systems | 341

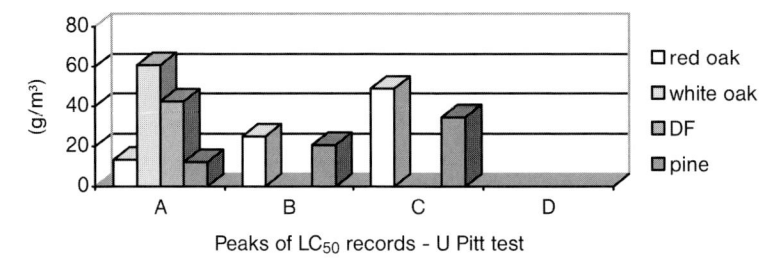

Figure 188a LC50 data of wood determined with the Ü-Pitt method

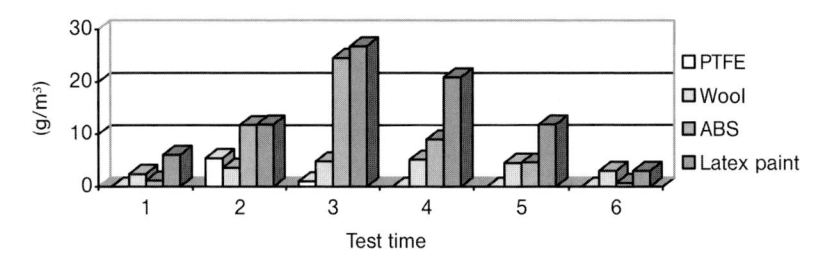

Figure 188b Influence of thermal stress (constant T and heating rate)

able the fire-gas concentration is developed in the course of the U-Pitt test method. The support of the ratings of the toxic potential on the temporal loss of weight supplies a further important argument for the missing proximity to reality of this combustion system. The detailed determined characteristics of the toxic potential, in mg/l, are used as universal material-specific characteristics considering the above demonstrated arguments. They are valid without limit, however, for those fire scenarios only that to simulate the combustion system is able. The data records of Fig. 188b show very clearly that depending on the experimental duration with this combustion system different effluent concentrations are released in the individual time periods. The produced streams of effluents produce within the stress range of 10 to 15 min peak values of the toxic potential.

The third group of decomposition systems contains, e.g. tube furnaces (see Fig. 97). In the case of the DIN procedure a simulation of all possible fire-risk situations was striven for. The DIN equipment (see Fig. 47c) is one of the best examined combustion systems for the generation of thermal decomposition products within the framework of biological investigations for the problem of the fire-gas toxicity. Prager [384] gives a comprehensive view of the most different studies with the DIN apparatus. Klimisch et al. [466] explain calibrating experiments with the decomposition products of fir wood that were produced in the primary air stream of 300 and/or 600 l/h at reference body temperatures of 300, 400, 500 and 600 °C. The CO and CO_2 content in the stream of effluents were determined as the dominating evaluation criteria. The calibration of the decomposition device is stipulated in accordance with the aid of the reference body, a metal strip with a welded thermocouple (see Fig. 47d), in the relevant standard. The exposure duration, as a rule there is a head-nose exposure mode, is 30 min. The utility of the method was verified in several laboratories. A corresponding experimental equipment was used in six more countries. In the UK (FRS) and Denmark (DAN Test) the method was used exclusively as a combustion system for the simulation of smoke release of different fire scenarios. In further national laboratories, e.g. Huntington/UK, VTT/Finland, University of Gent/Belgium, University of Paris/France as well as Kansas City/ FRC Salt Lake City/USA bioassay investigations were carried out [89, 296, 384, 403, 456]. Cocurrent and as well as partially counter-current tests were carried out in Paris [90]. In the majority of the investigations rats were employed as experimental animals. Also mice, rabbits as well as monkeys were used in special cases. A regular exposure was achieved in this case in specific series of experiments with rabbits by means of a forced breathing of the animals [90]. Edgerly and Pettett [691] describe extensive analytical measurements for the characterisation of the decomposition device according to DIN 53436. Within the framework of these investigations and the standardization activities at the SC3 of TC 92 arose substantial question formulations concerning a regular mass-burning rate, a corresponding dose relationship as well as a loss-of-weight measurement the focus with regard to the targeted standardization of the DIN equipment. The test conditions were varied in this case only within the framework of the standardized determination. The boundaries of basic possibilities of variation were not investigated. This model was submitted by the III among other things also within the framework of the investigation program of the PRC initiated by the FTC [14]. However, the two American procedures Potts Pot and U-Pitt were finally chosen. Opponents of the DIN model criticise, in particular, within the framework of the studies at the SC3 of TC 92, the unsure likelihood of ignition of the tested material as well as the missing weight loss instrument. Provided that the spontaneous ignition occurring in the case of high test temperatures then it is not sufficiently noticed, also onto a specific igniting appliance can be resorted after [471]. Purser [508] offers a further problem solution in which the test piece is moved instead of the stove. Investigations with the DIN decomposition method demonstrated that in the case of higher decomposition temperatures an uneven burning process of the strip-shaped test specimen occurred. In some studies it could be verified that by segmentation of the test pieces according to Figs. 189a and b an approximately continuous flaming combustion parallel to the stove movement occurs, but also a regular stream of effluents of identical concentration is released [471, 472].

The segmentation of the test pieces guarantees that an approximately regular stream of smoke over the duration of the experiment is produced. The graphical representations of the Figs. 190a and b demonstrate the effectiveness of the calibration setup at the reference material wood. Figure 191 illustrates that depending on the chosen test conditions, the O_2 content declines only slightly in the stream of effluents. The CO content in the fire gases differs, as Einbrodt and Jesse [693] have proved in extensive practical investigation, e.g. with cable test pieces, considerably with the variation of the test sample surface.

An analogous, regular decomposition process can also reached with the flexible PUR foam as well as with the EPS rigid foam. The graphical representations of Figs.

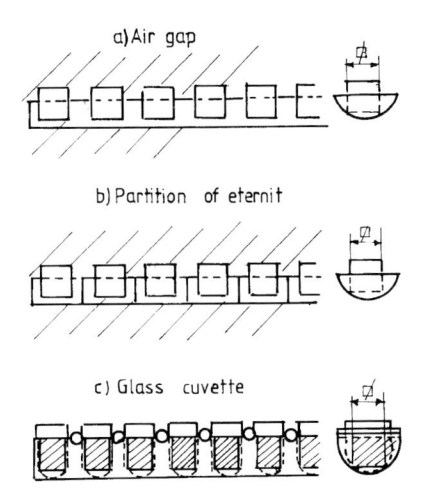

Figure 189a Influence of the thermal stress – constant temperature and constant temperature increase

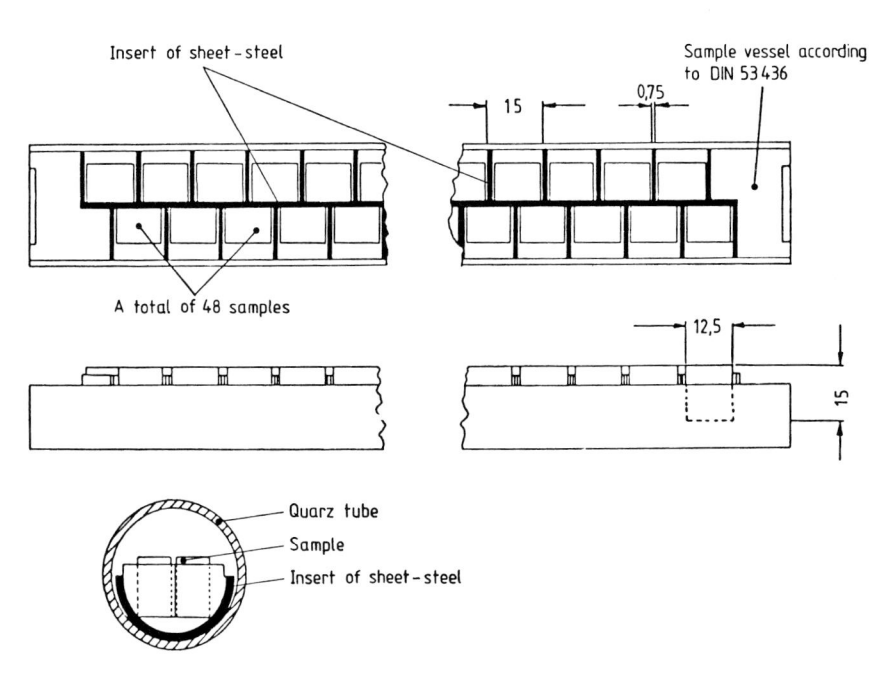

a) Air gap

b) Partition of eternit

c) Glass cuvette

Insert of sheet-steel

Sample vessel according to DIN 53436

A total of 48 samples

Quarz tube
Sample
Insert of sheet-steel

Figure 189b Segmentation of test samples for the DIN 53436

192a and b clarify using the example of flexible PUR foam and rigid EPS foam that the segmentation effect is generally applicable.

The variability of the procedure for the simulation of different CO_2/CO ratios was demonstrated likewise for the fir-wood reference material. An arbitrary CO_2/CO relationship can be adjusted through variation of the test parameters for the reference material wood (see Fig. 193). The graphical representations of Fig. 193 illustrate that

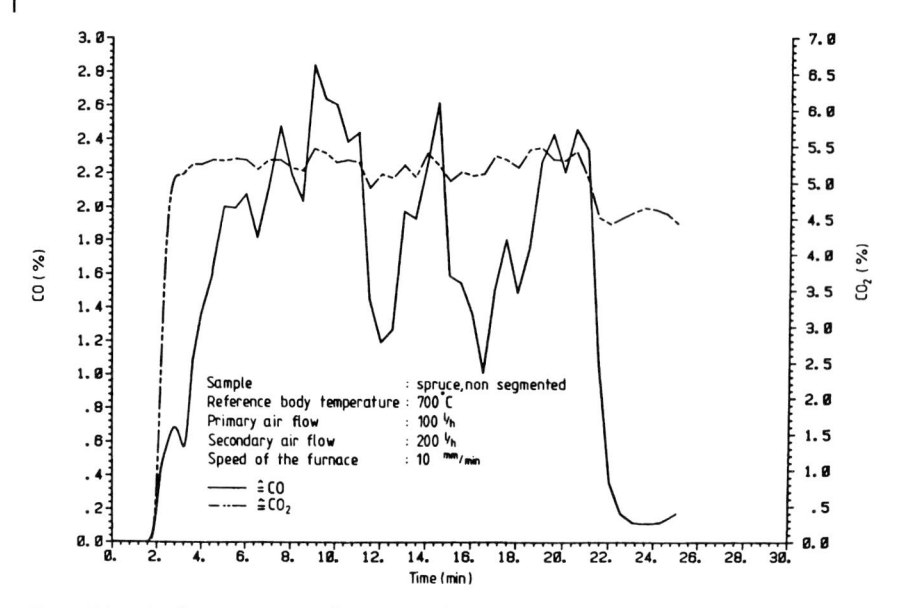

Figure 190a Smoke concentration of unsegmented test samples of wood

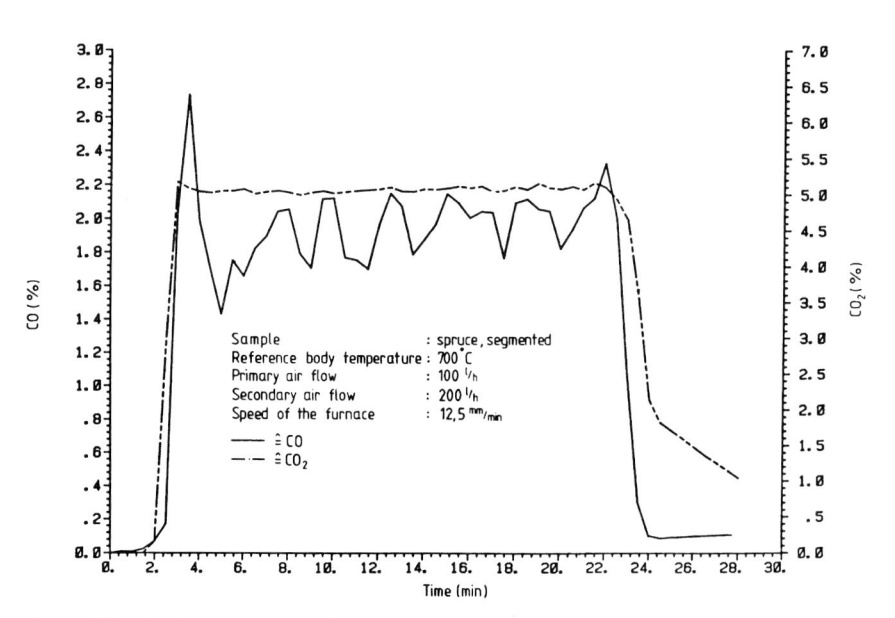

Figure 190b Smoke concentration of segmented test samples of wood

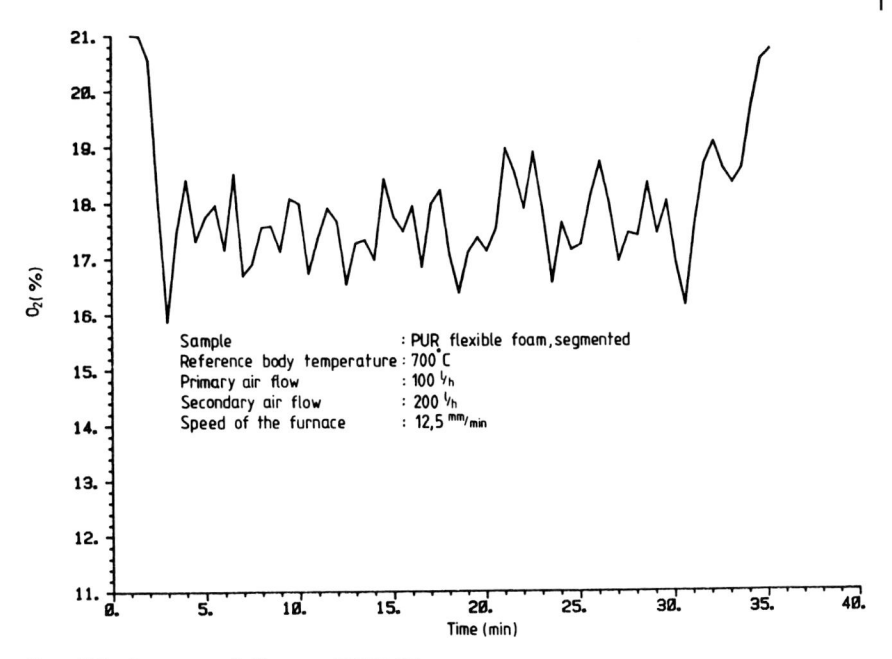

Figure 191 O_2 content of effluents – DIN 53436

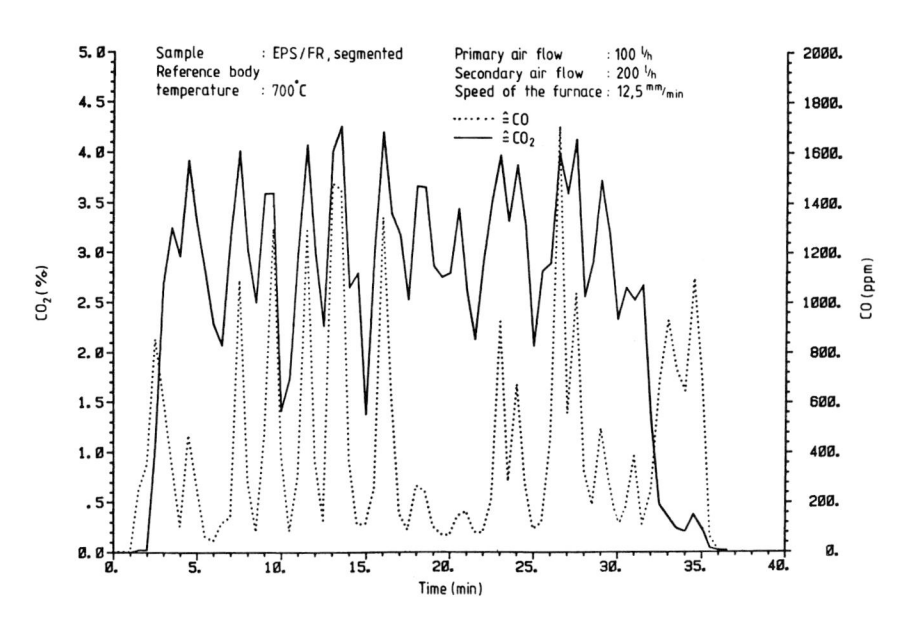

Figure 192a Smoke concentration of segmented EPS test samples

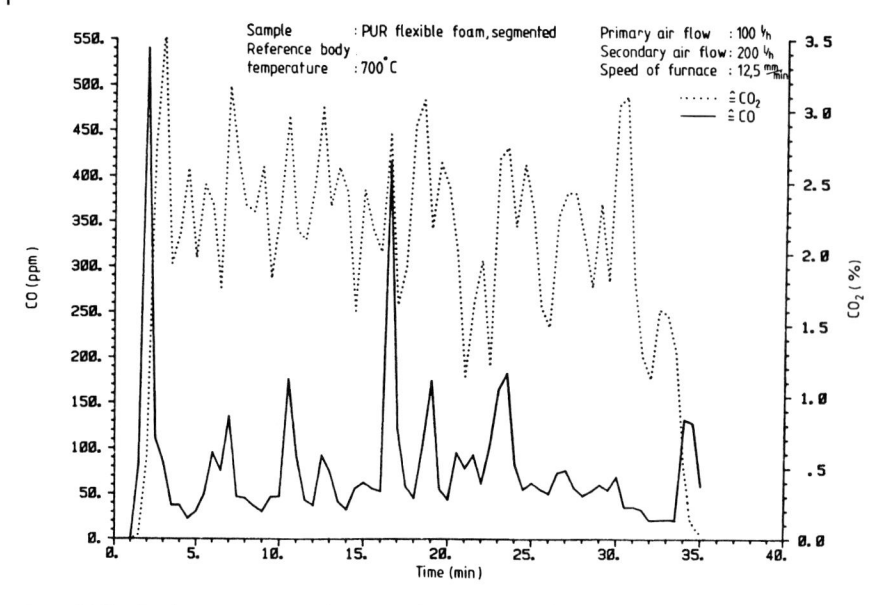

Figure 192b Smoke concentration of segmented PUR test samples

the streams of effluents of the essential fire scenarios can be basically simulated through variation of the test parameters of the DIN apparatus. Extensive studies, in which independent of the standardized conditions the basic possibilities of this equipment were examined, showed very impressively that this means is able to simulate every arbitrary fire situation.

The investigations illustrated that a decomposition temperature spectrum up to 1000 °C is attainable, e.g. through variation of the furnace temperature. The standard equipment allows, as shown by a different investigation, combustion with flames also without specific igniting appliances. In the case of correspondingly high test

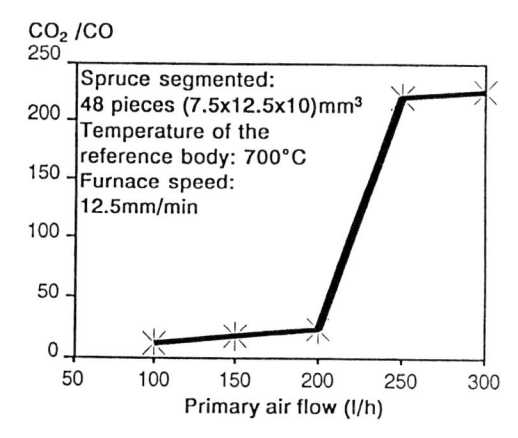

Figure 193 Influence of test parameters on the CO_2/CO ratio

temperatures spontaneous ignition occurs as demonstrated very clearly by Figs. 194a and b for fir wood. The graphical representations demonstrate that a relatively regular burning process with a sufficiently high test temperature is caused in the DIN tube furnace by spontaneous combustion.

The British experts noted in connection with the Furniture Regulation the evidence for the simulation of the self-sustained smoldering process of upholstery furniture. Through skillful selection of the furnace conditions also a smoldering process of flexible PUR foam samples could be simulated (see Fig. 17a). The graphical representations of Fig. 130a show the time–temperature relationship and Fig. 130b provides an impression of the carbonisation process of the flexible PUR foam sample.

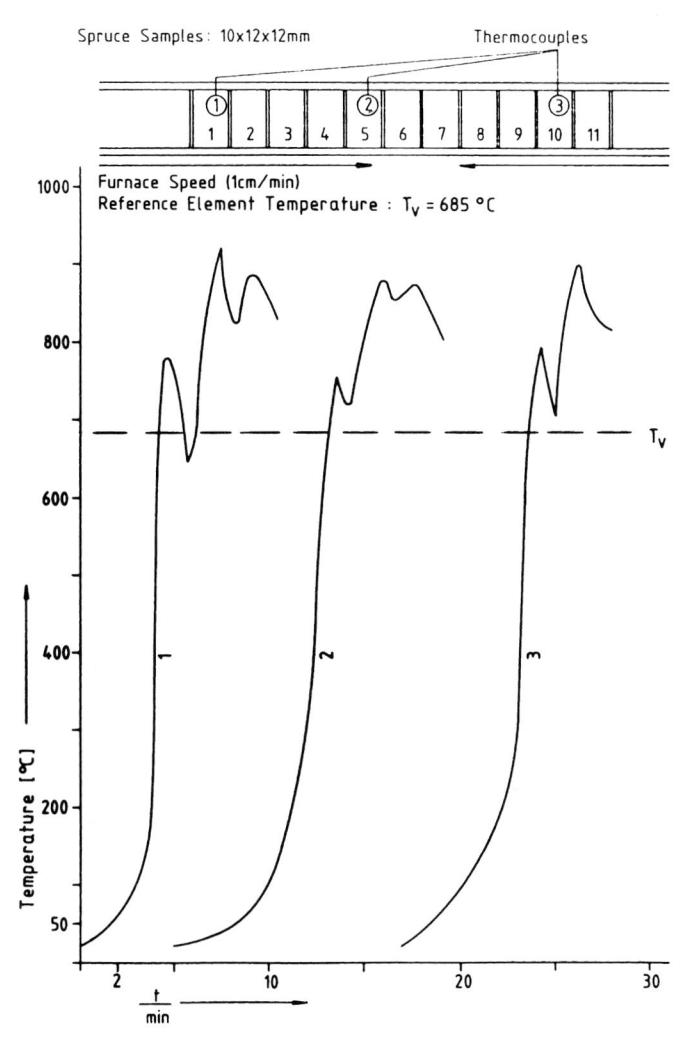

Figure 194 a Chronological temperature course – DIN 53436 wood tests

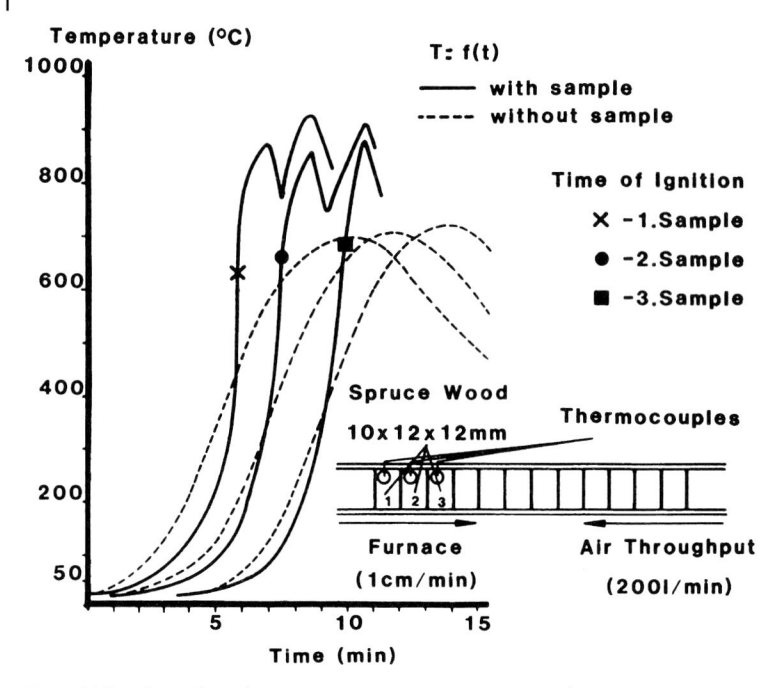

Figure 194b Chronological temperature course – DIN 53436 wood tests

The stream of effluents produced in the DIN equipment is used as a rule for the head-nose exposure mode (see Fig. 130c). The purposeful selection of apparatus-related parameters of the exposure conditions allows both the use of different amounts of secondary air and in the case of variation of the primary air the test animals are exposed to a stream of effluents of comparable concentration. The regulation of A2-classified materials requires whole-body exposure for the evaluation of the flue-gas toxicity in order also to allow an evaluation of the behavior of the exposed animals possibly from the toxicological viewpoint. Figure 195a illustrates very vividly that in the exposure chamber a regular flue-gas concentration does not appear before a longer test phase. While during the head-nose exposure mode, in accordance with Fig. 195b, an approximately constant fire-gas concentration exists, this is achieved in the whole-body mode for the simulation of the fire conditions after approx. 20 min. Jouany et al. [14, 90] examined among other things to what extent a parallel direction of furnace and air-stream movement the determined characteristics differ. These investigations showed that in the case of the direct-current principle due to the preheating parts of the test piece not yet decomposed that is improved flue-gas concentrations, more crucial data records of the toxic potential can arise. During the determination of different decomposition conditions, as they are propagated among other things by Purser in his IEC model [692], certainly relevant basic experiments are needed to clarify the facts.

The French combustion system (see Fig. 77c) has a comparable construction as DIN 53436. The longer testing-stove produces in the stream of air a sudden thermal

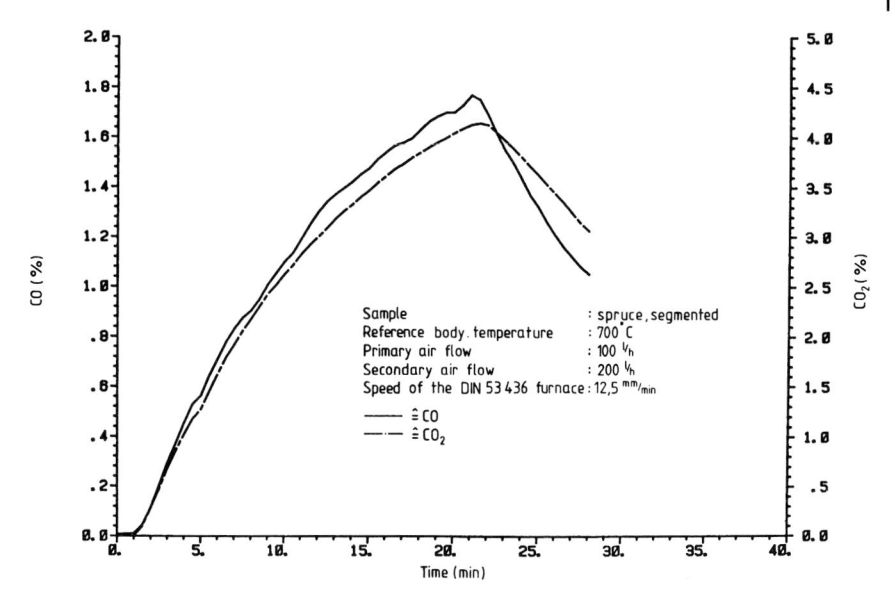

Figure 195 a Temporal course of the fire-gas concentration for the A2-classification – Whole-body exposure

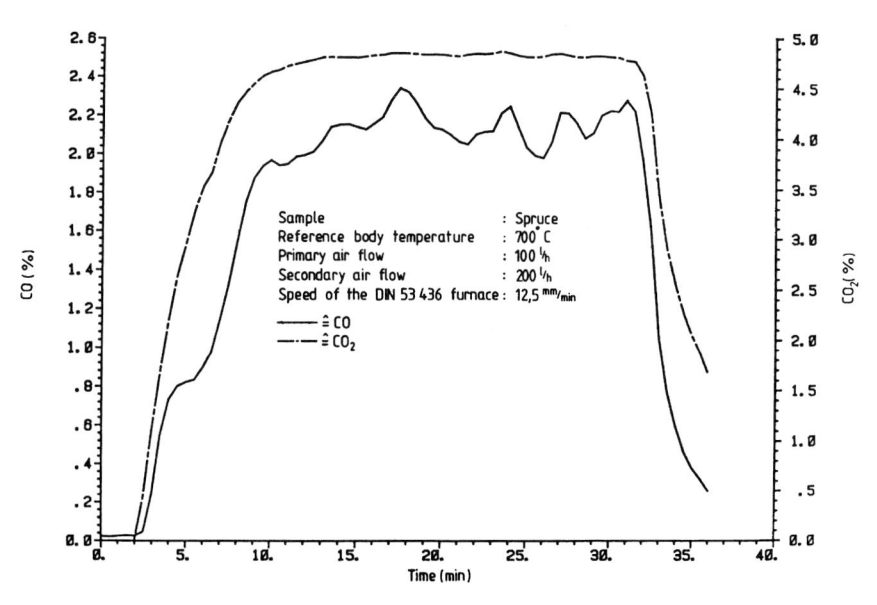

Figure 195 b Temporal course of the fire-gas concentration for the A2-classification – Head-nose exposure mode

stress of the entire surface of the test piece. The standardized procedure assumes a constant-mass basis. The decomposition of the test piece with a constant mass of 1 g occurs as a rule at 600 °C and 800 °C. The test duration is 20 min. The standardization of equal mass means that in the case of great deviations in the specific weight, also extremely different test surfaces will be thermally stressed. Especially in the case of high test temperatures, due to the sudden decomposition of the entire surface, an oxygen deficit will be possible. The inadequate air supply leads to a correspondingly low CO_2/CO relationship of the stream of effluents. The investigations of Sistovaris et al.[460] with the BIS/VCI equipment showed that by variation of the mass of the samples significantly different fire scenarios can be simulated at the same test temperature. The sudden thermal stress of the entire test piece, possibly variable smoke volumes, result in significant modifications of the concentration values of individual fire-gas components The graphical representations of Fig. 196 clarifies these deviations using the example of different groups of products. The influence of test pieces with uniform volume corresponds certainly to the more realistic material-related factors of a final application.

Whitely [694] examined, among other things, the practice-proximity of the French test method NF-T51. His results confirmed that with increasing testing temperature spontaneous ignition of the thermal decomposition products will occur. A serious steep drop of the smoke-density records occurs with the ignition of the decomposition products These investigations showed, however, also very vividly, among other things using the example of PEEK, that after the ignition at approx. 625 °C in spite of a still rising testing temperature only an insignificant modification of the value could be observed. The dependence of the material-related characteristics of the smoke potential of the test parameters was verified both for cumulative procedures, e.g. NBS Chamber, ISO Box and Gost 12.1, as well as through-flow systems as DIN 53436/437 and/or NF-T51. In the case of the question of an effective dose–effect relationship it is repeatedly referred to the advantage of the continuous loss-of-weight measurement. The loss of weight as a material-related characteristic of the fire-gas potential is, however, in the newest findings by no means uncontroversial. Since it is

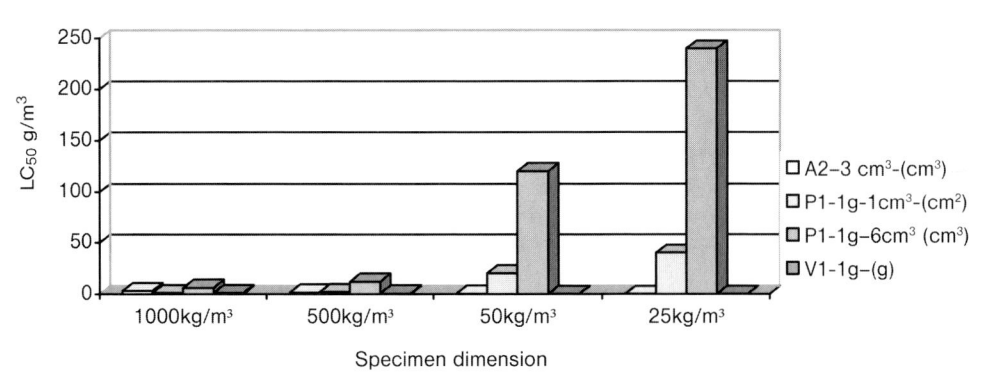

Figure 196 Influence of the test piece dimensions on the characteristics

not expected in any case that the released fire-gas composition as a rule remains constant about the temporal course of the fire occurrence, the different conditions depending on the general conditions of the procedure had to be taken into consideration. What for the pure pool fire, the combustion process of a liquid where an approximately constant fire-gas composition is guaranteed, must be denied for the normal case of a fire. Figure 116 clarifies that the decomposition products vary continuously both in the composition and in the concentration. With the DIN tube system it is guaranteed by the selection of the test parameters that per unit of time a fixed mass, an agreed volume or a specified surface area is decomposing. Figures 121 and 122a clarify using the example of the optical density D that the nominal concentration that is the tested mass or the tested volume or the identical surface area as a reference basis of the calculated values, allow more realistic estimates of a possible fire-gas danger in the case of a fire. In the USF method, which was developed at the University of San Francisco, the test material, as a rule 1 g, became thermally stressed at a constant test temperature or with a temperature increase of 40 °C/min in the temperature range of 200 °C to 800 °C. The results of Hilado et al. [695] are to be interpreted in such a way that the combustion system with constant temperature increase "40 °C/min" causes significantly more favorable characteristics of the toxic potential in comparison to constant test-temperature results. Specific investigations were carried out, among other things, without an additional air supply as well as with a limited supply of 1 l/min. The decomposition process is sped up by intensified ventilation, as the data records of Figs. 131d and e illustrate, and the reaction time of the experimental animals become reduced correspondingly, as demonstrated by the studies with the decomposition products of flexible PUR foam. These characteristics of the material-specific potential show that the modified ventilation conditions are basically mirroring the relevant dilution effects. This USF-B method, also mentioned as PSC procedure-1, is, among other things, the basis of the Bart (SFS-Subway) requirements.

In the case of the third group of decomposition systems in the predominant majority the entire test piece surface is suddenly thermally stressed by a constant intensity of radiation. Figure 56c clarifies the uniformity of the stress about the test surface using the example of the ignitability cone-shaped radiator in accordance with Fig. 74b. Analogous relationships must be reckoned on among other things at the cone calorimeter and its variants, as well as the modified NBS Chamber and the ISO Smoke Box. The numerical values of Fig. 127 show that in the case of high temperatures the combustion process progresses in particular under oxygen deficiency. An approximately regular, well-ventilated afterburning effect of the emerging decomposition products has to be awaited for products with carbonisation tendency after ignition occurred. Decomposition devices such as, e.g. US-Radiant, NBS-Chamber, modified NBS Chamber, ISO Smoke Box, Gost 12.1 Method and Cone Calorimeter are as the DIN 53436 Stove and the NF-T51 Tube Furnace basically suitable to supply the necessary functional material-related potential values. The comparability of the values is safeguarded when it is guaranteed by the selection of suitable test parameters that a stream of effluents of comparable concentration is released. Provided that the standardized test conditions start from a forced ignition, however, then the applicability of this test protocol is considerably limited. Packham [250] explains the results of relevant studies that were carried out at intensity levels of 25 to 50 kW/m^2 with wood. With carbonizing products, such as, e.g. wood, a clear combustion process takes place that leads to a significant reduced CO formation after ignition occurred. The CO_2/CO relationship is well above 40. Packham et al. [408] discuss fully this modified combustion system. This NIBS method is nationally among others under discussion for standardization. The NIBS test protocol release, like the former US Radiant Device, the thermal decomposition products under flaming conditions only. Reports on investigations with the thermal decomposition products of Douglas Fir, Southern Pine and annealed Hardboard are given in [191]. The data records of Fig. 197 provide a view of the thermal decomposition products produced under the thermal stress of 25 and 50 kW/m^2 and their characteristics of the toxic potency. The comparison of the data records illustrates that the serious differences between the smoldering condition and the flaming combustion are not measurable with this test procedure under standardized conditions. The higher test temperatures lead, with the chipboard, only to more unfavorable LC_{50} levels. The more favorable, larger values of the wood qualities are to be put down to the better combustion behavior of the effluent mixture. Alexeeff et al. [190, 697] report on studies with the decomposition products of PUR and PIR foam qualities in comparison to wood products. Under the selected external radiation of 25 kW/m^2 significant differences were determined. The evaluation was carried out from a medical viewpoint. Possible decomposition differences, caused by the modifications of the test parameters and their relevance to the actual fire conditions, was not dealt with. Alexeeff [699] reports on extensive studies with the decomposition products of various polymers. Douglas Fir was decomposed under the thermal stress of 25 and 50 kW/m^2 with and without flame impingement as reference material. The possible influence of the tested surface area was the target of the investigations. The results showed that the size of the surface area of the test piece is a dominating test parameter. It was also verified that the test form changes both the CO liberation and the loss of weight significantly. In the sequence of chips, slabs and cubes the reduced surface area leads to much smaller values. Alexeeff deduces due to his results that factors, e.g. presence of flames, test form and amount

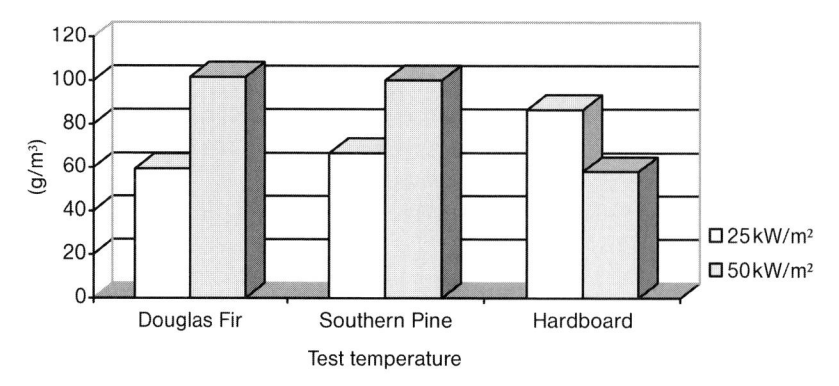

Figure 197 Influence of the intensity of radiation on the toxic potential

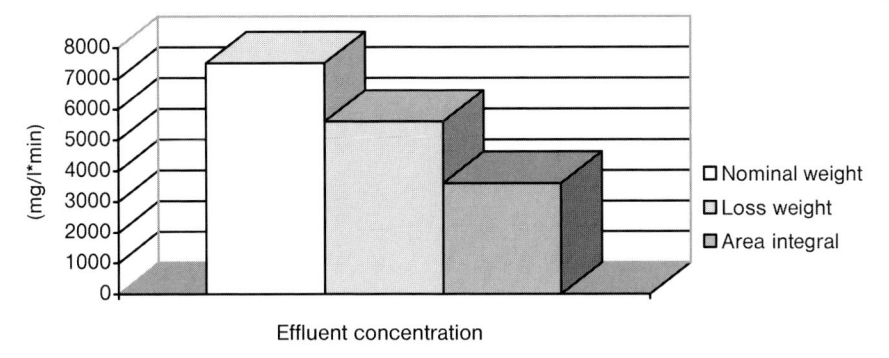

Figure 198 Influence of the reference basis on the rating of the toxic potential of PVC

as well as the rate of decomposition, play a special role for the determined fire-gas exposure profile. Within 3 min the thermal stress of 25 kW/m² leads to a weight loss of 90% at the flexible PUR foam, this was achieved with the rigid PIR foam due to the marked carbonisation effect first after 25 min. Provided that no corrections are made, serious differences in the fire-gas concentration must be taken into account. Using the example of the data records determined with PVC, it can be illustrated with Fig. 198 how the concentration values change significantly depending on the chosen reference basis.

The bioassay test results, the LC_{50} values, determined with the decomposition products of flexible PUR foams produced under flaming conditions and an additional thermal stress of 25 kW/m² and resulted in a concentration of approx. 25 g/m³. It was attempted to correlate the lab test data with the results of full-scale field trials at different exposure duration (5, 15, 20 and 30 min) A potential value of 40 g/m³ was determined under smoldering conditions. The rating determined in the Potts Pot procedure reached 26.6 g/m³. This rating comparison confirms that the lab test supplies, as a rule, more crucial values. In the crucible stove according to Gost 12.1, as a rule, thermal decomposition products are produced at test temperatures of 400 °C and 600 °C. The test pieces of dimensions 40 mm by 40 mm by 12 mm are thermally stressed in the decomposition chamber of the test equipment. The decomposition products are led into the 0. 51-m³ measuring chamber. By means of theoretical considerations the characteristics $Vg(m^3/kg)$, $Vt(m^3/kg)$ and $Vs(m^3/kg)$ are defined, where $Vs = Vt = 10Vg$. The value $Vg = 10\,m^3/kg$ is accepted as valid for all combustible materials. Only one exception exists, that is PE, which is 11 m³/kg.

As a permissible fire-gas amount for the characteristic Dm the boundary values of 20 and 200 are given. Figure 199 provide a view of some data records measured in the case of smoldering and flaming fires. The practice-proximity of the Gost method was confirmed by comparative measurements in the test chamber (0.51 m³) as well as in the testing room of size 19.3 m³ [699]. The determined test data confirm the relatively good agreement (see Table 118). On the evaluation of the material-related single results the partial considerable differences in the mass concentration are to be considered. The fire-gas toxicity according to Gost 12.1 is based on the thermal decomposi-

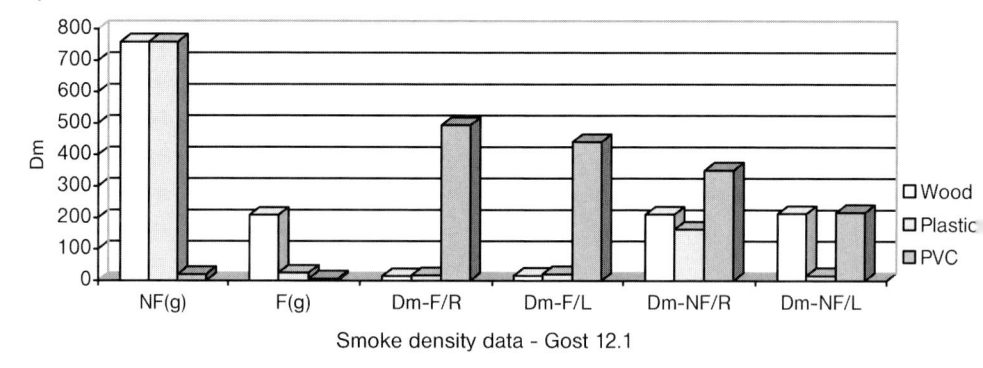

6 Use and Interpretation of PUR-Test Results Determined under Enduse Conditions

Figure 199 Influence of the sample mass and the volume of the measuring room

tion products produced at a thermal stress of 25 and/or 75 kW/m² (respectively 400 °C or 600 °C).

In the smoke-density test equipment, developed at the National Bureau of Standards (NBS), the vertically arranged test sample is thermally stressed at 25 kW/m² by an electrically heated radiator [700]. The influence of the decomposition products, produced with and without flame impingement, on the translucence of an optical system is determined and registered. The photometric system is vertically arranged in order to register possible fire-gas layers. In the case of cumulative systems such as, e.g. the NBS Chamber, the ISO smoke box or the Gost method a significantly different value is produced depending on the determined test conditions. The limitations of the test parameters carried out as a rule under the simulation conditions of the smoldering fire and the combustion with flames under reduced oxygen offer seems not only to be justified regarding the smoke-density temperature–time course due to the experiences, but also to the manageable costs of appraisal. The relevance of the combustion model NBS Chamber can be judged by means of extensive studies. The conditions selected for the NBS Chamber and its numerous modifications with 25 kW/m² with and without pilot-flame impingement allow, as a rule, an overview of the spread of the material-related smoke density potency values. The data records of Figs. 200a to 201 underline this statement. Malhotra [49, 86, 596] explains the possible influence of the imitated fire scenarios on the values by means of the round-robin test results determined within the framework of the standardization of ASTM-E662. Under both decomposition conditions a dramatic modification of the values occurred accordingly. If the NBS Chamber test results are taken as a basis, these test results determine, among other things, the admissibility of materials in the transport section, then the results achieved for the examined products confirm the above statements. Deviations were measured up to more than 600%. Through the determination on the smoldering fire conditions as well as the combustion with flames the two essential fire scenarios can be simulated. In the case of the smoldering-fire scenario deviations of the possible peak values must be taken into account. The studies carried out by Chien with a modified NBS Chamber at the Fire Research Center (FRC) of the Utah University showed that with the use of 25 kW/m² with and without pilot-flame impingement, e.g. possibly a differ-

6.1 Relevance of Combustion Systems

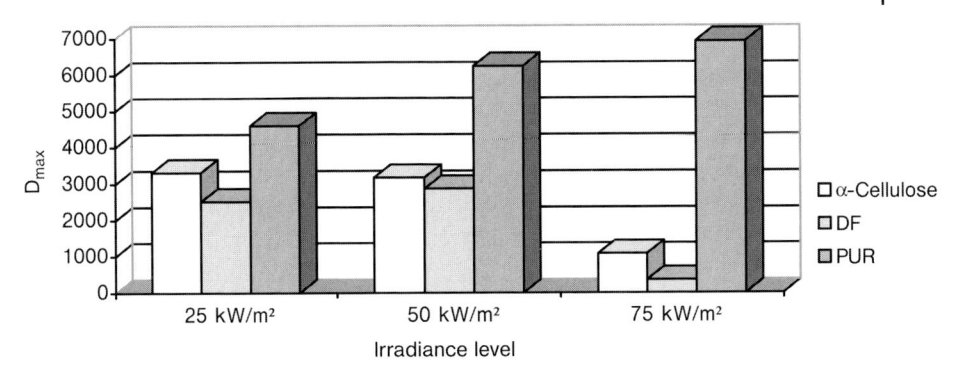

Figure 200a Influence of the test conditions on the smoke potential

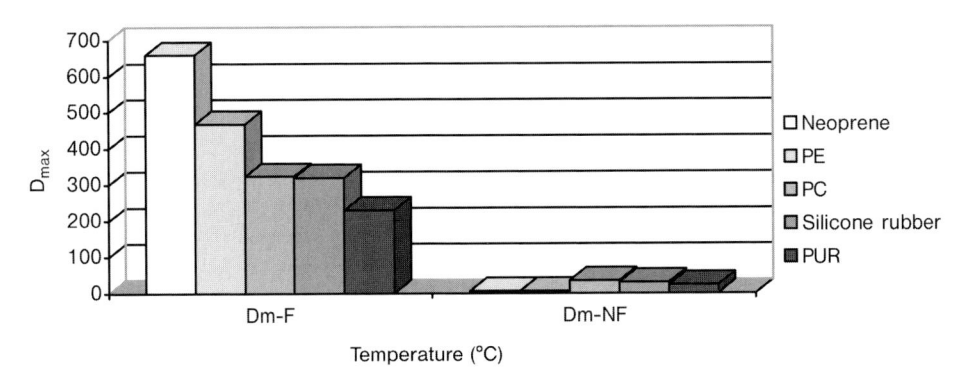

Figure 200b Influence of the radiation conditions on the MOD values

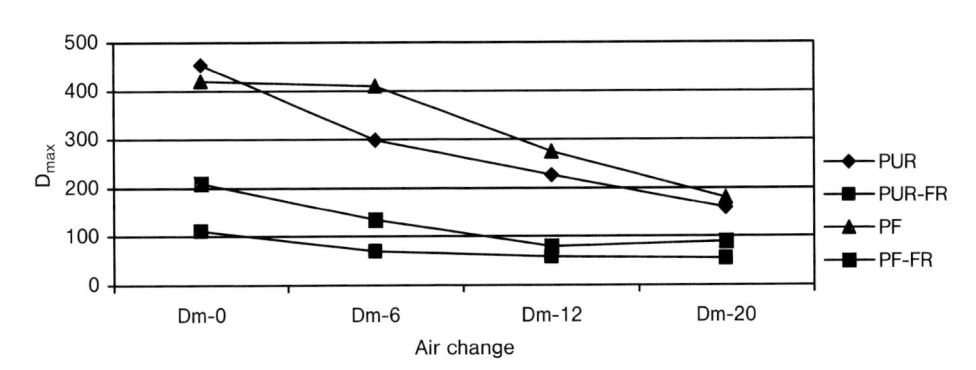

Figure 201a Influence of the ventilation on the smoke-density values – NBS

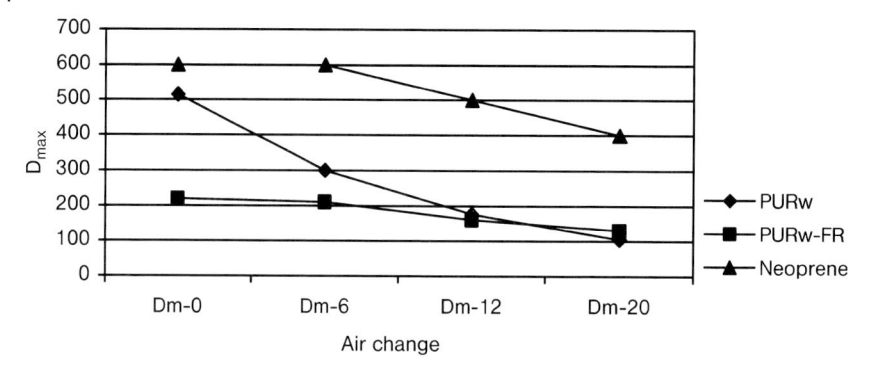

Figure 201b Influence of the ventilation on the smoke-density values – NBS

ent burning process into the test sample interior is initiated. The high intensity of radiation produce during the combustion with flames products with a strong carbonisation tendency. Gaskill [701] describes extensive investigations on the influence of the optical density through variation of the experimental parameters, e.g. changes of air numbers. The Dm values were determined at an intensity level $I = 25$ kW/m^2. The results (see Figs. 201a and 201b) are to be interpreted in such a way that increasing change of air numbers cause a significant reduction of the optical density D. The dilution effect is not directly proportional, however. In the predominant majority of the experiments a twenty-fold air change numbers lead to an average dilution of more than 40%. Gaskill [702] compares, among other things, the general conditions of laboratory test methods such as, e.g. ASTM – E666 and/or ATS 1000 and LRL ($I = 25$ kW/m^2, irradiated area $F = 6.56$ cm^2).

It was guaranteed by modification of the test conditions that different fire scenarios can be simulated. In Utah, investigations with different intensities of radiation could be carried out by modification of the radiator in the range of 10 to 80 kW/m^2. Analogous investigations with such a miniature ISO ignitability radiator were initiated, among other things, by Teichgräber and Topf [703]. The influence of enamel effects could be avoided by horizontal arrangement of the test sample. The first activities for the standardization of the modified NBS Chamber were practised at the SC1 of TC 92. The study of Hovde [280] in this connection led to the standardization activities at the SC1 of ISO TC 92 and to the resulting works at the SC4 of ISO TC61. The results, determined by Flisi [704] with such an equipment, clarify, as demonstrated by Fig. 202a, that through the horizontal arrangement of the sample in the Single Box Method in comparison to the NBS Chamber and its modifications such as, e.g. ATS 1000, in particular with melting products an essential increase of the peak values of the optical density D is possible.

This method being up for standardization enables analogous studies to the ISO Smoke Box. The data records (see Fig. 202b) demonstrate the good agreement of the results. In the Single Chamber Box the horizontally arranged test samples will be thermally stressed by a variable heat stress of 10 to 50 kW/m^2 as in the ISO Smoke Box. These numerical values clarify that the smoke-density data with increasing test-

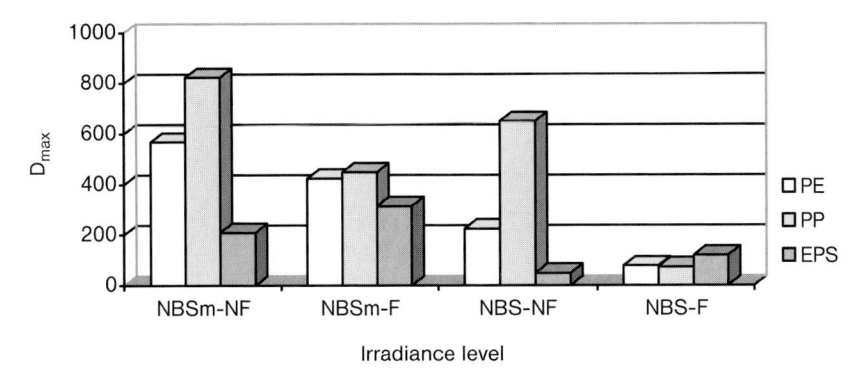

Figure 202a Influence of the test sample arrangement on the smoke density D (NBSm-modified single box method)

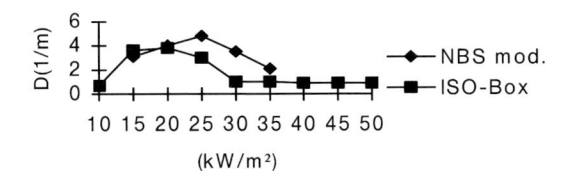

Figure 202b Influence of the intensity of radiation on smoke density

ing temperature reach a maximum at approx. 500 to 600 °C. The values of experience gained within the framework of a small round-robin at the SC1 of ISO TC 92, demonstrate the influence of the temperature on the results (see Fig. 203a). The characteristics of Figs. 203a and b illustrate that the material-related values were determined with a relatively low CO_2/CO ratio in the stream of effluents. By modifying the test parameters, it is guaranteed that risk-oriented values are provided. The value of the CO_2/CO ratio remains altogether below 20. The graphical representations of Fig. 203c clarify the decay of the CO_2/CO concentration.

Within the framework of the standardization activities at the SC1 of ISO TC 92 in particular the Scandinavian experts criticise the missing ignition source of the Dual Chamber Box (ISO Smoke Box). The characteristics of Figs. 203a and b show very vividly that the determined records are achieved in the case of critical CO_2/CO ratios. They are therefore suitable as a basis for legal arrangements as, for example, NEN6066 or numerical approaches. Contrary to the NBS Chamber, the modified Single Box allows similar to the Dual Chamber Box, also called the Munich Box, and the Cone-Calorimeter procedure, the examination of material composites. The data records of Fig. 204 show vividly that protective facings can change significantly the record levels of the current smoke-density potential, also by the testing of joint structures.

The cone calorimeter simulates a pool fire under variable preheating. For constant pool diameter the scale of the fire-gas liberation is exclusively determined by the combustion behavior into the test-product interior. The mass-burning rate is influenced decisively by the carbonisation tendency of the material to be tested. In the

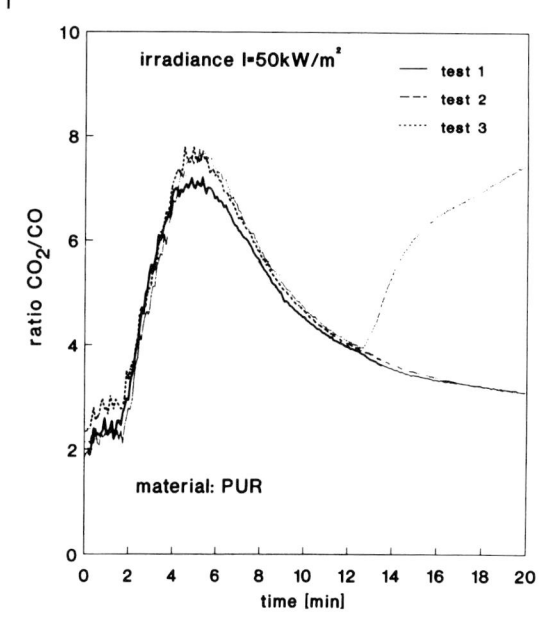

Figure 203 a CO_2/CO ratio in the ISO Smoke Box

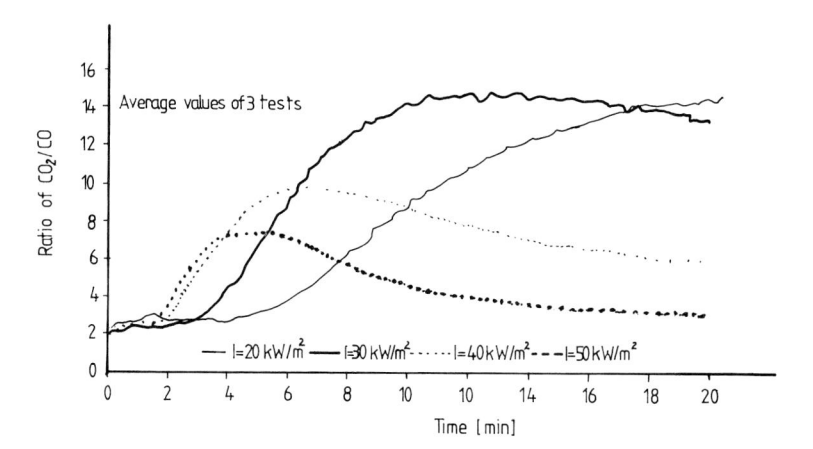

Figure 203 b CO_2/CO ratio in the ISO Smoke Box

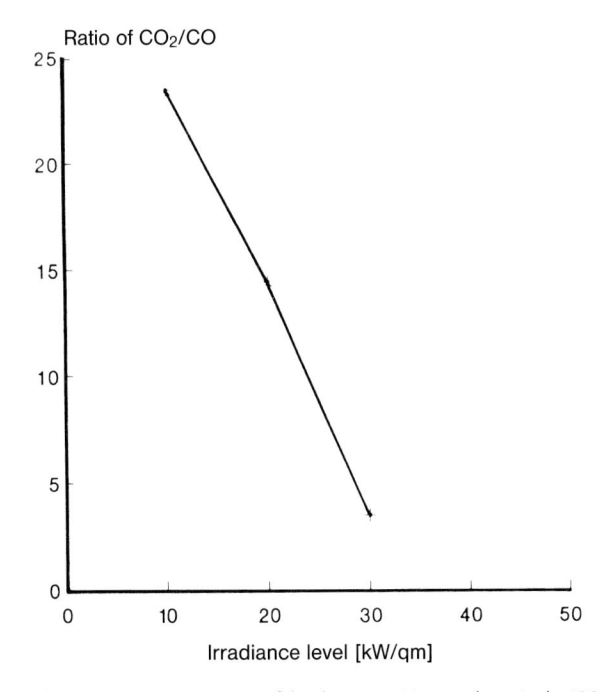

Figure 203 c CO₂/CO ratio of the decomposition products in the ISO Smoke Box

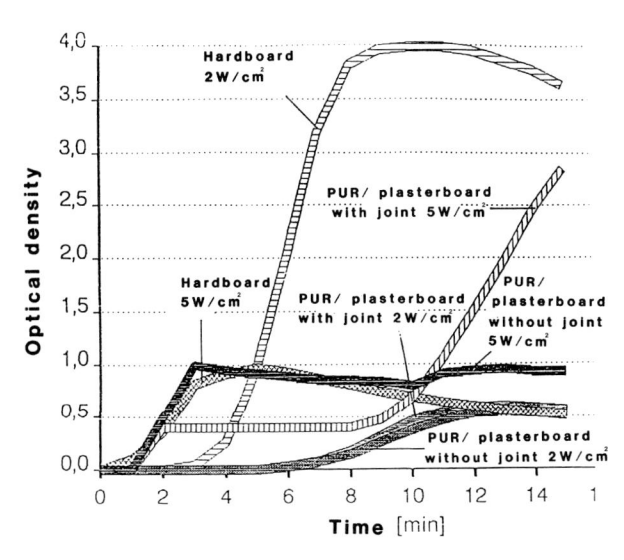

Figure 204 Records of the optical density of PUR sandwich elements

case of products with pool-fire conditions that is product melting in particular, these extreme test conditions determine, as the investigations of Oestman et al. [573] have illustrated in accordance with Fig. 122a the combustion with reduced oxygen offer and high values of the optical density D. The determination of risk related smoke density records by the cone calorimeter test procedere were decisively limited by adjusting the test parameters for the combustion with flames and high ventilation. In the case of products with marked carbonisation tendency the conditions were created to burn the decomposition products optimally. Babrauskas [705] offers the cone-calorimeter procedure that he developed as a realistic dynamic test is in his opinion suitable to assess the possible fire-gas danger in the case of a fire. Although on the one hand he is pretending that the correlation to full-scale field trial results was successfully verified, he clarifies on the other hand in [580] that this combustion system produces fire effluents that supply with cellulosic products, e.g. like wood in comparison to real-fire false CO-concentration data. The obtained CO_2/CO records differ significantly from those of real-fire situations (see Table 119).

Table 119 Corrected LC50 data according to Babrauskas

Test method	DFu	PURu	PVCu	DFk	PURk	PVCk
SWRI/NIBS	100	20	20	21	19	13
NBS Cup	40	18	18	21	9	16

Babrauskas et al. [506] compare the toxic potency of the thermal decomposition products of rigid PUR, PVC and DF produced with different combustion systems and under real-scale conditions. Being aware that the CO concentration of the effluents will be significantly influenced by the combustion parameters, a correction of the CO concentration is proposed with regard to the CO content of the tested material. It is quite evident that the goal of the authors was to get a better basis of acknowledgement for the NIBS test protocol. The data records of Fig. 205 and Table 119 clarify the

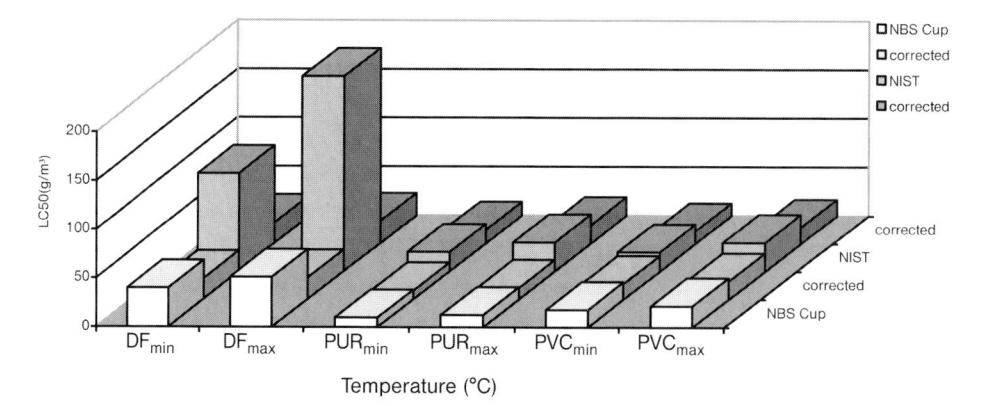

Figure 205 LC50 records patched as the basis of the carbon contact of the materials to be tested

attempts of Babrauskas to correlate the LC_{50} rating by means of a correction factor on the basis of the carbon part of the substance to be tested. Babrauskas thus corrected the characteristics of the toxic potential of the different products determined in the Potts Pot as well as in the NIBS procedure. The thus-corrected $I.C_{50}$ values were compared with the findings of full-scale field trials in the 1:1 scale. A detached analysis of the general conditions of these room experiments shows, however, that the simulated room-fire conditions do not reflect the course of an average living-room fire [705]. The combustion of the wall panelling was caused by a very specific ignition source, burning cribs of the material to be examined. This incipient fire can never be classified as a practical simulation of a common apartment fire. Circumstances that are completely unsuitable for a flue-gas evaluation due to the obviously well-ventilated combustion process analogous to the wooden crib fire as demonstrated by Fig. 206 become simulated. In the opinion of the CBUF consortium, a round-robin test with upholstery combinations reached basically an acceptable reproducibility. The data records of Fig. 207 provide a view of the statistically evaluated smoke-density characteristics measured with a padded chair as well as a double-seated settee. The appraisal "very acceptable" must be designated as extremely doubtful.

These results confirm the doubts about the risk relatedness of the values determined in the cone-calorimeter procedure for an estimate of possible fire-gas dangers in the case of a fire. The investigation results that were achieved with realistic upholstery combinations confirm the above testimony. All cushion combinations ignited within 16 s. The examined combinations resulted in no great differences of the ignitability risk accordingly. The smoke-density characteristics, the potential value per kg

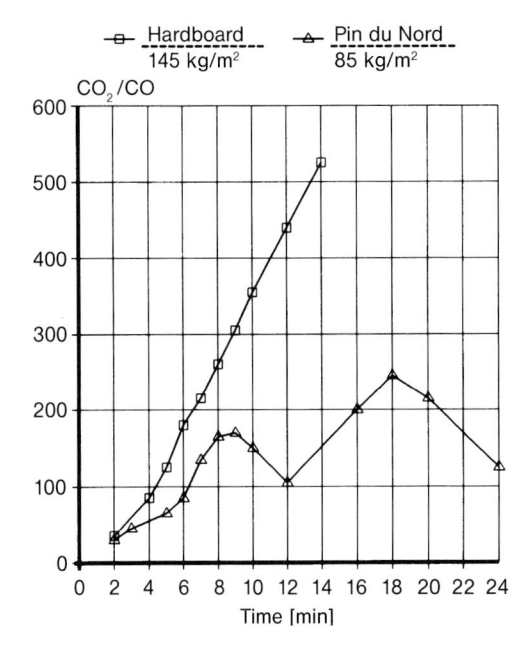

Figure 206 CO_2/CO ratio of the effluents of burning wooden cribs

and their divergencies, provide an impression of the quality of the cone calorimeter as smoke-density test equipment. The determined characteristics "r" and "R" of the mathematical statistic show for the reproducibility as well as for the repeatability considerable deviations. The data records of Fig. 207 demonstrate that these results do not already come into consideration as a basis for legal arrangements for reasons of reproducibility. The numerical values of the confirmation experiments show very vividly the difficulties concerning an acceptable reproducibility. The regular combustion process of the PES/FR-HR upholstery combinations that results also in the modified IMO/ISO Spread of Flame Test according to Fig. 159 might be responsible for the good agreement. The extremely strong differences in the value of the smoke liberation, clarify the difficulty to determine obligatory boundary values by means of the testing parcel proposed by CBUF.

Morikawa [395] reports, e.g. about a smoldering fire experiment with an upholstery layer from cotton flocks that shows clearly that the concentration of the effluent components acrolein and formaldehyde can vary strongly in the 25-m^3 large experimental room depending on the fire occurrence. In the studies with the U-Pitt-2 combustion system, basically a modified cone calorimeter, Caldwell and Alarie [706] examined the influence of the external intensity of radiation as well as the air supply for the CO_2/CO ratio in the decomposition products of wood. The results of Fig. 208 illustrate that the standardized decomposition conditions of the U-Pitt-2 method do not permit a realistic simulation of the different fire scenarios. The effluents of wood demonstrate, that the standardized decomposition conditions produce at the low intensity of 21 kW/m^2 a CO_2/CO ratio of more than 100.

The data records of Fig. 209 illustrate that such decomposition systems corresponding to the cone calorimeter, are up to now not suitable to simulate with the reference material wood the smoke production of real fires. The CO_2/CO relationship of the produced stream of effluents corresponds for cellulosic products as a rule to that of afterburner systems. The lethality as well as characteristics of the breathing capability were defined as evaluation criteria analogous to the U-Pitt method. These characteristics of the toxic potential are, according to the limitation of the combustion system, only of very limited use.

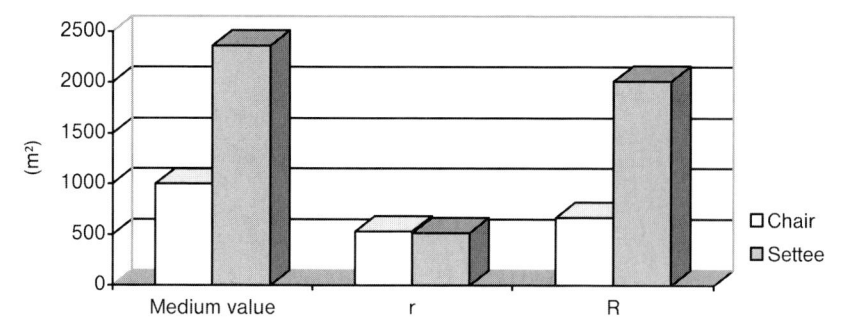

Figure 207 Reproducibility of the fire-gas liberation during combustion of upholstery furniture

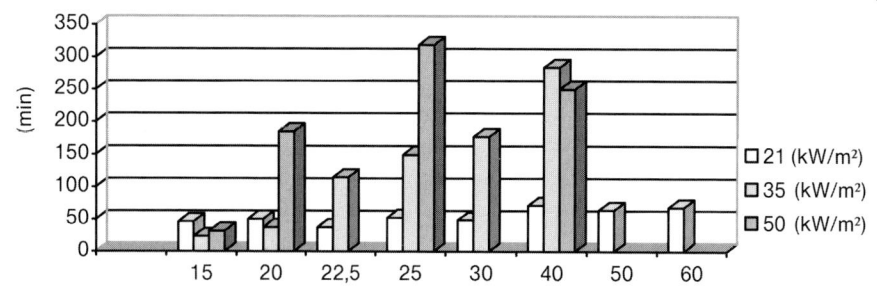

Figure 208 CO$_2$/CO ratio in the fire-gas stream produced in the U-Pitt 2 method

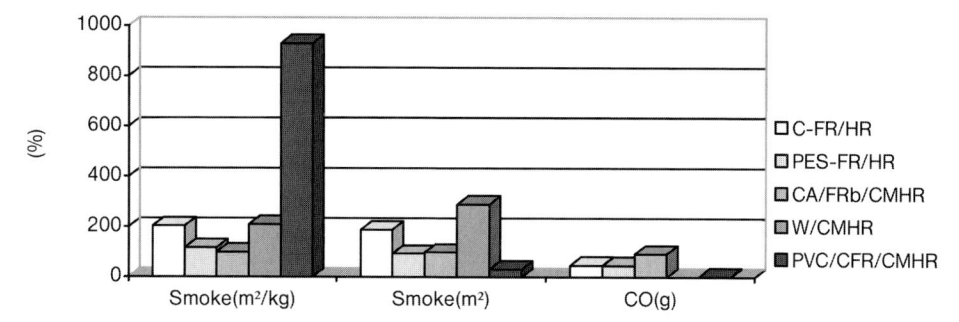

Figure 209 Reproducibility of the determination of the smoke-density potential in the cone calorimeter

The procedure RTB, an investigation project in Japan, corresponds likewise to the cone-calorimeter test method. The cone-shaped radiator is identical with the ISO ignitability equipment. The thermal decomposition products are also produced due to testing considerations in this procedure only with the simultaneous action of a pilot flame. According to Yusa [473] the fire-gas generation of the reference product wood could be simulated only at an extremely high ventilation that is also a very high CO$_2$/CO relationship. The characteristics determined in analogy to the cone-calorimeter test are of limited use also. Saito attempts with the box method (see Fig. 62c) to simulate a room-fire situation in an extremely reduced scale. The decomposition products are produced under two different combustion conditions. The different, standardized conditions are defined as follows:

A: Exposure chamber 125l, thermally stressed surface area 300 cm^2 and air supply 12 l/min

B: Exposure chamber 600l, thermally stressed surface area 462 cm^2 and air supply 12 l/min.

The fifth group of combustion systems contains combined methods, where primarily the resistance against defined ignition sources is examined (see Figs. 41b and 75a). These combined smoke-density test methods, e.g. the tunnel test according to ASTM-E84, the chimney-test method and the critical-flux test according to DIN 4102 and DIN

5510 as well as the modified fire-propagation test according to JIS A1231 are, due to their standardized test parameters, not able to provide functional characteristics of the smoke. The material-related smole density characteristics of these combined test methods and special procedures like the 3m cube of IEC TC 89 are linked to the respective fire scenarios. They base, due to the different contribution to the flame spread on quite different effluent concentration (see Table 120). This fact can be explained using the example of the chimney-test procedure according to DIN 4102 in detail. The data records of Fig. 148c show very vividly using the example of EPS slabs, to what extent modifications of the test parameters change the fire reaction behavior in the chimney test. The scale of the fire-gas formation is changed correspondingly. The fire disaster of the Düsseldorf airport showed that the enduse conditions can also change considerably the burning behavior and therefore the fire-gas formation [529, 707]. The B1-classified EPS slabstock materials failed in the chimney test as well as in the critical flux test due to the Al lamination. In the smoke-density test method according to IEC TC 89 the test samples, approx. 1 g cable pieces and/or upholstery parts are thermally stressed with different ignition sources (1000 ml methylated alcohol and/or 0.5 kg alcohol-soaked charcoal). The different contribution of the horizontally arranged test samples to the lateral flame spread will possibly also vary the extent of the effluent formation. The scale of the fire-gas liberation can be changed significantly through the orientation of the samples and the different sample construction, e.g. the variable cable assignment. Defined surface units will be thermally stressed without exception, as in the cone calorimeter, by parts of the combined test methods, such as, e.g. the NT 004 test or the fire-propagation test according to JIS A 1321. With this indeed an essential condition for the definition of nominal surface based material records is met not, however, the guarantee for risk-relevant results. Since these procedures are exclusively based on the combustion with flames, similar to the cone-calorimeter test, the applicability of the results is possibly very much limited. The listing of Table 120 provides a view of the usability of the individual decomposition models for the three most important fire scenarios, which are necessary for the assessment of the smoke danger. The combined procedures, like the chimney test and the critical-flux test in accordance with DIN 4102 and DIN5510 or the tunnel test according to ASTM-E 84 provide characteristics that are totally linked without exceptions to the simulated fire situation. Combined procedures such as, e.g. the XP2-Chamber that indeed allow a reference to the subjected mass-, volume- or area-related unit, are linked altogether to the combustion with flames. Even the variation possibilities of the external radiation such as, e.g. in the Early Fire Hazard Test according to AS-1530 or with the cone-calorimeter procedure supply, due to the forced ignition by flames, only characteristics of limited informative capability. Test methods such as the NBS Chamber in accordance with ASTM-E662 or ATS 1000, NF-T50 or the modified NBS Chamber or Gost 12.1 as well as the ISO Smoke Box, are in a position to supply in the same way material-related mass-, volume- or area-based potential values of the smoke density similar to dynamic systems like DIN 53436/37 or NF-T51. All these procedures enable through variation of the test parameters basically the simulation of the smoke generation of the different fire scenarios. Single-point measuring procedures provide statements that suffice for the checking production or supply conditions only. They are, however, not suitable in view of the functional dependence of the fire-performance potential for estimates of possible smoke dangers in the case of a fire by means of numeric approaches. These material-related records can be changed possibly crucially out of experience by the test conditions and as a result of the end-application definition. The data sets of Figs. 210a and b show very vividly that also so-called materials of the class A can possibly release considerable CO and HCN concentrations in the case of a fire. The evidence of such material-related potency values, which have been published by Szpilman and Nisted at DAN Test by means of the DIN equipment [708, 709] with an air flow of v = 100 l/h, can not be delivered, e.g. by the standardized room-corner test.

Table 120 CO_2/CO ratio of effluents of large scale tests and fire investigations.

CO_2/CO ratio	<10	10–20	20–40	>40
Developing stage of a fire	34%	29%	24%	13%
Flash – over stage	42%	11%	25%	22%
Real Fires [98]	50%	22%	11%	17%

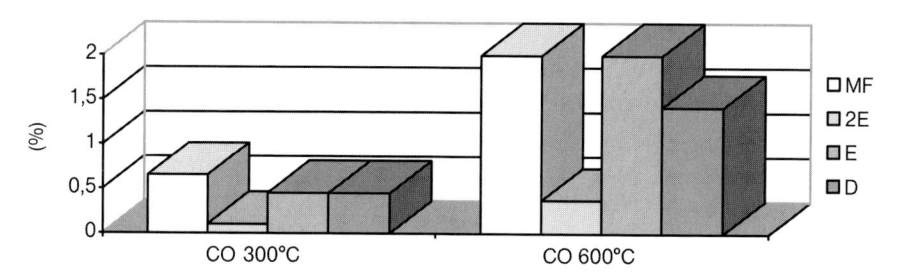

Figure 210a Influence of the test temperature on the CO liberation of Danish class A materials

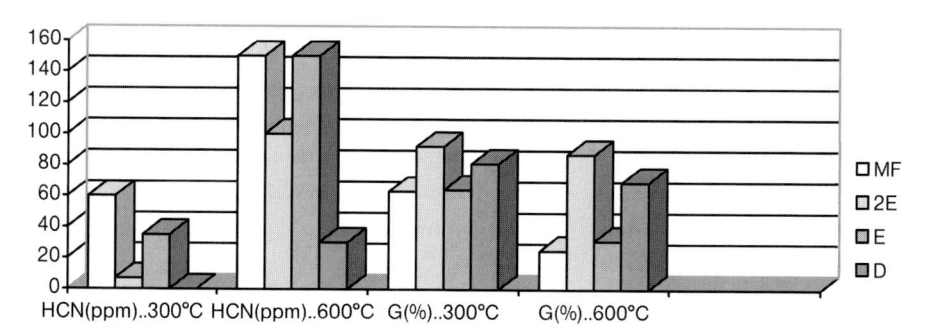

Figure 210b Influence of the test temperature on the HCN liberation of Danish class A materials

6.2
Relevance of the Evaluation Criteria

For the assessment of the relevance of evaluation criteria the testimony spectrum of the criteria must be carefully separated from the possible importance in the real-fire case. The individual criteria in themselves can provide relevant indications of the quantity and quality of different properties of the fire occurrence. They are nevertheless due to the chosen combustion system and the standardized test conditions not able to draw conclusions from the real fire course. The question to what extent laboratory test methods are suitable to simulate the determination of the different characteristics of the fire performance relevant to the risk was the subject of numerous research projects. Through comparative investigations with practical full-scale field trials on the one hand parallel and on the other hand obvious weaknesses of the laboratory test methods were sought. With the aim to improve the repeatability and the reproducibility of the results the testing and the evaluation criteria are oriented in the first place at the usability of the test method, i.e. the interests of the inspection technology, and not at the targeted protection destinations of the preventive fire protection. Foreseen from some exceptions, e.g. the small-torch procedure in accordance with Fig. 49a where only a fast, upright spread is supposed to be suppressed in the first 20 s or the Motor Vehicle Safety Standard (MVSS) 302 according to Fig. 72 where the horizontal flame spread is limited, all these test protocols require the temporal and local limitation of the flame spread altogether as proof of the resistance against acting ignition sources. How much it takes, possibly in the inspection technology, a precise determination of the evaluation criteria, can be demonstrated using the example of the small-burner procedure. The schematic representation of which (see Fig. 49a) clarifies that due to the vertical flame spread within the stream of decomposition products, failure is certified in the test although the surface spread of flame does not reach the marker at the end of the experiment. While products that combust completely after the observation period achieve the rating class B2, normally ignitable, insulants of the first category are classified also as core material of composites as inadmissible material class B3. The previous building law is guided by these test related misjudgements, which certainly do not satisfy the targets of the preventive fire protection in the substance. Using the example of the A2 classification of building products according to DIN 4102 as shown by Table 8 as well as of the room-corner test procedure after IS 9705 it can be illustrated that a judgement of evaluation criteria can be carried out only in relation to the different factors, determining the initiation and the fire development. Independent of the risk relatedness of the individual evaluation criteria both the compatibility of the individual characteristics and also the place value of the material-related potential value within the framework of the entire fire hazard must be considered sufficiently. The concentration data, underlying the definitions of the building authorities in accordance with Table 8 clarifies the erroneous overall concept for the estimate of possible dangers in the case of a fire. How far the end-application-oriented conditions endanger the scale of the danger in the case of a fire and not the flashover criterion of the room-corner procedure clarify the experiences of numerous risk-oriented large-scale field trials carried out in

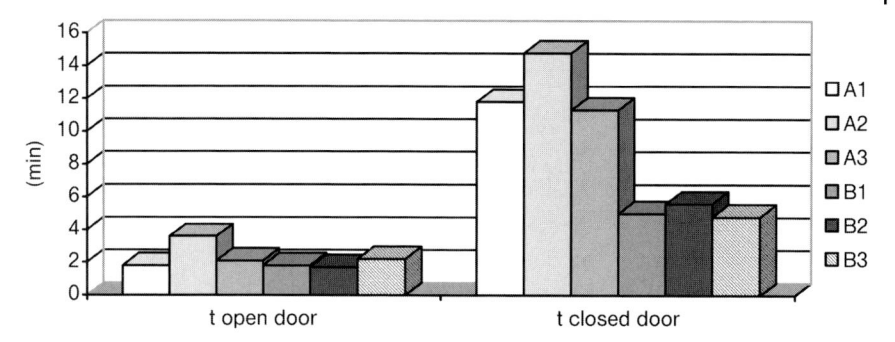

Figure 211a Influence of the ventilation on the scale of the fire-gas liberation

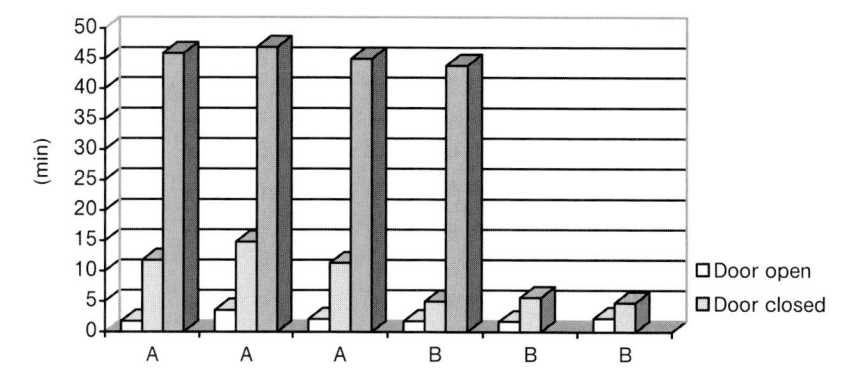

Figure 211b Survival time as a function of the exposure

the scale 1:1. The possible influence of the combustibility classification of wall panelling on the developing and fully developed fire in the dwelling field were examined, among other things, in the St. Lawrence research program [710]. The burning rate including the smoke liberation when open and closed doors were investigated (see Figs. 211a and b). The use of noncombustible wall panelling led with closed doors, to a significant increase of time to flashover occurred. For high ventilation, i.e. open doors, the influence of the combustibility rating seems to be of minor importance. Figure 211b illustrates that there was an analogous result for the time necessary for survival.

This fire situation, simulating a room-corner experiment, demonstrates that the classification targeted in IS 9705 does not mirror the situation of a real apartment fire, but rather the risk spectrum of an empty room, e.g. an empty kitchen in the construction and/or repair stage. The behavior of individual materials can be studied carefully indeed under ideal conditions, statements for security reasons are allowed, however, as a rule exclusively under observation of the general conditions of a real-fire occurrence. The results of this testing method do not form a realistic basis for security reasons with that for regulation in the dwelling field. This interpretation of the results of the St. Lawrence research program is, among other things, confirmed

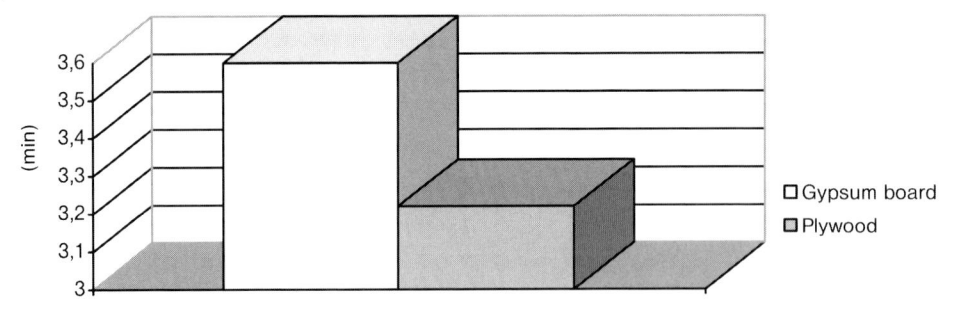

Figure 212 Influence of the wall panelling on f.o. conditions in a fully furnished room burn out

also by the investigation of McNeill [711], where the influence of the combustibility classification of the wall panelling on the flashover time in the trailor sector was studied in detail. The characteristics of Fig. 212 clarify that the fire performance of the content of the rooms investigated determined the scale of the fire. The time of the flashover situation is changed only slightly, as shown by the graphical representations of Fig. 212 by the use of gypsum board as wall panelling instead of plywood. The ceiling tiles consisted in both experiments of plywood boards.

6.2.1
Risk of Ignition

In the definition of the key data for the selected ignition sources the question is whether it can be guaranteed that the ignition source is able to simulate the initiation of the fire to be investigated. The undestroyed length as an evaluation criterion plays in the majority of the standardized test methods a not unimportant role. Since it is primarily a question of the proof of the resistance against ignition sources of daily life, an on-going differentiation is important for very specific simulated test cases only, but this over-differentiation obviously does not bring any increased safety levels. This over-differentiation does not allow any interferences on modified circumstances such as, e.g. caused by variation of the size of the ignition source, the exposure time, the ventilation or also the test configuration. The length of time to ignition t_i is the central criterion that is used for numeric approaches in particular concerning the fire-performance classification. The data records of Fig. 213a clarify the dependence of this characteristic on the external intensity of radiation. These results illustrate that the differences of the material-related characteristics reach more and more a negligible order of magnitude at high intensity of radiation. The investigations carried out within the framework of the cooperation of different industrial laboratories (I-ILDA) with the ISO ignitability equipment demonstrate this very well [712]. The characteristics of Table 121, determined by cooperation of I-ILDA with ISO TC92 for many products, demonstrate the approximation of the ignition times t_i with increasing intensity of radiation. An immediate ignition must be reckoned on always in the flame-action field of the flame impingement.

6.2 Relevance of the Evaluation Criteria

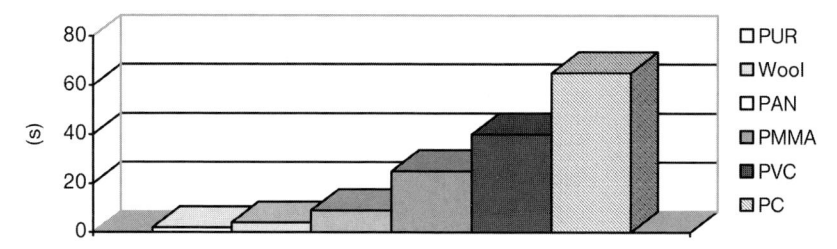

Figure 213 a Igniting risk at high intensities of radiation

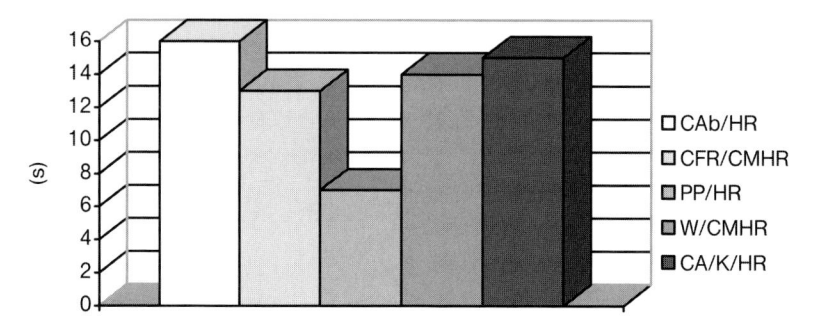

Figure 213 b Influence of the cushion contribution on t_i under the acting of the CBUF propane torch

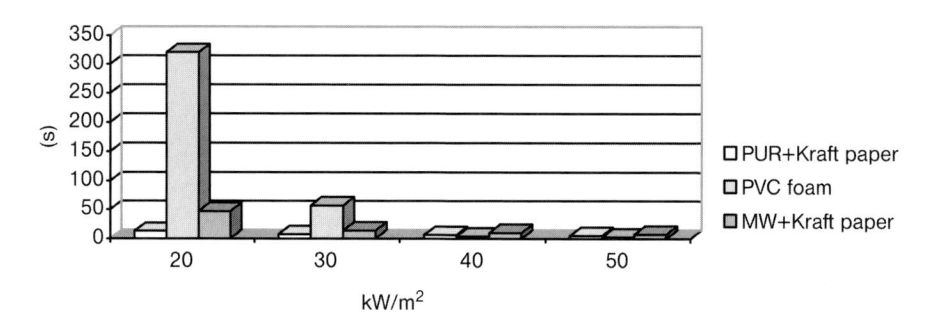

Figure 213 c Influence of the irradiance level on the risk of ignition

Table 121 Influence of the external intensity to the ignition time t_i (s).

Material	20 (KW/m²)	30 (KW/m²)	40 (KW/m²)	50 (KW/m²)
EPS	410	69	33	15
Nylon carpet	115	60	40	26
MOD	24	14	11	8
ABS	137	55	39	29
CA	94	47	30	19
PMMA	106	54	33	24
PC	–	464	116	65
PVC	402	107	58	40

Since the intensity of radiation of the common ignition sources, in particular the bunsen-burner flames achieve a very high level t_i does not seem of substantial importance to the igniting risk in the direct flame impingement area. The data records determined by cushion units in the CBUF program according to Fig. 213b confirm this statement.

The physical regularities of the horizontal surface spread of flame are detailed from the viewpoint of the CBUF consortiums using the example of mattresses. Experiences showed that the ventilation effect is considerably more effective. Figure 168a shows that the intensity of radiation can increase, e.g. approx. more than an order of magnitude in the ventilation direction. An effect that is not attainable through variation of the cushion combination as a rule. The surface finish, i.e. the top layer of a composite, can also gain the dominating role. In the case of melting products, the formation of a pool rules over the scale of the surface spread of flame. The properties of the substrate may play due to the peeling effect (see Fig.69c) a dominating role in the case of thermoplastic covers. The differentiated influence of carbonizing and melting fabrics was also verified in the CBUF program. The resistance against ignition sources of small and larger intensity can be modified possibly considerably by the use of flame retardants. The positive aspects of the fire-retardant additives with regard to an increased resistance against acting ignition sources and also against an obviously reduced fire risk were detailed in numerous publications [486, 523, 576]. Piechota [713] investigated, among other things, the influence of phosphorous- and chlorine-containing additives on the resistance of PUR foam qualities against various ignition sources. These additives do not cause seemingly any significant modifications of the chemical combustion process. The addition of melamine seems to cause an endothermic sublimation as well as a physical dilution of the decomposition products in the flame as well as an oxygen barrier. As further effective fire-retardant additives, among others, urea, aluminum hydroxide, ammonium phosphate and magnesium hydroxide are named. Through the trend to integrate electrical appliances/devices into sleeping and seat furniture, the primary ignition sources of the electrotechnical sector must be considered in the furniture field also. The electrical outfitting of beds includes a considerable fire risk [714]. As a possible cause of the often fatal fire misfortunes drive motors, underground laying of cables and erroneous moisture protection have been named among other things. The investigation carried out from the testing and research laboratory TÜV refer to a necessary re-equipment of the beds. The length of time up to the entry of the f.o. situation can among other things significant modified through the size and order of the igniting flame, the stress duration and the spatial conditions including room size, wall- and ceiling lining as well as the ventilation. The determined detailed material-related characteristics are particularly significant for the selected general conditions. Generalization of the conclusions is inadmissible due to the experiences from fires as well as realistic large-scale tests. A material-related classification of building products, as it is, e.g. recommended in the Eurific program [541] (see Fig. 27) for regulation, is therefore misleading. The room-corner test does not offer any risk-related prediction under standardized conditions for the fully furnished room.

6.2.2
Burning Dripping

The dominating criteria for the assessment of the burning dripping is the afterburn time of the dripping or dropping parts as well as their igniting capacity. The listing of Table 114 concedes a view of the igniting potential of dripping material. The decisive issues are, whether the burning dripping/dropping material is in the position to become a secondary ignition source for the environment. At the B2 test procedure, the risk relatedness must be questioned from the beginning, because the evidence of the ignition of a tissue paper as a pad within the first 20 s only, including 15 s impingement time, will be taken into account. If burning outflow should occur after 20 s this is no longer evaluated. The definition of a completely specified material for the proof of the ignition capability caused doubts about the risk relatedness. The partner material as proof of the ignition risk should be available as a receptor also in the practical fire case. The use of the tissue paper facilitates indeed the job of the testers, but does not safeguard, however, the information requirement of the preventive fire protection. The failure boundary is displayed as a rule due to its easy ignitability in a shorter period of time. Exery practice-relevant statement is led through this definition ad absurdum. The so-determined evaluation criteria lack any reference to the targeted destination accordingly. The limited afterburn time in the chimney test according to DIN 4102, which is supposed not to exceed 20 s, is faced with the primary flame intensity according to Table 114 for the neighboring field irrelevant since the igniting potential of the dripping burning material is not sufficient to cause ignition of the surrounding B1-classified materials. Full-scale field trials with vertical and horizontal glazings demonstrated this very impressively as illustrated by Figs. 149a and b. In France, the igniting risk by burning dripping is assessed by a specific procedure (see Fig. 54d). Although the procedure permits a more objective assessment of this side effect of a fire, the determined data records are also not suitable to carry out a risk-relevant evaluation, because the requirement profiles presuppose a resistance against ignition sources of essentially higher intensities than to be verified by the cotton material receptor. The afterburn time of the burning drops, which also decides the admissibility of upholstery combinations, corresponds also not to real fires. Neither in mass-transport systems like aeroplanes or trains nor in assembly rooms, will local limited burning dripping be able to become a secondary ignition source of sufficient intensity in the regulated flooring field. The resistance against ignition sources of much higher intensity is compulsory for these fields of application [714].

6.2.3
Relevance of the Criteria for the Evaluation of the Heat Release

Net calorific value
The net calorific value is certainly the classical criterion for the classification of the possible heat release. What seems to be justified with regard to the mobile fire load, to use the net calorific value per kg for all considerations of the preventive fire protec-

Table 122 Deviation of the heat release of upholstered furniture.

Outer covering	CMHR (35 kg/m³)	CMHR (50 kg/m³)	PUR (Standard)
Acryl	490	94	661
Cotton	145	674	630
Viscose	415	395	350
PP	482	402	430

tion, must be questioned as not risk relevant not only for the building-material application. If one takes the entitled room as a guide line, everything then points to volume- and/or area-related ratings. The data records (see Fig. 123) clarify that the observation of the risk-relevant dimension would lead without a doubt to another image. If than, as required in the expect report for the Dusseldorf airport fire disaster [1], material characteristics are disclosed from the built-in-state viewpoint, than the representations of the Figs. 99a to c demonstrate using the examples of different insulants that, presupposed an equivalent insulation effect, a completely different picture is presented by area-related ratings. While mass-based characteristics postulate a smaller heat potential for mineral insulating boards, demonstrate volume-based records a comparable insulation effect presupposed, quite equivalent values. The classification of cushion units carried out by Nelson in connection with the Dupont Plaza Fire [715] confirms the variation width of the heat release. These characteristics of Table 108 characterize, however, the difficulties of the authority to enforce regulations and taking into account these scientifically protected research results.

Fire-gas temperatures

Fire-gas temperature maxima were used, as illustrated by the data records of Fig. 148a, as an evaluation criterion for the heat-release potential by many laboratory test methods. These material-related data records, however, are not suitable for generalisation. The data records of Fig. 171 give a view of the boundary values of the fire-gas temperatures in different laboratory test procedures. The conversion of these fire-gas temperature levels in different fire scenarios is, as a rule, not possible. These values are linked to the limited combustion of the test procedures and can not be used as the basis for numeric approaches for the danger estimate in the case of a fire. In the case of the boundary values for B1- and A2-classified building materials these boundary values symbolize the scale of the heat release in the case of the impact of a burning wastepaper basket on a A2- or B1-classified wall panelling. If the tested material retains itself in the acute fire case according to the material classification to DIN 4102, then the stream of effluents includes a real igniting risk neither in the convective nor in the radiation component for the relevant environment, because even under the crucial chimney conditions no inadmissible surface spread occurs. The graphical representations of Fig. 148a illustrate that under realistic wall and corner configurations a much more favorable behavior is to be expected. With the action of ignition sources of raised intensity, e.g. burning upholstery furniture components, a stream of effluents of raised temperature must be reckoned on. The temporal process of the fire-gas temperature proved also in risk-relevant full-scale field trials with pieces of furniture units including fully furnished rooms as a meaningful evolution criterion of the heat-release potential. The temporal course of the effluents allows inferences to be made in the heat-release potential only with knowledge of the factors influencing the fire occurrence. Investigations in the scale 1:1, as, e.g. the III program [56, 104] or the study of BRMA/Europur [321, 677] within the framework of the targeted Furniture Directive [298] of the EG, illustrate that characteristics of the temporal process of the fire-gas temperatures is a prerequisite for estimating the possible smoke danger in the case of a fire. The graphical representations of Figs. 214a–c show very vividly that through variations of the outer covering and upholstery filling and through constructive measures the value of the stream of effluents can be changed considerably. The use of interliners causes a reduction of the temperature level from about 700 °C to 200 °C. As Fig. 214a shows very vividly, the use of interlayers, in this case a cotton fabric with fire-retardant additives, causes a considerably delayed combustion process and therefore reduced fire-room temperatures. These data records illustrate the different burning process due to variation of the composite. In the sequence standard polyether, FR quality, HR and impregnated quality a reduced burning rate could be observed. Composites with modified FR fabrics result in decomposition temperatures below 100 °C.

The feedback effect through the smoke layer arising during the fire process depends on the size of the room and the arrangement of the burning item. Mizuma et al. [714] verified in realistic full-scale field trials that the burning rate depend also significantly on the fire-load distribution. In the sequence middle of the room, wall and corner configuration an intensified burning process takes place. The interpretation of analogous results is not unambiguous with the studies of the CBUF program. The numerical values of Table 123 provide a view of the mass-burning rate after 10 min as well as the corresponding peak values. Strange to say the interliner combination gave the highest peak value. The large room (7.4 m × 5.7 m × 4.0 m) brought in comparison to the ISO room (2.4 m × 3.6 m × 2.4 m), which is taken as the more crucial scenario, only an insignificantly reduced fire course. The graphical representations of "Fig. 68" of the CBUF final report [545] clarify that in the larger room the intensity of radiation at the measuring point with <2 kW/m^2 deviates significantly, contrary to 10–14 kW/m^2 in the ISO Room. The unfavorable result of the combination with the interliner, usually classified as the most secure combination, has been not discussed in the CBUF report. Fire tests with fully furnished rooms were carried out by the International Isocyanate Institute (III) in the training center of the English Fire Fighters in Moreton in Marsh (see Fig. 215a) These graphical representations demonstrate that the findings of the RAPRA trials [100, 101] can be partly transmitted, as investigated in Moreton in Marsh [104] to other fire scenarios. Figure 215b clarifies the measurement of room temperatures carried out in the two-storey experimental plant. The significant modifications of the fire-gas temperatures that varied considerably due to the variation of the upholstery combination could be verified in the fire room, in the neighboring staircase and also in the upper floor. The representations of Table 124 provide a view of the findings of the CBUF program.

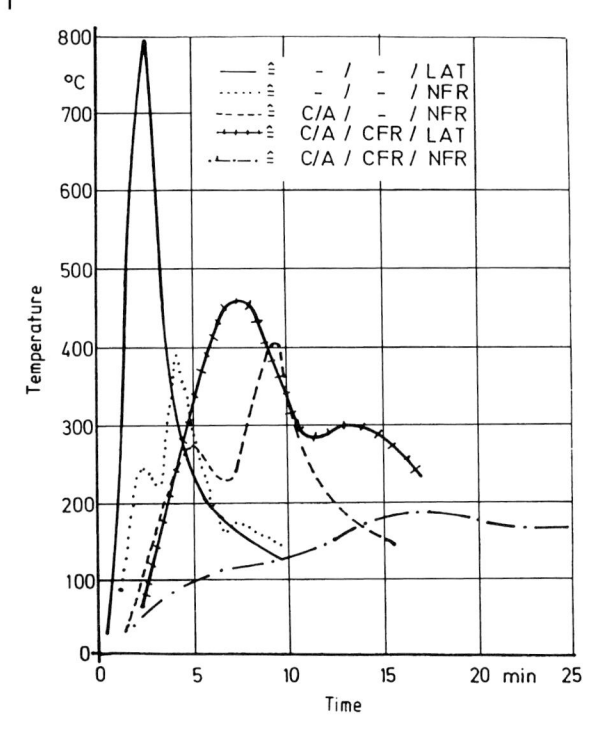

Figure 214a Influence of the upholstery combination on the heat release

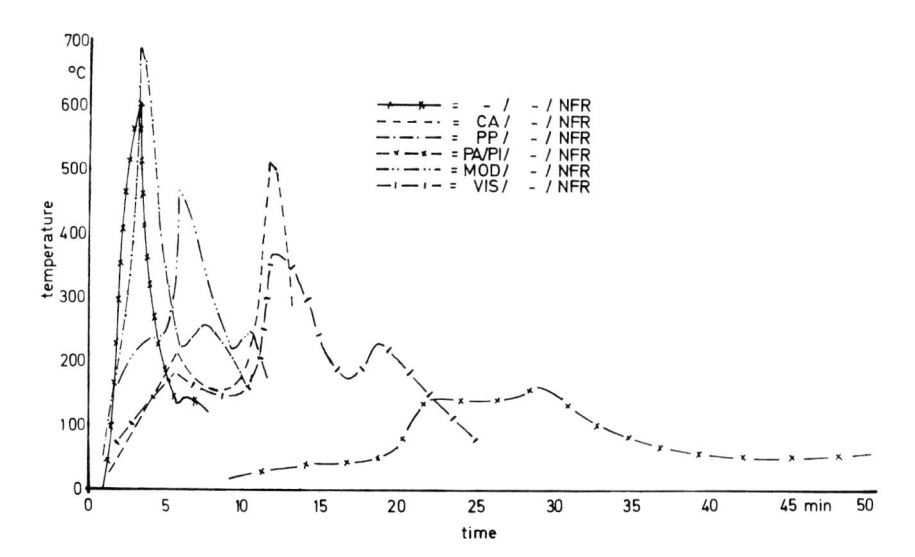

Figure 214b Influence of the upholstery combination on the fire-gas temperatures

6.2 Relevance of the Evaluation Criteria

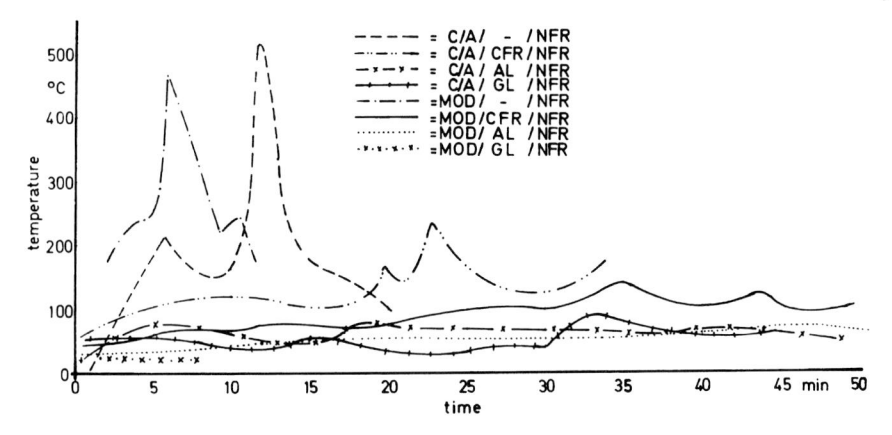

Figure 214c Influence of the cushion contribution on t_i under the acting of the CBUF propane torch

The furniture of the living-field produce unambiguously higher values of the heat release in a shorter period of time.

Smith [677] and Powell [715] reports on analogous investigations of Europur/BRMA, which were carried out at RAPRA and at WFRC in the late 1980s with regard to the planned European Furniture Directive. The temporal process of the fire-gas temperatures formed the determining criterion for the assessment of the heat-release rate also with these studies. The graphical representations of Figs. 179 and 180 clarify using the example of prominent values as, e.g. RHR peak, 60 s or 180 s value both questions concerning the reproducibility and the conversion of the findings on true-scale testing units in the Furniture Calorimeter as well as in the laboratory test units in the cone-calorimeter protocol. A detailed discussion is given by Creyf et al. [321] who planned, organized and financed these extensive investigations. These studies of the industry illustrates that even when the upholstery filling itself is not able to make any significant contribution, e.g. in the case of the application of rockwool as upholstery filling, both the heat and the smoke-release rate achieve comparable values. The material-specific records are seemingly only slightly changed (see Fig. 215c).

Heat release per unit area

During the last three decades, the peak values of the heat release per stressed area unit as well as the medium value per area unit gained acceptance as important evaluation criteria. They should be similar to the building product A2 classification, where alternatively a net calorific limitation value of approx. 4200 kJ/kg and a heat-release rate within the fire-resistance test (test duration of 30 min, 50 cm × 50 cm test piece) of 16 800 kJ/m² were standardized, allow the estimate of the possible contribution to the fire development. In the aeroplane field, boundary values of the possible heat-release rate on the basis of the OSU Chamber with 65 kW/m² and 65 kW min/m² had to be determined as there are in order to limit the possible contribution to the fire development in the case of a fire. These two characteristics are to be classified as

Table 123 Influence of the room geometry to the mass burning rate of upholstery combinations

Upholstery combination	Mass burning 600 s ISO (g/s)	Mass burning 600 s Test room (g/s)	Max. mass burning ISO (g/s)	Max. mass burning Test room (g/s)
4.1 CFR/PUR-HR	8,4	7,5	44,0	40,1
4.2 PES-FR//PUR-HR	8,8	7,4	39,1	28,5
4.3 CA/FRb/K/PUR-HR	12,5	11,3	46,0	52,0

Table 124 Fire load as a function of the use.

Application area	HRR Peak (KW)	Time to HRR Peak (min)
Dwellings	1278 ± 719	339 ± 278
Public areas	727 ± 465	490 ± 430

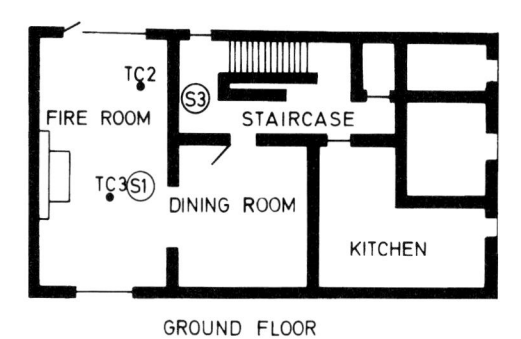

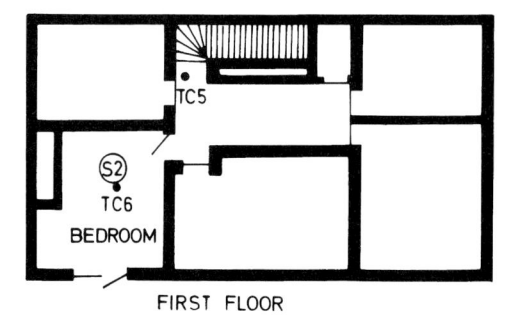

TC = Thermocouples, S = Sampling

Figure 215a Fire test house in Moreton in Marsh

6.2 Relevance of the Evaluation Criteria | 377

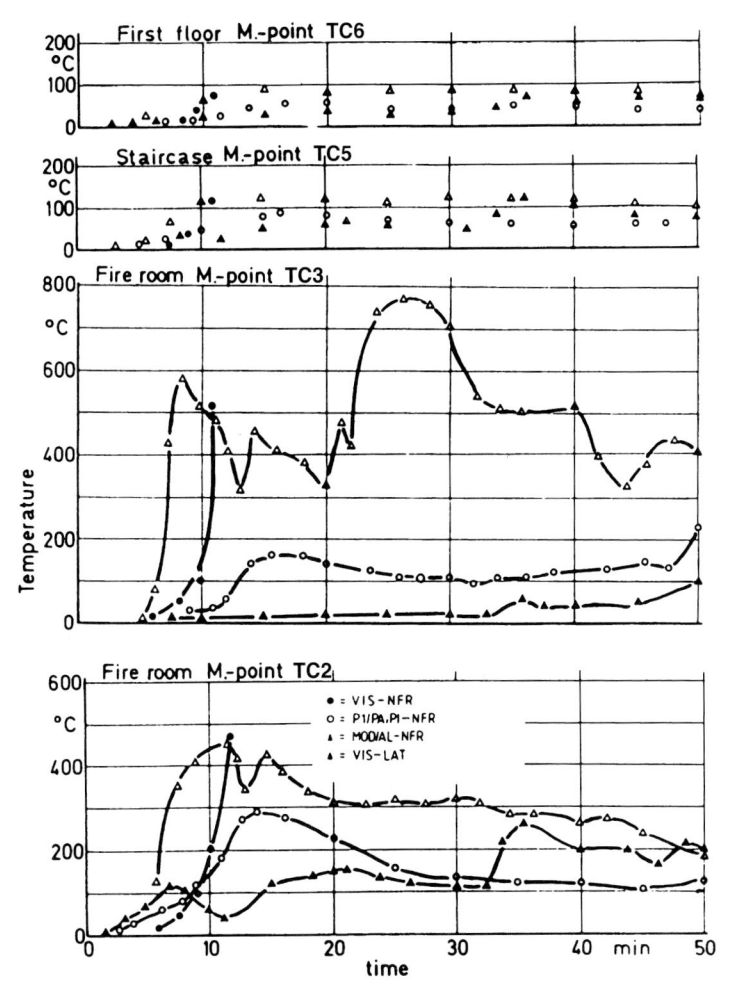

Figure 215 b Fire-gas temperature development of fully furnished room burn out

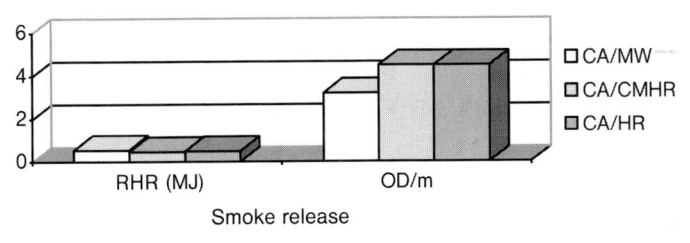

Figure 215 c Influence of the upholstery filling on the material-related characteristics

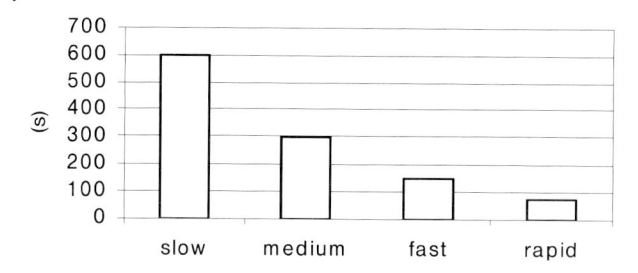

Figure 216 Characterization of the combustion behavior of room fires

practice-relevant records. If one compares the testing method for the limitation of the contribution at the thermal stress of ignition sources of raised intensity, then it turns out that in the aeroplane field the decision test procedure according to Fig. 67d, the vertical flame-spread test, does not allow to define adequate safety levels. Tran [716] demonstrated with different woods in the OSU chamber that the heat-release assessment by means of the fire-gas temperature measurement supplies approx. 25% to 30% lower values in relation to the oxygen-consumption definition, as is compulsory for the cone-calorimeter test. The investigation results showed, furthermore, that test pieces in a horizontal position lead, contrary to the vertical position, to higher heat-release values. After a testing period of 5 min, however, comparable levels were achieved for the total value. The data records of the heat release concerning the combustion process into the interior of the test sample, form a basis for numeric approaches whereas for the developing stage relevant parameters are orientation, composites, sandwich construction or ventilation and the surface spread of flame caused through them are to be duly considered. Two fire scenarios increase the scale of the nonlimited heat and fire-gas generation. On the one hand, this is the self-sustained smoldering fire that spreads due to exothermic reactions and, on the other hand, it is the flaming combustion enlarging to the fully developed fire. Sundström [179, 545] discusses, among other things, the differing burning behavior of furniture and furnishings in residential buildings and in the contract field. He also explains the activities at ISO TC 92 for the characterisation of room fires in accordance with Fig. 216. According to ISO CD13388 one distinguishes t^2 fires based on the mathematical relationship $Q = Q_o \ (t/t_o)^2$. The heat-release rate as the basic characteristic remains after 1 min at 1 MW.

6.2.4
Relevance of the Evaluation Criteria of the Smoke Potential

For the evaluation of a possible fire-gas danger in the case of a fire a list of different criteria were defined within the framework of the relevant standardization activities. Table 125a provides a view of the essential criteria for the classification of the thermal decomposition products released in the different procedures. Some data are offered by Table 125b. These criteria for the characterization of the quality and quantity of the produced stream of effluents are not by themselves altogether suitable to allow an evaluation of the possible fire-gas danger.

Table 125a Evaluation criteria of the smoke potential.

Smoke density

Criteria	Definition	Dimension
T	Transmission	% . % min
D	Optical density	1/m
Ds	specif. optical density	
k	Coefficient of extinction	m²/s
SAE,Ao	Area of extinction	m²/kg
V	Smoke volume	m³

Smoke toxicity

Criteria	Definition	Units
td	The exposure time that leads to death	min
ti	Time to incapacitation	min
CO-Hb	Carboxyhemoglobin content	%
Mortality	Death rate	%
LT50	The exposure time that lead to 50% mortality	min
RD50	The smoke concentration that leads to a 50% reduction of breathing	g/m³
EC50	The smoke concentration that leads to a 50% reduction of the reaxtion	g/m³
LC50	The smoke concentration that leads to a 50% lethality	g/m³
L(ct)50	The product from concentration and exposure time that leads to a 50% lethality	g/m³

Table 125b Smoke density classification

Classification	(% min)	D (1/m)	S (m)
DIN – 5510-SR1	<100	0,176	4,4
DIN – 5510-SR2	<50	0,079	7
DIN 4102	300	9,155	5

Smoke-density measurement

The selected criteria for the evaluation of the smoke release with regard to a possible reduction of visibility through thermal decomposition products do not consider, as a rule, the all-decisive factor of the effluent concentration. This elementary characteristic will even be ignored in the case of the evaluation criteria for the assessment of the toxic potency of thermal decomposition products. All currently employed evaluation criteria for the visibility reduction due to the fire-gas action are based altogether on photoelectric measurements of the luminous absorption through thermal decomposition products. There is a direct reference to the possible reduction of the visibility for all criteria, as illustrated by Fig. 109 and the relevant mathematical equations.

Transmission

The relationship of Fig. 109 illustrates that also the criterion transmission T (%), which is used, e.g. in Germany, Austria and in the Switzerland for the characterization of the smoke-density qualities, allows the direct reference to the possible visibility reduction in the case of a fire. The relevant characteristics of the optical density D

can be determined in the case of a cumulative system in accordance with Fig. 110 by means of the material-related transmission records. The nominal mass-, volume- and area-related concentration data of the thermal decomposition products are achieved by the test-piece dimensions of these cumulative test methods. Analogous numerical calculations are after the determination of specific general conditions absolutely possible with flowthrough dynamic systems, e.g. DIN 5510 or DIN 53436/37. The necessary conversion can also simply be carried out, for example, in the case of DIN 53436/37 by means of the transmission values assuming there is a regular stream of effluents. In the case of a flowthrough system like DIN 5510, the temporal process of the smoke-density data allows, due to the continuously changing of the fire-gas concentrations characterized through corresponding smoke-potency data, an equivalent definition. The transmission measurements taken within the framework of full-scale field trials with passenger coaches [80, 81] in the stream of effluents of upholstery seats demonstrate on the one hand, as shown by Fig. 129, the rapid decay of the transmission values in the cabin and on the other hand that these transmission values do not reflect the visibility in the bordering emergency exit. While in the stream of effluents of a burning upholstery unit in the effected compartment after the development of a fire a steep drop in the transmission was to be listed immediately, no crucial decay of the visibility could be observed in the neighboring escape route even after a fire duration of 5 min.

Temporal course of the transmission
A further, frequently used evaluation criterion is the temporal process of the transmission value in (% min). While the transmission values as an evaluation criterion are mainly assigned to cumulative procedures such as, e.g. the XP2 chamber, characterizes the temporal course of the transmission mainly dynamic procedures with continuously changing fire-gas concentrations. A relevant example here is the tunnel test according to ASTM- E84 (see Fig. 88), where the absorption is recorded as a function of the test duration t. Figure 217 illustrates the basic characteristics of the reference material red oak.

For these procedures also a computational definition of the characteristics of the optical density D as well as the relevant visibility can be carried out under the handicaps of specific assumptions. The relations represented in Fig. 218 are found, if a test duration of 10 min and an approximately constant fire-gas concentration are assumed.

These numerical data are coupled, as a rule, immediately to the simulated fire scenario of the test method and can not be transmitted without special evidence on other scenarios. These ratings are not suitable as basic values for numeric approaches for the estimate of a possible smoke danger in the case of a fire. If the general conditions of the different procedures are considered and the defaulted nominal effluent concentrations put into relation to the circumstances of a real-fire case, e.g. in a passenger coach, then considerable deviations of the fire-gas concentration for wall and ceiling linings as well as flooring materials will be released, as shown by Table 126.

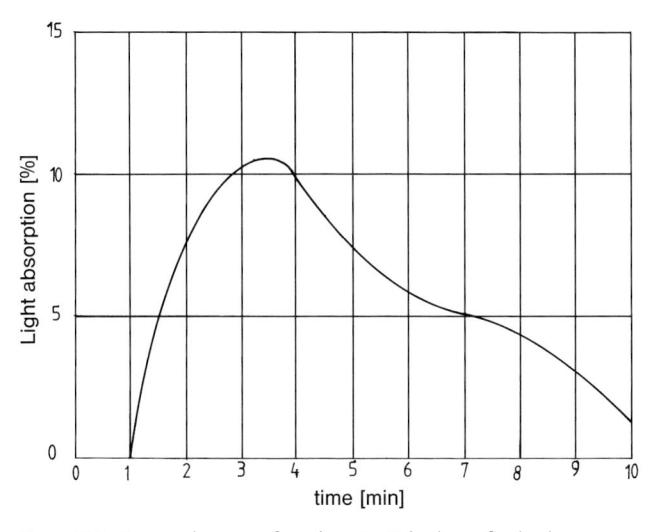

Figure 217 Temporal course of smoke potential values of red oak

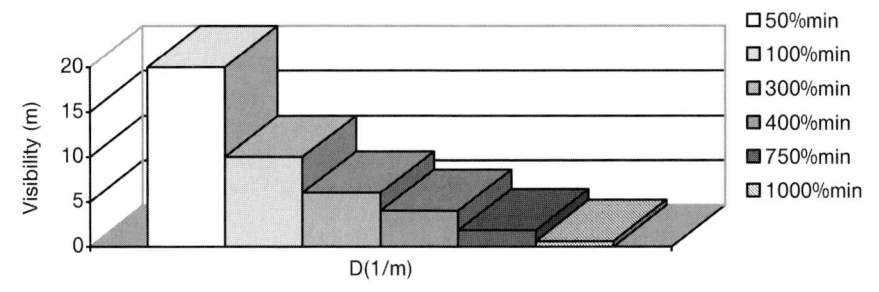

Figure 218 Relation of the records of the optical density and the visibility

Table 126 Influence of the upholstery combination on the optical density

Upholstery combination	t(min) for OD=0,2	OD (1/m)	Smoke (100m³) for OD=1
CA/PUR-NFR	1,25	0,7	3,07
PP/PUR-NFR	1,25	0,8	1,14
Viscose/PUR-NFR	8,5	0,3	0,71
Wool/PUR-NFR	3,5	0,4	1,32
PVC/C/PUR-NFR	2,5	1,9	5,0
Viscose/PES Fasern	3	1,2	1,77
Viscose/C Wadd./PUR-NFR	6,5	1,1	3,14
Viscose/CFR-Interl/PUR-NFR	8,5	0,3	0,75
Viscose/PUR-NFR	8,5	0,3	0,71

The boundary value 300% min for B1-classified flooring materials means, assuming a regular transmission, an optical density $D = \lg 10 \times 100/30 = 0.52$ and defaulted with that a range of visibility of approx. 2 m. The defaulted boundary values of the maximal permissible reduced visibility, e.g. the classes SR in the train field to mean thus

SR1 <100% min = $D = 1.51$, that is 1.0 m visibility,
SR2 <50% min = $D = 0.77$, that is 1.8 m visibility.

Optical density D – specific optical density Ds
The characteristics of the smoke potential as the optical density D, the specific optical density after 90 s "Ds_{90}", the maximum optical density D_{max}, the extinction coefficient k or the extinction area SEA are altogether in direct relation to the possible, relevant reduced visibility. The dominating evaluation criterion is the optical density D, which stands (see Fig. 111) in direct relationship to the visibility. The test parameters such as temperature and external radiation intensity as well as the air supply influence the value as much as the mass, the volume and/or the stressed test surface. Also, material-related characteristics, e.g. the open-cell structure and the material compound, modify the fire-gas concentration depending on the combustion behavior. These results of extensive studies document that the temperature effect is to be classified as a dominating factor. The results of fir-wood tests measured with the DIN equipment and illustrated by Fig. 112b and 219a underline the strong temperature influence. The characteristics determined in the modified NBS Chamber depending on the external radiation in accordance with Fig. 202b confirm this statement. These results demonstrate also that the records run through a maximum in the temperature range of 350°C to 500°C. With the start of flaming combustion, a sudden decay of the optical density occur. Unlike the temperature effect proved material-related differences mainly as less dominant. The characteristics of the optical density D for a chipboard quality, determined by means of the ISO smoke box as well as the modified NBS Chamber depending on the external radiation intensity, confirm that these records pass through a peak value depending on the test temperatures. An analogous behavior was verified, among other things, also by Czech studies (see Fig. 219a). The strong decay of the smoke-density data with the occurrence of flaming combustion was verified through numerous research studies with different decomposition systems [86, 367, 716].

The graphical representations of the Figs. 112c and 116a clarify also the strong influence of the test temperature, while Fig. 115 documents that, provided that the combustion with flames is mandatory, the variation of the external intensity of radiation modifies the characteristics of the smoke-density potential only slightly. Without mandatory ignition other circumstances result (see Fig. 219b). In particular, the results of PE and the chipboard clarify the strong decay of the optical density D after ignition occurs. This effect was, e.g. not verified for the products PMMA and rubber in the NBS Chamber. No self-ignition occurred up to a maximum radiation intensity of 40 kW/m^2 with many materials such as PS, PC, PVC and the tested wood samples.

Also, for the extinction coefficient k and extinction area SEA used as evaluation criteria for the smoke density in the official Japanese procedure according to JIS A 1321

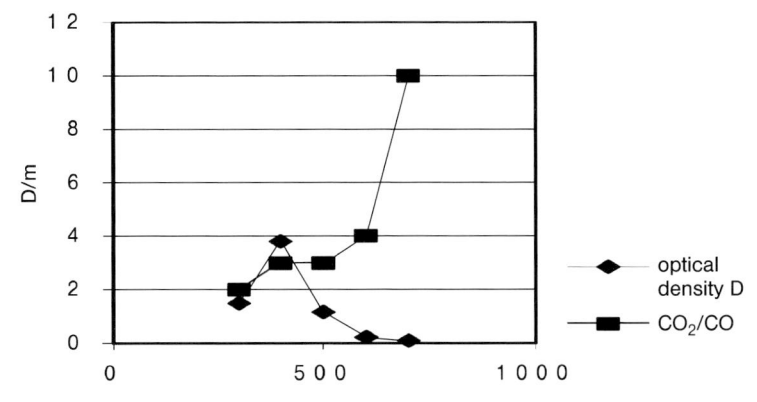

Figure 219a Influence of the test temperature on the records of the optical density D

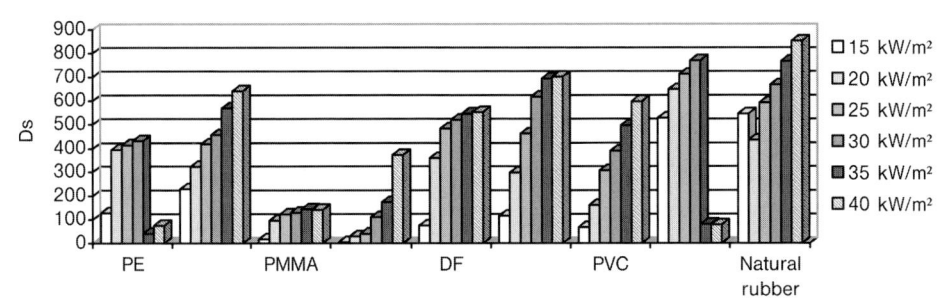

Figure 219b Influence of the external radiation on the characteristics of the optical density Ds

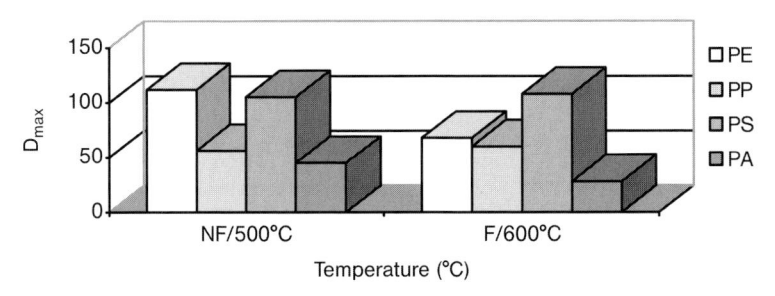

Figure 219c Influence of the external radiation on the records of the optical density D

there exists, as the mathematical relationships of Fig. 111 illustrate very vividly, a linear correlation with the optical density D, respectively, the visibility S. The graphical representations of Figs. 112c, 219c and 220a confirm both the significant influence of the test temperature as well as that of the concentration. These results clarify the passing of a peak value that seems to suit, as a rule, the temperature range of 450 °C to 550 °C. In the case of the concentration dependence both the primary air and also the pure dilution effect of the secondary air must be considered.

The selected criteria for the evaluation of the fire-gas formation with regard to a possible reduced visibility through thermal decomposition products do not consider, as a rule, that one all-decisive factor of the effluent concentration. The characteristics of the optical density D determined by means of the DIN equipment depending on the nominal concentration confirm this interpretation as demonstrated by Fig. 196 very vividly. A considerably larger volume of the flexible PUR foam must be involved, contrary to latex foam in the fire occurrence, to produce a comparable fire-gas formation and/or equivalent reduced visibility. An increase of the thickness of the test samples, i.e. a raised concentration, leads as expected to higher values of the maximum optical density D. The graphical representations of Fig. 220a show the approximately linear connections. The different carbonizing tendency of the product shows that the variation led to different effects. These data records show, furthermore, that mass-related test results may also possibly be changed as functional characteristics as well.

The graphical representations of Fig. 220b clarify that through variations of the primary air in the DIN 534536 system the characteristics of the optical density D of the stream of effluents are significantly modified. The reduction of the air supply leads to a strong rise of the value.

Gaskill [701] found analogous results in his studies with the NBS Chamber, as demonstrated by the graphical representations of Figs. 202a and b. The influence of the concentration could be verified, among other things, by the variation of the test-

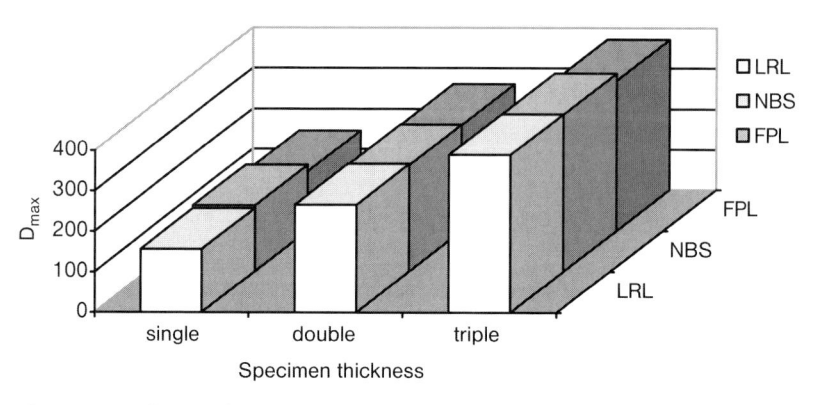

Figure 220a Influence of the records on the optical density D_{max}

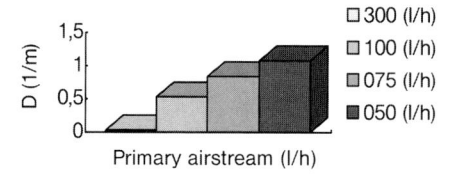

Figure 220b Influence of the primary air on the records of the optical density D

piece dimensions [281]. These significant deviations of the measured values underline the necessity to standardize the potential values and to require mass-, volume- and area-related characteristics. If these data are oriented on comparable fire-gas concentrations, so it is illustrated that in comparison to the common mass- relationship, volume- and area-relationships produce possibly extremely different values. Figures 202a and b underline, on the one hand, the strong influence of the test temperature and show, on the other hand, using the example of the thermal decomposition products of fir wood the strong dependence on the chosen basis. A comparison of the relevant diagram clarified how the values of the characteristics depend very much on the chosen dimension basis. Especially the material-related characteristics of Figs. 111a–116 demonstrate that material-related characteristics of the smoke potential require a careful interpretation. While the cone-calorimeter test results convey the impression that the dependence on the external radiation is negligible, the data set of Fig. 115 demonstrate the significant influence that is achieved on the smoldering fire and on the flaming combustion with reduced oxygen offer. The selected criteria for the evaluation of the effluent formation with regard to a possible reduced visibility through fire-gas formation do not consider as a rule the decisive factor of the smoke concentration. This elementary characteristic is not noted in the case of evaluation criteria for the assessment of the toxic potential of thermal decomposition products. The nominal fire-gas concentration of a B1-classified coating ($600 \, g/m^2$) produce in the Critical Flux Test according to DIN 4102 pt. 16 with respect to the maximum admissible surface destruction of $F = 1 \, m^2$, ventilation of 5.8 l/min and a test duration of 12 min a max. concentration of $840 \, g/m^3$, at a test period of 30 min, a max. concentration of $313 \, g/m^3$ is assumed. In the chimney test according to DIN 5510, a B1-classified material will produce a nominal load of max. $0.057 \, m^2$. Considering the ventilation of $10 \, m^3/h$, i.e. $0.50 \, m^3/3 \, min$, the nominal concentration lies at approx. $60 \, g/m^3$. To reach equivalent concentration values in the real-fire case of a passenger coach, approx. 127 and/or 9 kg at the fire occurrence have to be involved accordingly. These data clarify that it is necessary to align the ventilation handicaps for a testing method under the conditions of a real fire in order to guarantee such a real proximity. If one starts from a rectangular curve during the temporal process of the smoke-density records then depending on the transmission numbers visibility reductions in accordance with Fig. 218 have to be taken into account. These evaluation criteria are used in particular in the cone-calorimeter examination as well as in the experiment in the 1:1 scale. As shown by Figs. 155a and b, they form then also the basis of the judgement in the relevant full-scale field trials. The material-related characteristics of Fig. 115 clarify that the ratings themselves need a careful interpretation. The mass optical density (MOD), this potential value base on the tested sample mass, was used as an additional evolution criterion. Chien et al. [369] demonstrated (see Fig. 201b) by means of a modified NBS Chamber the strong influence of the radiation intensity to the mass-based characteristics of the optical density D (MOD) in the range of 25 to 75 kW/m^2. Material-related basic records for numerical calculation approaches are achieved under enduse conditions if the set of data were based on the tested sample volumes. Such material-related mass-based potential values are, as a rule, also only suitable for a very specific simu-

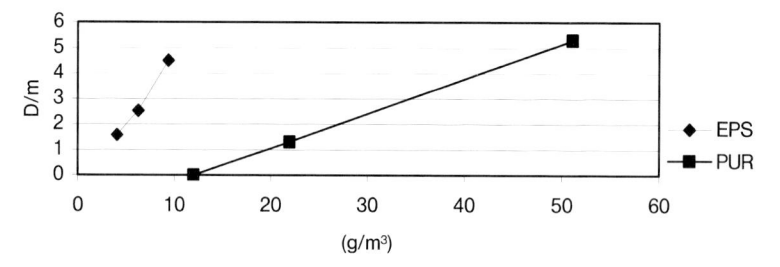

Figure 221 Influence of the concentration on the optical density D

lated fire situation. The records of Fig. 112c illustrate that the test-temperature variation changes the material-related smoke potential significantly. The data records of Figs. 221 and 11d clarify that the value depends on the concentration. The length of time up to obtaining the characteristics provides information concerning the burning behavior. This relevance was striven for, among other things, through the determination of these characteristics depending on the stress duration. The graphical representations of Fig. 217 clarify the time effect, which does not represent a material-independent characteristic. Characteristics, such as for example Ds_{20} that couple the material rating to a specific experimental duration are to be put into perspective by means of the values of edge parameters.

Smoke indices as an evaluation criterion

Sometimes it was also attempted to carry out a material-related smoke-density evaluation by means of specific index numbers. With the procedure NF-F16, which is based on NF-F51, a connection of the smoke-density records of two different procedures were carried out by means of the mathematical relation

$$F = S/m$$
$$V = D_{max}/m \ (t-to) \text{ and}$$
$$I = D_{max}/m$$

F is the smoke density index and S corresponds to the temporal integral of the optical density D. "D_{max}" in the maximum/peak value of the optical density D, m is the mass of the test piece and t correspond to the time space to D_{max} is reached and to represent the length of time until the stream of effluents reach the measuring point. If the combustion of the simulation models are realistic, then these material-related potential values can be used in connection with the knowledge of the concentration circumstances as basis records for numerical approaches for the estimate of possible dangers in the case of a fire. Detailed knowledge of the temporal process of the fire participation as well as the endangered room space is prerequisite. The schematic representation of Fig. 222 clarifies the basic connections of laboratory test results with the estimate of possible smoke dangers.

The graphical representations of this section clarify how much these potential values can vary within the framework of the test conditions. The deviations illustrate

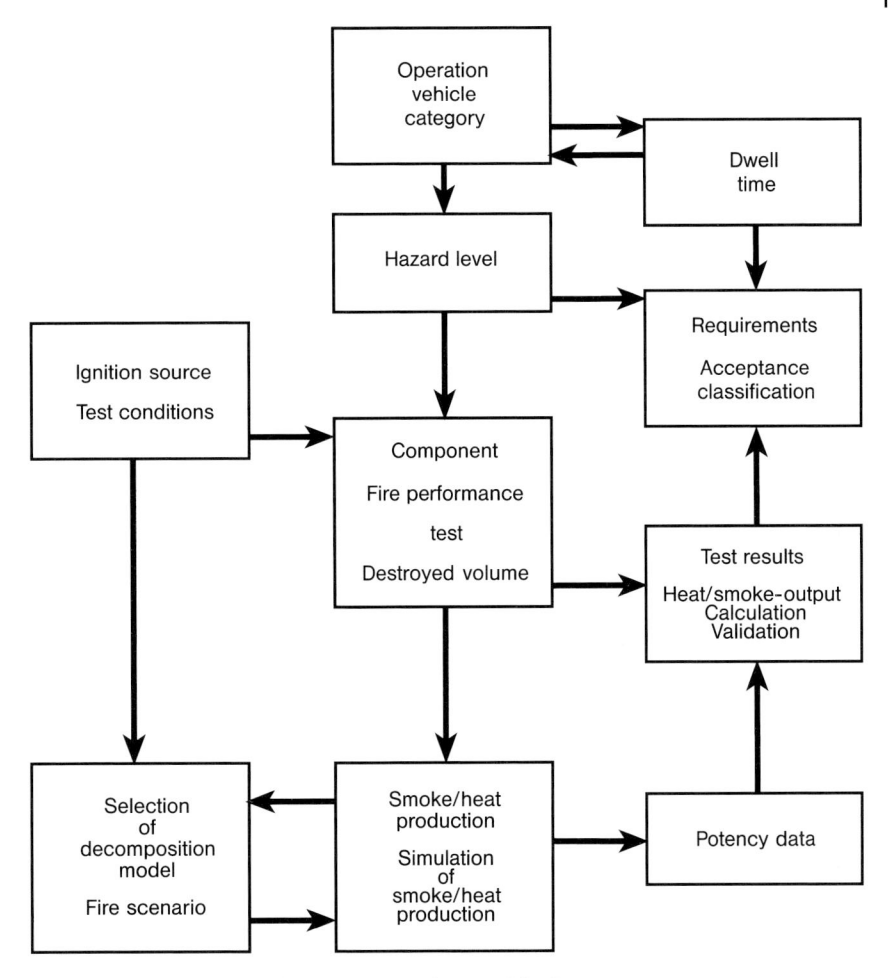

Figure 222 Basic possibility of a risk-relevant evaluation of the fire-gas danger

that the test conditions of the combustion system change the respective value substantially. The general conditions are to be adjusted to the essential fire scenarios to a large extent. In the case of a cumulative system such as that of the XP2 chamber, the corresponding data records of the optical density D and/or the relevant visibility range can be determined in relation to mass-, volume- and/or area-related concentration data of the effluents. If one starts from a closed system, e.g. the XP2 chamber, then the relevant characteristics of the optical density D can be determined by calculation from the transmission value by means of the mathematical relationship (see Fig. 109). Analogous calculation approaches are possible also with a flowthrough system such as, for example, DIN 53436/37 after the determination of specific prerequisites. In the case of the DIN procedure assuming a regular stream of effluents a corresponding conversion can be carried out, on the basis of the measured values. The practical conversion of these numerical values requires that the relevant concentration values can be assigned to the range of visibility.

Smoke volume

The fire-gas volumes are favored to be used as an evaluation criterion in the case of experiments in the scale 1:1. Since in such investigations the fire performance can cause decisive changes, the fire-gas volume criterion in (m^3) must be oriented as a functional characteristic on the general conditions of the fire course. A communicability on other scenarios is not admissible. The appraisal of the fire-gas volume as an evaluation criterion is the assignment to the fire occurrence is indispensable. Shern [717] discusses among other things by means of the investigation results of the operation program "School burning" their possible influence on the smoke-layer formation. Flisi [704] compares the results achieved within 19 different procedures. The characteristics of Fig. 202a illustrate very vividly that already a changed test piece orientation can lead to a substantially deviation of the value. The difficulty, to carry out relevant smoke-density investigations in the laboratory standard is explained also by Mickelson and Traicoff [718]. Such ratings exemplary represented in Figs. 182 and 183 are not to be classified as material-related potential values. They are significant for the simulated fire scenarios only. A communicability of the conclusions on other risk situations without separate evidence is inadmissible. Realistic fire tests in the 1:1 scale demonstrate that the possible smoke danger depends decisively on the smoke-layer formation, which is increased for its part from the fire course and the spatial circumstances. The graphical representations of Figs. 223a and b clarify the scale of

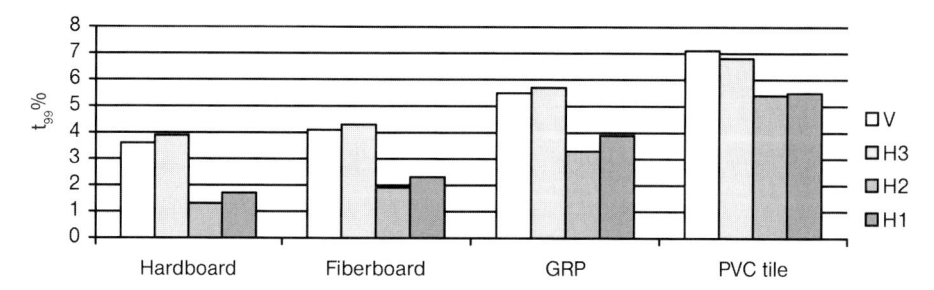

Figure 223a Influence of the room height on the smoke shift
>formation

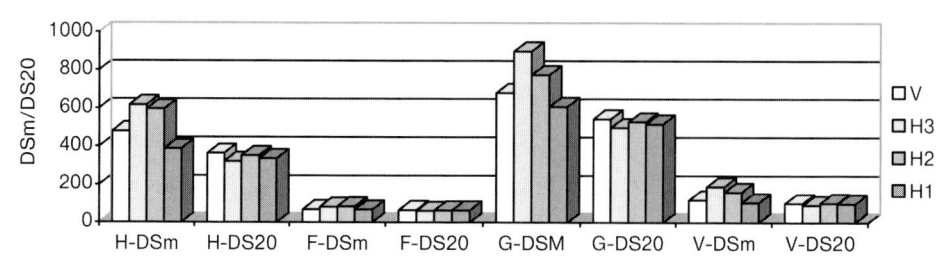

Figure 223b Influence of the room height on the smoke shift formation

the smoke-layer formation by means of the criteria Ds_{20}, Dm as well as the length of time to the transmission decay to 99% ($t_{99\%}$) due to the fire-gas action.

The practical relatedness of the cone calorimeter and the large-scale test supporting this special laboratory test, the Furniture Calorimeter Test procedure, was the subject of realistic comparative investigations. The smoke-density results of different upholstery combinations represented in Fig. 224a can be interpreted as follows, while with the experiments in the Furniture Calorimeter at the WFRC the chair, upholstered with the incombustible mineral-fiber insulation in comparison to the PUR-CMHR upholstery is showing a considerably reduced fire-gas formation and the PUR-HR upholstery produced a significant higher value. These differences could astonishingly not be confirmed in the room experiments at RAPRA. The mineral-fiber upholstery led in the realistic room experiment to a comparably heavy fire-gas formation.

The risk-relevant full-scale field trials of Mizuna et al. [714] demonstrate also the limited information value of the characteristics determined in the Furniture Calorimeter. These full-scale field trials with cushion units demonstrated that the arrangement of the piece of furniture in the room can possibly significantly change due to feedback effects the burning behavior. While the flame height in the case of wall and corner configurations achieved a comparable value, the time–temperature

Figure 224a Influence of the test conditions on the data level of the optical density D

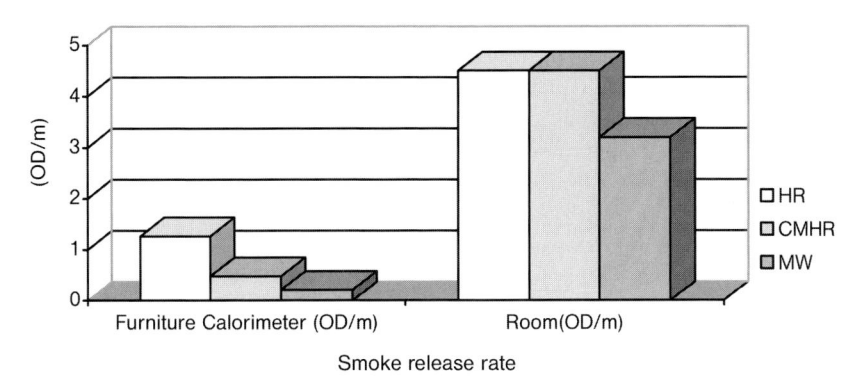

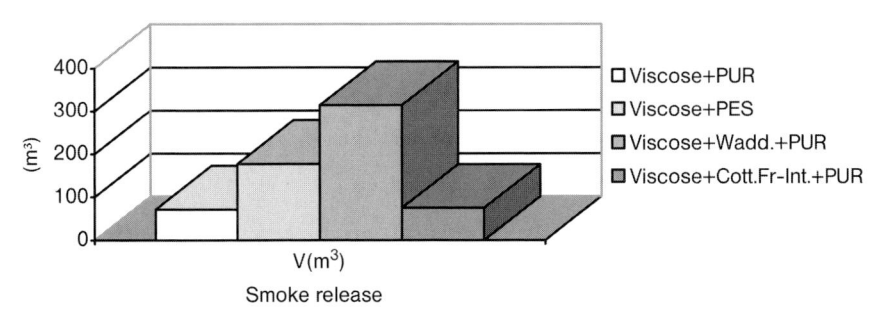

Figure 224b Influence of the upholstery combination on the effluent generation

Table 127 Influence of the upholstery combination to the smoke release

Upholstery combination	T Room (°C)	t Room (min)	OD/m Corr.	V K (m³/min)	T Corr. (°C)
Acrylic/PUR-NFR	706	2,5	1,7	37	491
Acrylic/HR	661	4	2,1	32	463
Acrylic/CMHR-35	490	6	0,8	18	375
Acrylic/CMHR-50	94	15	1,0	10	89
Cotton/PUR-NFR	630	3	0,9	3,4	313
PP/PUR-NFR	430	4	1,7	3,1	300
PP/HR	520	4	2,6	41	270

courses showed significant deviations. Significant different temperatures were obtained in the central line of the flames. Paul [525, 526] varied, among other things, in his studies both the cover material as well as the filling. The fire-gas liberation of the different cushion combinations was evaluated by means of the criteria optical density D and fire-gas volume. Both the length of time to attain the value $OD/m = 0.2$ and the peak value of the optical density D were significantly changed by the outer covering as well as by the chosen combination. The smoke volume with the optical density $D = 1$ showed likewise considerably differences. The listed data records illustrate that there is no homogeneous sequence for the different evaluation criteria. The length of time up to attaining a specific rating of the optical density, e.g. $D = 0.2$ provide information about the fire development. The information obtained by Paul demonstrate also with different upholstery combinations, see also the listings of Table 127, that both the upholstery fillings as well as the outer coverings can change the fire-gas formation significantly. Both the length of time up to obtaining specific smoke-density records as well as the potential values themselves, are changed significantly. Paul features the scale of the fire-gas formation by means of the characteristics of the optical density D and with the aid of the information about the smoke volumes. These results clarify that for the different criteria no direct connection exists and therefore also no basis for a homogeneous classification.

The data records of Fig. 224b [526] clarify the influence of the upholstery material on the scale of the effluent release. An intensified fire-gas liberation was observed in the sequence PUR, PUR with cotton interliner, PES fiber material and PUR wadding in realistic full-scale field trials. The evaluation of the smoke volume in (m³) was based on the rating of the optical density $D = 1$.

The characteristics of the smoke potential determined with the aid of the DIN equipment (see Figs. 123a and b) illustrate that estimates, as have been undertaken, e.g. by Ames [373] for the combustion of 2 kg material within a room of 100 m³, have to be based not only on the actual fire occurrence, but the detailed knowledge of the functionality of the smoke formation has to be taken into account as well. These characteristics confirm the estimates of Woolley and Ames [374] that during the combustion process of each 5-kg EPS and rigid PUR foam within a room volume of 1000 m³, the visibility range was reduced to 1 m (EPS-smoke – 1453 m³) and to 2 m (PUR – 502 m³). Figure 225 illustrates the connections between the optical density D, the range of visibility S and the fire-gas volume V.

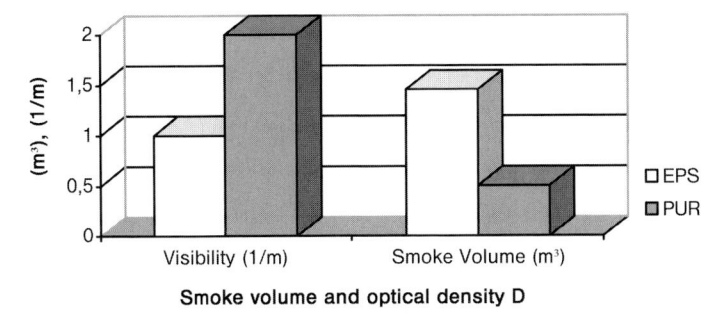

Figure 225 Fire-gas volumes as material-related characteristics

The conversion of the material-related data records of the smoke-density potential within the framework of a danger estimate presuppose that the range of visibility reduction can be assigned to the fire-gas concentration. Figure 109 demonstrates very impressively that material-specific transmission data are not in themselves suitable only as the basis for risk-relevant estimates of the smoke danger.

Assessment criteria for the toxic potential of effluents

Autopsy findings – experience values of the fire statistics

The relevance of the individual evaluation criteria, as they have been determined in the different national and international standardization committees, can be demonstrated only in agreement with the experiences of a realistic fire statistics. The basic findings for the evaluation of the flue-gas toxicity supplied extensive autopsy results for fire victims. The study of relevant fire statistics documents shows very vividly that the flue-gas action is to be classified as the dominating cause of death. The data records of Fig. 226a provides a view of some at the World Health Organisation (WHO) existing experiences concerning the pathologic investigations of fire victims. These data records in accordance with Fig. 226a show by means of statistical documents that for Japan, UK and USA the dominating role of the fire-gas action as a cause of death has to be confirmed. Burns as a cause of death can be verified only to a far smaller extent. The representations of Fig. 226b provide the information concerning the cause of fire deaths in USA, USSR and Belgium. While the American and Russian fire-statistics data confirm the dominant role of smoke, astonishingly, burns dominate in Belgium.

Nelson [723] reported, based on an intensive literature survey, that the effluent component CO plays the important role for the fire victims as well as for the nonfire victims in the case of a fire. CO and the combination of CO and others are the dominant cause of fire victims. The importance of burns is obviously of minor importance (see Fig. 226c).

The data records of Zirner [724] about Berlin experiences confirm these statements (see Fig. 226d). Zirner discusses, among other things, the necessary medical treatment of fire injuries. The heat release in the fire case can lead, as Stoll et al. [725, 726] report, e.g. through radiation on human organisms, possibly together with

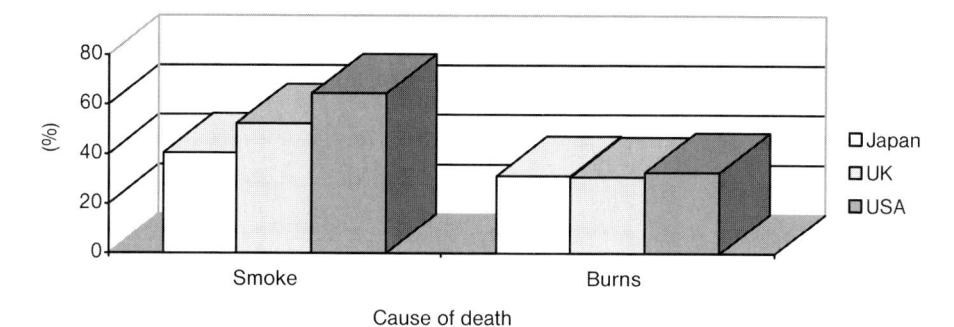

Figure 226a Autopsy results in UK, USA and Japan

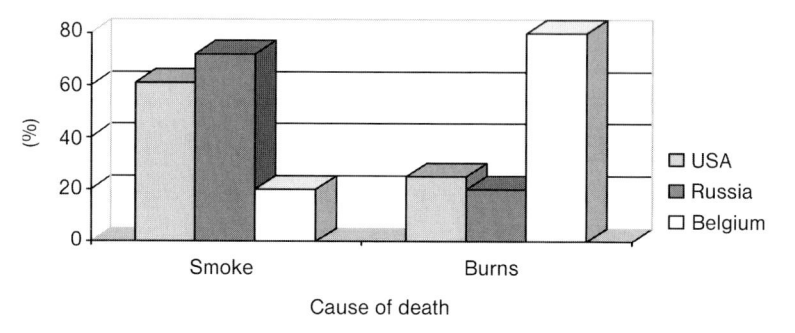

Figure 226b Autopsy results in USA, USSR and Belgium

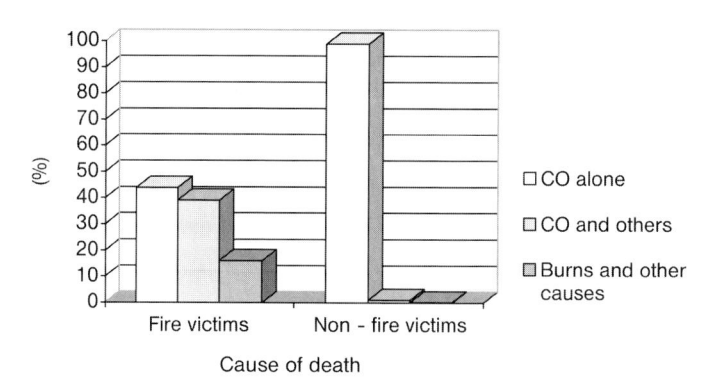

Figure 226c Fire victims – the role of the effluent component CO

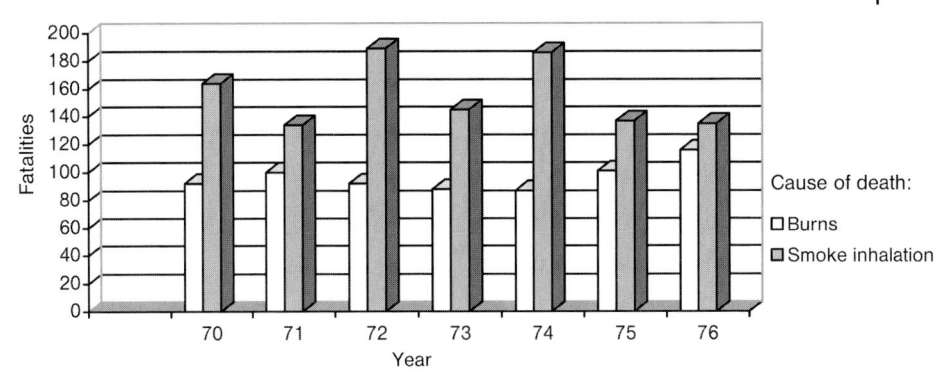

Figure 226d Berlin autopsy findings

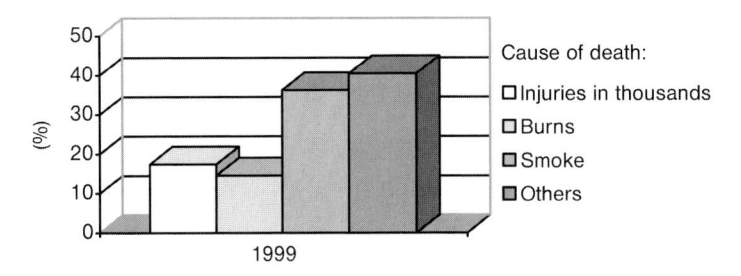

Figure 226e Korean fire statistic data

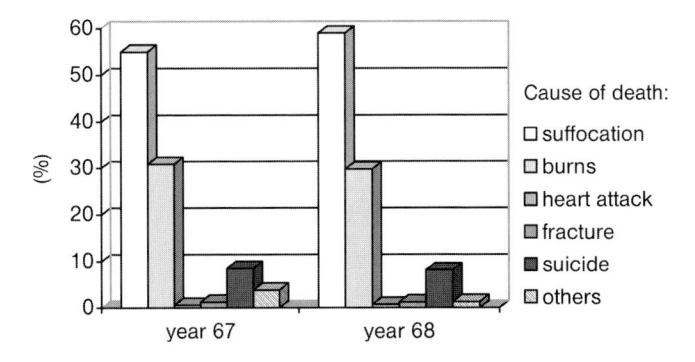

Figure 226f Japanese autopsy records

direct flame impingement to fire victims. The permissible exposure time for different intensity levels of radiation in relation to possible danger is pointed out. The data records of Fig. 227 provide a view of possible damages of the skin through external radiation action.

Halpin et al. [400, 727] and [728] reported the first comprehensive autopsy findings. The results of these joint investigations, carried out by the Johns Hopkins University and the fire department of Baltimore in Maryland, showed that CO is the

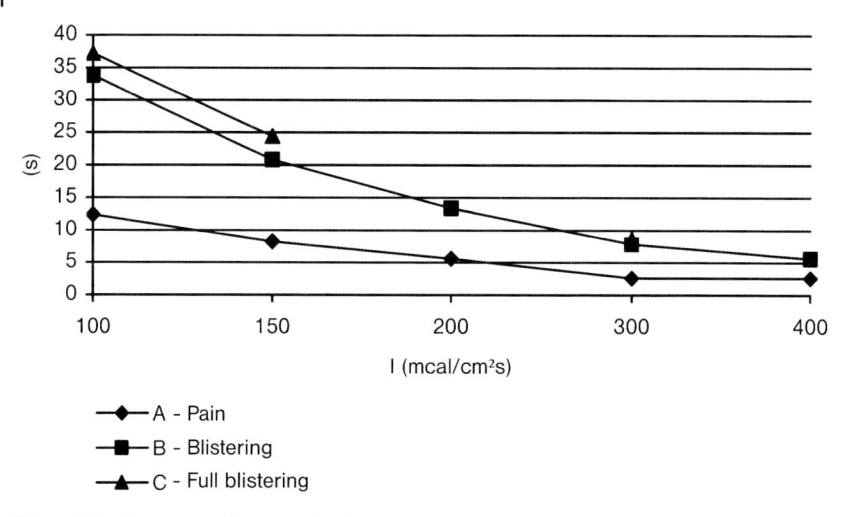

Figure 227 Skin reaction by external radiation

dominant fire-gas component. The results show, however, that due to the measured CO-Hb data and considering possible burns, additional parameters have to be paid some attention. The present results of the study (see Fig. 228a) are to be interpreted in such a way that to a considerable degree combined effects such as, e.g. CO uptake and illnesses are possibly responsible for the death of the fire victims. In the case of the interpretation of the Maryland study also fringe effects such as, for example, drug abuse and suicide are registered. The significant influence of alcohol consumption has been confirmed also in other studies. The research work of Harland and Anderson [426] in Great Britain gave in particular comparable results for the area around Glasgow. Noguchi [729] confirmed the strong influence of the alcohol component in the same way for the area of Los Angeles as Gerson [730] did for the Ontario region. Jones [116] also explains the fire-gas problem in general and the connection of fire victims and alcohol consumption. Teige et al. [731] report on autopsy results in Norway. These investigations of the Oslo area, however, did not confirm the strong influence of alcohol consumption. The raised danger potential in connection with diseases was confirmed through Kishitani [732]. Kishitani reports on relevant Japanese studies that underline the dominant role of the CO uptake. Clark and Ottoson note in [733] that, e.g. heart illnesses are able to cause possibly fatal poisoning at considerably lower CO-Hb values. Kulgemeyer [734] reported on autopsy findings in Aachen within the framework of a dissertation. These findings (see Fig. 228b) demonstrate that, depending on the nature of the smoke mixture, significant deviations of the CO-Hb values are to be observed. The effects of the exposure with the thermal decomposition products of wood were compared among other things with the experiences by the inhalation of car-exhaust fumes.

Luepke and Schmidt [735] report on the combined effect of CO and HCN uptake. CO-Hb values, determined for a room fire with two fatalities, did not suffice in themselves for the interpretation of the lethal effect of the combustion gases. Tsuda et al.

6.2 Relevance of the Evaluation Criteria | 395

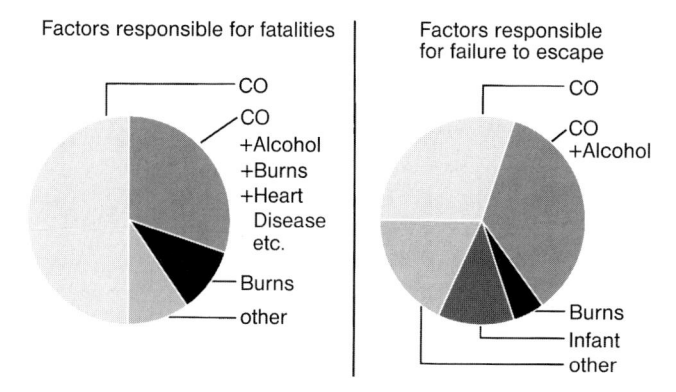

Figure 228a Maryland autopsy results

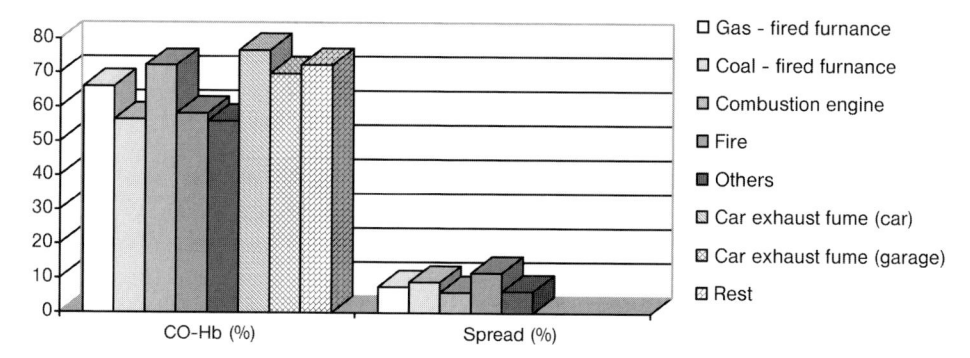

Figure 228b Aachen autopsy findings

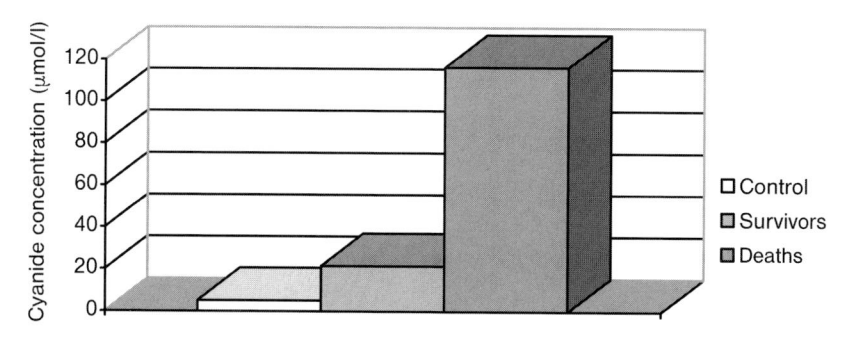

Figure 228c Fire victims of the Paris area – Blood cyanide concentration

report in [736, 737] on relevant Japanese studies. The action of the effluent components CO and HCN was found in particular in approx. 75% of the dwelling fires; for example the autopsy results at a hotel fire showed a CO-Hb level of >60% and a HCN value of >2.0 µg/ml. The medical investigations of fire victims illustrate according to Tsuda that in the case of the majority of fire victims, there is a damage

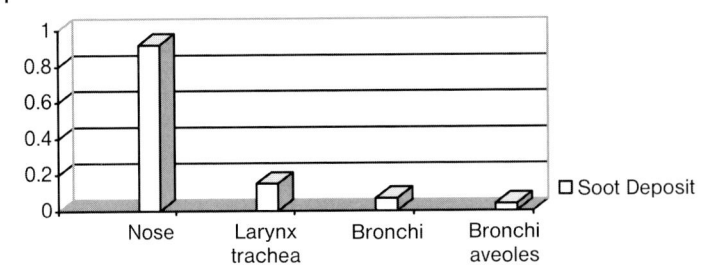

Figure 229 Soot deposits in the respiration tract of fire victims

through hot gases in the respiratory tract. The listings of Fig. 228c provide an overview of autopsy results concerning the blood cyanide concentration of the Paris area. Figure 226e show comparable Korean findings. Japanese findings are illustrated in Fig. 226f.

Milla explain in [738] like Harland et al. [426] their autopsy findings. The determined violation in the respiratory tract can be caused, in their opinion, basically both through hot air and irritants. The graphical representations of Fig. 229 illustrate the findings of the Harland investigations and provide information about possible soot deposits in the respiratory tract.

Zirkia [422, 739] examined the influence of the flue-gas temperatures. He used dogs as experimental animals. Buettner [740] handles the influence of extreme temperature action. The possible danger to human life due to external radiation is discussed by means of the relevant literature in detail. The complexity of the evaluation of a possible smoke danger in the case of a fire requires a rating of the various parameters. The material-related potential values as mass-, volume- or area-based characteristics have the necessary informative value only by considering the performance-influencing factors such as, e.g. orders, orientation, distribution, ventilation and feedback effects. The different evaluation criteria for the material-related potential value of the thermal decomposition products stand altogether in direct relation with the possible acute danger to human life. The majority of these cases provide information about the mortality behavior of the exposed animals. Their usability for the evaluation of the acute toxicity of the thermal decomposition products is mandatory, combined with the general conditions of the decomposition and exposure model. Only if it is guaranteed that decomposition and exposure conditions have a sufficient accessibility for the fire occurrence to be evaluated, can these characteristics be used as base values for danger estimates in the case of a fire. An unambiguous dose–effect relationship is necessary to perceive estimates as represented in the scheme of Fig. 229.

Concentration of individual fire gases

The know-how on the acute inhalation toxicity of pure gases was the starting point of first considerations for the danger in the case of a fire. With the evaluation of the possible fire-gas risk due to the acute toxicity of the smoke components themselves there exists very obviously a danger of misinterpretation (see Fig. 230). Figure 230 provides

6.2 Relevance of the Evaluation Criteria

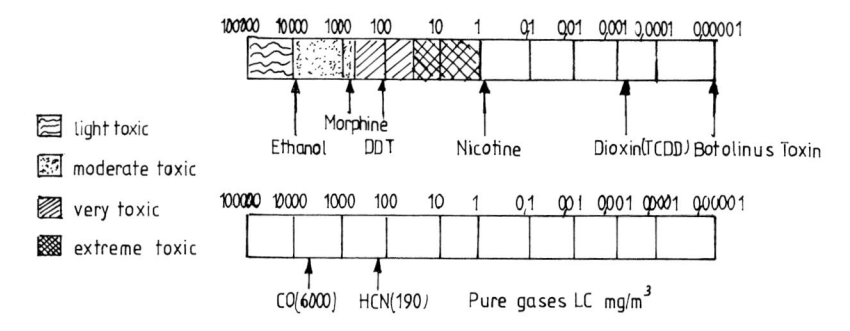

```
100000  10000  1000  100  10  1  0,1  0,01  0,001  0,0001  0,00001
```

- light toxic
- moderate toxic
- very toxic
- extreme toxic

Ethanol — Morphine — DDT — Nicotine — Dioxin(TCDD) — Botolinus Toxin

```
100000  10000  1000  100  10  1  0,1  0,01  0,001  0,0001  0,00001
```

CO(6000) HCN(190) Pure gases LC mg/m^3

Relation of some characteristic LC_{50} material data

	Material	LC_{50} g/m^3
1	PUR(f)	17
2	PUR(r)	5
3	EPS	27
4	Pine	17
5	PVC	17
6	Wool	5
7	Cotton NFR	25
8	Cotton FR	8
9	Mineral wool (ΔW)	35

Figure 230 Toxic potency data – poisons – gases – effluents

a comprehensive view on the classification of the acute inhalation toxicity of pure gases as well as fire-gas mixtures in relation to known toxins. The graphical representations illustrate the scale of the possible danger through fire-gas components and fire-gas mixtures in relation to common toxins. Misinterpretations can be shown very impressively using the example of the effluent component SO_2 (see Fig. 231a). While in different orders, a high danger potential is being formulated, recent bioassay test results of Pauluhn [741] demonstrate the obviously serious deviations to the actual danger potential, e.g. in the transportation field. The regulation carried out in the field of aeroplane construction through the FAA, in the ship-building field through the IMO as well as in the railroad field, e.g. in France, Portugal and Spain in accordance with Fig. 231b demonstrate, as these data records illustrate, the different meanings. Although it is undisputed in the expert committees that there is altogether an additional effect mechanism for the essential fire-gas components, this knowledge state is not a component of the relevant requirement profiles. Furthermore, the rating of the single components is not by any means homogeneous as this listing demonstrates.

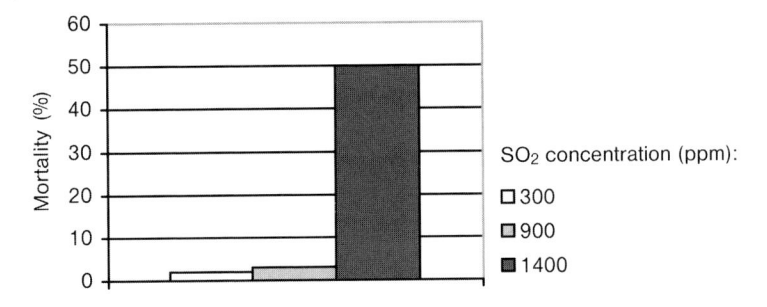

Figure 231 a Characteristics of the toxic potential of SO_2

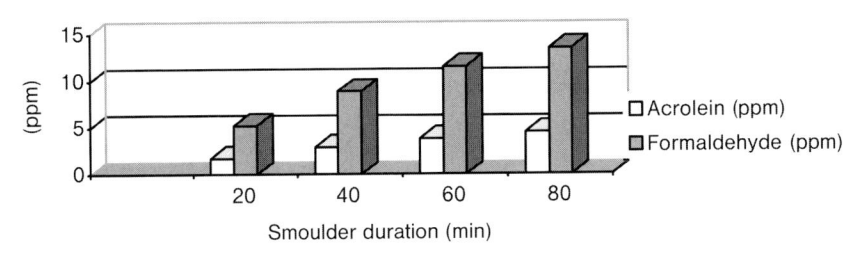

Figure 231 b Aldehyde concentration of effluents

Engler et al. [742] treat in detail the problem of health damages through combustion products and fire areas. There is a detailed discussion concerning the main components and the problematic nature of the acute and the chronic toxicity. Questions concerning the teratogenicity, the mutagenicity as well as the cancerogenicity are discussed by means of the present findings from different studies. The occurrence of symptoms seems to depend on the dose as well as the sensibility of the exposure in any case. Safeguarded findings about the most important components are dealt with just the same as the present fundamental knowledge about the human toxicogical danger potential as, e.g. polycyclic hydrocarbons or the highly chloridized dibenzodioxins. The variety of the subjects and the missing explanation of the combined effect mechanism are demonstrated. In the public discussion on the possible danger to human life through fire-gas action the dioxine problem is repeatedly addressed such as, e.g. at the Düsseldorf Airport Disaster. Unity exists in the expert committees that in fires of residential buildings and assembly places no relevant concentration of these particular fire-gas components has to be expected. Dioxin is the comprehensive term for 75 allied chemical substances that differ strongly in their poisonousness. The correct denotation is dibenzo-para-dioxins. The "Seveso poison" dioxin is designated 2,3,7,8 – tetrachlordibenzo-p-dioxin (TCDD). A detailed discussion on the dioxin and furan problem with regard to possible dangers through the exhaust furnace of incinerating plant is reported. The scientific consultants of the federal medical chamber (Bundesärztekammer) require a real discussion instead of politically motivated posturing after the event. Einbrodt et al. conducted tests with effluents and arreas produced with the DIN tube system to investigate nonlethal affects. Pieler and

Figge report the first relevant investigations results of the thermal decomposition products of fir wood produced with the DIN 53436 equipment. The test system was designed so that the resulting condensate was also introduced in the diluted stream of effluents and the estimating of possible chronic effects was done via the Ames test [742, 743].

Evaluation criteria by means of bioassay testing
The criteria for toxicological evaluation of thermal decomposition products are based altogether on analytically determined concentration data of the essential fire-gas components as well as their rating by means of animal-experimental investigation results. For the assessment from the toxicological viewpoint relevant mortality records as well as reaction and behavioral criteria are used. An often-used criterion for the expert evaluation of animal-experimental investigation results is the CO-Hb content in the blood of the exposed animals. Possible reactions of the exposed animals like groping, hearing, pain reactions are often used, but not standardized, which complicates the possibilities of comparison of the data records. The material-specific characteristics of the toxic potential, possibly the mortality, the CO-Hb content in the blood of the exposed test animals or the exposure duration up to the attainment of specific reduced activities or characteristics as EC_{50}, LC_{50} and Lt_{50} are not suitable to carry out classifications in themselves.

Time to reaction reduction including death
Material-specific characteristics of the criteria t_i and t_d allow the direct comparison of the results, e.g. within identical series of experiments only. The thermal stress is one of the possible influential parameters. The possible influence of the heating rate was demonstrated by Hilado [690] via the USF method. The graphical representations of Fig. 232a to c illustrate that the variation of the thermal loading can significantly change the composition as well as the concentration and from this it follows also different reaction times of the test animals. The exposure with the thermal decomposition products of ABS polymers produced at 800°C causes, in comparison to the constant heating rate of 20°C/min in the temperature range of 200 to 800°C considerably reduced reaction times. The higher temperature levels lead obviously to an increased effluent concentration and therefore to a significant reduction of the reaction times (see also Table 130). These material-specific characteristics have a statement value in relation to the in each case selected test conditions only. A direct comparison with the analogous characteristics of other combustion systems is as a rule not admissible. Only if it is guaranteed that there are equivalent concentration circumstances, can a direct comparison of these numerical values be justified. The material-related differences of the exposure duration up to the death of the exposed animals have to be related presumably to different fire-gas concentrations arising depending on the heating rate. The assessment of the characteristics has to consider that the examined products depending on the carbonisation tendency produce, on the one hand, different concentrations and, on the other hand, burn at different speeds after the self-ignition.

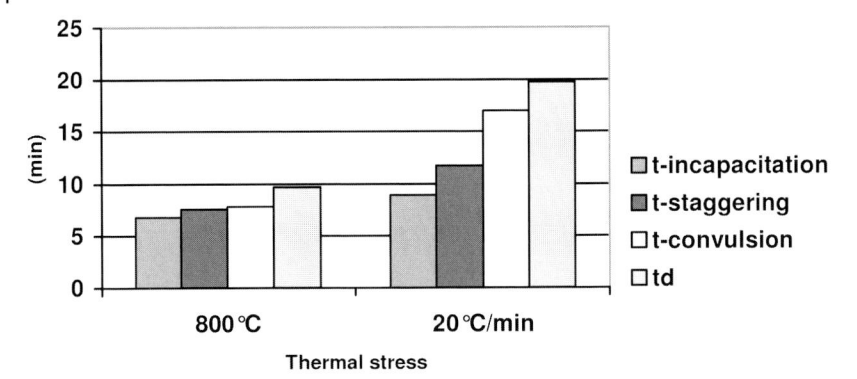

Figure 232a Influence of the test temperature on the toxic potency of wood

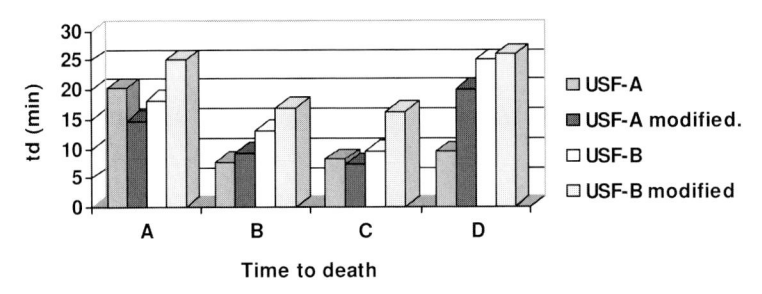

Figure 232b Influence of the ventilation on the toxic potency of effluents

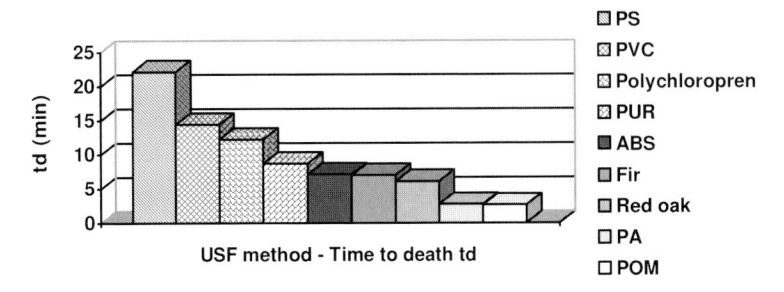

Figure 232c Toxic potency records of various plastics

Mortality and CO-Hb values

The mortality of the test animals is an essential criterion of the biological investigation results. The characteristics determined by Oettel [463] and Herpol [745] in accordance with Tables 129 and 130 confirm the passing of a maximum value depending on the test temperature. The supplementary measurements of the loss of weight provide information about the different scale of the thermal decomposition of the tested products.

Table 128 Influence of the test temperature to toxic potential

Material	300 °C	300 °C	400 °C	400 °C	500 °C	500 °C
	Mortality (%)	CO-Hb(%)	Mortality (%)	CO-Hb(%)	Mortality (%)	CO-Hb(%)
Cork	0	17	28	69	100	86
Rubber	0	<15	100	78	100	75
Lamb wool	17	<15	100	19	100	27
Pine	25	32	100	86	100	85
Leather	100	67	100	69	100	63
PVC	83	25	92	45	100	70
PA6	75	–	0	<10	100	30
EPS	0	<15	0	<15	100	72
PE	0	<15	100	79	100	85

Table 129 Influence of the test temperature onto the mortality

Material	400 °C	ΔG(%)	500 °C	ΔG(%)	600 °C	ΔG(%)
PVC (10 g)	0	95	3	62	100	86
PE (10 g)	94	61	94	100	100	100
Plywood (9,6 g)	25	76	89	97	0	97
Plywood (10 g)	0	45	94	71	0	98
Beech	72	78	54	99	17	99
PIR (3,4 g)	0	30	96	92	0	97

Table 130 RAITK Determination of characteristics of the toxic potential according to DIN 53436

Test method	PE		PS	
A : 200–800 °C	40 l/min	16,68		19,04
B : 600 °C	0 l/min	7,95		10,34
C : 600 °C	16 l/min	5,31		4,09
D : 600 °C	40 l/min	4,08		3,08
E : 800 °C	0 l/min	3,57		5,99
F : 800 °C	16 l/min	2,91		3,88
G : 800 °C	40 l/min	5,11		6,11
H : 600 °C + recirculation		4,88		>> FE

The mortality data provide information about to what extent and/or when the selected conditions lead to the death of the test animals. When the comparative juxtaposition of the different criteria of the toxic potency is contrary to the smoke-density determination to consider not only the temporal process of the flue-gas concentration but to consider the exposure duration in addition as a further actuating variable. Following DIN 53436 the mortality records were determined from Oettel et al. [463, 746] and Kimmerle [93, 88] as illustrated by Tables 131 and 132 depending on the test temperatures. These graphical representations provide an impression about the RAITK values, which list the lowest-test temperature in the case of mortality.

Table 131 RAITK Determination of characteristics of the toxic potential according to DIN 53436

Material	RAIT (°C)	CO-Hb (%)
Fir	>300	29
Cotton	<250	69
PUR	>300	39
Wool	<300	17

Table 132 Material-related records of the toxic potential

Material	Mortality	CO-Hb (%)
Cotton FR1	38	74
Cotton FR2	100	56
Neopren	75	21
PES Fibers	100	14
PUR	75	44
Wool	100	32

The data records of Tables 130 and 131 confirm the results of the numerous autopsy findings [402, 425, 727, 729, 747–749] that state that the CO-Hb content does not always suffice by itself to interpret the cause of death. As a rule, further components such as, e.g. HCN, HCl, NOx or also aldehydes become effective next to the dominating CO concentration. The research studies carried out with the DIN equipment show, as the results of Table 131 illustrate, that the CO-Hb values are suitable only in connection with additional information for the interpretation of the toxic potential.

Without doubt, the CO-Hb content is the dominating evaluation criterion for the toxic potential of the combustion gases. In [750] and [751] methods are described for the proof of the CO-Hb level in blood. Possible modifications e.g. aging effects are discussed. No modification of the CO-Hb level within 2 to 4 days are expected, provided no death occur. The graphical representations of Fig. 124d clarifies in which scale the CO-Hb content of exposed persons is reduced by subsequent clinical processing [152]. These figures illustrate how the time factor itself already causes a recovering stage. The mortality figures published by Klimisch et al. [466] provide an impression about the reproducibility of this characteristic depending on the test temperature. The animal-experimental results of a small round-robin exercise (laboratories A – B – C) (see Fig. 233a) show the relatively good agreement of mortality numbers and CO-Hb content of the decomposition products of fir wood. Increasing test temperatures produce a rise of the mortality numbers. Significantly reduced values are achieved, as demonstrated by the graphical representations of Fig. 233b by variation of the air supply from 100/100 l/h to 200/200 l/h and/or 400/400 l/h. The data records of Fig. 132a underline the influence of the test temperature on the lethality of the decomposition products. The influence of the testing temperature was determined, among other things, also with different decomposition products produced by means of the Gost equipment. The mortality figures pass a marked peak depending

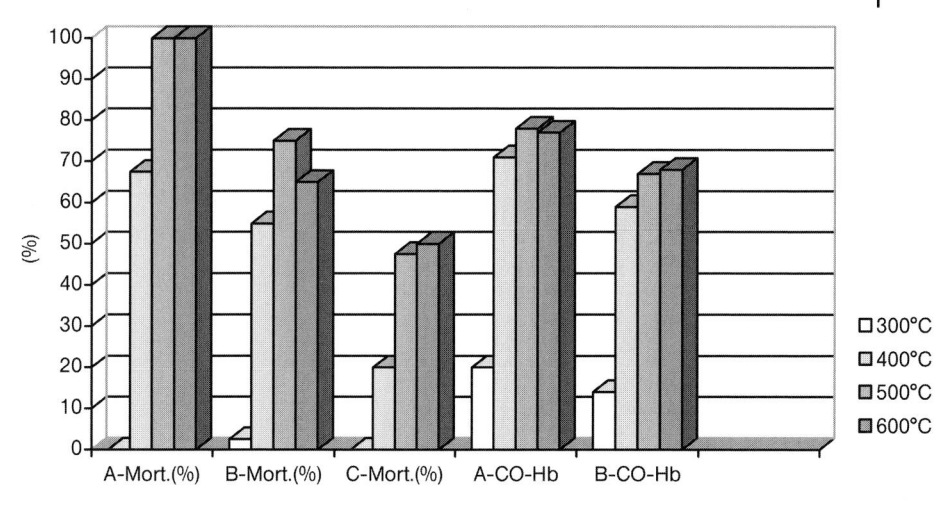

Figure 233 a Influence of the test temperature on the CO-Hb values caused by the exposition with effluents

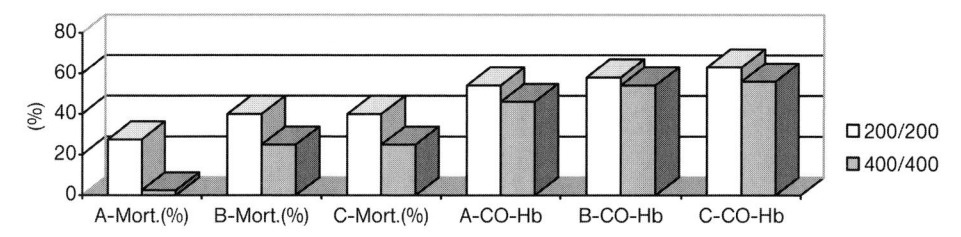

Figure 233 b Influence of the ventilation on the toxic potency of effluents

on the test temperature (see Figs. 117 and 235a). The data set of Figs. 234a to c demonstrate the influence of the exposure time on the CO uptake, which is also a function of the activity (see Fig. 234b). The CO-Hb content plays a decisive role in the expert assessment of the acute toxic potency of the thermal decomposition products of A2-classified building products. Too high CO-Hb values are, according to Einbrodt [412], responsible for the failure with the decomposition products of the predomi-

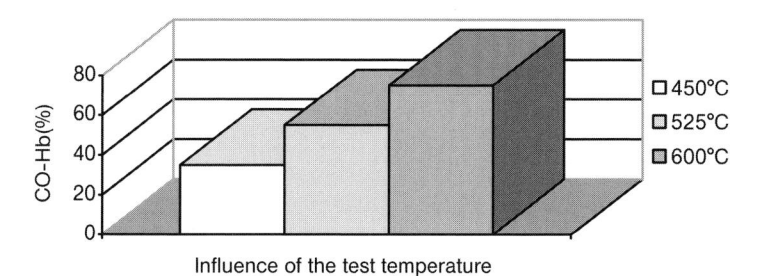

Figure 234 a Influence of the test temperature on the CO-Hb values caused by the exposition with effluents

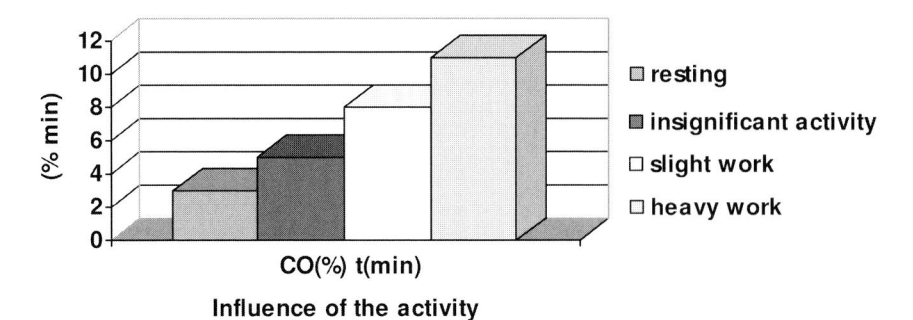

Figure 234b Influence of the activity on the CO uptake

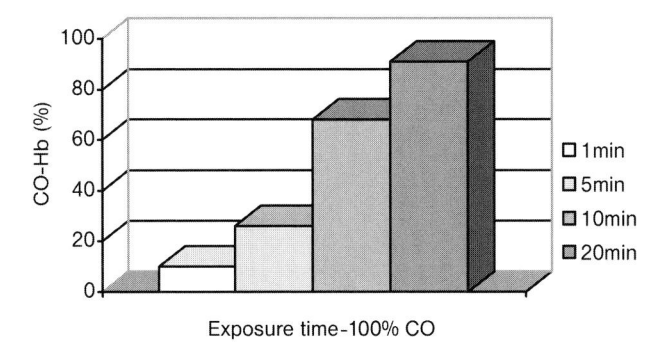

Figure 234c Influence of the exposure conditions on the CO-Hb level

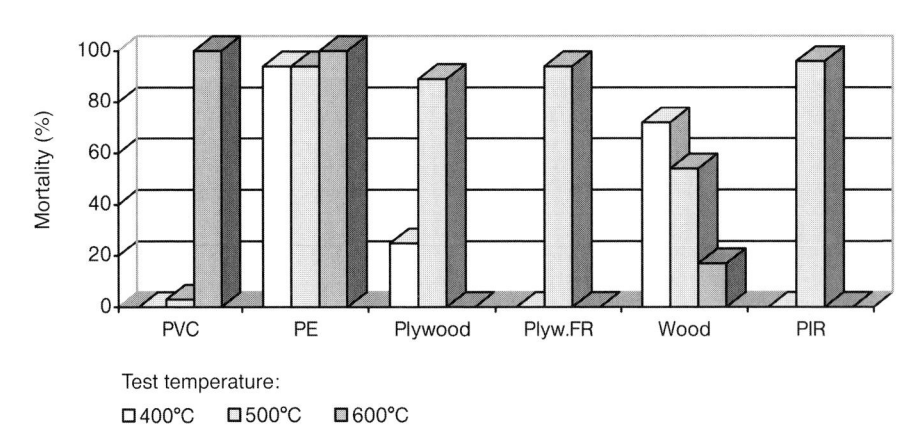

Figure 235a Influence of the test temperature on the response of the test animals

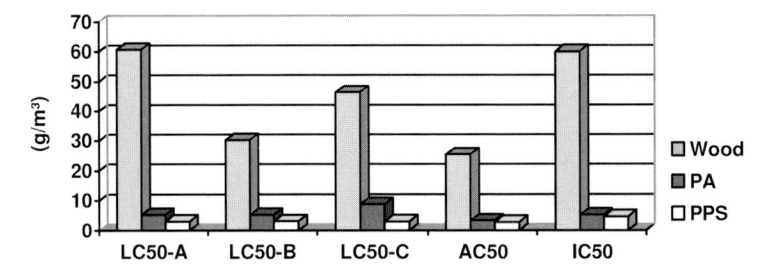

6.2 Relevance of the Evaluation Criteria | 405

LC50-A   End of test
LC50-B   trained animals, 14 days observation
LC50-C   untrained animals, 14 days observation

Figure 235 b   Influence of plating on LC_{50} data

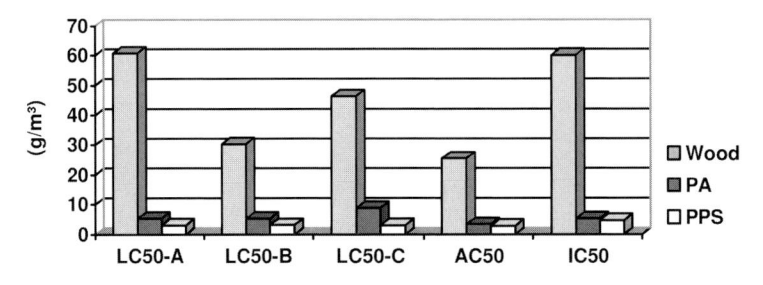

LC50-A   End of test
LC50-B   trained animals, 14 days observation
LC50-C   untrained animals, 14 days observation

Figure 235 c   Influence of training to the reaction behaviour

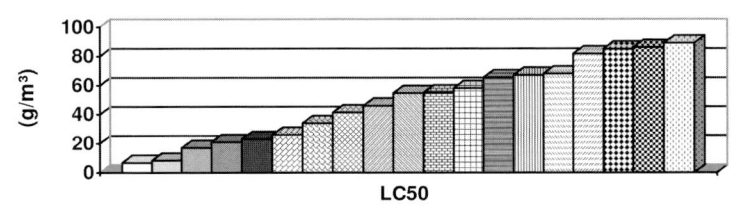

1	Durethan B 30 S	11	Tedur KU-1-4623
2	Durethan TP KU 2-2227	12	Durethan BKV 30N1
3	Novodur L3FR	13	Apec HT KU 1-9354
4	Novodur P3T	14	Makrolon 6030
5	Tedur KU-1-9510-1	15	Makrolon 3103
6	Novodur P2M	16	Apec HT KU 1-9359
7	Bayblend T65MN	17	Makrolon 9415
8	Pocan B4235	18	Pocan TP-KL-7913
9	Bayblend FR90	19	Makrolon 6465
10	Makrolon 6555		

Figure 235 d   Records of the toxic potency of various plastics

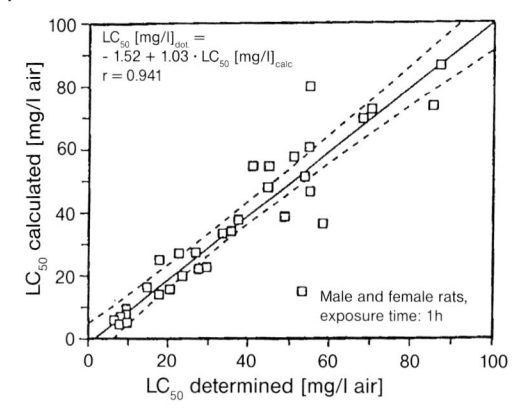

Figure 236 Correlation between calculated and biological determined LC_{50} data

Table 133 Toxicity of the Pyrolysis Products of Semi-rigid Polyurethane Foams

Sample	Temperature (°C)	Concentration in air		Deaths Out of 20
		CO (ppm)	CO-Hb (%)	
Per Volume				
1	400	1000	41,7	0
	500	1400	57,8	4
2	600	3000	62,0	0
3	500	1600	49,7	0
	600	2500	56,0	1
Per Weight				
1	350	3000	63,2	0
	400	7000	71,2	20
2	350	1300	34,0	0
	400	3500	68,9	6
3	350	2000	61,2	0
	400	6600	71,5	9

Tests with equal volume (300 by 10 by 5 mm) and equal weight (1,2 g per 100 mm)

nant number of incombustible materials. With regard to the increased rate of breathing of children, the CO-Hb content must not exceed the boundary value of 35%. The importance of the CO-Hb issue was also testified by the studies of Fleischmann (see Fig. 133) [481].

Influence of the test parameters on the toxic potential LC_{50}

The most frequently used criterion for the assessment of the toxic potential of thermal decomposition products from the toxicological point of view is the LC_{50} rating determined as a rule on the basis of biological investigations. Nearly all of the selected test protocols by SC3 Toxic Hazard in Fire of TC92 provided LC_{50} data. The cone calorimeter is the only exception. This material-specific characteristic is usually determined as mass-, volume- and/or area-related (g/m^3, cm^3/m^3 and/or cm^2/m^3). The exposure duration of 30 min, agreed at the SC3 of TC 92, suits the rating LC_{50} as a basis in the majority of the investigations. The most advanced test protocols for bioassay testing were the DIN tube system and the Potts Pot procedure [47, 93, 185, 406, 752–754]. Increasing temperature leads obviously to a somewhat less positive LC_{50} record. After reaching a peak value, the level drops again. These characteristics are to be interpreted in such a way that an application of Haber's rule to the gas mixture of different decomposition models does not appear opportune. Prager [464] report complete investigations with the DIN equipment. Table 134 provides information about the results of the small round-robin exercise with the Potts Pot Procedure. These tests were conducted by eight laboratories using the thermal decomposition products of wood (DF). Each one of the participating eight laboratories produced an entirely lower level of toxic potency for the thermal decomposition products of wood produced under flaming conditions. The graphical representations of Figs. 134–137 as well as the listed date of Tables 95–100 demonstrate to what extent variations of the test device and the test parameters possibly influence the test results. Increasing temperature leads obviously to a somewhat less positive LC_{50} data. Anyone of the classification criteria is inseparably combined with the fire scenario simulated by the decomposition and exposure conditions. The necessity to analyse the edge parameters also in the case of a criterion as the LC_{50} demonstrate, for example, the characteristics determined by the U-Pitt protocol. The test results LC_{50} (g), Lt_{50} (min) and $L(ct)_{50}$ (g min/l) demonstrate that this test procedure does not guarantee comparable test parameters without significant deviations. The material-specific LC_{50} data records determined in accordance with the U-Pitt protocol are by no means based on a standardized exposure duration of 30 min. These characteristics of the U-Pitt protocol must be linked to obviously different exposure times. The comparison of some material-relevant potential values, which were determined with different test methods indicate partially considerable deviations. The deviations in the temporal process due to the thermal loading as well as the exposure duration are obviously the cause of the difficulties of a comparative evaluation. The functionality of these characteristics is documented by innumerable investigation results. The comparison of the data records, demonstrate that, e.g. the U-Pitt and the NIBS methods [755, 756] supply values that deviate strongly from those of the DIN and the Potts Pot methods for products with marked carbonisation tendency, e.g. with cellulosic products. An analysis of these different data records clarifies that only by considering the concentration circumstances, may the relevance of these characteristics be interpreted. The newest findings of the SC3 Toxic Hazard in Fire, which allow a computational determination of the LC_{50}'s ratings, will surely increase with high probability the acceptance of this evaluation criterion. The data records of the toxic potential of Kimmerle et al. [931], Sakai, Le Moin as well as Michal [757–759] (see Tables 135–137) show in a very informative way that for comparable fire-gas concentrations the different test protocols supply equivalent LC_{50} data, if one carries out an approximation of the exposure times according to Haber's Rule. Significant deviations in the potential value result, if the decomposition conditions are modified by fillers, components or fire-retardant additives. The graphical representations of Figs. 237a and b show to what extent LC_{50} data will be changed by the addition of fillers (see Fig. 135c), flame-retardant additives (see Fig. 135d) or, e.g. metal plating (see Fig. 235b).

Table 134 Potts pot Method – Round Robin Results.

Lab.	LC_{50}-NF	LC_{50}.F
1	16,7	35,9
2	27,6	45,3
3	26,8	28,0
4	24,0	29,6
5	25,9	38,4
NBSa	20,4	41,0
NBSb	22,8	39,8
8	18,5	29,8
Medium	X=22,8	X=40
Value	S=16,7–27,6	S=28–45,3

Table 135 Comparability of the characteristics of the toxic potency

Material	Temperature (°C)	NF-X70-100 LC_{50}-10 min (g/m³)	NF-X70-100 LC_{50}-30 min (g/m³)*	DIN 53436 LC_{50}-30 min-600°C (g/m³)
Cotton	500°C	215	72	24
	850°C	112	37,5	
Silk	500°C	46,6	15,5	–
	850°C	13,8	4,6	
Wool	500°C	55	18,3	6,5
	850°C	19,4	6,5	
PAN	500°C	21,6	7,2	6,0
	850°C	14,6	4,9	
PA	500°C	31,8	10,6	10,0
	850°C	14,8	4,9	

Table 136 Comparability of the characteristics of the toxic potency

Material	DIN 53436 LC_{50} 30 min determined (g/m³)	DIN 53436 LC_{50} 10 min calculated (g/m³)	NF-106 LC_{50} 10min determined (g/m³)	NF-106 LC_{50} 30min calculated (g/m³)
PAN	6	18	23,8	7,5
PA	10	30	34,3	11
PUR	17	51	56,7	19
ABS	17	51	50,7	17
PC	50	150	141,8	47
PVC	17	51	208,9	70

Table 137 Comparability of the records of the toxic potential

Material	LC_{50} 20 min *) measured (g/m³)	LC_{50} 30 min *) calculated (g/m³)	DIN 53436**) LC_{50} 30 min$_{600°C}$ (g/m³)
PA	66,7	44	10
PUR flex	41,0	27	17
PUR rig.	22,0	15	7
FIR	27	18	18
Plywood	16,7	11	–
UF	13	8,7	–
MF	8,7	5,8	–
PAN	8,9	5,9	6
Cotton	33	22	24

The test concepts of the biological assessment demonstrate that the outcome of analytical measurements can not be used for predicting the possible threat to life. Different kind of animals like mice, rats, rabbits, dogs or even baboons have been used, to get information about breathing behaviour, motor activity, leg flexion shock avoidance, pole-climb avoidance, escape response, swimming activity or respiratory depression. The overwhelming majority of bioassay tests were conducted with rats as test animals. Because animal testing is very expensive, various attempts have been undertaken to reduce the number of tests e.g. by training of the test animals (see Fig. 235c). The pressure to look for alternative procedures was the basis for the development of mathematical modelling using chemical analysis for prediction of toxic potency data. The graphical illustrations of Fig. 236 demonstrate the achieved progress, which enables to reduce bioassay testing for at least to validate the toxic potency data gained by calculation.

These material-related differences, as they are documented by these graphics, are to be put into relation to the deviations made by the technical experimental variation in order to guarantee risk-relevant interpretations. The findings, gained at the SC3 of TC 92 Toxic Hazards in Fire, that relevant LC_{50} records can be determined also with the aid of analytically determined concentration data of the essential fire gas components on behalf of the FED concept by means of the Finney Formula, should improve the importance of this evaluation criterion LC_{50} decisively. The establishment of mathematical models is based on the ground works done by the WG5 "Prediction of toxic effects" of SC3 of ISO TC 92 "Toxic Hazards in Fire". The FED (Fractional Effective Dose) model developed by Hartzell et.al. [408] has so far proved to be the most successful, since it offers a practical approach to the problem. The toxic potency (LC_{50}) of the material tested is calculated by means of the analytically determined concentration of the relevant components of smoke using the following equation

$$FED = \frac{[CO]}{LC_{50}CO} + \frac{[HCN]}{LC_{50}HCN} + \frac{C_i}{[LC_{50}C_i]} \ldots$$

where the values for concentrations of smoke components correspond to the integral ct values for the duration of the experiment. The LC_{50} data for the pure gases determined in biological tests are familiar from the literature.

A further approach is the N-Gas model of NIST, reported e.g. by Levin and Gann [394] which takes into account the interaction of four different fire gases (CO, CO_2, HCN and low O_2).

$$\text{Total FED} = \frac{m[CO]}{[CO_2]-b} + \frac{[HCN]}{d} + \frac{21-[O_2]}{21-LC_{50}O_i}$$

Pauluhn [741] developed a mathematical model for the DIN tube 53436 decomposition apparatus. His model is based on the Finley equation and enables the LC50 value to be calculated using the analytically determined concentration data.

$$\frac{1}{LC_{50}} = \frac{ppm.of.toxic.effect_1.mg.component^{-1}.lair}{LC_{50}.of.toxic.effluent}$$

The problematic nature of the dose-effect relationship is in this case important. A detailed analysis shows very clearly, that checking bioassay test data are still necessary for the validation of the calculated material-related LC_{50} characteristics. A computational determination of the relevant LC_{50}-ratings for different requirement profiles could be carried out, as e.g. in the ATS 1000 by means of the FED-concept starting from the additive effect mechanism of the different fire gas components, including irritants. The graphic representations of Fig. 236 provide information about the chance to reduce bioassay testings to a great extent by calculation.

An evaluation criteria coming frequently for the application in connection with the LC_{50} value is the "$c \times t$" product. The graphical representations of Fig. 237 permits a view of relevant material- and test-specific data records, as they have been reported by Hartzell et al. [756], Alexeeff [190] and Purser [757, 508] (see also Fig. 238).

Figure 237 provides a view of the $L(ct)_{50}$ data of different plastics and natural products. Alexeeff et al. [698] refer to the good agreement of the critical $L(ct)_{50}$ values with the ALE (approximately lethal exposure) characteristics. The data records of Fig. 138a clarify using the example of a PIR foam quality, on the one hand, the good agreement of the values and illustrate, on the other hand, that the combustion behavior at the higher intensity of radiation obviously leads to a smaller toxic potential of the effluents.

With this evaluation criterion $LC_{50}t$ the direct comparison of bioassay test results is facilitated in analogy to Haber's Rule. Equivalent decomposition conditions with comparable concentration circumstances are the basis for the direct comparison.

Influence of the test parameters on the toxicity-index classification approach
The classification system proposed for the evaluation of a possible smoke danger in the case of a fire is based on a list of different basic data. Both analytically determined concentration data of essential fire-gas components and bioassay results form the basis of these evaluation systems. The thermal decomposition products of wood were used, furthermore, worldwide as a reference material. There was already a list of efforts to classify the smoke risk of the examined products by means of individual

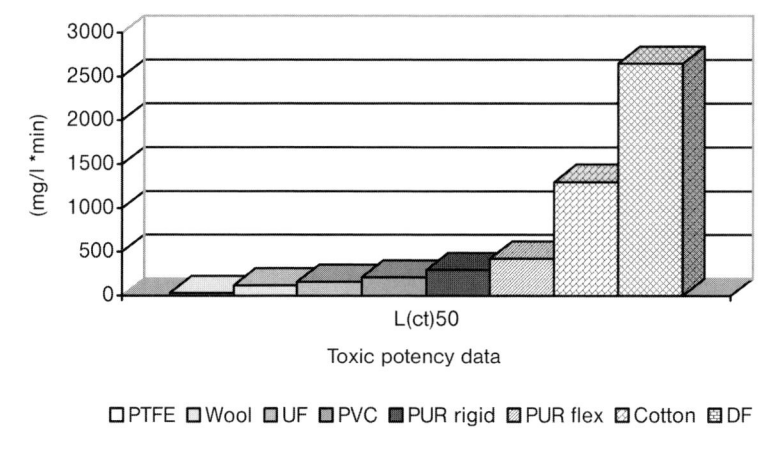

Figure 237 Characteristics L (ct) 50 of the toxic potential

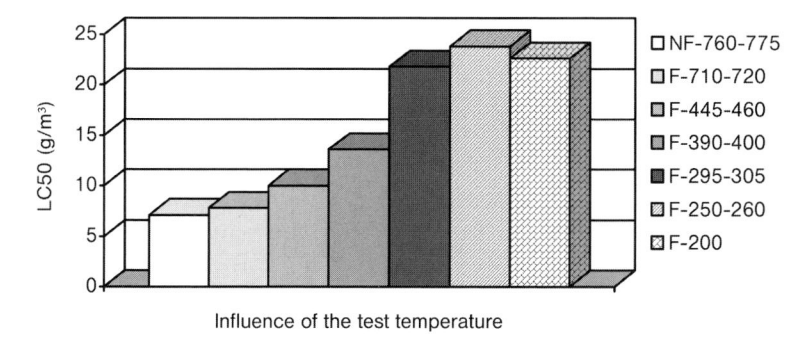

Figure 238 Influence of the test temperature on the toxic potency

laboratory test results. The material-related characteristics were classified by means of mathematical equations with the determination of the corresponding indices. Ausobsky [520] chose the CO-value of the thermal decomposition products of fir/spruce wood as a basis for his classification system. He set the different fire-gas components of the produced thermal decomposition products in relation to the CO value of the wood test run. The research studies of Jouany [220] clarify among other things the obvious weak point of this special classification system. The graphical representations of Fig. 240a demonstrate that also with the wood material next to CO even more fire-gas components become effective. Jouany could verify that not only the LC_{50} data records of PVC but the wood records as well could be reduced significantly also through the commitment of a water veil as a filter.

The numerical values of Table 138b provide a view of the relevance of the index system in accordance with NES 713. The different material-related fire-gas formation caused by the bunsen-burner impingement disclosed from the beginning a universal classification. By comparing these index figures that are based on the concentration data determined under standardized conditions and projected on $100 \, g/m^3$ are set in relation to the corresponding material-related LC_{50} characteristics, it can be demonstrated very vividly that they are neither suitable for the appraisal of the possible fire-gas risk in the case of a fire nor for the assessment of the toxic potential.

Sumi et al. [759] propagated a toxicity index already at the end of the 1970s by means of the mathematical relationship.

$T = t \times V/W = ct/cf \times W/V$
T = toxicity index
t = LC_{50} 30 min
ct = measured concentration
cf = LC_{50} 30 min concentration
V = air volume
W = sample volume

Table 138a Comparability of the records of the toxic potency

Material	LC50 20min *) measured (g/m³)	LC50 30 min *) calculated	DIN 53436**) LC50 30 min-600 °C (g/m³)
PA	66,7	44	10
PUR flex	41,0	27	17
PUR rig.	22,0	15	7
FIR	27	18	18
Plywood	16,7	11	–
UF	13	8,7	–
MF	8,7	5,8	–
PAN	8,9	5,9	6
Cotton	33	22	24

Table 138b Relevance of the NES 713 Index Numbers of the toxic potential.

Material	NES 713 index	LC_{50} record (g/m³)
PA	7	5
PVC	7,2	17
PUR	12,5	5
Wood		18

The analytically determined concentration data are put in relation to the LC_{50} data records of the pure fire-gas components. Sumi et al. assumed an additive-effect mechanism. The weaknesses of calculated characteristics can be illustrated in relation to the corresponding characteristic LC_{50} values using the example of the Sumi indices [74, 452, 760]. The partly serious differences for the relevant animal-experimental investigation results are illustrated by the graphical representations of Figs. 240–240c. The characteristics of Fig. 240b provide a view of the corresponding studies of Martin [499]. These determined Sumi indices demonstrate the different toxic potential of the examined products.

6.2 Relevance of the Evaluation Criteria | 413

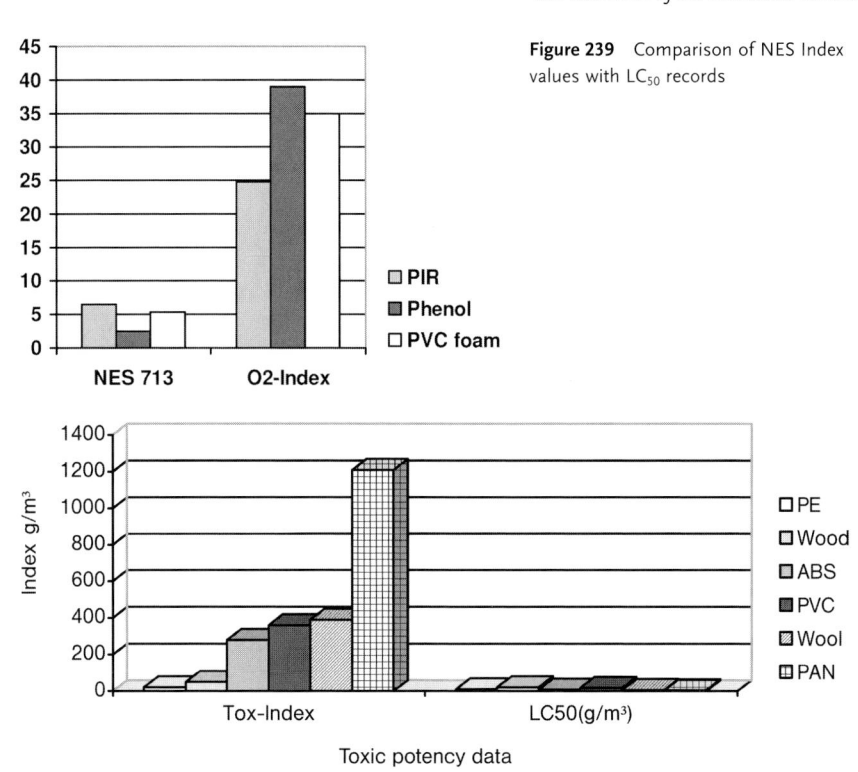

Figure 239 Comparison of NES Index values with LC_{50} records

Figure 240a Relevancy of Sumi Indices in relation to $LC50$ data

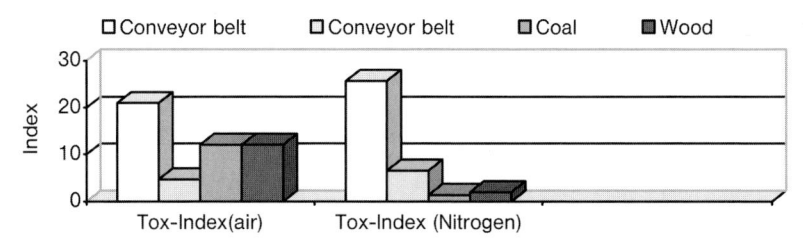

Figure 240b Influence of the oxygen availbility on Sumi indices

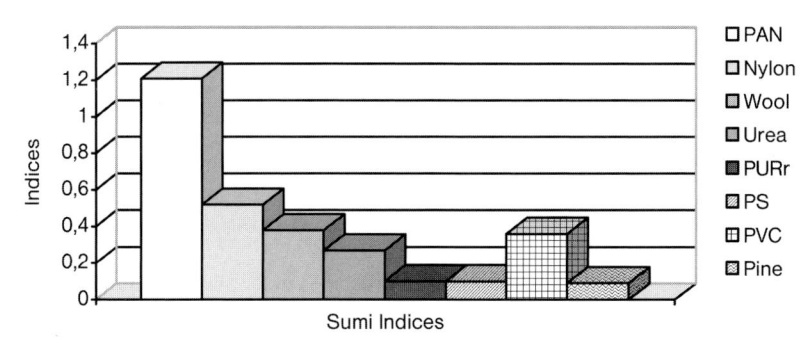

Figure 240c Sumi Indices of different products

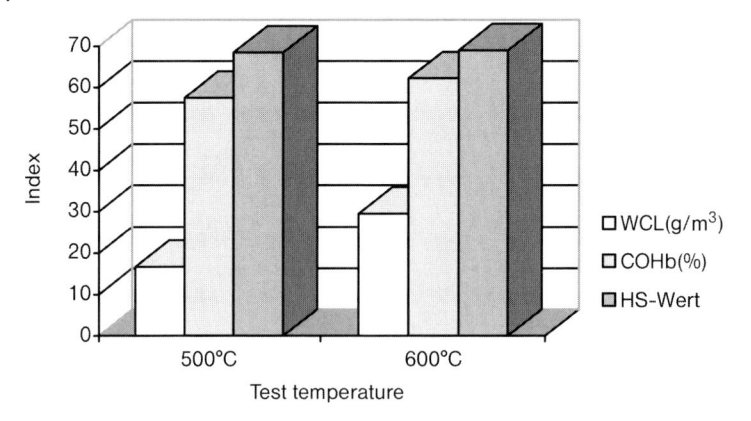

Figure 241 Influence of the oxygen availability on Sumi Indices

These graphical representations clarify that the level of the indices can be modified significantly by variation of the decomposition conditions. The oxygen offer at the decomposition process leads to a significant rise of the index level in the case of the reference products coal and wood. Sumi and Taylor [761] explain their newest considerations. Starting from the above formula, Sumi developed the so-called dynamic toxicity index

$T = V/A \times t \times Cs$, where
A = stressed surface area
t = test duration and
Cs = LC_{50} 30 min of the respective fire-gas component

Sumi refers on the uncertainty of laboratory test results for the characterization of the effluent toxicity and recommends corresponding reservations for legal actions. The decisive weakness of his system is certainly the chosen combustion protocol that excludes the determination of risk-relevant potential values. The classification system of RATP is based on analytically determined concentration levels of the most important fire-gas components. These characteristics are set in relation to the toxic potential of the pure fire-gas components by means of mathematical relationships.

$ITC = \Sigma\ ti/CCi$

Where ti means the concentration values in the stream of effluents produced at 600 °C and 800 °C and CCi means the corresponding characteristics of the toxic potential in accordance with Fig. 114 that do not lead within the exposure duration of 15 min in the biological test to any health injuries. An additive effect of the individual components is likewise supposed.

A list of classification systems is also based directly on bioassay test results. The assessment by means of an additive mechanism of the individual fire-gas components corresponds to the newest state of the art of the scientific findings [89]. The knowledge that next to carbon monoxide even more components became effective in the stream of effluents, was used to develop multiple specific classification systems.

In this case it was attempted, independently of the achieved default value of the toxic potential, the LC_{50} record that already considered these additional components for itself due to the additive-effect mechanism, to use the existence of further components, as, e.g. HCN, to evaluate as an additional negative element. This special processing of further fire-gas components contradicts the widely accepted additive-effect mechanism of the individual components. In the classification system of Gost [78] that was discussed in the transportation field, e.g. in the IMO [230] and also in the railway field a correction of the measured LC_{50} records by means of the determined CO-Hb values via the mathematical formula was defined.

$$\text{Itox} = \frac{10^4}{LC50/1 - ML \times 10^{-2}}$$

where
LC_{50} = lethal concentration for 50% of the exposed animals
ML = smallest measured CO-Hb values

The graphical representation of Fig. 241 illustrates that as a rule decomposition temperatures of 500°C to 600°C were used. The comparison of the data records illustrates that the temperature-dependent material-related differences are obviously suppressed due to this correction. The data records of Fig. 241 demonstrate how significantly the record levels of the toxic potential can become changed. A detailed analysis shows very vividly that the correction link, the CO-Hb value in the blood of the test animals can be changed significantly by variation of the test temperature.

This special processing of further fire-gas components in comparison with CO contradicts the indicated present additive-effect mechanism of the most important fire-gas components. Sand et al. [480] propagated also on the occasion of the FRC symposium in Salt Lake City a more unfavorable rating system in the case of additional acting fire-gas components. This specific classification system in accordance with Table 139 considered rating next to the LC_{50} 30 min in addition the content of the CO-Hb value and the RAITK/Tc value in accordance with the DIN tube system. A detailed analysis of this implementation approach demonstrate that from the very beginning an unfavorable rating for nitrogen-containing materials was intended, although this approach had been refused already before in numerous scientific studies. Herpol et al. [403, 762] propagated in accordance with Table 140 a toxicity index on the basis of bioassay test results. The mathematical assignment was based on the relationship

$$Tr = \frac{\Sigma\ Ki \times mo}{\Sigma\ Ki}$$

where

mi = mortality (%)
ti = exposure duration up to the death
To = best result, all animals are alive after 30 min
$T100$ = worst result, all animals are dead after 6 min
Ki = correction factor; K_1 (6min) = 5, K_2 (12 min) = 2.5, K_3 (18 min) = 1.66, K_4 (24 min) = 1.25, K_5 (30 min) = 1

Table 139 Classification scheme according to Sand et.al.

Class	Critical temperature T_C	LC_{50} 30 min (g/m³)	CO – content
1	$</>$ 400 °C	$>a$	significant
2	$>$ 400 °C	$<a$	significant
3	$<$ 400 °C	$>/<a$	significant
4	$</>$ 400 °C	$>/<a$	sublethal

a = 10–30 g/m³

Table 140 Rating of the toxic potency according to Herpol

Class	Value Ki
1	0–0,60
2	0,60–20
3	20–40
4	40–66,5
5	$>$ 66,5

Alarie et al. [92, 762] propagated a classification of the fire-gas toxicity in accordance with Table 141 by means of LC_{50} data records determined according to the U-Pitt protocol within the framework of the PRC- program [14]. They selected also the toxic potential of the decomposition products of wood as a basis of the classification system.

Alarie [763] report on the toxic-potency data LC_{50} (g) and the Acute Lethal Health (ALH) values determined according to the mathematical equation

where

$$[A\,L\,H] = \frac{K \times D}{T \times LC\,50}$$

T = test temperature (K) at 1% weight loss
LC_{50} (g) = the concentration, where 50% of the test animals are dead
K = the thermal conductivity (W/cm K)
D = density (kg/m³)

Table 141 Classification of the toxic potency according to Alarie

Class	Relevance	LC50 (g)	LC50 (g/m³)
1	much more toxic than wood	0,2–2	0,33–3,3
2	more toxic than wood	2–20	3,3–33,3
3	as toxic as wood	20–200	33,3–333
4	less toxic than wood	$>$ 200	$>$ 333

Assessment of the classification criteria of the smoke corrosivity

The pH value was one of the most used criterion for assessing the corrosive risk in the case of a fire. The standardisation of risk-relevant test methods has been initiated within ISO TC 61 Plastics and in IEC TC 89. Corrosion-relevant criteria are the goal of these investigations.

6.3
Relevance of the Requirement Profiles

The effectiveness of material-related requirements is discussed in detail in relation to the various test and classification criteria. In this section it will be stated using relevant examples that it is not the excellent reproducibility of a test that has the decisive impact but the possibility of this test procedure simulating sufficiently the risk situation and to reduce effectively the extent of damage. The final conclusions based on national findings cannot be transferred to other national recommendations. Detailed studies of national experiences demonstrate that life customs as well as environmental conditions will influence fires to a great extent. These facts have to be taken into consideration by setting national requirements. The nationally and internationally agreed requirement profiles concerning the combustibility are components of the different national legislation. The noncombustibility demands as they are stipulated in the building regulations, e.g. [764–766] are supposed to guarantee in particular that in the case of a fire these products do not provide any contribution to the initiation and spreading of the fire and the personal safety and rescue and extinguishing measures. Again and again, as by the Düsseldorf airport fire [1], increased safety is ascribed in the preventive fire protection in the case of a fire by the material specific classification "noncombustible". Fire protection technical orders are based, as all orders for security reasons, basically on negative experiences of the community. In the face of fire disasters such as for example the Charing Cross Fire [623], the Düsseldorf Airport Fire [1] or the Kitzstein Tunnel Fire [767] the issue about punctual modifications of the existing fire protection orders is repeatedly raised. On such occasions it becomes clear to the community how much the preventive fire protection is necessary. The fire protection orders are based also in the field of application of furniture and furnishings, transportation and electromechanical engineering, basically on the basic demands of the structural fire protection (see Table 1). The experience of numerous fires and risk-relevant full-scale field trials in the 1:1 scale, both in the building and furniture or transportation sector had demonstrated that pure material-specific fire-performance grading as a rule does not allow a risk-oriented evaluation of the measures of the preventive fire protection. The fire-protection regulation is based worldwide as a rule on laboratory procedures that are supposed to allow the assessment of the resistance against the acting of simulated ignition sources of the daily life. It has to be accepted, that material-related classification makes the task easier to implement relevant safety requirements by regulation and by the insurance industry. According to the Steinhoff [768] technical rule, which does not have any textbook character, but rather includes prescriptions for the practice already serves its purpose in a sufficient way, if it covers about 80% of the conceivable cases. In order to be able to cover the destination of the preventive fire protection evaluation are to be carried out in the application-oriented state. By means of some examples it is supposed to be demonstrated that more of the political appeasement of fear is used for modifications from requirement profiles as a rule by far in connection with essential fire disasters than the effective reduction of the available danger potential. For itself it would be desirable to scrutinize the fire-performance demands whether a sufficient safety is guaranteed in the case of a fire:

- Questions concerning the risk of ignition are in particular to be answered for the built-in state and for possible repair stages.
- The problematic nature of the possible fire spread must be clarified both for the operating stage and for the repair stage.
- A possible fire-gas danger in the case of a fire must be shown for the different fire scenarios depending on the respective fire development.

In order to make a pertinent, risk-relevant assessment for the regulator possible, general conditions were already given, e.g. in the 1970s in the German Technical Standards Committee FNM 851 "Coordination of the standards for the testing methods for fire performance". These guidelines allow principally a risk-oriented judgement of the exceptionally complex fire-performance testing and with this to select a fitting procedure for the possibly different danger situations. Analogous determinations were elaborated in numerous committees of the international standardizing organisations, e.g. ISO TC 61 or IEC TC 89 as well as in the industrial associations like VKE, APME or SPI. These handicaps should allow clarification of to what extent the planned testing and evaluation criteria are suitable to provide dependable predictions of the targeted destination. The selected samples illustrate that carried out quick shot solutions the general conditions elaborated by the experts for the determination of relevant fire-performance testing and classification systems are not considered due to last but not least reasons of time. The testing and evaluation criteria for mattresses, modified in the UK as a result of the HMS Sheffield fire [769] during the Falklands war, demonstrate very vividly that unsuitable means were decided. The modified bunsen-burner test was neither raked to the targeted protection destination nor allowed the definition of more relevant test and evaluation criteria for analogous fire scenarios. This specific igniting risk by missile action could not, without doubt, be simulated by means of a small bunsen-burner action. The decomposition products of smoldering bed linen represent a possibly fatal danger Smoldering bed linen as a fire cause was responsible for the Taunton sleeping car fire and the Scandinavian Star fire but did not, however, lead to intensified end-application-oriented testing and evaluation criteria [5, 770, 771].

The modified Danish orders due to the Scandinavian Star fire concerning possible fire-gas dangers in the fire case were not based on the smoldering fire risk, as the up to now valid combustion system based on DIN 53436, but rather on the cone-calorimeter and/or room-corner test configuration, a decomposition model, completely unsuitable for the smoldering fire risk.

How much the regulative determinations for the issues of possible initiations of a fire obviously claim general conditions that stands opposite to the conversion of these protection destinations, should be clarified by means of two examples that concern both ranges of application, the rigid PUR foam as well as the flexible PUR foam.

1) The demand of the supreme construction supervision in Germany in the building section "Insulants to be tested for themselves" and
2) The determination of the British Furniture Regulation that basically requires a fire performance evaluation of PUR upholstery fillings themselves. The required testing of compounds with polyester fleece causes for the expert due to the "peeling effect", (see Fig. 166), a pure material-specific evaluation of the PUR upholstery filling.

The modified order level in the respective editions of the guidelines over the use of combustible materials in the building construction field clarifies the arbitrariness as a basis of the respective regulation. The comparison of the editions of 1968, 1970/72 and 1978 document using the example of insulants, demands on the roof insulation, the missing direct reference to the experience from fire disasters and relevant large-scale testing. Using the example of the insulant orders in the guidelines on the use of combustible materials in the building construction field it can be documented that unrealistic orders of the preventive fire protection are not to be corrected as long as "insulants to be tested for themselves" or the normative condition for the classification of the burning dripping in accordance with DIN 4102 is adhered to. The requirement profiles onto the insulants offers as a special example that clarifies the findings from fires and relevant large-scale tests do not lead to the modification of insufficient test and evaluation criteria. The transfer of the findings leads as a rule to a perfectionism of the effected testing method and not to substantial modifications. In the criterion of 1972 of the guidelines it is called under Chapter 7.0 Roofs:

"7.1 Below a roofing, which must be classified according to §.1 BO NW [203] resistant against flying brands and radiant heat (Hard Roofing), insulant layers may consist also from materials easy to ignite (class B3) if the entire construction is sufficient resistant against flying brands and radiated heat according to DIN 4102 p. 3 Chap. 8. The insulant layer must be protected against ignition in an all-round way. Roof surfaces of more than 1000 m² have to be partitioned through stripes of at least 1 m width from incombustible materials partitioned [204, 205]." In the reworked edition of the guidelines from 1978 the wording of Sect. 7.1 is as follows: "7.1 Insulants below the roof cladding must be for themselves only tested at least as normally ignitable (class B2). If roof surfaces with combustible roof cladding or combustible insulants (class B2) borders on rising walls with apertures, these roof surfaces must be protected then up to a distance of at least 5 m with a shift of at least 5 cm thick from incombustible material, e.g. coarse grid edition. To the inspecting of specific roof surfaces of multistorey buildings must be in whole face protected against ignition correspondingly."

The findings that the use of incombustible represents an unrealistic measure for the limitation of the fire spread across the roof was taken into account in the 1978 edition, not, however, the fact that classification according to DIN 4102, which takes a surface evaluation as the basis for classification and is therefore unsuitable for the proof of the protection destinations for the built-in state required from the LBO. These unrealistic material-specific rating remained components of the regulative and assurance-technical demands, although Zorgman [244], TNO Delft, demonstrated already in 1976 on the 5th IBS in Karlsruhe the obvious weakness of the system by means of risk-relevant full-scale field trials. These illustrations demonstrate that the steel-trapezoidal roof insulated with the incombustible rockwool fiberboard can produce a fast fire spread on the neighboring roof surface units due to the surface layer and the wind conditions. This spread did not occur with the roof constructions isolated with different PUR grades.

The positive behavior of the PUR insulation of the steel roof construction was confirmed for the Pentan blown rigid PUR quality with analogous test runs at the FMPA Leipzig [772]. All these positive results did not change anything at that time to the regulative demand "Insulants to be tested for themselves only". If, for example, the real admission requirement for insulants in Germany are considered in more detail, it shows then that the requirements for composites as laid down in DIN 4102 were not implemented. A corresponding transfer was carried out obviously only there, where, as for example at the core material PUR/PIR rigid foam a more favorable classification was reached through the use of protective linings. The Düsseldorf airport fire demonstrated very clearly that by using a favorably classified insulating core, the possibly negative effect of the outer covering was not taken into account. Without doubt a comprehensive competence/expert knowledge is condition for the implementation of risk-relevant, application-oriented test and evaluation criteria in law form. Even expert committees succumb to the temptation to interpret pure material-related classification as risk relevant. Hence it is called, among other things, in the final report of the set-in expert group for this situation in Section 5.45 [773] "The DIN 4102 testing and evaluation standard demands that building products are to be judged fundamentally application-oriented that means that the influences that results from their application are to be considered at the testing and evaluating of their fire performance. In order to form the classification clearly, the standard plans that from all possible applications conceivable in the practice the most unfavorable conditions for the fire performance are simulated. More favorable application conditions may be put only then as a basis if the manufacturer can guarantee that these more favorable application conditions are available in the practice." This determination is in conflict with the evaluation of the noncombustible classified mineral-fiber layer in the suspended ceiling according to Sect. 5.44 [773] "Under that fire conditions these products do not burn or only into negligible extent (small amounts of organic components) They do not add to the evolution of smoke and toxic gases." This inconsistent interpretation of the final report of the expert commission speaks for itself. This unlimited evaluation of the noncombustibility does not correspond very obviously to the possible danger potential in the case of a fire.

The English Furniture Regulation of 1968/1978 [774] is a second, significant example of the abnormal behavior of administrative bodies. The Upholstered Furniture (Safety) Regulation [775] was introduced in 1980 in Great Britain and required for upholstery seatings the passing of the cigarette test according to BS 5852 part one. Furniture that did not pass the BS 5852 part one match-type flame test had to be labelled. More stringent requirements for furniture used in domestic buildings were established in 1988 with the Furniture and Furnishing (Fire) (Safety) Regulation [776]. "Upholstery components, which pass the relevant tests, can be used with other components. These demands require that the filling materials and the covering meet certain criteria in respect of their resistance to ignition. If one compares the agreed levels of fire safety for upholstery combinations, quite different test and evaluation criteria have to be met." The most important issues are: The agreed material-related fire performance properties allow an interchange of the various components if they pass the relevant tests, The manufacturers are therefore in the position to classify their products. One exception is the cigarette test, where the composite always has to be tested and evaluated.

As far as safety reasons are concerned, the defined classification demands (see the following wording) raise some substantial doubts:

- "No furniture, to which this regulation applies shall include upholstery that does not pass the cigarette test."
- "This test has to be conducted as described in BS 5258 part 1 and disclosed in schedule 4 of this regulation as composite."
- "Ignitability test for polyurethane foam in slab or cushion form …."
- "…An ignition source crib 5 is placed in position."
- "……provided that the resultant mass loss is less than 60 g, the foam passes the ignitability test."
- "Ignitability test for Polyurethane foam in crumb form"
- "Paragraph 3: This test is than carried out in accordance with BS 5852 part 2 using ignition source 2 as specified therein."
- "Ignitability test for latex rubber foam."
- "…. The test is then carried out using ignition source 2 specified therein"
- "Ignitability test for nonfoam filling singly."
- "…. The specimen comprising the filling material to be tested and the specified cover fabric shall be tested with ignition source 2 as specified in BS 5852 Part 2"
- "The cover fabric used according to paragraph 2 shall be made of 100 % FR polyester fabric."

In reality, due to the peeling effect (see Fig. 81) of the thermoplastic nature of this fabric, the PUR upholstery filling will be classified as a pure material property. A further differentiation is settled for the outer covering.

The test for the stretched covers (Part II of Schedule 5) shall be conducted in the same way, except the filling has to pass the ignitability test in part 1 of Schedule 1 (crib 5) instead of nonfire-retardant polyurethane foam.

There are substantial doubts that these requirements will inevitably mean a greater focus put on life safety. It is quite questionable whether these demands have been motivated by safety reasons only. The differentiation of defining partly composite (cigarette test) and partly pure material test and evaluation criteria (match test and crib 5 test) seems to be more or less politically motivated disregarding safety aspects. The findings of real fires and realistic large-scale trials demonstrated very clearly that the filling material latex foam obviously contains a higher degree of hazard in the case of a fire. Latex is more easily ignitable, and the heat and smoke release, once ignited, will be, due to the flame spread, much more hazardous. Numerous results had been available worldwide and some of those large-scale test have been initiated and financed by the Government/Home Office itself [56, 777–783].

Although innumerable risk-related full-scale field trials with upholstered furniture, carried out also in the task of the responsible authorities in the UK (DOT), demonstrate that pure material-specific characteristics are completely unsuitable to classify possible dangers in the case of a fire, the requirement profiles become predominantly aligned by pure material-specific characteristics. In this basic attitude of the first edition of 1968 nothing was changed by the revision of 1978. The flexible PUR foam is, through the regulation, without doubt more unfavorably assessed and classified than latex. The missing relation to the practice becomes clear using the example of regulated outer coverings. The peeling effect of thermoplastic fabrics guarantees that the PUR foam upholstery filling will be tested by itself. Polyester FR fabric that suppresses the smoldering risk in the cigarette test procedure without doubt significantly must be tested according to the regulation in the match test in combination with the PUR polyether foam grade inadmissible by law. The pure material-related test and evaluation criteria for every combination in the standardized cigarette test for filling stuff, would have ruled out a high percentage of cotton wadding and latex foam [100, 784, 785] but no polyurethane foam. Damant and Prager demonstrated that more than one cigarette in one test is needed to start a cigarette-induced smoldering fire in bare polyurethane foam [128, 786]. FR polyester fabric as outer covering would minimize or rule out the risk of a cigarette-induced smoldering fire. In combination with CMHR polyurethane foam the resistance against match-type flames could be awaited. The requirement to test this cover material in combination with the nonfire-retardant polyurethane standard foam, which is legally not allowed to be used in the furniture production must be politically motivated. There are no arguments of fire experts available.

6.3.1
Relevance of Full-scale Field Trials

The assessment of the relevance of full-scale field trials must be based in particular on the objective of the studies. In fact two basic scenarios are to be differentiated
  Fire development and limited fire spread and
  Fire development and unlimited fire spread including the smoke action.

All usual, national fire-reaction test methods simulating the origin of a fire, are linked to the fire scenario with limitation of fire spread, providing at the present general conditions a limited testimony field for the evaluations are clearly decisive. The reference of full-scale field trials in the scale 1:1 is frequently used as a safeguard of the scientific interpretation of the test activities in the laboratory scale. The evaluation of the relevance of full-scale field trials presupposes the answers of some basic questions:

- Is the selected ignition source realistic?
- Meet the dominating general conditions such as, e.g. fire-load distribution and ventilation conditions in real circumstances?
- Bear in mind the planned interpretation of the results sufficiently the limitation of the carried out fire test and/or the experimental course and
- Will the aspired classification be linked to the target?

It has to be differentiated between the two essential fire scenarios "Risk of ignition and limited contribution to fire spread" and "Unlimited contribution and smoke evolution combined with that". Numerous large-scale tests were carried out both with flexible and rigid PUR foam qualities to clarify their possible contribution to the surface spread of flames as well as to the release of thermal decomposition products and to produce risk-related findings for the preventive fire protection in the case of a fire [56, 111, 244, 321]. In the case of the first scenario it is to be guaranteed that all-important factors, such as ignition source size, order and exposure time, fire-load distribution, material compound and ventilation are duly considered. With the second scenario it must be guaranteed for general obligatory statements that the temporal process of the fire-gas formation and the exposure time allow relevant characteristics for a risk-oriented interpretation of the smoke dangers. With the interpretation of the effluent concentrations it must be guaranteed that these are material-relevant characteristics that are suitable as basic data for danger estimates in the different fire scenarios. In the field of fire science currently the status of a risk-relevant full-scale field trial [223, 545, 615] is given in particular to the ISO room-corner test developed at ISO TC 92. The results determined in these test procedures demonstrate severe material-related differences [237]. The standardized frame conditions such as room size, orientation and distribution of the fire load as well as the ventilation must be duly considered for the interpretation of the results. The statement for the room-corner experiments according to IS 9705 remains limited to a small empty room, e.g. a kitchen and a spot-like ignition source of high intensity, e.g. a broken gas pipe, and an equivalent ventilation aperture. The communicability of the results onto other room sizes, different ignition sources as well as differing ventilation conditions is not permitted without special evidence. This restriction is, in particular, valid for the second fire scenario of the unlimited spread with the connecting flashover situation and the fire-gas dangers caused by this. The standardized room-corner protocol is accordingly completely unsuitable to default material-specific assessment of the smoke danger with regard to the reduced visibility and health injuries through acting effluents. The aspect of the possible fire-gas danger due to the fully developed fire, was the central point of the majority of the studies, with first priority in the furniture field. The findings from numerous full-scale field trials with upholstery furniture and composite building products show in a very informative way that the fire protection material-related classification of the piece of composites can be changed by the material compound both in the positive and in the negative sense decisive in the furniture and building section. Extensive, risk-relevant investigations of the surface spread of flame were carried out in cooperation with the official testing station of NW in particular with facade linings at the testing station of the Bayer AG in Krefeld-Uerdingen [787-

789] as shown by Fig. 242a. The relevant studies show that also with incombustible facade linings flaming ignition can occur on the curtains at the floor above the fire room due to the intensity of radiation of the hot gas stream. This effect has also been observed in real fires [790]. The primary destination of these investigations was the checking of the resistance of facade linings against ignition sources of different intensities up to the hot gas stream of a fully developed room fire. The distribution of the fire load of about 25 kg/m^2 in the fire room is illustrated by Fig. 242b. Within relevant studies, it could be proven that the thermal stress as documented by Fig. 243 by means of a wooden crib fire on the facade linings could be simulated within an experimental duration of 10 min. This experimental duration correspond to the standardized testing time of the by law defined laboratory test, the chimney test according to DIN 4102. The room fire lead to a thermal stress over a time of 30 min.

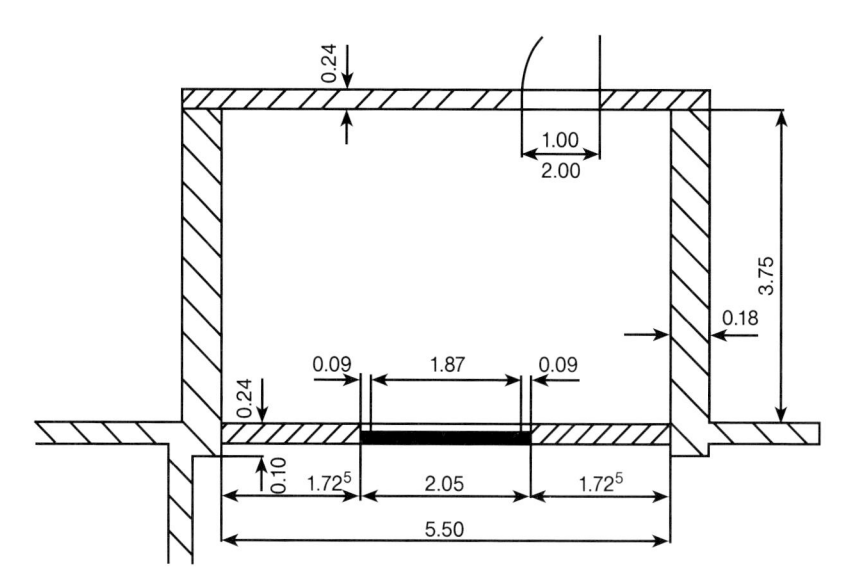

Figure 242a Plan of the test house in Krefeld-Uerdingen

Figure 242b Fire load for the facade test

6.3 Relevance of the Requirement Profiles | 425

Determination of the Fire Load

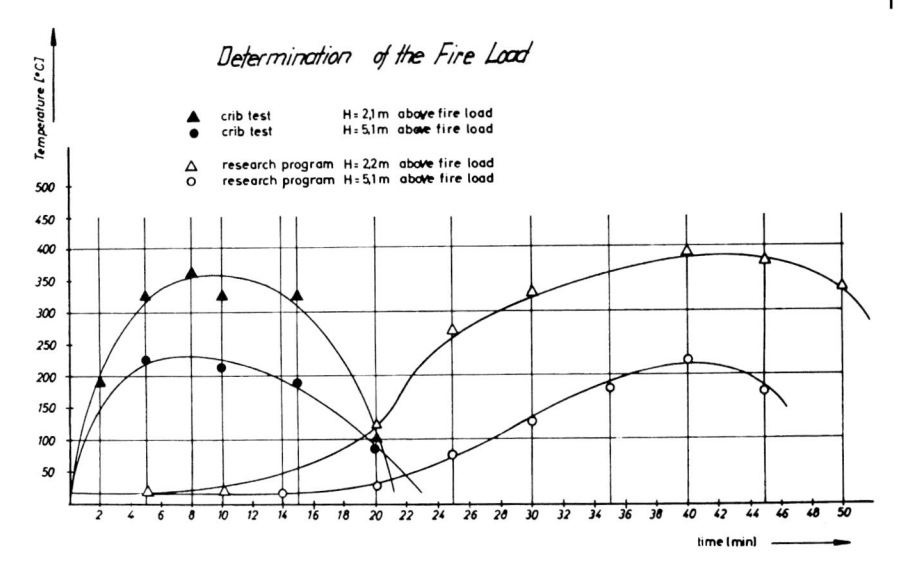

▲ crib test   H = 2,1 m above fire load
● crib test   H = 5,1 m above fire load
△ research program   H = 2,2 m above fire load
○ research program   H = 5,1 m above fire load

Temperature [°C] vs time [min]

Figure 243   Simulation of the thermal stress of the facade during a room fire

Figure 244a   Facade test with a PUR steel-sandwich element

Figure 244 b Facade test – large sclae trail at the FMPA Leipzig

The possible influence of the PUR core on the contribution of the facade sandwich elements to the surface spread was examined several studies. In order to verify that panels, applied before B3-classified core materials were prohibited, do not represent any inadmissible risk in the initial stage of a fire case, a 1:1 experiment was conducted [788]. Figure 244a illustrates that even when a limited portion of the B3-classified PUR-FR core material is exposed to the impact of the burning crib, the burning process within the flame-exposure area was of limited duration and effected a limited surface area only. In Krefeld-Uerdingen as well as in the Leipzig testing station analogous results were obtained with additional B1-classified facade elements, which were manufactured with frigene and pentene blowing agents [791-794]. The aim of the official Leipzig test runs was also to provide practical data to demonstrate that the proposed "E" classification of the Eurific program, based on the ISO corner test protocol of PUR steel-sandwich elements does not reflect the reality [795]. The worst rating of fire performance of that proposed European classification system of building products for PUR sandwich elements is without doubt quite contrary to the findings of realistic and meaningful well-published large-scale, test results. The tests, conducted in Laue near Leipzig, used a corner and wall configuration. The size of the test facade was about $80\,m^2$ consisting of seven elements each 1.16 m × 10 m. The ignition source, a 40-kg wooden crib, designed to simulate the thermal stress of a room fire, was located outside. In the corner test, the crib was positioned against the panels, where the protecting steel facing was partly removed as shown by Fig. 244b simulating a construction or repair situation. These tests have shown that the selected thermal stress of a room fire will lead to a very limited flame spread only. These panels performed satisfactory under full size fire tests. The variation of the blowing agent does not obviously change the fire performance in the early stage of a fire. The fire performance of metal-faced PUR panels was part of a special program conducted by the University of Stuttgart [796]. The fire tests were performed in a real building under construction. Figures 245c and d illustrate the extent of damage of the facade with the limited contribution to fire propagation in the case of a room-fire attack.

A series of tests have been run to study the fire behavior of thermoskin composites. The test program, conducted with a B1-classified EPS foam core and two different PUR/PIR foam grades demonstrated that the composite material is totally responsible for possible variations of the flame-spread behavior of the facade linings. In the case of the B1-classified EPS core, a chimney effect was observed in the corner position, due to the melting processes. Decomposition products evolved at the top of the test building as illustrated by Fig. 246a. Figure 246b illustrates the thermal course of the EPS-core facade test. With one of the PUR/PIR combinations a smoldering fire spread to the top of the test house in the corner position. A limited spread in wall and corner positions could be stated for the second combination (see Figs. 246c and d) [797].

Figure 245a Facade test – real 1:1 test at the FMPA Stuttgart

Figure 245 b Facade test – real 1:1 test at the FMPA Stuttgart

Figure 246 a Facade test with an EPS-class B1 composite

Figure 246b Facade test with an PUR-class B2 composite

Figure 246c Facade test with an PIR-class B1 composite

430 | *6 Use and Interpretation of PUR-Test Results Determined under Enduse Conditions*

The objective of the ventilated facade test runs was to investigate the fire performance of a B2-classified PUR insulation for ventilated facades. The aim was to cover both the "during construction" as well as the "in-use" conditions and to test whether PUR foam grades that just failed in the chimney test according to DIN 4102 behave possibly similarly to B1-classified foam grades. These tests, organized and financed by the III, were performed at the Bayer testing station in cooperation with the official testing station of NW. The fire load was defined based on the experience of earlier tests [787]. A wooden crib (1000 mm × 500 mm × 400 mm) consisting of stripes and weighing 40 kg was used for testing the unclad foam as well as the ventilated facade. The shape of the crib, but not the weight, was changed to get a much more severe flame attack at the ventilated facade test run. Figures 247a and b illustrate the test arrangement and the damage done in the ventilated test run. In both tests a limited contribution to the spread of flame was investigated. The bare foam, which just failed the chimney test, behaved similarly to a B1-classified grade. In both tests with the ventilated system, wall and corner position, an ignition of the decomposition products above the top of the cladding was observed. Under the more severe conditions of the corner position a longer-lasting burning process had to be recorded. A vertical

Figure 247 a Ventilated facade test

Figure 247 b Damage of tested ventilated facade

spread had to be expected with the B2 grade. The extent of the spread is significantly influenced by the size of the ignition source and by construction parameters like the size of the gap between cladding and insulation board. The influence of the size of the gap was one objective of relevant investigations with ventilated facades at SWRI in San Antonio [798, 799] the findings were that the nature of the insulants can influence the effect of the size of the gap. The use of horizontal barriers, e.g. horizontal support of the cladding will stop the fire propagation behind the cladding.

How necessary the crucial questioning of the general conditions during the generalization of such pseudo scientific conclusions is has to be illustrated by means of some examples.

a) The ISO room experiment for the fire-performance classification of a flashover situation.
b) The Rockwool-steel trapezoidal roof-fire tests demonstrating the contribution of roof insulation to the heat and smoke formation.
c) The NBS-room-fire experiments for the determination of realistic material-relevant characteristics of the toxic potential
d) The computational determination of material-related LC_{50} characteristics by means of analytically determined concentrations of fire-gas components in 1:1 scale field trials.

To a) The definition of the flashover situation in room fires in accordance with ISO 9705 (see Fig. 58e) negotiates serious differences in the contribution to the fire spread of dwelling fires. While the legislator and the specialist of the preventive fire protection are being advised of significantly different safety levels in accordance with the room-corner test, illustrate realistic large-scale tests with fully furnished rooms very clearly that these material-specific significant differences are only representative for empty, small rooms [322]. Scandinavian studies have shown that serious deviations of the values are possible depending on the room size. In fully furnished rooms these material-specific fire-performance deviations are actually negligible. Large-scale tests [800] have shown, as demonstrated very clearly by Fig. 212, that even in the case of significant differences in the fire-performance classification of wall panelling, as are given for wood and gypsum board, the time to flashover is only slightly changed. The influence of fire load and ceiling tiles as well as ventilation conditions was obviously decisive for the burning characteristics of the fully furnished room, as illustrated by Fig. 211a.

To b) The investigation program of Rockwool at the University of Gent with different trapezoidal roofing constructions demonstrate the latent danger of false interpretations. The influence of a specific ignition source, a burning wooden crib, was examined in this case for the ignition and burning behavior of trapezoidal roofing partitions of a very small roof. By a purposeful selection of the primary ignition-source intensity was guaranteed that indeed in the case of the roof constructions with EPS and PUR insulation fire penetration and spreading across the roof and a corresponding smoke formation occurred, which did not occur with the mineral-fiber-insulated roof partition due to the better fire-resistance behavior (see Figs. 248a and b). The drastically smaller smoke formation documented by the Gent test series as well as the follow-up test series of the Rockwool Company was published without any reference to the realistic large-scale tests at TNO Delft simulating a small warehouse fire with a fire load of approx. 25 kg/m^2 [801, 802]. These results, already published [244], demonstrated that the incombustible classified rockwool insulations means only a delayed fire penetration. The aim of the TNO investigation, initiated and financed by the III, was two-fold:

- to provide data to assist the development of a realistic classification system and
- to demonstrate the limited contribution of the insulating material to the lateral spread of flame across the roof.

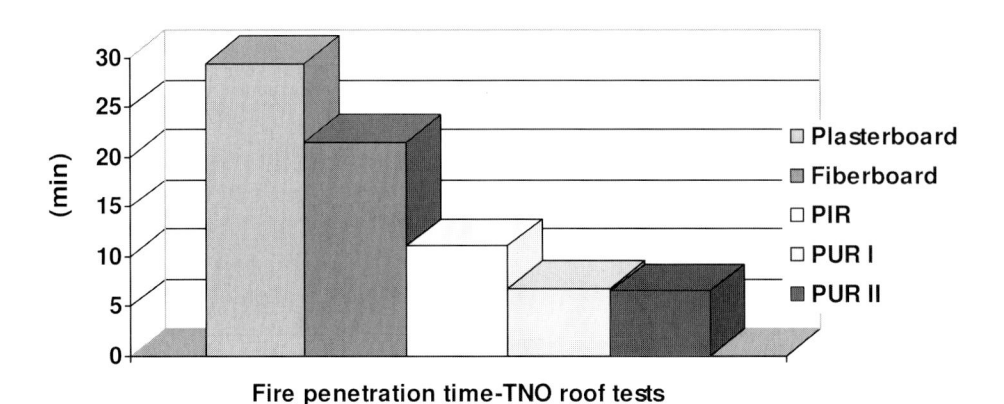

Figure 248a Fire-penetration test results at TNO

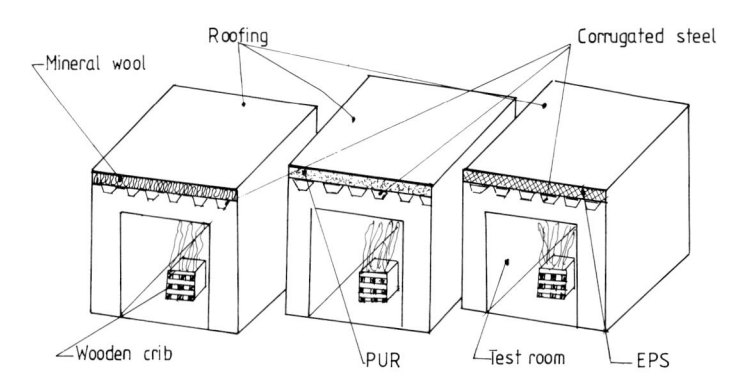

Figure 248 b Fire-penetration tests conducted by Rockwool

The selected fire scenario was not the impact of a single fire source but at least a fully developed fire in a small storage room. Figure 66a illustrates the arrangement of the fire load in the fire room and the test house. The fire load consisted of wooden cribs arranged by sticks with an air volume ratio of 1:1. The fire, once ignited, spread to the neighboring cribs. To get the desired time–temperature course within the fire room, as illustrated by Fig. 66b, the Thomas formula was used to calculate the necessary ventilation [244]. During the test runs, visual as well as video observations were conducted. The key conclusion of the test program was: After the fire penetration occurred, the fire performance of the top layer of the roof construction and the wind direction will have the most important effect on the lateral fire propagation. The Leipzig trials initiated and financed also by III, confirmed the TNO findings for the PUR insulation. The fire in the fire room spread at first, after ignition occurred, along the fire load before any significant ignition or penetration of the roofing partitions occurred. The fire spread across and within the construction appeared not to be influenced by the properties of the insulating material. In addition, the Leipzig trials confirmed the influence of the construction details to the fire-penetration time [792, 793]. Fire propagation to the roof partition of the neighboring room occurred as demonstrated by Figs. 249a and b with the incombustible rockwool only. This test was conducted under a significant wind input. After the fire penetration in the fire room occurred, a very fast fire propagation across the roof took place in the direction of the wind. Figures 249a and b show an extremely strong fire-gas formation of the roof insulated with mineral fiber, just after the fire breakthrough of the roof partition. Due to the wind influence lateral surface spread of flames occurred obviously on the neighboring roof section of the test house, a spread that could not be observed at the other constructions with PUR, PIR, EPS and PF insulation [244]. Figure 249c shows very vividly that, e.g. with the B2-classified PUR spray system Baymer DS1 ® the roof-fire was limited to the roof partition of the fire room. The fire occurrence came to a standstill at the roof also without any partitioning measure at the separating wall below the roof [803]. There was only one special composite in the TNO series, the vemiculite board, which released decomposition products, which led to a "flashover" in the second room, as demonstrated by Fig. 250, which was observed by means of video recording only. This special effect could be confirmed by an additional second test run with this special, American composite material [244].

To c) While the interpretation of the NBS-room fire experiments [705] postulate a realistic classification of the material-relevant stream of effluents of Douglas fir, rigid PUR foam and rigid PVC under real-fire conditions (see Fig. 251) indicate a careful consideration of the individual test parameters and the interpretation of the test results of Table 140 as well as to doubts about the generalization of the conclusions. This investigation illustrate according to Babrauskas [804, 805] realistic risk-oriented tests. Due to the very specific ignition source, cribs consisting of stripes of the materials to be tested, different concentration data of the effluents, changing during the fire course as well, are the result. Therefore, from the beginning, differing general conditions of the developing stage of a fire are created, which exclude, on the one hand, a comparative evaluation of the fire performance and, on the other hand, a classification of the acute toxicity of the produced stream of effluents. Material-related fire-gas component are made indeed, but already the concentration of the stream of effluents of the burning ignition source, consisting of the experimental material in crib form, is based on an irrelevant fire scenario. The burning behavior of the cribs varies from material to material and with that the produced stream of effluents of the ignition source that generate a different thermal stress at the wall panelling. Taking into account these limiting and basically inadmissible parameters it is then understood, as a natural reaction, that the results (see Table 142) of such studies are to be confronted with a list of reservation and restrictions with regard to the interpretation on legal orders. These characteristics are as a rule not suitable for legal actions.

Figure 249a Smoke formation at the roof construction insulated with Rockwool

6.3 Relevance of the Requirement Profiles

Figure 249 b Smoke formation and flame spread at the roof construction insulated with Rockwool

Figure 249 c Fire damage at the roof construction insulated with a PUR spray system

Figure 250 Deflagration in the neighboring room at the TNO test series

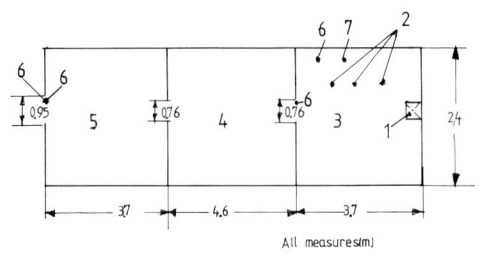

Figure 251 Room fire NBS – Test arrangement

1 Fire load
2 Sprinkler
3 Fire room
4 Corridor
5 Target room
6 Thermocouples
7 Probe for static pressure

Table 142 LC_{50} records and CO_2/CO ratio of effluents

Material	LC_{50} – Potts Pot	LC_{50} – NIBS	LC_{50} – 1 : 1	CO_2/CO total	CO_2/CO yield
Douglas Fir0	41–51	100–200	> 70	13–16,7	13,3–16,7
PUR rigid	10–13	20–30	30–40	12,5–21,5	12,5–22
PVC	18–22	20–30	35–45	0,7–4,0	0,7–3,8

To d) All general conditions of the fire-gas generation and distribution must be obviously considered carefully in the case of the computational determination of material-related LC_{50} data by means of analytically determined concentration values of full-scale field trials in the 1:1 scale. The investigation program with fully furnished rooms [56, 103, 104] initiated and carried out by the III in the training center of the British Fire Department in Morton in Marsh (see Figs. 215a and 252) allowed temperature and concentration measurements of the released stream of effluents, not only in the fire room but also in the bordering staircase as well as in the upper floor of the experimental test house. The graphical representations of Figs. 253a to c illustrate the temporal process of temperature and concentration of the essential fire-gas components CO, HCN and NO_x. In the case of the comparison of the concentration data of the different studies it appears that the dilution effect of the stream of effluents is not only determined from the physical regularities (dilution effect with the square of the distance). It also depends considerably on whether a stable smoke layer can develop and expands without turbulence. Also, the scale of dilution must be considered at the danger estimate accordingly depending on the spatial circumstances. In the numeric calculation of the LC_{50} records by means of analytically determined fire-gas component concentrations, it is to be noted that an unambiguous assignment to the fire load is not given as a rule.

a) – c) Variations of the ignition-source intensity, the ignition source action as well as the ventilation will among other things change the fire behavior including the fire-

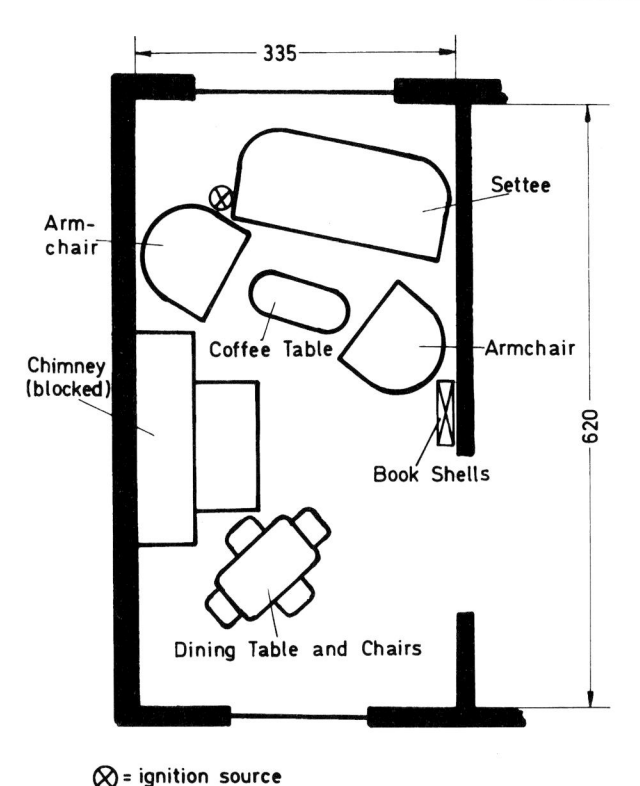

Figure 252 Test house Moreton – fire room

⊗ = ignition source

gas formation possibly significantly. During the estimate of possible smoke dangers by means of analytically determined records of fire tests in the 1:1 scale it is to be absolutely ensured that material-relevant characteristics do not give a reason for misleading interpretations. The area-related records of insulants instead of the usual mass-related characteristics, are, according to [806], certainly a step in the right direction to risk relatedness. Risk-related characteristics require an end-application-related orientation that is, e.g. a comparable insulation effect that must be selected as the basis for comparison. The different thicknesses of the test pieces are to be duly considered. The area-related rating will accordingly reach a significantly higher value. For dwelling fires, the respective fire load part of the materials must be, furthermore, set in relation to the entire fire load showed by practice-relevant investigations. A weighting of the single fire loads must be evaluated from that to the respective fire participation in the course of the fire. Material-specific characteristics, as bare values for estimates of possible smoke danger in the case of a fire, are to be classified only then as reliable if mass-, volume- or area-based data records are determined. For the concentration data analytically measured and registered in full-scale field trials shown in exemplary fashion by Figs. 253a–c, this specific feedback with the temporal restraint of the different combustion behavior is dropped as a rule (see Table 143).

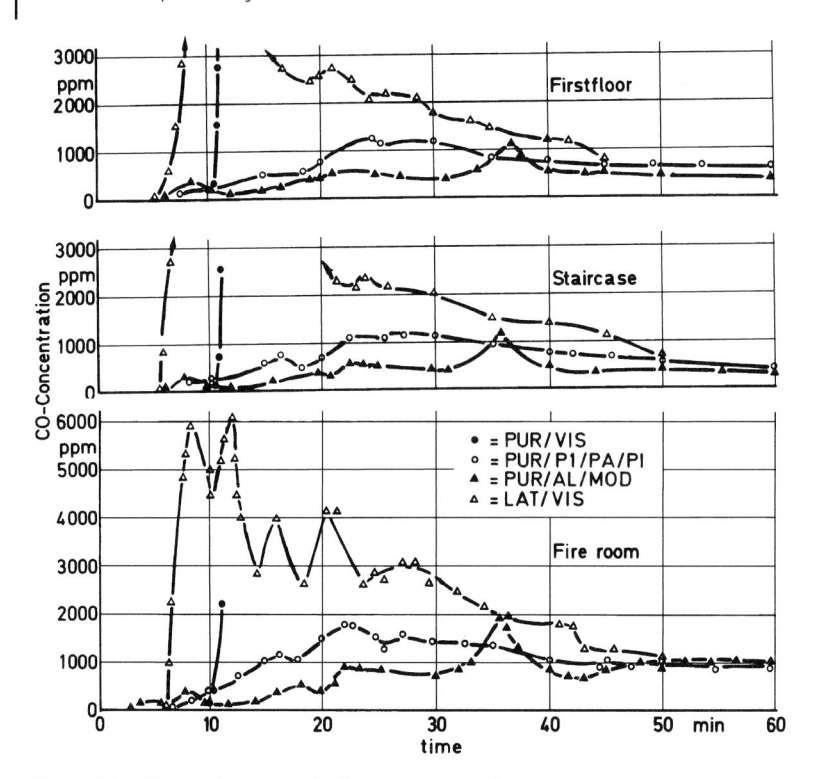

438 | *6 Use and Interpretation of PUR-Test Results Determined under Enduse Conditions*

Figure 253a Temporal course of the fire-gas stream at the Morton trials CO

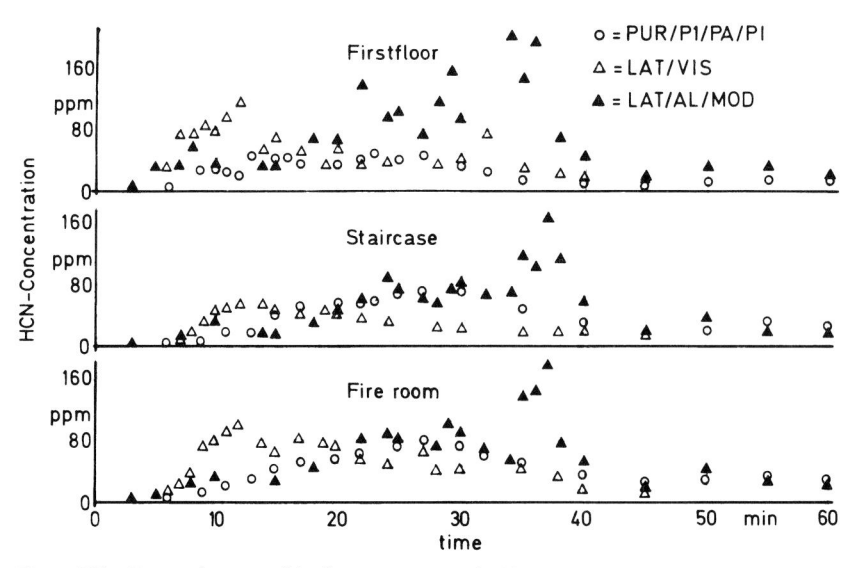

Figure 253b Temporal course of the fire-gas stream at the Morton trials HCN

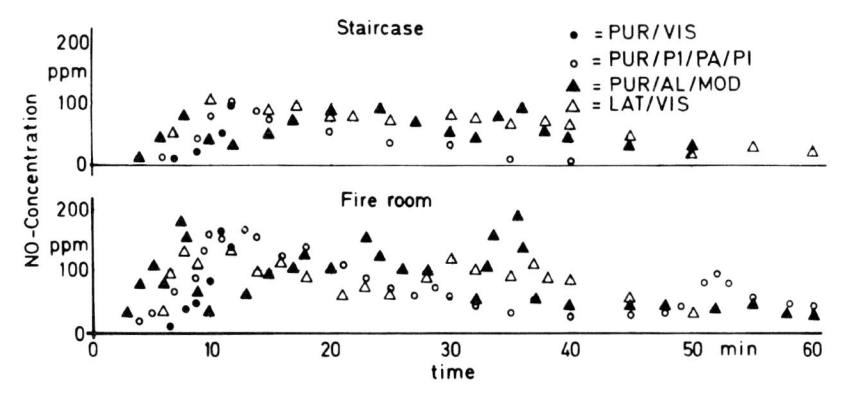

Figure 253 c Temporal course of the fire-gas stream at the Morton trials NO$_x$

Table 143 Toxic potency data – Bioassay test results and calculated data

Material and test protocol	LC$_{50}$ (g / m³)	LCt$_{50}$ (g min / m³)
PUR flex. – DIN $_{600}$	11,0	330
PUR rig. – DIN $_{600}$	2,1 calculated	63 calculated
PUR rig. – 600	29,0	870 calculated
PUR flex. – 600	6,6	198
PUR flex. – 500	7,5	225
PUR flex. – 400	22,5	960
TPU-FR – DIN $_{600}$	26,0	995
FPU – LS 1090	0,03 calculated	

The objection against pure calculation, by means of large-scale tests, as they are recommended by Purser [692] and backed by TR 9122 part 6, can be illustrated by means of the published LC_{50} records of polyurethanes All comparative studies in this connection, in which the toxic effects of the thermal decomposition products produced in laboratory and large-scale tests were opposed, have illustrated that in laboratory tests altogether more significant effluent concentrations are released [782–784]. LC_{50} data, calculated on the basis of analytically determined concentration data, are exclusively oriented on the temporal course of the simulated fire. All combustion systems that permit simulation of the fire-gas formation of real-fire situations basically allow statements about possible smoke dangers to be made. In accordance with Fig. 120 the comparability of these characteristics is guaranteed on the basis of equivalent concentration data. Purser [692] discusses toxic-potency data, LC_{50} and LCt_{50} data of polyurethanes (see Table 144). The thermal decomposition records were produced under the conditions of the three most important fire scenarios. The stream of effluents released by the three essential fire scenarios "smoldering fire", "combustion with flames" and "fiercely burning" as defined by Purser within the standardization activities of IEC-TC89, are readily distinguished with regard to the fire-gas concentration seriously. Purser assumes for his calculations based on 1:1 scale investigations

Table 144a Borderline values of smoke requirements

Combined methods	Borderline of admission	Area of applicability
Critical Flux Test	300%min	Germany
Tunnel Test ASTM E84	< 650 (100% = D Eiche)	USA
Early Fire Hazard Test AS 1530	D= 0,0082–4,20	Australia
Nord Test 004	T= < 10 (< 50) – < 30 (< 95)	Sweden
Smoke Test JIS A1321	CA = < 30 , < 60 , < 120	Japan
DIN 5510	50%min – 100%min	DB, Germany
Specific laboratory test methods	Borderline values	Area of applicability
XP2 Chamber	T=15% (D=0,235)	Germany
	T=< 10% , < 50% , > 50%	Austria
	T=< 50% , < 90% , > 90%	Switzerland
NBS Chamber	Dm 50–100–200	FAA
		Germany, ATS 1000
Ore 14 (mod.NBS Chamber)	> 300%min,	UIC
	150%min,< 150%min	
NF-X70		France
Gost 12.1	Dm50 , < 500, > 500	Russia
ISO Smoke Box NEN 6060	D= 10–5,5–2,2	Netherland
DIN 53436/37	T= 30% (D =	Germany

Table 144b Relevance of the national smoke density requirements

Test method	Concentr. (g/m³)	Class 1 D (1/m)	Class 2 D (1/m)	Class 3 D (1/m)	Class 1 Visibility (m)	Class 2 Visibility (m)	Class 3 Visibility (m)
XP2/Ö-3800	51,5	0,153	0,153–1,0	> 1,0	5,0	1,2–5,0	< 1,2
Switzerland	51,5	< 1,0	1,0–3,33	> 3,33	> 1,2	0,45–1,2	< 0,45
Nord.Test -004	–	< 0,2	0,2–1,3	–	4,5	1,9–4,5	–
JIS A 1321	–	< 0,52	0,52–1,0	1,0–2,0	> 2,5	1,2–2,5	0,7–1,2
JIS A 1201	–	< 0,2	0,2–1,0	1,0–2,4	> 5,0	1,5–5,0	0,8–1,5
Ore 14	–	0,4	0,15–0,4	<0,15	> 4,5	2,5–4,5	< 2,5
Gost 12.1	–	0,48	0,48–4,73	> 4,73	> 2,2	> 0,4	< 0,4
NEN 6060	39	> 2,2	2,2–5,5	5,5–10	0,7	0,4–0,7	0,3–0,4
DIN 4102							
A2-XP2	51,5	0,235	–	–	> 4,0	–	–
A2-DIN 53436	18	0,484	–	–	> 2,5	–	–
Radiant Panel	11	0,466	0,118–0,466	< 0,118	> 6,2	2,2–6,2	< 2,2
Chimney test-B1	–	0,58	–	–	> 1,8	–	–
DIN 5510	–	0,969	0,45–0,969	–	> 2,5	1,2–2,5	–

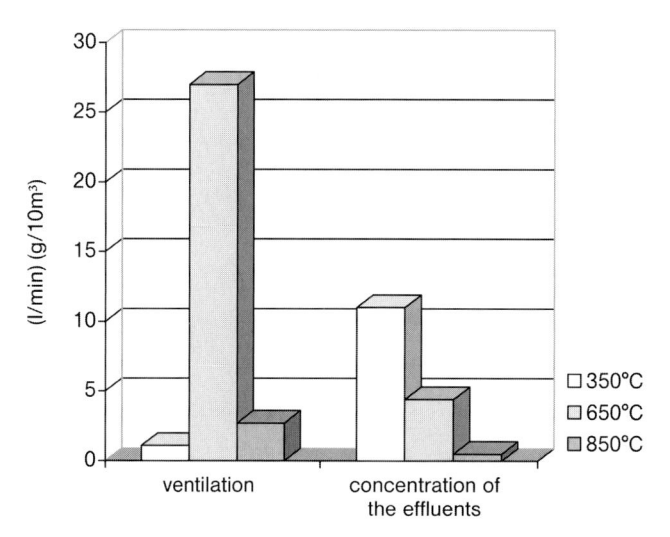

Figure 254 IEC fire scenarios for the determination of characteristics of the toxic potency

that next to CO and HCN further material-specific fire-gas components become effective and cause a significantly more unfavorable value (see Fig. 254). The listings of Table 144 demonstrate that the LC_{50} data records determined worldwide with the various different combustion systems do not back the Purser calculation. Even the most unfavorable values give evidence of lower acute toxicity than the calculated PUR records of Purser. This fact elucidates the necessity to confirm unfavorable calculated toxic-potency data at least by one bioasssay test result.

6.3.2
Relevance of the Investigation Results Concerning the Cause of Fires

Relevance of the fire-statistics data
A risk-relevant interpretation of the fire investigation results requires detailed fire statistics. The German intention, to establish a scientific fire statistic system, did not succeed during the last decades. The targeted, very detailed requirement profiles, about which Seidel [809] reported already at the 3.IBS 1967 in Eindhoven, could obviously not be implemented by the fire fighters on site. That also in the case of a very detailed fire statistics, as, e.g. that of the Home Office, UK, necessary care is to be taken, is supposed to be demonstrated by means of two examples. The first example is an interpretation of the trend concerning the number of fire victims caused by furniture fires on behalf of the numerical values of the years 1955 to 1973 (see Fig. 255a), mentioned also by the experts [809].

The graphical representations of Fig. 255a illustrate the danger of misinterpretation of fire-statistics data if one compares the predicted trend with the actual findings of the later decades (see Figs. 46 and 73). Here it becomes clear that it is necessary to consider very carefully the interpretation of possible trends of fire initiation and consequential fire damages. The second example clarifies that not the mathematical

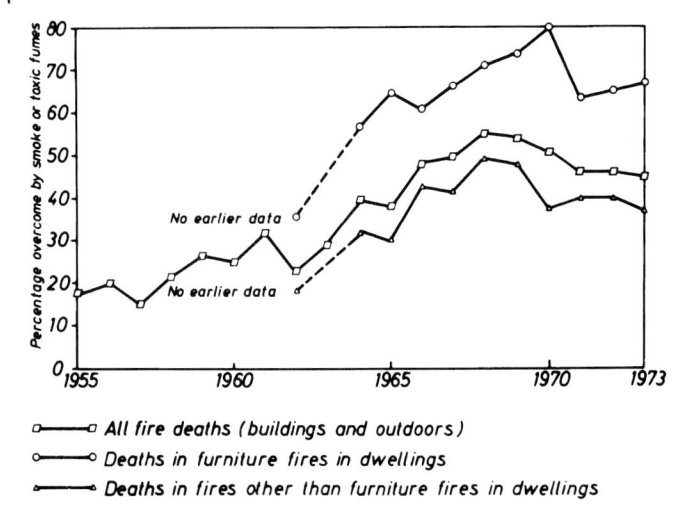

6 Use and Interpretation of PUR-Test Results Determined under Enduse Conditions

□────□ All fire deaths (buildings and outdoors)
○────○ Deaths in furniture fires in dwellings
●────● Deaths in fires other than furniture fires in dwellings

Figure 255 a UK fire-statistics data

Cause of deaths in furniture fires

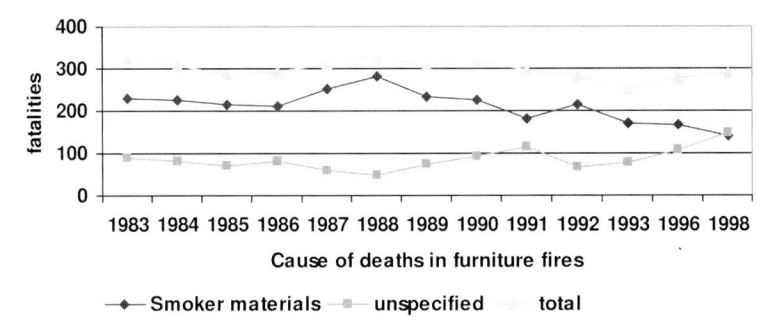

──◆── Smoker materials ──■── unspecified ······ total

Figure 255 b UK fire-statistics data – Dwelling Fires

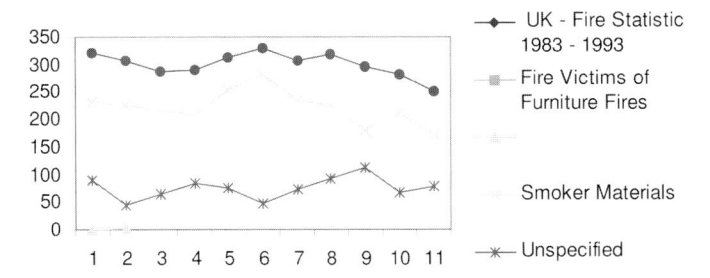

──◆── UK - Fire Statistic 1983 - 1993
──■── Fire Victims of Furniture Fires
       Smoker Materials
──✶── Unspecified

Figure 255 c UK fire-statistics data – Furniture Fires

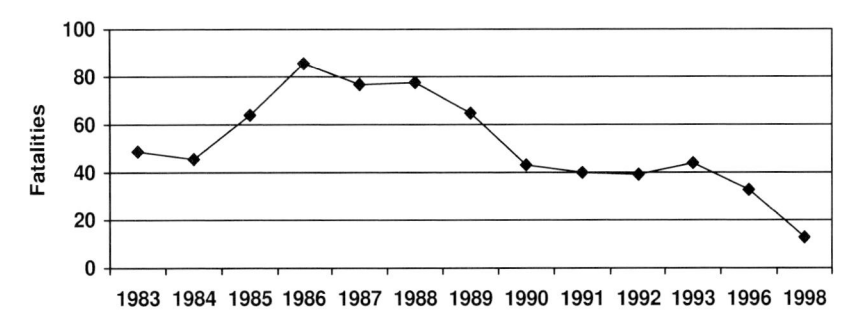

Figure 255 d

understanding but rather an excellent expert knowledge of fire science is required. The interpretation of fire-statistics data concerning dwelling fires carried out by the University of Sussex [810] on behalf of DTI (Department of Trade and Industry UK 1998) can be taken as an indication that very obviously politically required aims about the efficiency of the UK Furniture Regulation were setting the trend. The graphical representations of Figs. 255b and c show very clearly that the connection shown in the expert report of the University of Sussex is to be categorized more than speculatively vivid from the furniture fires and the implementation of the Furniture Regulation as well as the numbers of fire victims reduced by that. If it is considered that in the decisive period the part of the dwelling fires "not nearer specified" varied strongly, then the trend concerning the furniture fire and the estimated number of fire victims possibly avoided by the implementation of the Furniture Regulation must be interpreted differently. These graphical representations demonstrate the outcry of Colin Todd [811] "Lies, lies".

Furthermore, it is emphasizing that the experts of the University of Sussex deny a significant role of the intensified use of smoke detectors for UK running in parallel. The characteristics of the graphical representations of Figs. 35c and d clarify, on the other hand, the worldwide predicted and verified reducing effect of fire-alarm systems on the number of victims of dwelling fires. Smoke detectors are used as required worldwide.

Relevance of the analysis of fire processes and conclusions
Provided that there is or possibly is a participation of polymeric materials at spectacular fire cases the press, and experts, report on EPS, PUR or PVC fires and their possible material-specific dangers, e.g. the so-called PUR-fires, for example under the participation of flexible PUR foam systems as the Glasgow Fire [627], the Manchester Woolworth Fire [635] and the Leisure Park Fire [7] or under the participation of rigid PUR/PIR foam grades as the Saint Laurant du Pont Fire [626], the Kansas City Fire [14], the Köln-Deutz Gymnasium Fire [812], the fire case "Nürnberger Bund" [813, 814] the Tank Fire at the harbor Duisburg and different mine fires [815]. This focusing on the material-specific characteristics, for example, the heat and smoke liberation data is, as a rule, not risk relevant. In the Glasgow fire case, the barred windows

of the building, an old whiskey factory, were mainly responsible for the fact that the rescue and fire-fighting measures achieved little success. For the Woolworth Fire at Manchester [635], the official investigation showed that the inappropiate storage of the upholstered furniture caused in particular a decisive contribution to the fire spread and thus limited the rescue and extinguishing measures. With the Leisure Park Fire [7] in the realistic full-scale field trial, to the surprise of the fire experts of FRS, there was a very small contribution of the PUR mattresses to the fire development and in particular to the fire-gas liberation. "In this fire it is difficult to understand fully the source of the large amounts of smoke. During the early stages of fire the polyurethane foam in the crash mats burnt clearly with little smoke. Increased smoke would be expected as the temperatures inside the rig became higher and other fuel was involved". This full-scale test demonstrated that not material-specific characteristics but end-application-oriented composites and the local factor, e.g. ventilation and preheating effects influence the fire course essential. While the strong fire-gas formation during simulated warehouse fires [816] raised the profile of smoke from PUR foams in the UK, these realistic findings in the full-scale field trial concerning the furniture field have not been taken into account within the framework of the Furniture Regulation.

In the fire disaster of Saint Laurent du Pont, indeed the material-specific fire performance of the applied PUR spray system stood in causal connection with the catastrophic fire [4]. The court investigation [626] demonstrated that the rigid PUR spray system was applied as wall and ceiling lining by the dance-hall operator without permission. The licence was obviously hurt and the responsible building authorities deceived. Furthermore, an increasing number of emergency exits were barred due to commercial considerations. The fire case in Kansas City led to the FTC Consent Order and to the PRC 5-million dollar program [14]. The danger of misinterpretation caused by verbal definition of classification criteria and the faulty interpretation of the results coming out of small laboratory test methods, like the ASTM D 1692 procedure, were one of the consequences. The investigation of the gymnasium fire at "Köln-Deutz" [812] showed that the applied, bitumen-paper-laminated PUR boards of the ventilated facade had to be classified as class B3 and therefore were installed accordingly against regulation. In the fire case "Nürnberger Bund" the PUR core material of the sandwich elements for the initiation of the fire as well as for the fire spreading obviously played a minor role [813, 814]. Uncoated surfaces were responsible for the fire initiation and spreading in the mines fires under participation of PUR spray systems. While the front of the weather seal was protected by cementous coatings, the reverse side with the jute fabric, was untreated [815].

A detailed analysis of the Düsseldorf Airport Fire Disaster [707, 817] reveals that official investigations also, experts conclusions included, do not guarantee a risk-related assessment at all. The nonobservance of the fire-damage experience was, e.g. in particular substantially documented by the statement in the expert report to the Düsseldorf airport fire [1] "Under real fire conditions these products do not burn or only in negligible volume (small amounts of organic components) they do not contribute/add either to the evolution and toxic gases". The latter determination is all the more incomprehensible, nevertheless, it is to be read in the appendix of the expert report in the analysis of the Köln/Bonn airport among other things under 10.2.3: "Under the ceiling suspended ceilings are installed that are provided with a layer of laminated mineral wool. At the fire protection analysis (Sect. 10.2.1 put out that this layer of the incombustible classified material rated class B2. The operator intends therefore to replace the normal ignitable classified layer by an incombustible one." For experts this is a questionable issue. The testing modalities of the insulant examination of themselves concerned as a matter of cause also the side effect "drops burning from." This classification noted at the Düsseldorf EPS compound within the retesting of the MPA NW Dortmund must be assessed under the premise of the underlying B2 examination of the insulant themselves according to DIN 4102. According to the definition of the test and evaluation criteria this compound had to be classified as "no burning dripping". A detailed study of the testimonies for the airport fire showed after Prager and Sasse [707] that all the interpretations of the experts of the initiation and the spreading of the fire raised doubts. Prager supports this in his expert opinion [817]. The reproaches of the prosecution of representatives of the Düsseldorf professional fire department, negligent arson with resulting death, are before the background of the basic demands of the structural fire protection, laid down in §18 of LBO, NW [818].

Structural plants are to be ordered, constructed and maintained that the origin and the spreading of a fire will be effectively prevented and extinguishing operations in the case of a fire and the rescue of people and animals are possible and Materials that are still easy to ignite after the processing and the fitting must not be used to evaluate in the case of establishment and change of structural plants. A thorough analysis of legal requirements at the time of the project leads obviously to other conclusions, as those of the prosecution and their experts. In the expert panel called by the region government, the background, the fire occurrence and the consequential damages were examined in detail and conclusions were made without any guilt assignment for a new fire-protection concept. The final report for the Prime Minister of NW Rau [1] and testimonies during the court case such as the drafts of the expert comments for the prosecution forms the substantial basis of this document. The expert panel was of the opinion that the application of the Alu-faced EPS boards was against the regulations. The Alu-faced EPS slabs that were already applied in the first construction phase for the ceiling insulation, has been used as well in the third phase approved by the construction supervision, as noted by Fisher [819] on the 30.07.01 at the court. The noncombustibility classification, stipulated in the licensing procedure and quoted by the experts of the prosecution repeatedly as a proof of the inadmissible apply do not support this interpretation. The passages of the planning and building permission clearly do not include any reference to the insulation and/or the lining of the ceiling while contrary to this under 4.6 of 4.0 components (walls, pillars, ceilings and summers), wall and ceiling linings in the staircases must consist of incombustible materials, the ceiling tiles are cited explicitly. The finding is that neither the building advisory determinations nor the requirement profiles of the planning and building permission procedure contradict the application of the Alu-laminated EPS ceiling lining. The requirement in 4.21 of the building permission "The fire load density must be held as low as possible" is without doubt a main item of approaches against the two representatives of the BF Düsseldorf. What appears to be very superficial, obvious and risk related at an examination does not prove itself in a differentiated consideration by any means more than without limit correctly. In the case of the foamglass-insulation a raised fire load density must be calculated due to the usually used bituminous fixing in accordance with Fig. 256a in comparison to the EPS insulation [820]. Figure 256b provides information of the fire load involved at the Düsseldorf Airport Fire. The data of the rockwool in the suspended ceiling were assumed [707, 817], because the official report [1] does not contain any informative figures. In comparison to reports on other fire disasters like the Kings Cross Fire (see Fig. 256c) or the Beverly Hills Fire, many open questions are not answered.

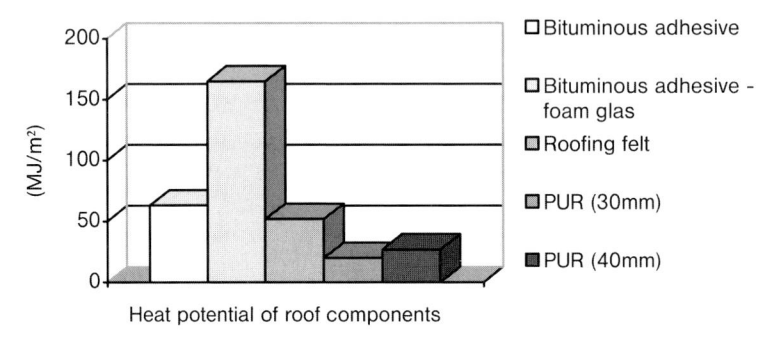

Figure 256a Fire load of roofing constructions

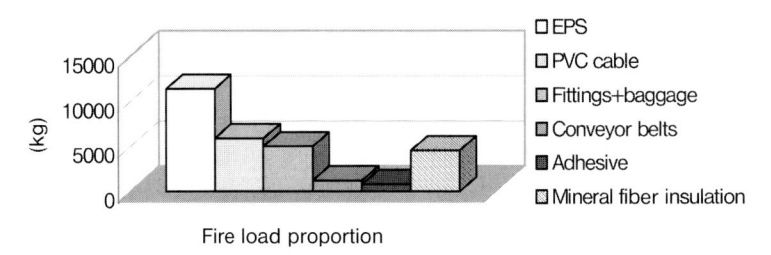

Figure 256b Fire load of the Düsseldorf Airport Fire

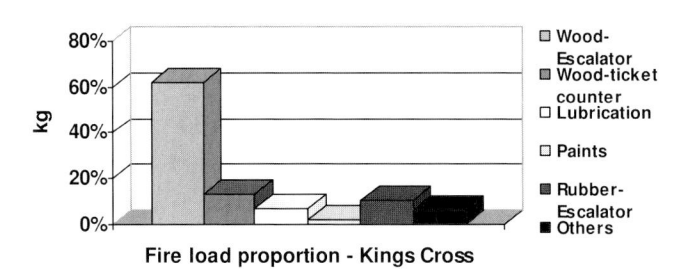

Figure 256c Fire load of the Kings Cross Fire

With the mineral-fiber insulation the fire-load comparison on the usually used weight basis in accordance with Fig. 99 led to wrong conclusions. If the volume-based material characteristics are compared and, furthermore, the area-related insulation effect duly considered, it is to be recognized that, as shown by Figs. 99a to c that other relations are relevant for the risk estimation. The incombustible alternative products do not obviously cause, in this particular case, any decrease of the fire load. The admissibility of the built-in Al-laminated EPS board products must be given due to its material-specific classification according to the testing and evaluation criteria of DIN 4102 valid at the time of planning and permission of the building construction. The testing technology was oriented on the handicaps of the supervision compartment consignment "Insulants tested for themselves". In Sect. 5.2.3 of the guidelines [821] it states: "At buildings, except for multistorey buildings insulants in the interior of external walls must be classified for themselves at least normally ignitable (class B2), at multistorey buildings at least difficult ignitable (class B1)" and in Sect. 7.1 it is called "Insulants below the roofing cladding must be classified for themselves normally ignitable (class B2)."

Prior to the former date, it was usual testing practice to dissolve coatings and covering foils and to subject the core material, the insulants themselves, to the testing modalities. The uncertainties in the existing theory that at the determination of the fire-performance classification of composites, sandwich elements with combustible foamed core materials are documented as well by the comments of Prager at the relevant committee (PA III) for combustible materials at the Institute for building technique (IfBt) [211, 212]. Prager illustrates through this that positive effects of incombustible linings like metal sheets, were confirmed in the risk-relevant fire trials in the scale 1:1 both in facade test runs and in the case of steel trapeze roof constructions at the TNO Delft [244]. These findings, confirmed by the later test runs at the FMPA Leipzig [243, 793, 801], were contrary to the predominant opinion of the relevant standardizing committees. The fear that easy ignitable core materials become a danger in the case of a fire in spite of the screening, protecting effect of covers and coatings, determined the former way of thinking. Possible negative effects of coverings and covers were part of the wording of the relevant standard, DIN 4102, but were at that time not subject to regulation. The built-in material, the Al-laminated EPS lining must be classified according to the former test and evaluation criteria, i.e. classified as class B1 and no burning dripping. The prosecution assumed that due to the hot work at the expansion joint, in accordance with Fig. 257, opened-up joint and came into contact with the elastic drainage bands below, which as a result of their material were rigidly touching in the middle of the joint. "As demonstrated in the fire test requested by the commission, this material was moderately flammable. They ignited and the heat energy released caused the neighboring polystyrene foam boards to melt. Further rigid foam boards detached themselves from the reinforced concrete ceiling. Liquid formed on the Al-facing. The adhesive bonding on the Al-facing facilitated the rapid spread of fire. The energy released caused the foam material in adjacent areas to melt and become detached from the reinforced concrete ceiling. Parts of the seams on the aluminum facing opened up and liquid polystyrene flowed/dripped onto the mineral-fiber insulation in the suspended ceiling, and burning droplets emerged through porous sections at irregular intervals into the arrival area level". This does not conform to the list of observations, there was no observation confined to the burning droplets. "These glowing particles moved with the wind" reported one of the witnesses to the court [822]. This is an indication that glowing particles of the binder or of dust instead of burning droplets were observed. The temporal course of the fire in the false ceiling/suspended ceiling can be estimated by means of the testimonies only very coarsely. It is undeniable that at 15^{47} a flashover in the false-ceiling field and a local collapse of the suspended ceiling with deflagration occurred. The deflagration was documented also by the smoke emerging above the joint. If one analyzes the temporal course of the fire, it started at 15^{25} with the ignition of the water-draining bands. The listings of Table 145 clarify the obvious inconsistencies. If one considers that according to the testimonies at 15^{45} a large glowing area of the mineral wool insulation of the false ceiling was increasing from $1\,m^2$ to $20\,m^2$/ca $100\,m^2$ within a short period of time, then it can be assumed that the chronological fire course accepted by the fire experts (see Fig. 258a) does not reflect the actual fire occurrence. Prager [817] clarifies that the fire development as it is assumed by the experts and the prosecution would have led to a free burning of the joint within the original stage of the fire and therefore also to an unhindered emerging of the "black" polystyrene smoke. But black smoke and rubber odor were noted, however, by different witnesses just at the time of the flashover situation at 15^{47}. The documented occurrence of light-colored smoke indicates a smoldering fire at the starting point. If one excludes a failure of electromechanical products than falling welding beads must have caused the smoldering fire. Prager et al. [817] assumed that the smoldering fire started and spread, as shown by Fig. 258b, within the mineral-fiber layer and was enhanced by dust deposits and cables. Melting polystyrene may have added later on to the developing flashover situation.

This interpretation became backed by the effectiveness of the smoke detectors. Provided that the installed smoke detectors respond only to large smoke particles, then it is understandable why the smoke detectors sounded first just after the flashover situation, when the polystyrene tiles become involved. If the polystyrene tiles had been involved at an earlier time of the fire occurrence, then the smoke detectors would have become effective obviously before the flashover situation. The smoke liberation of EPS, demonstrated by Klingelhöfer on the occasion of the symposium of the airport fire [823], documented indeed the fire-gas potential of EPS boards, but this large-scale trial did not simulate without doubt the suspended ceiling situation of the airport fire but simulated instead a pool-fire situation, which does not correspond to the real-fire disaster. This significant variation was neither mentioned nor discussed at the symposium. This fact clarifies the necessity of an exact simulation of the important fire parameters in large-scale field trials to obtain, on the one hand, meaningful results and, on the other hand, to avoid any misinterpretations by the regulatory people, who are not aware of the possible deviations. In the case of the third complex, the smoke liberation that was responsible for the death of the fire victims, the prosecution assumed that the CO concentration of the effluents released from the EPS boards was responsible for the fire victims. The HCN concentrations, determined within the framework of the autopsy investigations, (see Table 146) point at the participation of N-containing materials.

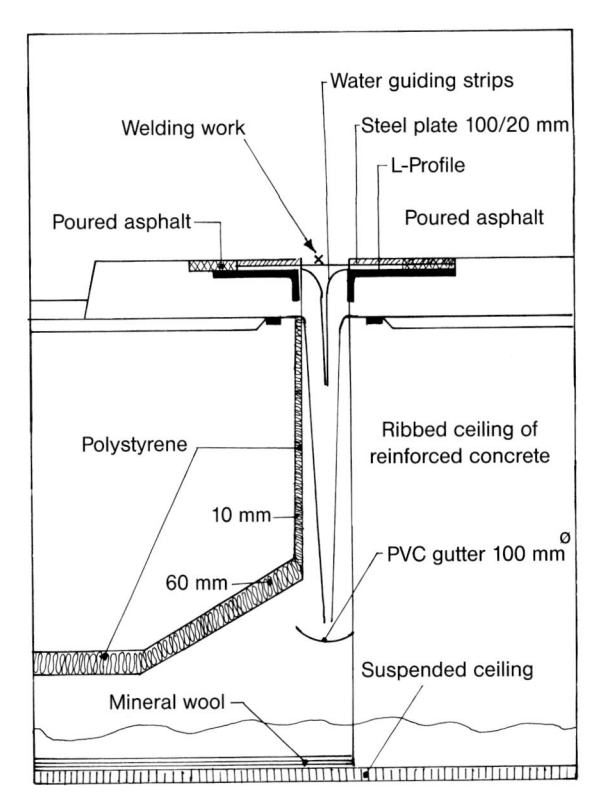

Figure 257 Joint of the ceiling construction of Düsseldorf Airport

Table 145 Observations on the course of the fire

Time	Observations
13^{00}	start of welding work around the expansion joint
1531	taxi driver reports sparks in the ceiling area at the flower shop
1533	individual glowing sparks observed
1535	considerable production of light-colored smoke around expansion joint
1540	electricians determine that the electrical equipment is not responsible for sparks
1543	smell of smoke, smoke visible at upper edge
1544	airport fire brigade orders welding work to stop
1545	smell of smoke from burning rubber
1547	fireman sees glowing fire in ceiling area – c.1 m^2, affected area increases rapidly to ca. 20 m^2, partial collapse
1550	deflagration around the expansion joint at workplace
1555	smoke escapes from suspended ceiling in lounge area
1556	smoke detectors triggered in underground level 1
1600	fire door closes
1605–1619	spread of smoke, calls for help from AF lounge and lifts
1700	fire victims recovered
1920	fire under control

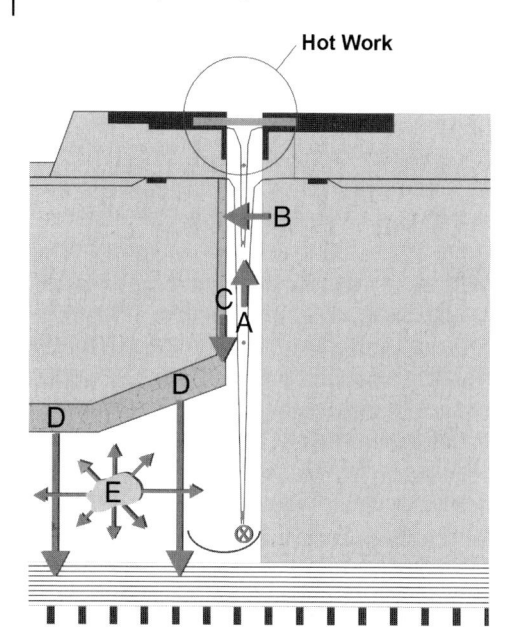

450 | *6 Use and Interpretation of PUR-Test Results Determined under Enduse Conditions*

Figure 258a Expert interpretation of the Düsseldorf airport fire

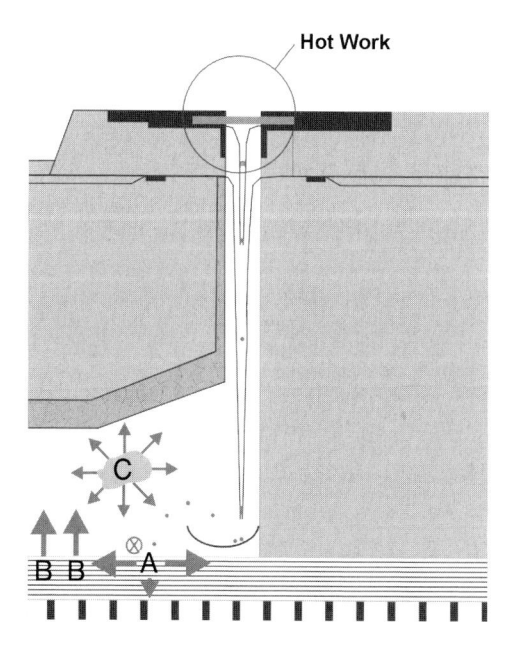

Figure 258b Interpretation of the airport fire by Prager

Table 146 Blood analysis

Place of victim	CO-Hb (%)	HCN (mg/l)
Stairwell lounge	82	0,6
Lift 41	50	0,4

HCN plays indeed, as confirmed by the witness Dalberg [824] in the Düsseldorf Airport Fire Case [1], a subordinate role only, because the CO intoxication caused the death of the fire victims already. This evidence is, however, an important indication that a N-containing material also participated in the CO liberation, possibly responsible to a not unimportant degree. A detailed analysis of the facts, observations as well as the relevant expert literature, shows very vividly that doubts exist in the interpretation of the prosecution. The EPS insulation and the PVC cable material can not be used for this. The mineral-fiber layer is the only N-containing material in the suspended-ceiling field. The relevant expert literature recognizes that depending on the decomposition temperature considerable parts of the effluent components CO and HCN are possibly released. Figures 210a and b illustrate the values published by the DAN test [208, 209]. Increasing decomposition temperatures lead to significant higher levels of CO and HCN concentration data. Babrauskas [825] reports on the characteristics of the toxic potential of insulants. These data records of Figs. 259a to c demonstrate that the mineral-fiber material contains, due to the binder content, a high toxic potential in comparison to EPS.

Doubts about an unprejudiced cause of fire investigation are certainly feasible due to the analysis. The demand of Schneider [826] to accept expert-reports only of qualified experts. He declared at the 5th anniversary of the Düsseldorf Airport Fire, that fire safety concepts presuppose a knowhow, that must be required by the nomination of experts. This demand is absolutely relevant for fields of safety. An aspect that both to heed with regard to the relevance of the interpretation of the public prosecution of the valid building, regulation orders or also the testing and evaluation criteria under enduse conditions.

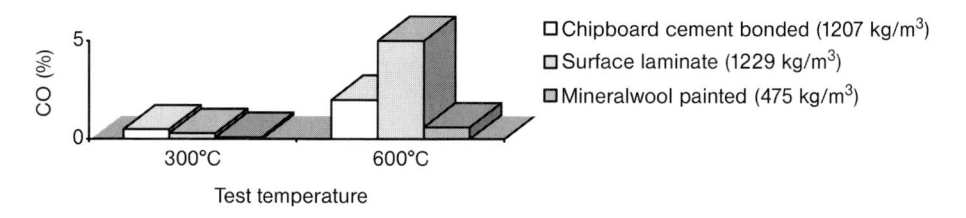

Figure 259a Toxic potential of mineral fibers

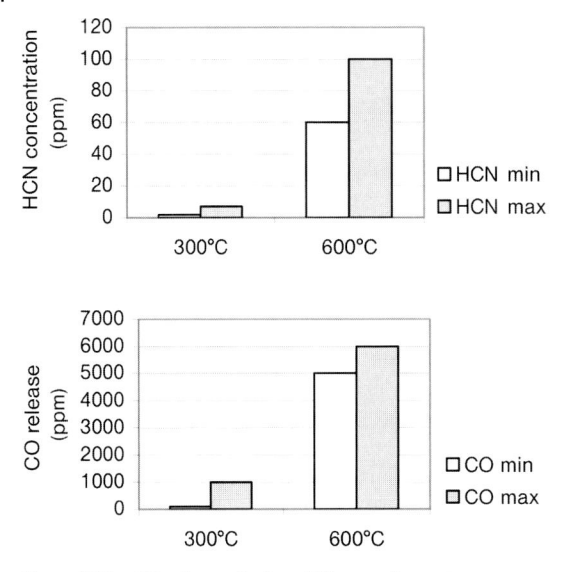

Figure 259b CO release of mineral fibers at thermal stress

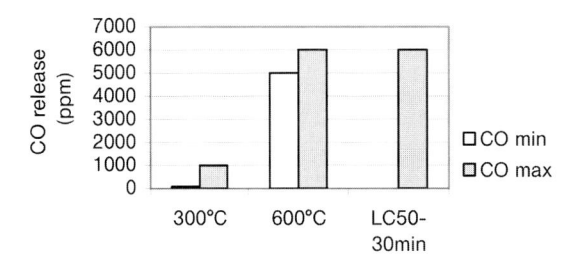

Figure 259c CO release in relation to weight loss

6.3.3
Orders Concerning the Risk of Ignition

The intention of all legal requirements concerning this important issue of preventive fire precaution is the aim to guarantee a sufficient protection against fire hazard according to the basic requirements laid down in Table 1. The goal to establish an acceptable resistance against ignition sources of the daily life, e.g. glowing particles or flames of variable levels of intensity (match-type flames, candle flames or torch flames) is a constituent part of these legal requirements in the different application areas. Experimental results have shown that pure material-specific characteristics do not guarantee without exceptions the targeted protection destination. The risk of ignition by ignition sources of small intensities can be significantly reduced by flame-retardant additives. This effect is without doubt relevant for match-type flames. [827–829]. Experience of the worldwide fire-statistics data has shown that match-type flames play a minor role as ignition sources of today in comparison to smoldering cigarettes. Risk-related cigarette tests have demonstrated that the use of flame-retardant additives for cellulosics obviously does not reduce the risk of a smoldering fire of such composites. On the other hand, such flame-retardant additives doubtless increase the risk of a possible medical damage by the effluents [830–833]. The B1 classification (DIN 4102) or the 25 rating (ASTM –E84) are principally a quality label of the fire performance of building products [834, 835]. Whether the requirement of B1 or rating 25 guarantees a higher level of safety, is a meaningful question. Do we need official certificates or is the self-certification of the industry, similar to the UFAC procedure in the United States of America, an acceptable alternative? Such kinds of official directives are a perquisite for the motto "Competence and Responsibility". B1-classified fabrics had, e.g. been required for upholstery seatings in smoker cinemas by law [764, 765]. The B1 classification, which is quite meaningful, e.g. for fire curtains in theatres or cinemas, is obviously for fire experts meaningless for upholstery furniture. In the late 1970s, a B1 classification was gained for CMHR qualities. Since the research work failed to improve this positive rating for sample thickness larger than 20 mm, these certificates were returned to the PA III by the Bayer AG, to avoid any misinterpretation by merchandise actions. The standardized classification "noncombustible" according to the 750 °C furnace test as well will not guarantee a safe unlimited use of this material. Various studies have demonstrated that the incombustible classified Rockwool can, on the one hand, initiate in connection with the chemical nature and the concentration of the binder a self-sustained smoldering process and can provide a contribution to the surface spread of flames due to the surface treatment (see Figs. 18a to c) [1, 68, 322]. Pure material-specific characteristics do not guarantee without exception the targeted protection destination. The positive ranking in the 750 °C furnace test does not reflect neither the possible tendency to a selfsustained smoldering process nor the possible failing. The rockwool boards classified in the furnace test as incombustible break down obviously as shown by Fig. 260, possibly in subordinate testing methods for incombustible materials.

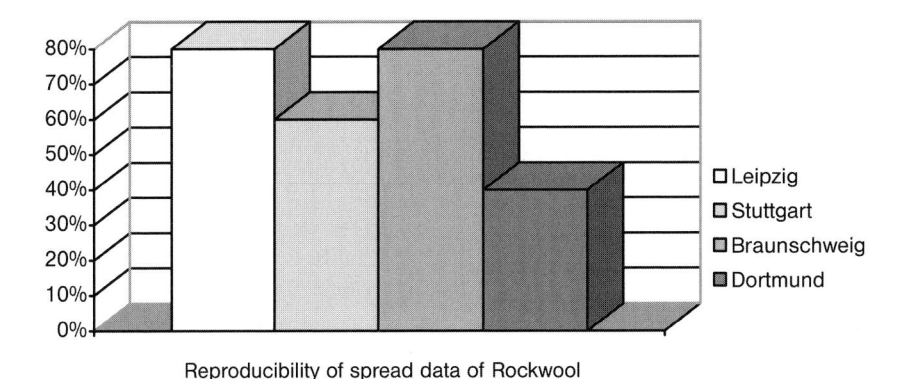

Figure 260 Control of the A2 classification of Rockwool material

Furthermore, it is to be considered that the fire-protection classification in the furnace test is not able to evaluate comprehensively possible fire-protection deviations of the classification caused by coatings, paintings or lamination, i.e. the built-in state. The experience has shown that there is possibly a severe influence of coatings or laminations on the development and spreading of fires [836–845]. The possible danger remains unconsidered furthermore, for rockwool through the scientifically verified smoldering-fire tendency that depends on the kind and amount of binder. An analogous discrepancy to the targeted protection destination exists also at the demand of the insurance industry. The standard VdS 2211 is an example. Section 41 Roofing materials and roofing constructions gives the advice: "Materials tending to glowing (the testing report and/or admission) are supposed not to be set in as materials in roof cavities". It has to be taken into account that, on the one hand, specific test methods were not standardized and, however, on the other hand, the existing, extensive scientific studies concerning the smoldering fire risk with this addressed protection destination are obviously not realized. From the viewpoint of the preventive fire protection a completion of the fire-performance examination would be welcomed with regard to the fire risks through hot works. The expert report of Jagfeld [846] demonstrated that rockwool insulation boards tend to smolder under the DIN 4102 roof test conditions as well. The danger of welding beads as an ignition source was set under evidence further through the Düsseldorf airport disaster. Regulative orders exist only in the United States of America [847]. Insulants on cellulosic basis that are currently favored also in Germany as nonpolluting alternatives, according to Fig. 84, must be judged in the USA with regard to the possible igniting risk by means of a cigarette test as well as concerning the possible contribution to the surface spread in the modified Radiant Flux Test according to Fig. 41c [848]. The listings of Table 145 show that the permissible limited flame spread is always a component of the different nationally and internationally stipulated requirement profiles. Only very specific test procedures, such es e.g. the Motor Vehicle Safety Standard according to Fig. 72 or the small burner torch test according to DIN 4102 (see Fig. 49a) stipulated after defined time space an unlimited flame spread. The precise knowledge of the application-oriented igniting risks and the present environmental impacts must be assumed for risk-relevant orders.

Resistance against the impact of ignition sources of low intensity
The requirement profiles concerning the resistance against glowing cigarettes are, as documented by the various national fire statistic records, in accordance with real life risk situations. While for the American requirements [848] a risk-relevant covering and variations of the ignition source application find sufficiently observation, these findings remain unconsidered, for example, at the relevant ISO and CEN standards due to the small basic studies at BSI. A detailed analysis of the fire statistics show very vividly that the cushion compound and the modified ignition source location in accordance with Fig. 140b can lead, e.g. to significantly modified results. The requirement profiles do not consider, as a rule, either this finding or the fact that a sufficiently secure rating of the tested upholstery combination is certainly not guaranteed with the standardized "two" test samples required by the relevant ISO and

CEN standards only. Modified, more economical test parameters would allow, e.g. smaller test pieces, or more than one test per sample and therefore also a more risk-relevant evaluation. The data sets of Tables 60a and b provide information about the influence of various test parameters concerning the nature and the construction of the composite on the cigarette test results with various upholstery combinations. The relevance of the cigarette test has been investigated in various studies [849–853]. The proof of the resistance against burning matches is part of all essential, national preventive fire-performance requirement profiles. The match flame as a usual ignition source of daily life is simulated in numerous national and international specifications and standards by means of gas flames. Both in the building section and in the furniture and furnishing field as well as the transportation sector and in the electromechanical engineering field the simulation with a burning match in accordance with Fig. 142b is striven for. The listing of Table 149 illustrates the extent of the deviations of the different requirement profiles. The existing deviations are obviously a result of the different evaluation criteria. Detailed analysis of the present findings about the combustion of matches provided finally that the test duration of 20 s, fixed in the English Furniture Regulation and was stipulated at CEN on 15 s. The requirement profiles are based altogether on a limited contribution to the flame spread. The definition of the material classification B2 according to DIN 4102 forms the only exception. The proof of the normal flammability class B2 means, by definition, a limited contribution to the flame spread in the stream of effluents within 20 s including 15 s impingement time. After the 20th second an unlimited spread of flames across the surface becomes permissible by definition. Figure 255d illustrates the importance of the burning-match ignition source in relation to the category negligent arson through children playing.

Resistance against ignition sources of raised intensity
The requirement profiles in the transportation field and for electromechanical products also mainly support bigger flames (see Table 110 and Fig. 145). The failure of electromechanical products is, as the data records of the fire statistics show very vividly, the starting point of many fire cases. The requirement profiles are supported by testing assessment on the basis of national and international specifications. The decisive evaluation criteria are based on material-specific and device-particular testing methods that are supposed to guarantee that, e.g. the electromechanical product does not become, in the case of a fire by burning droplets, a secondary ignition source to the environment. The requirement profiles are mainly oriented for electrical products on material-specific characteristics [854–858]. In exemplary fashion for this are, in particular, the insurance technical rating as they are distributed and brought to validity worldwide by the Underwriters Laboratory (UL). Material-related characteristics facilitate this without the stipulation of classification tests or the more low-cost UL control system. Device-specific demands as they have been intensively discussed and partially standardized at the relevant IEC committees are, for this purpose, comparably of less importance. Device-specific fire-performance requirements require naturally a considerably higher knowledge of the testers, controllers and admission committees. The resistance against ignition sources of raised intensity

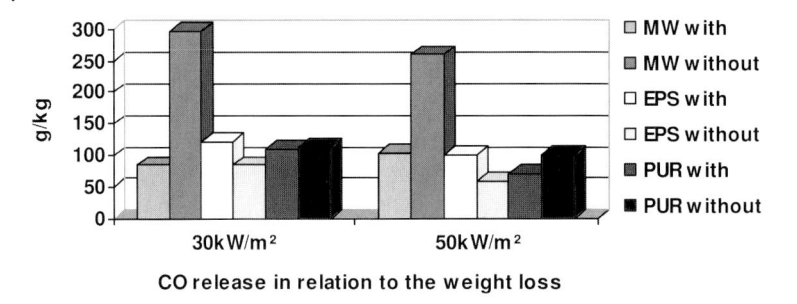

Figure 261 Toxic potency of effluents of steel sandwich elements with and without facings

(CO release in relation to the weight loss, at 30 kW/m² and 50 kW/m², for MW with, MW without, EPS with, EPS without, PUR with, PUR without)

like candle flames, burning paper or blow pipes and burning items hedges the possible igniting risks. Standardized ignition sources of raised intensity are simulated, last but not least, due to a better reproducibility of the thermal loading by gas burners. The listing of Table 108, 110, and 111 illustrates the bandwidth of the energy supply of these different standardized ignition sources. Partially serious differences exist at the ignition sources of raised intensity. Examples are, e.g. the NIST burner, according to Fig. 145c in the furniture field and the sand-bed burner, according to Fig. 58f, in the simulation of a wastepaper basket fire, e.g. for the SBI test procedure and the sand-bed burner according to Fig. 58f, in the ISO Room-corner test. The targeted protection destination should be exactly defined at the regulative orders and the scale of possible deviations clarified there to possible modifications and operation variation of the test parameters. The general goal of these test protocols is still the limited contribution to flame spread [859, 860, 861].

Orders concerning the characteristic burning dripping
The requirement profiles concerning the side effect burning dripping are in many cases of different application fields a component of the fire-performance classification (see Table 114). These requirement profiles guarantee that materials that give off burning drops in the case of a fire, will not or only partially come to application in specific application areas. The different conversion of fire experiences into the respective national system of rules clarifies the comparison of the German Guidelines about the use of combustible materials in the building construction with the original guidelines of the regional (cantonal) fire assurances of Switzerland. In the first edition of the guidelines of the regional fire assurances of Switzerland [208] the test and evaluation criteria of the side effect burning dripping were a substantial component. This old Swiss testing and evaluation criterion has been the starting point for the different analogous national requirements, among other things, for those of DIN 4102 also. Due to the positive experiences during the evaluation of current fire cases that indicated that this aspect of fire-performance examination and evaluation of materials was of no significant value, this aspect was deleted on the revision of the Swiss Guidelines. The different imitators in the various standardization working groups without exception did not follow this consistent step. The German Guidelines for the use of combustible materials in the construction field are a classical example of this. During the repeatedly carried out revisions of the require-

ment profiles efforts were concentrated altogether on the perfectionism of the test-oriented procurement and not on the conversion of experiences from fires and realistic large-scale field trials with facade linings, glazings and roofing materials. Examples are the large-scale test with insulated PVC faccade elements as well as the facade test with horizontally and vertically arranged PC glazing elements on the subject of burning dripping [547, 553, 862]. The obvious weak points of the relevant standardized testing and evaluation criteria of DIN 4102 were not eliminated. How much, e.g. the expert knowledge of the legislator is a necessary condition in order to stipulate risk-relevant requirements clarifies the relevant demands in the guidelines about the use of combustible material in the construction field in Germany [863]. In the relevant building codes of, e.g. NW [864, 865], the relevant demands are as follows:

- 5.2.1 In buildings with more than 1 storey, materials that give off burning droplets are not allowed
- 5.4 Lining of materials that may give off burning droplets are not admissible
- 6.2 Linings of ceilings can be manufactured from at least normally ignitable materials class B2, provided they do not give off burning droplets.

The latter requirement 6.2 demonstrates the nonexistent relation to the aspired goal of protection against the danger of burning dripping. According to the standardized definition the B2-classified material can produce "burning rain" after the 20-s benchmark. The relevant material classification is to be carried out with the defined test protocols (chimney test and small-burner test) for difficult to ignite (class B1) materials and normally ignitable (class B2) materials in accordance with DIN 4102 [866, 867]. After that, the following state of affairs exists for class B2. If the tested materials ignite by burning drops the subordinate tissue paper within the 20-s match flame test, the impinging time of 15 s included, it is classified as "burning dripping". If the ignition of the tissue paper occurs, however, just after the 20th second, then the material will be classified by definition as "not burning with flaming droplets". This technical test classification means without doubt a quite different protection definition as worded in the guidelines to the building regulation.

In the Epiradiateur Test an automatic down-grading into the most unfavorable class M4 is carried out on proof of burning dripping in the special test procedure according to Fig. 54d. In the transportation field as well as for electrical appliances this side effect of a fire is used in order to limit the admissibility. The requirement profile including the side effect burning dripping becomes in practice predominant for electromechanical products in accordance with UL subject 94 analogous to Fig. 170 in order to control the admissibility of materials. This specific quality classification is carried out among other things also in the transportation field for railway vehicles. The afterburn time of burning droplets forms predominating the basis of the individual rating. The listing of Fig. 80a–b and c is to be interpreted in such a way that only there, when the energy liberation of the burning droplets shows a higher energy potential as the torch flames coming in the appropriate environment to the effect for the material classification, does an assessment of the danger seem to be justified [80, 81, 868]. Only when the burning dripping from electromechanical products becomes effective as a primary ignition source, and only then, such requirement profiles appear justified from the viewpoint of the preventive fire protection.

6.3.4
Demands Concerning the Heat and Fire-Gas Liberation

The different national requirement profiles concerning the heat and fire-gas release are mainly component of the fire-performance classification of building products and materials. With some exceptions the limited fire spread forms the basis of the judgement/evaluation altogether.

Heat-release rate
In the case of the predominant number of requirement profiles the heat release is regularized by means of the limitation of the fire-gas temperature level in the combustibility testing. Benjamin [869] gives an overview of the differences of the relevant studies in this connection. The heat release is influenced by the extent of the surface spread of flame. As a rule parallel measurement occurs accordingly to the relevant building products/materials test methods as, e.g. the chimney test according to DIN 4102 and/or DIN 5510, the Epiradiateur Test according to NF-P92-501, the Tunnel test according to ASTM-E84 or the Early Fire Hazard Test according to AS 1530. The graphical representations of Fig. 60b provide a view of the assignment of the indices to the heat-release rate in (kJ/m^2) according to AS 1530. Area-related data are defined, among other things, in the Nord test NT-004, in the Vlam Overslag Test according to NEN 3883 as well as in the fire-propagation test according to JIS A1321 being based on BS 476 pt. 6. These determined test results clarify the permissible deviations of the material classes A, B and C.

In the course of the European Harmonisation a classification of the heat release was proposed according to Table 27a based on the Cone Calorimeter Protocol. Also, in the SBI test procedure the fire-gas temperature levels are components of the targeted European material classification, as illustrated by the listing of Table 149. All these procedures supply, with the exception of the Early Fire Hazard Test, no material-specific potential values. With the Early Fire Hazard Test the measured data will be possibly significantly modified through the upright position of the test specimen; possibly caused by the sample orientation, e.g. through different enamel behavior. These material data are without exception only valid for the simulated fire scenario. In the A2 classification in accordance with DIN 4102 alternatively to the modified 750 °C furnace test procedure the possible heat-release contribution will be required by the boundary data $\leq$ 4200 kJ/kg (Hu – DIN 51900) as well as $\leq$16 800 kJ/m^2 (DIN 4102 pt. 8 ETK demand, 30 min). In France the heat release of noncombustible classified materials (class M0) is determined likewise by the permissible net calorific value. The permissible heat release is limited on 2500 kJ/kg. In addition the criteria of the building material class M1 are to be verified. In the furniture sector, as far as the building material testing methods are valid, these boundary effluent temperature levels are to be considered. Examples are, among other things, the evidence of class B1 according to DIN 4102 for upholstery outer coverings according to the VO of NW [765, 766] (Regulation of Assembly Rooms).

The demand of VdS 2216 "basically fire loads in the roof should be hold as small as possible" does not correspond to reality. The use of insulants on an inorganic basis is to be striven for, with priority instead of insulants of organic basis (e.g. rigid foams). Though the usual net calorific value assessment on a weight basis is incomprehensibly overlooked that at the risk-relevant evaluation, in particular considering the necessary insulation effect, the incombustible rockwool offers obviously no advantage in this connection. Figures 99a to c demonstrate that under realistic, comparable test conditions rockwool boards have a comparable fire-load potential [322, 872]. Exceptions include, among other things, the mattress test according to CAL TB 121 for high safety areas. In this examination in accordance with Fig. 60b the fire-gas temperature level must not exceed the boundary value of 260 °C. In the upholstery seat examination in accordance with the FAA requirement the fire-gas temperature level also determines, among other things, the admissibility of the upholstery combination. For large area units like wall-lining materials, boundary values have been fixed by the FAA which limit the scale of the heat release. These definitions suppress the use of materials that exceed, during the testing of test pieces with outer dimensions of 150 mm × 150 mm in the OSU Chamber according to Fig. 73b under the heat stress of 35 kW/m², the fixed boundary values of 65 kW/m² and 65 kW min/m² at a testing time of 5 min. Some deficits exist in comparison to the cone-calorimeter examination due to the vertical orientation of the test piece.

Smoke formation
The protection destination to be covered by the requirement profiles becomes defined through the basic demands of the preventive fire protection "Rescue and extinguishing actions must be guaranteed". The visibility-reducing effect of the resulting thermal decomposition products in the case of a fire is the reason that this is not guaranteed [871]. The determinations for the material class A2 are in this connection a very vivid example. The graphical representations of Fig. 262 clarify the existing differences in the requirement profiles for the burning with flames and the smoldering fire. A material specific value of 2.5 m range of vision was settled on for the burning with flames in the XP2 chamber and for the smoldering fire a minimum range of vision of 4.0 m was agreed upon by means of DIN equipment, which has to be verified in the smoldering temperature range of 250 °C to 550 °C DIN 53436/37. Although the range of vision reduction is the dominating danger parameter in the opinion of the preventive fire protection it is the fire-gas toxicity evaluation that must be determined according to Fig. 263 with a 30-fold concentration of the thermal decomposition products [502, 836]. Ries [872] required at the VdS seminar "air port fire Düsseldorf-5 years later" with regard to the possible smoke danger in the case of a fire, e.g. at tunnel fires, to define the new protection targets. The possibility for the self-rescue must be, in his opinion, more clearly conceded, i.e. the visibility-reducing effect has to be risk-related classified of decisive importance for the perception of the escape possibilities and the realization of rescue operations is the scale of reduction of the visibility.

The data records of Tables 144a and b give a view of the different national smoke-density requirements for building products and materials. In the case of a more detailed consideration of these values it appears that the selected classifications are neither able to provide information to make escape ways sufficiently safe nor to guarantee the length of time necessary for the perception of the rescue possibilities.

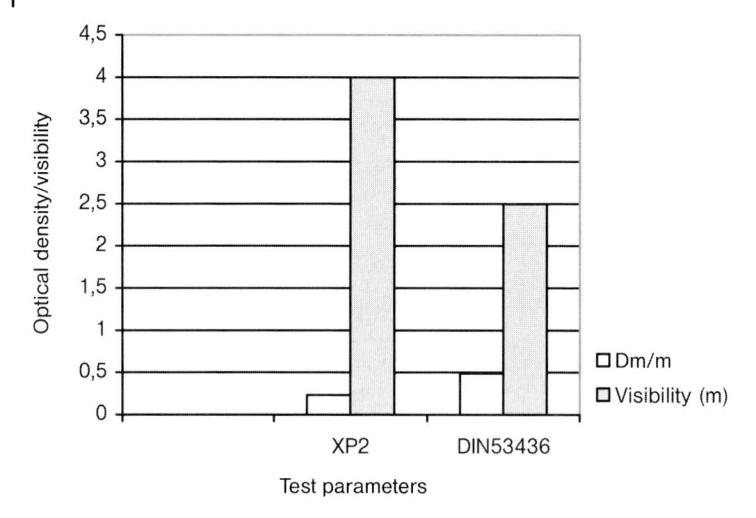

Figure 262 Borderline values of the optical density of the effluents of A2-classified building products

Tables 144a and b provides the equivalent values of the optical density D as well as the corresponding ranges of vision compared with the fixed classification values [873]. In the comparison of the classification systems it appears that the reference of the classification to the reduction of visibility was very different. As single-point data they are not suitable without doubt to be used as boundary values for security reasons. Since the underlying combustion systems can simulate altogether only a completely specific fire scenario in the developing stage of a fire, these material-specific keys are not suitable for computational estimates of the possible visibility reduction in the case of a fire as base values. Also a specific smoke-densities test protocol, like the XP2 chamber, does not supply, due to the general conditions, any material-specific potential values. The listing of Table 149 provides a view of the possible smoke-concentration records in relation to the FAA requirements.

Table 149 Deviations of the smoke requirements FAA

Requirements Ds	Requirement profile	Optical density D (1/m)	Visibility S (m)
15	Within 15 min for mains / cables	0,19	4,5
100	Within 90 s/ 4 min for textiles	1,263	1,2
200	Within 90 s and 4 min for upholstery	2,525	0,6

Ds will be determined by the mathematical relationship $Ds = \dfrac{V}{L \cdot A} \cdot D$, at what

Ds = the specific optical density
V = the chamber volume (m³)
L = the light path (m)
A = the surface area subjected to the thermal stress (m²)
D = optical density value at the photometer

The above testimonies have also validity for the record "% min". The listing of Tables 144a and b clarifies that the grading carried out on the basis of the data records % min, do not really provide any risk-relevant differences concerning the possible reduction of visibility in the case of a fire. The % min records were calculated by assuming a constant transmission (T-value) and multiplying by the test duration. With such a medium T value the corresponding numerical value of the optical density D, by means of the respective light path length in 1/m, as well as the range of visibility S were determined. The permissible boundary values in the Critical Flux Test of 300% min (Germany) as well as 250% min and 400% min (Switzerland) will provide under the assumption of an approximately constant stream of effluents and a testing period of 10 min an optical density D of 0.25/m and/or a range of vision S of 2 m. The analogous permissible value in the railroad field were agreed in Germany in accordance with DIN 5510 with class SR1 = 100% min and SR2 = 50% min. At observance of the stream of air of 11 m^3/h that is 0.55 m^3/3 min at a test duration of 3 min, and the test-piece volume, these boundary values indicate an optical density and a range of visibility reduction $D = 1.5$ and $D = 0.77$ as well as $S = 1$ m and $S = 1.8$ m. In the UIC classification system Kaminski [840] favors the classes 1 to 3 with boundary values of $T1 > 300\%$ min, $T2 = 150–300\%$ min and $T3 < 150\%$ min. Table 149 clarifies the requirement profiles ordered by the FAA that are valid worldwide and were defined on the basis of relevant companies specifications [271, 272, 874].

For the time being these requirement profiles are valid also for the naval field [843], as far as no risk-relevant test and evaluation criteria are agreed by the relevant ISO committees (ISO TC 61, ISO TC 92). The listed records of Table 144 demonstrate deviations of the range of visibility under use of the characteristics of the optical density an estimation of the possible reduction of visibility can occur in the case of a fire risk relevantly. These examples confirm the experiences from fires that the reduction of visibility is by far the greater danger of smoke since this effect arises, contrary to the toxic effect, immediately without temporal delay.

The different national requirement profiles concerning the fire-gas toxicity in the case of a fire in the different fields of application are based both on pure analytically determined concentration data of essential fire-gas components and animal-

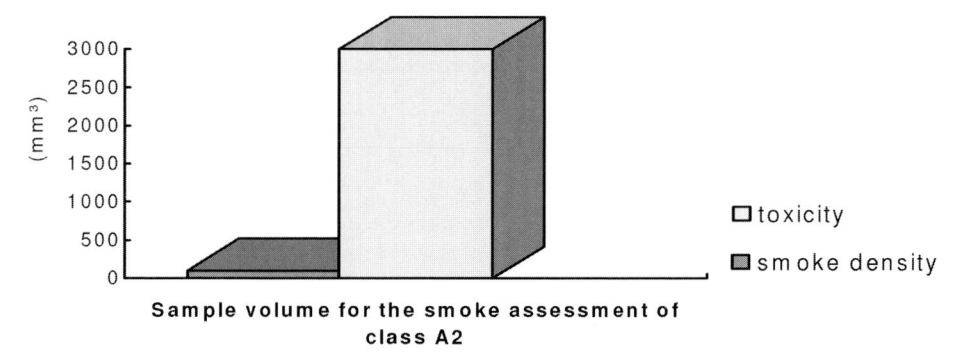

Figure 263 a Smoke requirements class A2

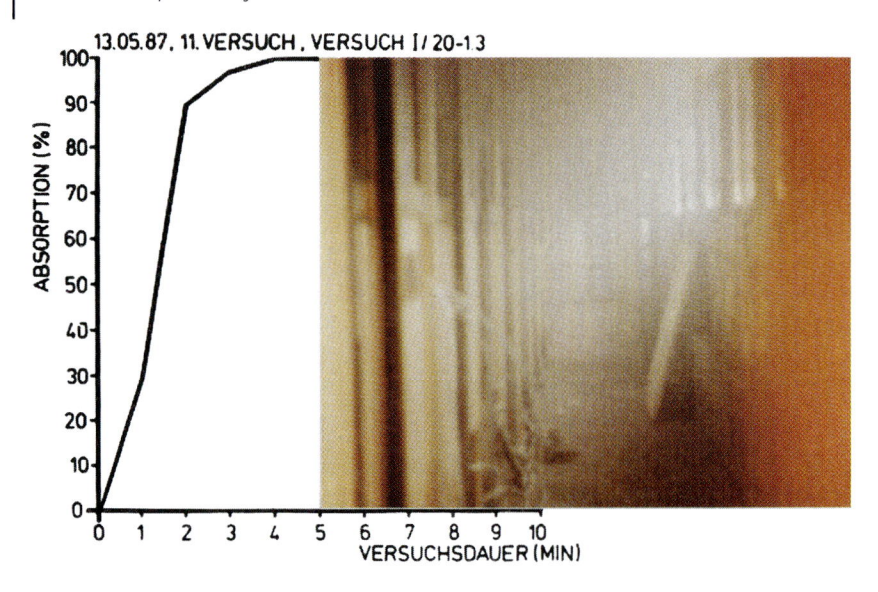

Figure 263b Smoke density records in the cabin in relation to the visibility in the gangway

experimental investigation results. Orders concerning the acute fire-gas toxicity of the thermal decomposition products were stipulated as an emphatically high priority in the building sector. The natural product wood was, in the majority of the requirement profiles, defined as reference material. The findings of the studies of the SC3 of TC 92 Toxic Hazard in Fire illustrate that the determination of maximal-permissible concentration data of essential flue-gas components as classification parameters do not guarantee, as a rule, sufficient safety against possible fire-gas dangers in the case of a fire nor creates the prerequisite for material-related potential values as a basis for danger estimates. Figure 263b demonstrates the relation of the smoke density in the cabin and the visibility in the rescue way [81]. The danger of misinterpretation of material-specific potential characteristics of the toxic potential illustrates the French-Nitrogen-Chlorine-Law (see Fig. 264, as a result of the fire disasters of Saint Laurent du Pont [626] and the crash of the Varig plane 1971 near Paris [630]. In accordance with this law, which was inserted as a priority measure at the EG [876], the nitrogen and chlorine content is to be determined in the stream of effluents produced at 700 °C with synthetic products. With reference to the concentration data that cause death within 15 min, the permissible boundary values were defined as 5 mg/m^3 for nitrogen and 25 mg/m^3 in the encircled room for chlorine. These values were based on the assumption that only about 10 % of the nitrogen and/or chlorine content are participating immediately in the fire occurrence, i.e. the fire-gas liberation. Neither the application-oriented modifications in accordance with Fig. 264 nor the quoted calculated examples obscure the fact that the French legislator chose a wrong approach. The missing relation to the practice becomes in particular manifested through the fact that the proven dominance of the fire-gas component

CO in the fire case was not taken into account during the definition of the law text. The numerical values are to be interpreted in such a way that one wanted to complicate presumably primarily the application of specific material groups and/or to suppress the application, i.e. striving for a political solution. Under the assumption that, e.g. a blended fabric 350 g/m^2 consisting of 25.1 parts of acrylic fibers and 74.9 parts of wool, then there is only 1/3 of the existing HCN potential danger to be taken into account, because the remaining 2/3 parts given off by the wool does not fall in the range of this law. A risk-relevant assessment of the possible fire-gas danger will therefore not be possible.

The requirement profiles for incombustible materials of class A2 in accordance with DIN 4102 are based on the expert evaluation by means of animal-experimental investigation [877, 878]. The requirement profile was agreed upon in the 1960s by representatives of the construction supervision under cooperation with toxicologists and representatives of the scientific and economic fields as follows: The thermal decomposition products produced at a reference body temperature of 300 °C and 400 °C with the DIN equipment in the stream of air of 100 l/h diluted to a total airstream of 300 l/h, induce within a 60-min exposure and an after observation time of 14 days no deaths. The medium CO-Hb content in the blood of the exposed animals must not exceed 35 %, considering the increased danger for children, and no toxicological objectives of other kinds must exist [879, 880]. Prager referred in Würzburg [836] to obviously weak points of these fixed requirements for A2-classified materials. The fire disaster at the Düsseldorf airport confirmed that the defined fire-performance requirements of class A2 materials are not sufficient for the aspired aims. The graphical representations of Figs. 262 and 263 clarify that the fixation of the test temperature to 300 °C and 400 °C is not able to meet the variety of real fires. Indeed the smoldering fire stage was recognized as a dominating danger situation, but the spectrum of the possible smoldering fire was not estimated correctly. After recent studies [97, 322], the self-sustained smoldering fires of rockwool reach, depending on the binder and as demonstrated by Fig. 18 considerably higher tem-

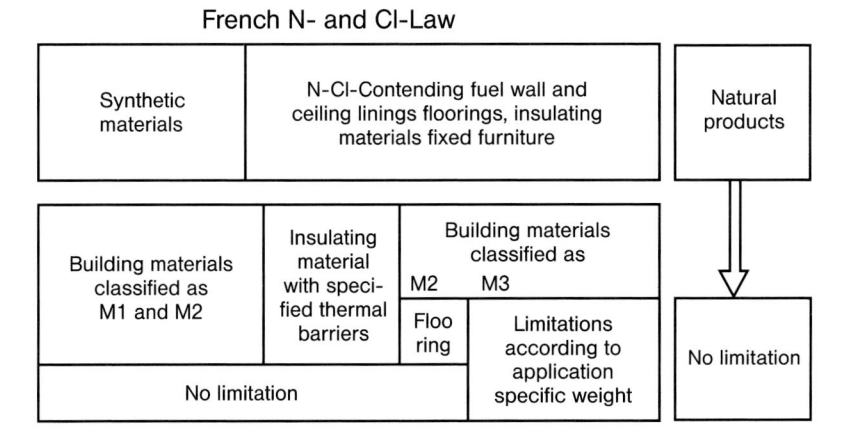

Figure 264 French Nitrogen-Chlorine Law

perature levels. Relevant research studies illustrated that considerably greater concentration of the relevant fire-gas components, in this case CO and HCN, are released at the higher test temperatures [708, 709]. It is to be assumed that these requirement profiles for these incombustible classified materials have been agreed primarily out of politically motivated reasons and only secondly from safety-conscious considerations. The transfer of these testing and evaluation criteria onto combustible materials becomes at least in the Düsseldorf airport fire, required repeatedly.

Within the framework of a workshop of the relevant German standardizing working group for the determination of thermal decomposition products and their toxicological examination could, using realistic examples, convincingly demonstrate that B1, difficult ignitable, classified material, provided that they show an appropriate behavior also in the current fire case under the relevant spatial relationship, do not produce risk-relevant fire-gas concentrations [125]. Prager and Sand demonstrated by Figs. 265a and b the circumstances [882].

Example 1: 1-mm thick wall panelling of PVC (density 1.25 kg/m^2)

If the danger estimate uses a LC_{50} value of 17 g/m^3 as a basis, this means taking a room volume of 400 m^3 and an exposure time of 10 min in accordance with Haber's Rule, an equivalent concentration requires a burned mass of approx. 21 kg, i.e. about 16.5 m^2. Such an extent of the surface spread under the action of a burning wastepaper basket can not be assumed for a B1-classified PVC lining.

Example 2: Wool carpet, 9.7 mm thick (about 2.7 kg/m^2)

For a B1-classified flooring material a maximal area of ca. 0.4 × 1.7 m^2 will be destroyed, i.e. a maximal fire load of 1.08 kg. If one bases the danger estimation on a LC_{50} value of ca. 7 g/m^3, the maximal dangerous fire load of the endangered room of ca. 400 m^3 will be ca. 5.7 kg. To produce an equivalent concentration, approx. 21 m^2 of the floor covering must participate in the fire occurrence. Taking into account as well an exposure time of 10 min, the burning area needed will be, according to Haber's Rule, approx. 6.3 m^2, which can not be expected with a B1-classified floor covering under the impact of a burning wastepaper basket (see Fig. 265b). Provided the products behave in the actual current fire situation like a B1, difficult to ignite material, than a primary flue-gas danger can be excluded. If the 10-min data are used as a basis for the danger estimate for the different fire scenarios as in cabins of ships and railway carriages, small places like kitchen and other residential rooms, cockpits, ship parlours and assembly places, then equivalent flue-gas concentrations can be estimated with regard to the different LC_{50} material-related data for an exposure time of 10 min.

These values clarify that for a crucial danger situation a disproportionately large amount would be necessary. Starting from the material-specific potential value a concentration value is estimated for the endangered volume considering the test amount participating in the fire occurrence. Provided that within a given time and spatially limited burning is guaranteed, there will be, as a rule, no dangerous concentration. The data records clarify by means of some examples that only under spatially cramped conditions, for example the cockpit, ship or railway cabins, crucial circumstances are possible. Tables 150 and 151 provide an overview of the relevant requirement records. In the furniture field a relevant requirement profile was deter-

6.3 Relevance of the Requirement Profiles

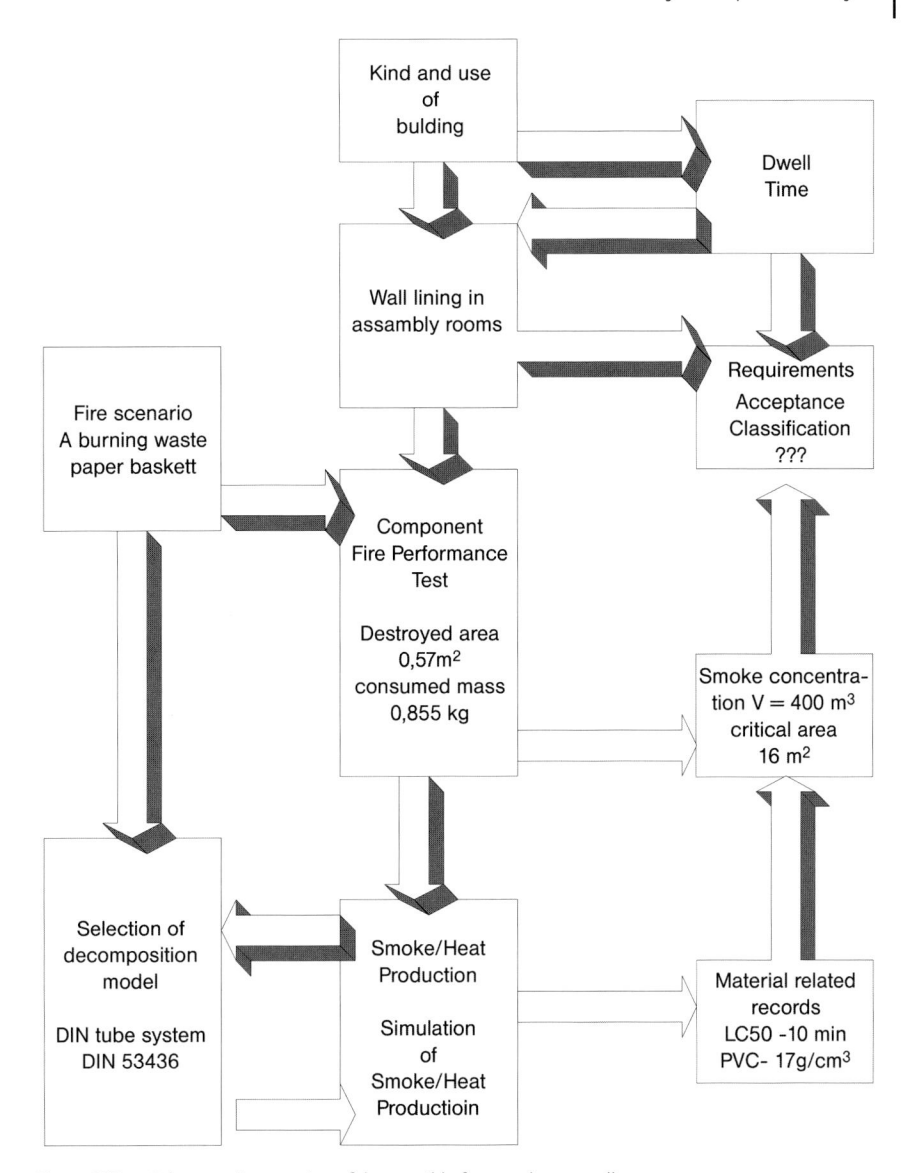

```
                            ┌─────────────────┐
                            │  Kind and use   │
                            │       of        │
                            │    bulding      │
                            └────────┬────────┘
                                     │         ┌──────────┐
                                     ├────────▶│  Dwell   │
                                     │         │   Time   │
                                     ▼         └──────────┘
                            ┌─────────────────┐
                            │  Wall lining in │
                            │ assambly rooms  │
                            └────────┬────────┘
                                     │         ┌──────────────┐
                                     │         │ Requirements │
  ┌──────────────┐                   │         │  Acceptance  │
  │Fire scenario │                   │         │Classification│
  │A burning waste│──────┐           │         │     ???      │
  │paper baskett │      │            ▼         └──────────────┘
  └──────┬───────┘      │    ┌─────────────────┐
         │              └───▶│    Component    │
         │                   │ Fire Performance│
         │                   │      Test       │
         │                   │                 │
         │                   │ Destroyed area  │
         │                   │    0,57m²       │
         │                   │ consumed mass   │
         │                   │    0,855 kg     │    ┌──────────────────┐
         │                   └────────┬────────┘    │ Smoke concentra- │
         │                            ├────────────▶│  tion V = 400 m³ │
         │                            │             │   critical area  │
         │                            │             │      16 m²       │
         ▼                            ▼             └──────────────────┘
  ┌──────────────┐           ┌─────────────────┐
  │ Selection of │◀──────────│   Smoke/Heat    │
  │decomposition │           │   Production    │
  │    model     │           │                 │    ┌──────────────────┐
  │              │           │   Simulation    │───▶│ Material related │
  │DIN tube system│          │       of        │    │     records      │
  │  DIN 53436   │           │   Smoke/Heat    │    │   LC50 -10 min   │
  │              │──────────▶│   Productioin   │    │   PVC- 17g/cm³   │
  └──────────────┘           └─────────────────┘    └──────────────────┘
```

Figure 265 a Scheme – Assumption of the possible fire-gas danger wall lining

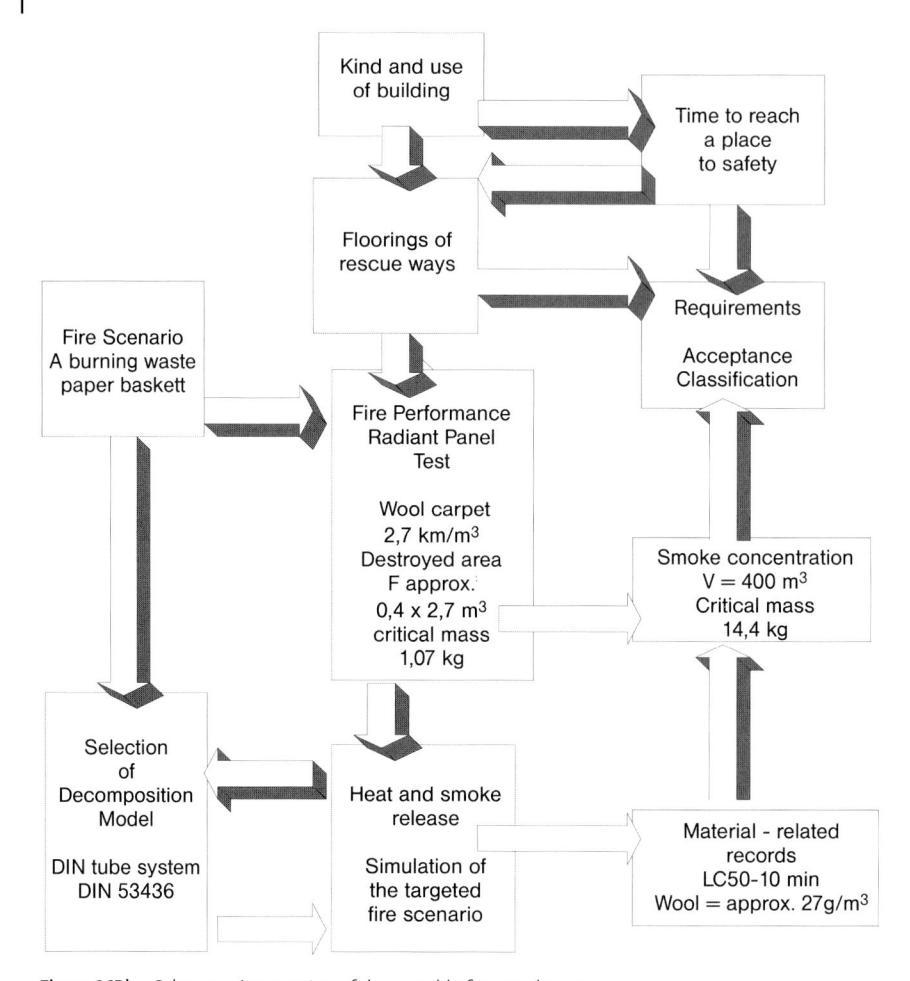

Figure 265 b Scheme – Assumption of the possible fire-gas danger flooring material

6.3 Relevance of the Requirement Profiles | 467

```
                    ┌──────────────┐
                    │Transportation│
                    │    Field     │
                    └──────┬───────┘
                           │          ┌──────────────┐
                           ├─────────▶│Time to reach │
                           ▼          │  a place to  │
                    ┌──────────────┐  │    safety    │
                    │  Upholstery  │◀─└──────┬───────┘
                    │  furniture   │         │
                    │              │         │
                    │ PUR flexible │         ▼
                    │density 30 kg/m³│  ┌──────────────┐
                    └──────┬───────┘──▶│ Requirements │
┌──────────────┐           │           │ No dangerous │
│Fire scenario │           │           │ concentration│
│A burning paper│──┐       │           └──────────────┘
│   cushion    │  │       ▼                   ▲
└──────┬───────┘  │ ┌──────────────┐          │
       │          └▶│Fire Performance│         │
       │            │     Test     │          │
       │            │UIC paper cushion│        │
       │            │Kerosine burner test│     │
       │            │e.g. weight loss│  ┌──────────────┐
       │            │not more than 10%│─▶│ Test Results │
       │            └──────┬───────┘   │Heat/Smoke-Output│
       │                   │           │ Calculation  │
       │                   │           │  Validation  │
       ▼                   │           └──────────────┘
┌──────────────┐           │                   ▲
│  Selection   │           │                   │
│      of      │◀──────────┤                   │
│Decompositioin│    ┌──────────────┐           │
│    Model     │    │ Smoke/Heat   │           │
│              │    │ Production   │  ┌──────────────┐
│  DIN 53436   │    │              │─▶│Material related│
│  Tv = 600 °C │    │  Simulation  │  │    record    │
│              │    │      of      │  │              │
└──────────────┘    │  Smoke/Heat  │  │ LC50-30 min  │
                    │  Production  │  │PUR flex = 17g/m³│
                    └──────────────┘  └──────────────┘
```

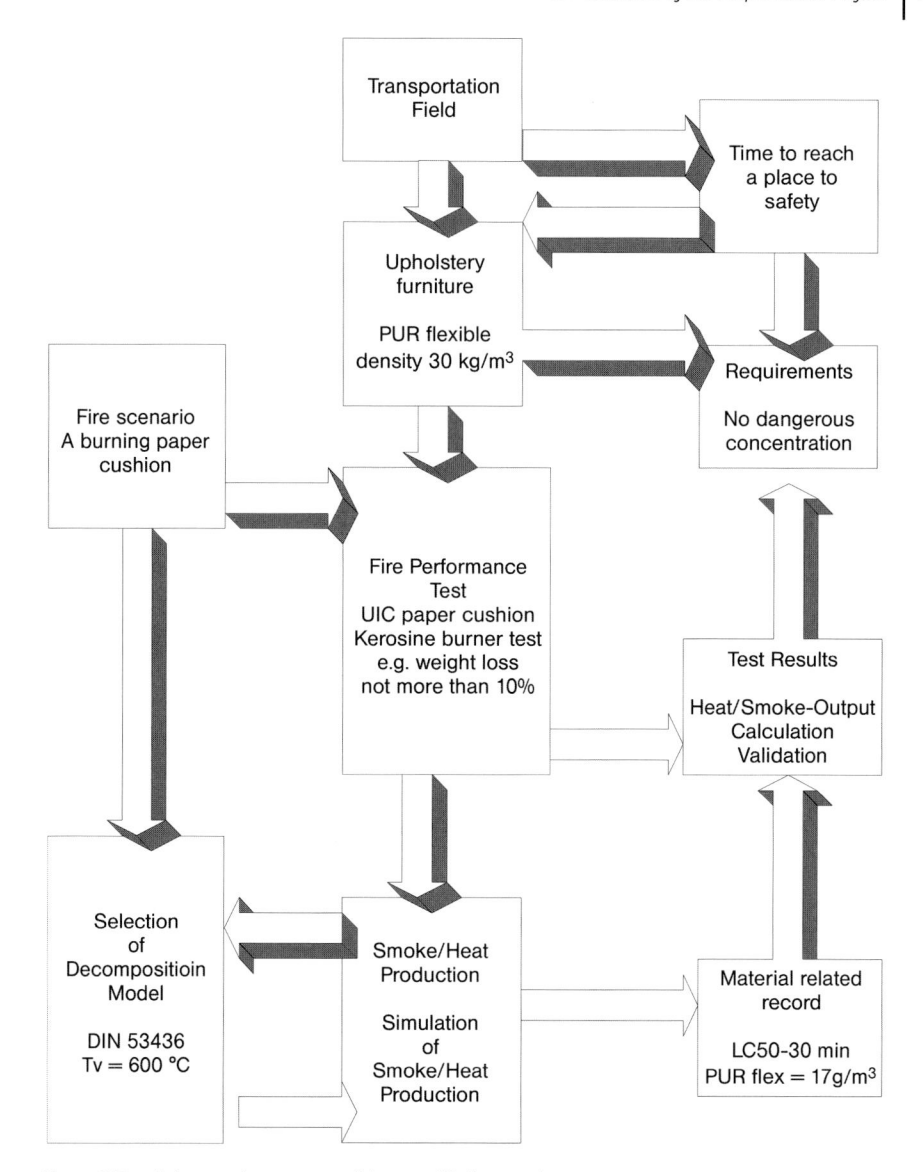

Figure 265c Scheme – Assumption of the possible fire-gas danger upholstery furniture

mined only in connection with the mattress examination according to CAL TB 121 (see Fig. 68a) of decisive importance for the perception of the escape possibilities and the realization of rescue operations in the extent of the reduction of the visibility. The schematic representation of Fig. 265c shows using the example of the UIC-paper cushion test that, provided the cushion combination supplies only a correspondingly limited contribution to the fire occurrence, then also in the real-fire case no dangerous reduction of the visibility occurs.

The results of numerous risk-relevant investigation as well as the findings from fire disasters, e.g. the Düsseldorf airport fire confirm that the results of lab-testing methods are not suitable to carry out a classification of the possible fire-gas danger in the case of a fire. The scale of the danger is determined decisively from the fire development. An estimate of the danger potential is possible on the basis of the material-specific potential values only. The results of extensive studies with the simulation of numerous different fire scenarios document that material-specific smoke potential values are functional characteristics. Such material-related potential values are to be determined as absolute data, such as, e.g. the net calorific value and the decomposition temperature or the functional characteristics of the heat and fire-gas release depending on the test temperature and the air supply. While estimating possible dangers, knowledge is accordingly necessary in particular about the variation width of these values depending on temperature and air supply. In the furniture field a requirement was set only in connection with the mattress examination in accordance with CAL TB 121 (see Fig. 68a). Beneath the limitation of the permissible fire participation of 10% of the mattress mass, a maximally permissible CO concentration in the stream of effluents of approx. 1000 ppm became prescribed. A more detailed consideration of this requirement profile analogous to Fig. 161 shows very clearly that risk-oriented orders require a reference onto the spatial factors of the fire room and the possible concentration circumstances combined with that. In the case of the currently targeted furniture directive at the DGIII Industrial Affairs is next to the igniting risk in particular with the second essential requirement the fire-gas problem responded [883]. The present testing consideration strives in analogy to the California law [884] initiative for mattresses, a problem solution for upholstered furniture a limited mass loss in the post-ignition field.

In the transportation field the predominant numbers of the relevant requirement profiles are based on analytically determined records. The permissible boundary values of the concentration of the essential fire-gas components were listed in Tables 150 and 151. If one compares the different data records of Tables 150 and 151 it appears then that, e.g. in the case of HCl fire-gas component there are in particular significant deviations, although doubts still exist. The data records of SNCF form an exception. In the SNCF requirement the accessibility to the biological test is ensured. As a combustion system the tube furnace is stipulated in accordance with NF-T51. The stream of effluents produced in the temperature range of 600°C to 800°C and guided in the approx. 400-dm^3 sized measuring room, collected and analysed. The fire-gas toxicity of the stream of effluents is determined. The boundary values are supposed to guarantee that the produced fire-gas components cause no harmful effects (ILDH values for a 20-min exposure) within an exposure duration of 20 min.

An expert assessment of the thermal decomposition products was arranged in California, among other things, according to Jenkins as a result of the subway fire under the bay of San Francisco [885, 886]. The thermal decomposition products produced with a temperature increase of 40 °C/min by means of the USF test protocol (see Fig. 176) must not influence the reaction behavior of the experimental animals inadmissibly. The death of the exposed animals must not occur within $t_d = 25$ min. There is no doubt that the boundary value t_d is, in this general form, very obviously not suitable as a classification criterion for the acute material-related fire-gas toxicity. Different requirements have been agreed in the naval field. Doubts exist about the risk relatedness of the requirement profiles in the naval field at SC3 of ISO TC92 in particular at the highly distributed NES 713 (see Fig. 74c). The thermal decomposition products of insulants on sea-going vessels under a German flag are to be evaluated from the toxicological viewpoint by means of animal-experimental investigation results. The decomposition products, produced by means of the DIN 53436 apparatus, had to be assessed in relation to the analogous decomposition products of wood and cork. The listing of Fig. 266 gives information about the recent Danish requirements concerning the acute fire-gas toxicity. The requirement profiles based on the DIN tube system DIN 53436 have been modified by the Danish authorities due to the disastrous fire of the ferry boat Scandinavian Star [887]. The cone-calorimeter and the room-corner test were agreed as additional decomposition models. The permissible CO concentration was significantly reduced. The boundary values determined for the room-corner experiment are only 10 % of the, up to now, valid data sets, which has been determined in the stream of effluents produced at a reference body temperature of 400–600 °C with the DIN Equipment in an airstream of 100 l/h. Being aware

	Requirements DSB Danish type approval IMO Res. A. 653 (16) (Bulkhead, Ceiling and Deck Finish)			
	DIN 53436 T = 400 °C, 600 °C L = 100 l/h (different information (dated 92/93))			Room corner ISO/DIS 9705
	Concentration [ppm]			
CO	40 000	40 000	20 000	2000
CO_2	600 000	600 000	600 000	60 000
HCl	2000	2000	2000	200
HCN	1000	500	500	50
HBr	500	500	500	50
HF	300	300	300	30
NOx	500	500	500	50

if danish certification is still valid
DIN 53436 provide potency data (24 g/m³) (24 cm³/m³ p = 1000 kg/m³)
Room corner provide data for the specific fire scenario emply small room and specific gas burner flame impingement

Figure 266 Danish requirements concerning the possible fire-gas dangers in the case of a fire on ships

Table 150 Validation of limiting values of smoke components

Smoke gas components	Maximum concentration (ppm)	
	After 1.5 min	After 4 min
Hydrogen fluoride (HF)	30	30
Hydrogen chloride (HCl)	50	500
Hydrogen cyanide (HCN)	100	150
Sulfur dioxide (SO_2)	50	100
Carbon monoxide (CO)	3000	3500
Nitrous gases ($NO+NO_2$)	50	100
Calculated LC_{50} (g/m^3)	Cellulosic products	N-containing products
Finney Formula (see Fig.263a)		
ATS 1000	130	4
NES 713	4	0,8
NF-T51	87	45
DIN 53436 – LC50 (g/m^3)	10	18

Table 151

	Fire gas toxicity Fire gas components							
Component	(maximum allowable concentration)						ISO TR 9122: LC_{50}	
	NES 713	ATS 1000		SNCF/RENFE	SI 755	DSB (DIN 53436)		
	(C_f) [ppm]	90s [ppm]	4min [ppm]	(ILDH) [mg/m³]	(C_f)	(C_f) [ppm]	5min [ppm]	30min [ppm]
HF	100	<50	<100	17	---	300		>1774
HBr	150	---	---	150	---	500		
HCN	150	<100	<150	55	350	500	250-400	170-230
NO_x	250	<50	<150	---	200	500		100-200
SO_2	400	<50	<100	260	500	---		
H_2S	750	---	---	---	---	---		
NH_3	750	---	---	---	---	---		
HCHO	500	---	---	---	65	---		
CH_2CHCO	400	---	---	---	---	---	500 - 1000	50-135
HCl	500	<50	<150	150	500	2000	12000-16000	2000-4000
CO	4000	<3000	<3500	1750	9000	20 000	12000-16000	2500-4000
CO_2	100000	---	---	90000	---	600 000	>10%	>9%

that the room-corner test protocol is not in the position to produce material-specific potential values and, on the other hand, is not able also to simulate the quite important smoldering fire situation of the DIN protocol, this law modification can be designed primarily as a political, tactical move for the appeasement of the ignorant public. The smoke formation was e.g. at the Düsseldorf Airport Fire the most important issue. The smoke toxicity of the fire effluents will be significantly changed by variation of temperature, concentration and exposure time as illustrated in Fig. 267a. By backing the assessment of the smoke hazard mainly on the comparison of very specific gained material-related records shown by Fig.267a, one neglects that smoke development can be changed to a great extent e.g. by the way of burning (see Fig. 267b), which influences the smoke density, as shown in Fig. 267c. Fig. 268 provides some material-related smoke volume records, calculated for stoichiometric combustion.

The State of New York requires, as a result of some spectacular fires, a toxicological assessment of the thermal decomposition products of materials of the furniture field and electromechanical appliances. Since Dec. 1986, these toxicological records, LC_{50} data of the thermal decomposition products determined with the U-Pitt test protocol (see Fig. 169), have to be registered and made public [888–890]. This test protocol is obviously not able to risk relevantly simulate the thermal decomposition products of the most important fire scenarios [89, 376, 465].

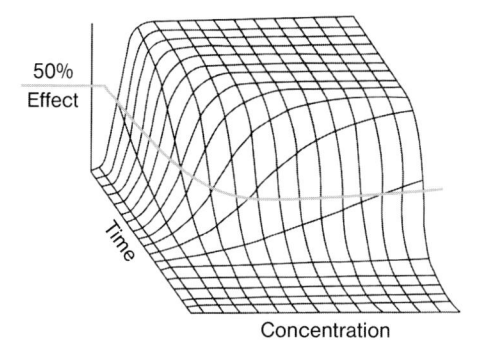

Figure 267a Concentrationtime relationship of the toxic potency of effluents

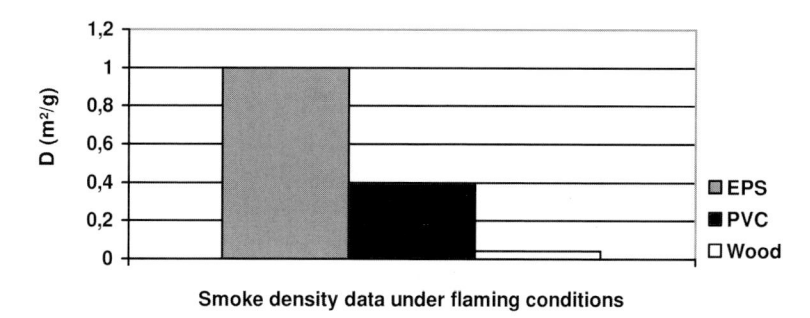

Figure 267b Material related smoke density data

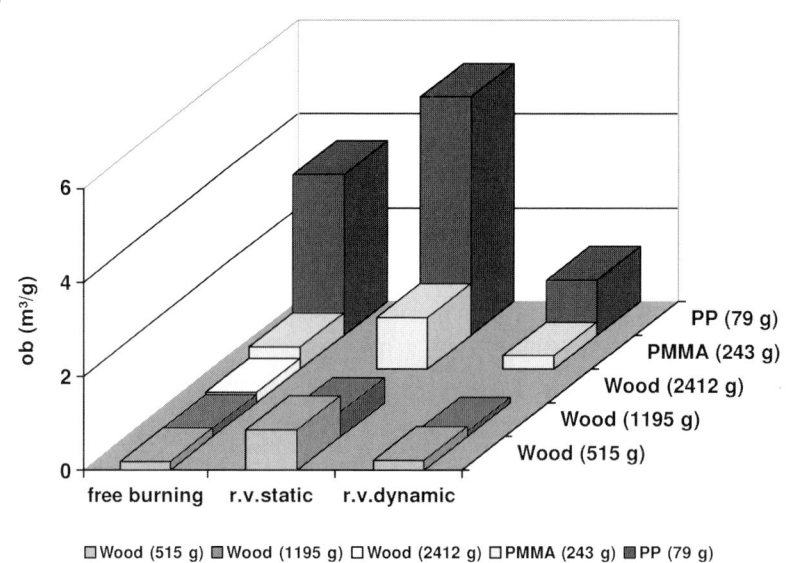

Figure 267c Smoke density data influenced by the way of burning

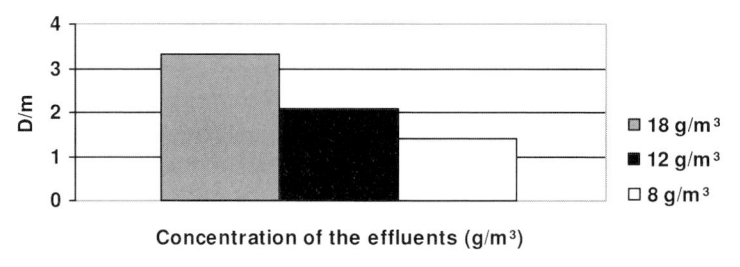

Figure 267d Influence of the concentration on the optical density

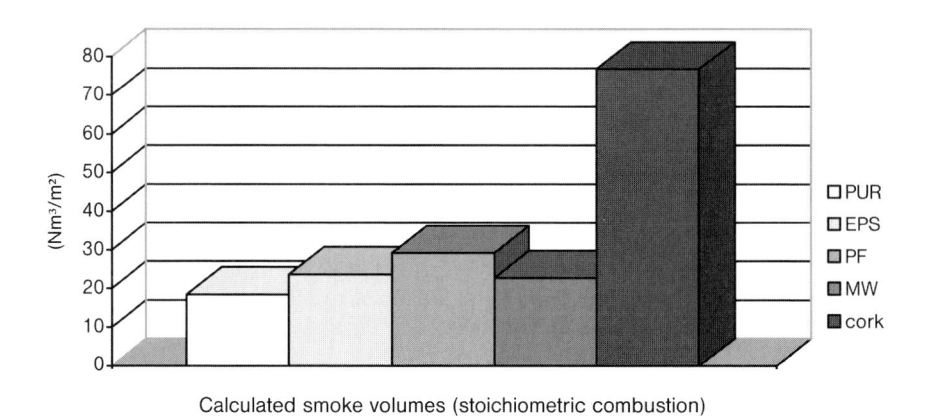

Figure 268 Calculation of smoke volume for stoichiometric combustion

Summary

Every facet and the relevance of the fire-reaction testing and classification systems are discussed in detail using the example of flexible and rigid polyurethane foams as well as their basic components. A detailed discussion is given about the various key parameters of ignitability, flame spread, burning dripping and heat and smoke rate in laboratory, bench and large-scale tests. The possible risk of fire initiation and development and the hazard created by the heat and smoke formation is discussed on the basis on meaningful fire-statistics data. It is shown that pure material-related records can be used for the estimation of the possible damage in the case of a fire. The key issue "risk of ignition" is linked to the resistance against ignition sources of daily life.

Chances and risks of the use of fire-retardant additives are one of the important issues. The increased resistance against the impact of flames due to the flame retardency reduces the risk of ignition significantly. This circumstance is a real advantage, no doubt, but there are disadvantages as well. While the use of flame-retardant additives increases the resistance against flames, for example the impact of flames against the ceiling tiles, no advantage is given in the case against the impact of a smoke layer. In such a specific fire scenario, the decomposition temperature of the applied material will be valid. This temperature level is not increased but might be slightly decreased by the use of flame-retardant additives. It has also been proven, that the use of flame-retardant additives, like, e.g. halogenated or brominated compounds or others, does not decrease the risk of a self-sustained smoldering fire in upholstery combinations. These fires, as proven by the worldwide fire-statistics records, are responsible for the largest proportion of fire victims. Fire-statistics data concerning fire disasters like Star Dust, Charing Cross, Du Pont Plaza or Düsseldorf Airport have demonstrated that instead of pure material data, environmental conditions and the enduse combinations are responsible for the fire development and therefore for the heat- and smoke-release rates. The use of flame-retardant additives produces, as demonstrated by various large-scale test series, an increased risk of smoke formation. These decomposition products lead to a stronger reduction of visibility and to increased health problems The characteristic LC_{50} value of the toxic potency will be decreased significantly.

It is demonstrated that test and classification parameters are settled as a compromise by the members of the standardisation committee. Therefore, large-scale tests do not have to be in line with the standardized conditions. Large-scale tests may deviate, to get a better answer due to safety reasons. The outcome of a survey of different

Polyurethane and Fire. Franz H. Prager and Helmut Rosteck
Copyright © 2006 WILEY-VCH Verlag GmbH & Co. KGaA, Weinheim
ISBN: 3-527-30805-9

national and international standardized fire-reaction tests reveals that most of these decomposition models are not in line with the recommendations of the SC3 of TC 92 Toxic Hazard in Fire. Realistic fire scenarios were chosen, e.g. for the A2 test procedure according to DIN 4102. Realistic test and classification parameters have been standardized for smoke-density measurements. As far as the smoke-toxicity orders are concerned, not only the test parameters differ from reality. The limitation to a smoldering process up to 400 °C has been agreed, which differs from reality.

The smoke-density requirements settled by the harmonized SBI-test are without doubt linked to this very specific fire scenario and therefore not valid for an overall classification. The limitation to the room-corner fire scenario, initiated by a single burning item, is doubtless a hindrance to assess a realistic hazard-related classification. The "zero" classification reached by definition at the records for smoke formation of $RSP_{AV} < 0.1 \, m^2/s$ and $TSP < 6 \, m^2$ is therefore not relevant for most fire scenarios, where the impact of a smoke layer produced by a local flashover has to be taken into account.

The reproducibility and repeatability are key issues for test methods becoming standardized. The outcome of a round-robin exercise is usually filtered by ruling out outliners according to the relevant ISO standard 5725-2. The regulatory people, convinced by the corrected deviation, do not know that possibly their own national laboratory was ruled out as an outliner. Additional test stations without the round-robin instructions will possibly lead to much greater deviations.

The regulator should be aware of the total deviation measured in the round-robin exercise before setting the legal orders. A further example for political decisions in the standardizing field is the cigarette test. Out of financial reasons only 2 test runs are required for classification. The very expensive test rig, which is no doubt meaningful for research work, determining the degree of smoldering, is not necessary for the classification system, where the extent of smoldering is extremely limited. Being aware of the limited extent of the self-sustained smoldering process, the dimension of the test rig could be significantly shortened. Then many more test runs could be performed at the same cost.

Surprisingly, the risk of ignition by hot work, glowing contacts and glowing tobacco is usually not part of the testing and classification system in the field of building materials. The tendency of materials to a self-sustained smoldering process is totally neglected by the harmonized European Fire-reaction Test Methods. Experience of large-scale testing, which provides positive information with respect to fire reaction, e.g. burning dripping, is not taken over by the standardizing committees. The committee's actions are dominated by the members of the national test stations. The latter also dominate, as a rule, the international standardization work. The A2 classification according to DIN 4102 is a good example to demonstrate the nonexisting relationship of the test parameters to the possible hazardous situation in the case of a fire. The classification of the polyimide foam as incombustible insulation material PA III nr. demonstrates also the nonexisting relation to the hazardous situation in the case of a fire. The differing test parameters for the smoke-density testing and the toxicity assessment of the decomposition products of A2-classified materials is a further example of the lack of relation to reality.

The aim of ad hoc decisions for fire prevention after fire disasters is to calm the public. Such ad hoc solutions never provide risk-related solutions. The new Danish requirement for the assessment of the toxic potency of the thermal decomposition products in the naval area is a typical example. The cone-calorimeter decomposition model, relevant only for the flaming combustion with high ventilation, was agreed instead of the DIN Tube Furnace, which has been used for simulating smoldering conditions. This activity explains/demonstrates the need to define the framework of the aim by the regulator for the testing industry.

References

1 *Weinspach PM,Gundlach J, Klingelhöfer H-G,Nitschke K, Ries R, Schneider U.* Kommissionsbericht NW 14. April 1997
2 *Sarkos* DFVLR-Conference Köln 1987
3 *N. N.* Fire Prevention Nr. 266 Jan. 1994 p. 7
4 *Purser D.* Fire and Flammability March 1992
5 *N. N.* Figaro 26/2/71-Time 16/11/70
6 *N. N.* Fire Prevention 142 p. 12
7 *N. N.* Report On Investigation- 18/04/1984
8 *N. N.* Daily Telegraph May 19 1980
9 *Bardet E.* Int. Conf. of FR Textiles Selcta 97
10 *Duden* Dudenverlag
11 *Emmons H. W.* Scient. American 231 (Jul. 1974) 1 p. 21–27
12 *Emmons H. W.* J. Of Heat Transfer (1973) May p. 145–151
13 *Thomas P. H.* Fire Prev. Sci. And Technol. Nr. 23 p. 3–7
14 *PRC* PRC-The Final Report April 1980
15 *Einhorn I. N.* Conf. Ser. Salt Lake June 1973
16 *Baillet C, Delfone L,Lucquin M.* The Comb. Inst. Europ. Symp. 1973 p. 148–152
16 *Ryan J. E.* ASTM-STP 502 p. 11–23
17 *Brown W. Tipper C.* The Combust. Inst. Europ. Sympos. 1973 p. 137–141
18 *Rasbash D. J.* Intern. Sympos. Edinburgh 1975
19 *Paul K. Clearly W. Noisey L* Rapra Project A1320–1983
20 *Kuchta J,Furno A. Martindill G.* Fire Techn. 2–1969 p. 203
21 *Blazowski W. ,Cole R. Mcalevy R.* Combust. Inst. 14. Symps. Pittsburgh 1973
22 *Hailer-Hamann D.* Bild Der Wissenschaft (1977) Nr. 11 p. 88–94
23 *Becker W.* Chem. Ing. Techn. 52 (1980) Nr. 2 p. 162–163
24 *Merzhanov A. G. Averson A. E.* Combust. And Fame 16 (1971) p. 89–124
25 *De Ris J, Orloff L* The Combust. Inst. Europ. Symp. 1973 p. 153–158
26 *Gross D.* Fire Safety J. 15 (1989) p. 31–44
27 *Prager F. H.* Becker/Braun. Kunststoffhandbuch Bd. 1
28 *Thomas P. H.* The Combust. Inst. 9. Symp Pittsburgh 1963
29 *Kordina K. Bechthold R. Ehlert K. Wesche J.* Schriftenreihe des Bundesministeriums für Raumordnung,Bauwesen und Städtebau Nr. 4 037
30 *Kirkby L. L. Schmitz R. A.* Combust. and Flam. 10 (1966) Sept. p. 205
31 *Roberts A. F. ,Quince B. W.* Combust. and Flame 20 (1973) . p. 245–251
32 *Sheinson R. S. , Williams F. W.* The Combust. Instit. Europ. Symp. 1973
33 *Bartels H. ,Lowes P. ,Michelfelder4 S. ,Pai B.* Combust. Inst. Europ. Symp. 1973 p. 680
34 *Hastie J. W.* Poymer Conference Salt Lake City 1973
35 *Seeger P. G.* Vfdb-2 (1970) p. 91
36 *Modak A. T. , Croce W.* Combust. and Flam. 30 (1977)
37 *Warren P. C.* Spe J. 27 (1971) 2. p. 17
38 *Zukowski E. E. Kulota T. ,Catchen B.* Fire Safety J. 3 (1980/81) p107–121
39 *Friedman R. J.* Fire and Flamm. 2 (1971) p. 240–256
40 *Friedman R.* Fire Res. Abstr. and Reviews 10 (1968) P1–8
41 *Gross D. Loftus J. J.* Nbs. Report, NBS. Project Report
42 *Karwaller S. J.* Fire Journal 1976 p. 66
43 *De Ris J.* Combust. and Sci. Technol. 2 (1970) p. 239–258
44 *Roberts A. F.* The Combust. Inst. Europ. Symp. 1973 p. 159–64
45 *Thomas,Bullen,Quintiere,Mccaffrey* Combust. and Flam. 38 (1980) p. 159–171
46 *Heskestad G.* The Combust. Instit. Pittsburgh 1973
47 *P F. H. Prager, Troitzsch J.* Carl Hanser Verlag 1990
48 *Prager F. H.* Brüssel/88 Fire and Flamm. Bul. 10/nr. 5
49 *Malhotra M. L.* Fire Prevent. Sci. and Techn. Nr. 17 p. 24

50 *Van Kreveler W*. J. of Appl. Polym. Sci. 31 (1977) p. 269–292
51 *Kashiwagi T. ,Summerfield M*. Comb. Inst. 14. Int. Symp. Pittsburgh 1973
52 *Friedman R*. ASTM-STP 614 1976 p. 91
53 *Steingiser S*. J. Ff. 3 (July 1972) . p238
54 *Roberts A. F*. Insulation Nov/Devc. 1974
55 *Schuhmann J. G*. J. Of Cell. Plastics
56 *Prager F. H. Darr W. Wood J*. Cellular Polymers 3 (1984) p. 181–194
57 *Moor L. D*. NBSIR-78–1448 03/1978
58 *Cooksey P. N*. The Internat. Fire Chief 46 (1980) Nr. 9 p50
59 *Wilde D. G*. Jff 11 (Oct. 1980) p. 263
60 *Starrett Th. S*. Jff 8 (Jann. 1977) p. 5
61 *N. N*. Rpff Bulletin 1982 Vol. 3 Nr. 4/5 p. 2
62 *Bryan J. L*. Fire J. March, (1982) p. 37
63 *Williams F. A*. 16. Sympos. Combust. Inst. 1977
64 *Belshaw R. L*. Fire Safety J. 10 (1986) P19–28
65 *Demers P*. Fire J. 3/1979
66 *Malhotra M. L*. Fire Note 12 Oct. 1971
67 *Vandevelde P*. Fire and Mater. 4 (1980) 3 P. 157–162
68 *Quintiere J*. The Combust. Inst. 15. Symp 1974 p. 163
69 *Tewarson A. Pion R. D*. Combust. and Flam. 26 (1976) 1,p. 85–109
70 *Kashiwagi T* Jff Cons. Prod. Flam. 1 (Dec. 1974) p. 367–389
71 *Christian W. J*. Fire J. Jann. 1974 p. 22–28
72 *Hurd R. King R*. Cellular Polymers 10 (1991) Nr. 1 P.
73 *Tewarson A* Fire And Mat. 4 (1980) 4 P. 185–191
74 *Pagni P. J*. The Combust. Inst. 15. Symp. 1974
75 *Fang H*. Jff 7 (1976) P. 368
76 *Martin St*. Stanford Research Inst. Univers. San Francisco 1979
77 *Quintiere J*. 15th Sympos. on Combustion Aug. 25–31 1974
78 *N. N*. AIA Tarc Project 210–9 Final Report 28. 2. 83
79 *Krasny J. E*. NBS IR 87–3509 April 1987
80 *Prager F. H., Cabos H. P. Fischer H. Rosteck H*. Fire and Materials 18 (1994) P. 131–149
81 *Cabos H. P. Fischer H. Prager F. H. Rosteck H*. VFDB 2/88
82 *Von Grimbergen M. Reybronk G. von de Voerde H*. Zbl. Bakt. Hyg. I Abt. Orig. B 155 (1971) p. 123–139
83 *Rashbash D. J. Pratt B. T*. Fire Safety J. 2 (1979/80) P. 23–37
84 *Malhotra H. Rogowski B. Raftery M*. Qmc Sympos. 1982
85 *Berman A. M*. Qmc Sympos. 1982
86 *Rasbash D. J*. London 1966 Conference Flame Resistance.
87 *ISO TC 92* TR 9122
88 *Kimmerle G* Salt Lake City 1974
89 *Klimisch P. Doe J. Hartzell G. Packham St. Pauluhn J. Purser D*. J. Of Fire Sci. 5 (1987) March/Apr. p. 73–104
90 *Jouany J. M. Boudene C. Truhout R*. Salt Lake City 1976
91 *Hartzell G. Galster W. Farrar D. Hileman F. Blank T. Pedersen S. Williams S*. PRC Conference 1978 Tuscon
92 *Alarie Y. Anderson R*. PRC-Conference 1978 Tuscon
93 *Kimmerle G. Prager F. H*. Kautsch. u. Gum. Kunst. 33 (1980) 5 p354
94 *Beer F*. Brandschutz März 1970
95 *Malhotra M. L*. Cp 2/1980
96 *Bilina H*. Brandschutz/Deut. Feuerwehrztg 11 (1971)
97 *Wiendl S*. Diplomarbeit Bergische Universität Aug. 1996
98 *Burgess W. A. Treitman R. D. Gold A. Havard School Of Public Health*. NFPA Grant 1979
99 *Hillenbrand L. J. Wray J A*. Battelle Columb. Laborat. July 1973
100 *Humphries K. J. Prager F. H. Wilson W. J*. Rapra 1 Report to III 1972
101 *Dawson J. W. ,Prager F. H. Wilson W. J*. Rapra 2 Report to III 1974
102 *Wood J. F. Prager F. H. Wilson J. W*. Rapra 3 Report to III 1977
103 *Dawson J. W. ,Prager F. H. Wilson W. J*. Moreton-1 Report to III 1974
104 *Wood J. F. Prager F. H*. Report to III, PRC Project Rp 75–1–13
105 *NBS 1:1*
106 *N. N*. Report of The Tribunal of Inquiry Stardust Fire Dublin 14/02/81
107 *Emmons H. W*. Fire Techn. 192 (1983) Nr. 5 P. 115
108 *Fiala K. Dussa K*. DFVLR-Jahresbericht 1986 p. 39
109 *Stuckey R. E*. NASA TM – X – 58141
110 *Takita T*. Rail Internatioal July/Aug. 1977
111 *Bußmann R, Einbrodt H. J. Prager F. H. Sasse H. R*. J. Of Fire Sci. 10 (1992) P. 44
112 *Smith E*. Fire Techn. 8 (1972) P. 231–245
113 *Emmons H. W*. Fire Safety J. 12 (1987) P. 183–189
114 *Morikawa et. al* Bul. of Jap. Ass. of Fire sci. 41 (1992) p. 9
115 *Chandler S. E*. BRE-Cp 66/76
116 *Jones J. C*. Fire J. July 1982
117 *Banks J. Rardin R. D*. Fire Techn. 18 (1982) Nr. 3

118 *N. N.* Home Office Statist. Div. 1993 ISSN 0260–3098
119 *Alekhin E. Buschlinski N. Kolonietz J. Sokolov S. Wagner P.* VFDB 4/2001 p. 176–192
120 *CIB* CIB Conference Tsuchuba City 1976
121 *Prager F. H.* Fire and Flamm. Bulletin 10/88 Nr. 5
122 *Wilmot RTD* VFDB 1/2000
123 *Mulder S.* Fire Prevention 263 Oct. 1993
124 *N. N.* Fire Prevention 326 Nov. 1999
125 *N. N.* Home Office Statistical Div. 1985
126 *N. N.* Fire Prevention 263 Oct. 1993
127 *N. N.* FPA April 2002 p. 5 DTLR
128 *Prager F. H.* PUR Worldcongress Nice 1991
129 *Prager F. H.* 5. IBS Karlsruhe 1976
130 *N. N.* Home Office Statist. Div. 1985
131 *N. N.* Fe&Fp April 2002
132 *N. N.* Fire Prevention 329 Feb. 2000
133 *N. N.* Fire Prevention 328 Jan. 2000 p. 4
134 *The Loss Prevention Council* Libery of Fire Safety ISBN – 090216844–8
135 *N. N.* Fire Prevention 286 Jan. /Febr. 1996
136 *N. N.* Fire Prevention 306 Jan/Febr. 1998
137 *Drysdale D. D.* Fire Prevention Nov. 2000
138 *NFPA* NFPA Jahrbuch 1978
139 *N. N.* Fire Prevention 292 Sept. 1996
140 *N. N.* Home Office Statist. Div. 1995
141 *Yuill C.* Jff 1 (1970) P. 312
142 *Einhorn I. Newman M.* Uni Utah 1975
143 *N. N.* Fire Prevention 220 June 1989 p. 11
144 *Christian J. T.* Salt Lake City 1973
145 *Lundberg D. Petersen K.* Sintef Rapp. Stf25,A82008
146 *DTLR* FEJ&FP April 2002
147 *Anderson R. Willets P. Cheng K. N. Harland W. A,.* Fire and Mat. 7 (1983) Nr. 2, p. 67
148 *Halpin B. Radford P* Fire J. 5/1975 p. 11
149 *Berl W. G. Halpin B. M.* ASTM-STP 614 1976 p. 26
150 *Anderson R. A. Harland W. A.* Proc. of Europ. Meet. of instmass. of Forensic Tox. 1979
151 *Harland W. A. Woolley W. D.* Fire International 66 p. 37–42
152 *Anderson R. A. Harland W. A.* Fire and Materials 3 (1979) Nr. 2 p. 91
153 *Harland W. A. Anderson R. A.* QMC 1982
154 *Christian S. D. Nolan* Flame Retardents Conference 85
155 *Paul K. T.* Flame Retardants Conference 85
156 *Chandler S. E.* Fire Prevention 246 Jann/Febr. 92 P. 14
157 *Christian S. D.* J. of Fire Sci. 3 (1985) p. 310
158 *Christian J. F.* Salt Lake City 1973
159 *N. N.* Fire prevention 303 Oct. 1997
160 *N. N.* Fire prevention 340 Jan. 2001
161 *N. N.* Fire Prevention 328 Jann. 2000 P. 4
162 *Arvidson T* Fire Prevention nov. 2000
163 *Doherty B.* Fire Prevention 317 Febr. 1999 p. 26
164 *Clarke T. B. Hirschler M.* J. of Fire Sci. 9 (Sept. 1991) p. 406
165 *Lowry Et. Al.* J. of Forensic Sci. 30 (1985) Nr. 1 p. 59–78
166 *UK* Home Office Statistic Division Sept. 1993
167 *Curtis M. H. LeBlanc P. R.* Fire J. July 1984 p. 33
168 *N. N.* Fire Prevention 239,May 1991 p. 28
169 *N. N.* Fire Prevention 339 Dec. 2000
170 *DTLR* Statistical Div. 1994
171 *Briggs P* Report to APME
172 *N. N.* Fire Prevention 326 Nov. 1999 p. 6
173 *N. N.* May 28. 1977
174 *Herzog R.* FAZ 18. 01. 1999 Nr. 14 p. 50
175 *N. N.* Fire Prevention Nr. 181 July/Aug. 1985 p. 5
176 *N. N.* May 28. 1977
176 *Wood P. G.* Fire April (1973) p. 567 Frs Note 953
177 *Phillips A. W.* Fire J. May 1978. p. 69
178 *Jin T.* Jff. 12 (1981) 4, p. 130
179 *Sundström B.* EEC/QMC – Confer. Luxemburg
180 *Emmons H. W.* Abstracts And Review 16 (1968) Nr. 2 p. 133
181 *Curtat M.* CSTB Magazine 124 (1999) p. 18–20
182 *Benjamin J. A. Adams C.* NBS – IR 75–950
183 *Denyes W. Quintiere J.* NBS – IR 73–199
184 *Potts W. Lederer T. Quast I* J. Of Comb. Toxic. 5 (4) 1978,P. 408–433
185 *Birky M, Paabo M, Levin B,Womble S, Malek D.* NBS – IR 80–2077
186 *Benjamin J. A.* Fire Safety J. 7 (1984) P. 3–9
187 *Wallace D* Fire and Mater. 16 (1992) P. 77–94
188 *Alarie Y.* PRC-Project Conference Tucson 1978
189 *Barrow J Alarie Y* Polymer Conference Salt Lake City 1976
190 *Alexeeff G, Lee Y, White J, Putas N.* J. Of Fire Sci. 4 (March 1986) P. 100
191 *Alexeff G. Packham St.* J. Of Fire Sci. 2 (Jul/Aug. 1984) P. 306
192 *Babrauskas V. Levin LC* Nist-Techn. Note 1284 Jann. 1991
193 *Kolman. Voorhees,Osborne,Einhorn.* J. Fire Sci. 3 (Sept/Oct. 1985) P. 322
194 *Smith,Crane,Sanders,Abbott,Endecott* Polymer Conference Salt Lake City 1976
195 *Skornik W. Robinson S, Dressler P* . J. Of Comb. Toxic. 3 (1976)

196 *Skornik W. Dressler P. Robinson . S.* Salt Lake City 1976
197 *N. N.* Life Sci. Research Rep. 8 (1978) Nr. 3,July
198 *Hilado C. J.* Fire J. Nov. 1979 P. 69
199 *Hilado C. J. Huttlinger P. A.* J. Of Consum. Prod. Flamm. 7 (Dec. 1980) P229
200 *Alarie Y.* J. Fire Sci 11 (1993) The Forum P. 457
201 *Alarie Y. Caldwell D. J.* J. of Fire Sci. 8 (Jann/Feb. 1990) P. 23
202 *N. N.* G. V. NW 7. 03. 1995
203 *Gaedke, Böckenförde, Temme* Werner Verlag 7. Auflage
204 *Rößler W.* Carl Heymann Verlag 1971 Köln Et. Al.
205 *N. N.* Bayr. Gesetz U. Verordn. Bl. 20/7. 82 Nr. 18
206 *N. N.* Ministerialblatt . Amtlicher Anzeiger
207 *Hertel H.* Mitt. IfBt 1994 Nr. 2 und 3
208 *N. N.* Wegleitung Feuerversicherung
209 *N. N.* Mitteil Ifbt 9 (1978) H. 4 S. 121–126
210 *Jagfeld P. Frech P.* Agf-Bericht 16 Sept. 1971
211 *Prager F. H.* PAIII Vorlage 6. 5. 1980, IfBt Berlin
212 *Prager F. H.* PAIII Vorlage 15. 11. 1979, IfBt Berlin
213 *Schmid P.* Vfdb 2/1984
214 *Hoffmann R* Fire Prevention 150 P. 21
215 *Schmid H. R.* Textilveredelung 23 (1988) nr. 9 p. 301
216 *Hoffmann R.* Fire Prevention 150 P. 21
217 *N. N.* Journal Official nr. 7 0. 01. 1976/77
218 *Capron R.* Prev. Securite Suppl. Sci. et Techn. 95/23
219 *Jouany J Boudene C. Truhout R.* Arch. Mol. Prof. 38 (1977) Nr. 9 P. 751–772
219 *Boudene C. Jouany J. Truhout R.* 16. Jour. De Groupm. Fran. Lyon 1975
220 *Damant G.* Fire J. May/June 1988–13. Conf. San Francisco
221 *Daniel H, Henin J. Lefevre B* J. Of Fire Sci. 7 (March 1981) P. 81
222 *Minne R* ASTM STP 502 1972
223 *Wickström U.* Eurific Symposium Kopenhagen
224 *Wickström U.* Eurific Report
225 *Tolley,Nadeau, Reymore, Waszeciak, Sayigh* Polym. Conf. Ser. Detroit 1966
226 *Hildebrand Ch.* FMPA Leipzig
227 *Hildebrand Ch.* Plaste und Kautschuk 19 (1982) Nr. 10 p. 773
228 *Süß W.* Plaste und Kautschuk 19 (1982) Nr. 10 p. 771
229 *Hildebrand Ch. Süß W.* Vb Wissensch. Techn. Beilage 1/1973

230 *UDSSR* Imo Standard,Fb/353 19. 1. 1984
231 *Standard UDSSR* Imo,Iso Tc 92 Wg12–9,Wg4–177
232 *Harmathy T. Z.* Combust. and Flam. 31 (1978) P. 265–273
233 *Harmathy T. Z.* Combust. and Flam. 37 (1980) P. 25–39
234 *Ramsay G. C.* Int. Sympos. Edinburgh 1975
235 *Saito F.* NBS-Conference 15. Oct. 1980
236 *Saito F.* BRI- Res. Paper Nr. 65 March 1976
237 *Petterson O, Magnusson S. E.* Nordest Project 34–75 jan. 1977
238 *Blachere E* Eg-Report III 3197/88
239 *Wickström U,Göranson U* SP Report 1988:1
240 *Mikkula E.* Fire and Materials 17 (1993) p. 47
241 *Todd C.* Fire Prevention 337 Aug. 2001
242 *Hildebrand Ch.* CIB WG 14 1988
243 *Hildebrand Ch.* Bauzeitung 49 (1995) $^{1}/_{2}$
244 *Zorgman H.* 5. IBS Karlsruhe 1976
245 *Finegan F.* Fire Prevention 327 Dec. 1999
246 *N. N.* Fed. Reg. 38 Nr. 133 1973 Notice 7
247 *Damant G. Nurbakh S.* AFNA Conference !983
248 *N. N.* CAL TB 116 and 117
249 *N. N.* New York Port Authority 1970
250 *FAA* RPFF- Bulletin 5. 12. 1984
251 *Palmer K. N. Taylor W. Paul K. T* FRS/UK CP 3/1975
252 *N. N.* Versammlungsstätten VO NW 1976
253 *N. N.* Versammlungsstätten VO Hessia
254 *Rumberg* VDI-Ztg. 113 (1971) Nr. 4 März
255 *N. N.* Prefecture De Police Mai 1981
256 *Whoolley, VandeVelde, Moye, Klingelhöfer, DuPont, Blume, Briggs, Thomas, Twoney* Council Directive 89/106/EEC
257 *N. N.* Fire International 79 p. 13–16
258 *EEC – DG3* 83/189/EWG 23. März 1983
259 *Ramsay. G. C. Dowling V. P.* CSIRO ISBN 0643 035605
260 *Phillips W. A. Bartos K. Benisek L.* IWS Report CCD-32A 1981
261 *Goldsmith A.* IIT Res. Inst. Fire Techn. Rep. J. 6152/69
262 *FHA* Doc 3–3 Notice 2 Fed. Reg. 34 Nr. 249 1969
263 *N. N.* Ford SKM-99 p. 9500a
264 *N. N.* Opel 261
265 *N. N.* Tl-VW 1010
266 *N. N.* Str. VO
267 *FAA* Fed. Reg. 34 Nr. 135 12. 08. 1969
268 *FAA* Fed. Reg. 50 Nr. 73 1985
269 *FAA* Fed. Reg. 51 Nr. 139 1986
270 *FAA* Fed. Reg. 55 Nr. 71 1990 14CFR Pt. 25. 121
271 *Airbus* ATS 1000. 001 1989
272 *Boeing* BMS 8. 226r

273 *IMO* Resolution A 163
274 *Robertson A. F.* Fire and Materials 6 (1982) Nr. 2 p. 68
275 *Ulrich G.* Hansa-Schiffahrt-. 125 (1988) Nr. 22 S. 1415
276 *N. N.* California Assembly Bill 594 12. 08. 1983
277 *Damant G.* Flame Retardents 92
278 *Briggs P.* Fire Safety J. 20 (1993) P. 341
279 *Heskestedt A. W. Hovde P. J.* Fire and Materials 23 (1999) 4 p. 193–199
280 *Hovde P. J* Interflam 93
281 *Seader J. D. Chien W. P.* JFF 5 (1975) p. 151
282 *Best R.* Fire J. May 1980 p. 51–53
283 *Fernell D* Fire and Flammability Bulletin 10/1990 p. 5
284 *N. N.* New York Port Authority 1970
285 *N. N.* Fire Prevention 339 Dec. 2000
286 *N. N.* Prefecture de Police Paris Mai 19891
287 *DB* DV 899/35
288 *Deischl E.* Eisenbahningenieur 18 (1976) p. 41
289 *N. N.* ETZ – B26 (1974) M123
290 *Flatz J. Pohl D. Rickling E.* ETZ Report 1
291 *Reymers H.* Mod. Plast. 47 (1970) 9 p. 78
292 *Finger* Ieee July 1986 Vol. 2 Nr. 4 P. 24
293 *Schwarz K.* ETZ-B 14 (12962) p. 273
294 *Patten G. A.* Modern Plastics (1961) July P. 119
295 *Bottin M-F.* J. of Fire Sci. 10 (March/April 1992) p. 160
296 *Purser D. Fardell P. Rowley J. VollamS. Bridgeman B. Ness E.* Interflam 95
297 *Oertel G.* Hanser Publishers, Munich Vienna New York 1994
298 *CEC* Council Directive 82/501/ 24. june 82
299 *CEC* Amentments; 87/216 and 88/610
300 *CEC* Directive on industrial pollution prevention
301 *Kourtides Et. Al.* J. Ff. 7 (1976)
302 *Koseki H. Hoyasoka H.* J. Of Fire Sci. 7 (Jul/Aug. 1989) P. 237
303 *Carvel R.* FPA Aug. 2002 p. 41
304 *Pehersdorfer H.* Brandverhütung (1996) 1 p. 8
305 *Einfeld W. Mohler B. Morrison D. Zak B. D.* Scandia Laboratories USA 1985
306 *Fiala R.* VFDB 3/84 p. 95
307 *Gefahrst. VO* Bundesgesetzbl. 26. 8. 1986, 5. 6. 1991
308 *BImschG* Bundesgesetzbl. 15. 3. 1974, 22. 5. 1990
309 *Bützer P* Swiss Chem. 14 (1992) Nr. 1 p. 7
310 *Habermaier F.* Brandschutz, Deutsche Feuerwehrzeitung 10/86
311 *Marlair G. Prager F. H. Sand H* Fire and Materials 17 (1993) p. 91–102
312 *Marlair G. Prager F. H. Sand H.* Fire and Materials 18 (1994) p. 17–39
313 *Sand H. Marlair G* PU – World Congres 1992 Aken
314 *Allport D. C. Gilbert D. S. Outterside S. M.* MDI and TDI: Safety, Health and the Environment . John Wiley& Sons. LTD
315 *Rhone – poulenc* III Report nr. 11013
316 *Rhone – poulenc* III Report nr. 11014
317 *Klebert W. Prager F. H. Müller B.* III Report Nr. 11019
318 *Carter D. A.* Chem. Eng. Res. Des. 67 (July 1989) p. 348
319 *Halpaap W.* VCI 1/1987 Symposium Frankfurt
320 *Halpaap W.* VFDB 4/1981
321 *Creyf H. Hurd R. King D.* Cellular Polymers 14 (1995) 4
322 *Prager F. H.* Vogelbusch-Erg. Lfg. 3/99
323 *VdS* VdS 2053 1988–12
324 *N. N.* HM Inspectorate UK 1976
325 *VdS* VdS 2049 1994–09
326 *Sand H.* PU World Congress 1992
327 *Patten G. A.* Modern Plastics (1961) July P. 119–122,180
328 *Ryan J. E.* ASTM-STP – 502 p. 11–23
329 *NW* Ministerialbl. NW Nr. 57–6. 6 1978
330 *Ohlemiller Et. Al.* J. F. F. 9 (Oct. 1978) P. 489
331 *Tanaka* Cib Symposium Tokyo 1979
332 *Ohlemiller Et. Al.* J. Of Fire And Flamm. 9 (1978) 10 P.
333 *Ohlemiller Et. Al.* J. Of Cons. Prod. Flam. 5 (June 1978) P. 59
334 *Ohlemiller Et. Al.* J. F. F. 9 (Jann. 1978) P. 5
335 *N. W.* Ministerialbl. NW Nr. 57 6. 6. 1978
336 *Damant G. H.* J. F. F. /Cons. Prod. Flamm. 2 (Jun. 1975) P. 140
337 *Yuill C. H.* Jff 1 (1970) P. 312
338 *Jach W.* Schadensprisma 2/78 S. 23
339 *Bayer AG* PU – 51003 1. 10. 1971
340 *Lehnen A,Meckel. L* Chemiefasern/Textilind. 27/29 (1977) P. 267
341 *Haas P. F. Prager F. H.* PUR World Congress Nice sept. 1991
342 *Knipp U. Prager F. H. Wittbecker FW. Weber H. R. Wegenstein H.* VB 4/92 11 (1992) h. 4 s. 1
343 *Szabat Et. Al.* J. Fire Sci 8 (March 1990) P. 109
344 *Madaj E. J. Jasenalz J. R.* 33. Ann. Pur Techn. Market. Conf. Sept 90p177
345 *Prager F. H.* IfBt, PAIII 15. 11. 79
346 *Prager F. H.* IfBt, PAIII 6. 5. 1980
347 *Dowling K. C. Feske E.* Utech 94
348 *Hanusa R.* SPI Conference 1982
349 *Tewarson H.* Combust. And Flame 19 (1972) P. 101

350 *Thuresson P*. SP Report 1964–61
351 *Gebhard M*. Vogelbusch 71 Erg. Lfg. 3/99
352 *Scudamore M. Briggs P. Prager F. H*. Fire and Mater. 15 (1991) P. 147
353 *Tewarson A*. Fire And Materials 4 (1980) 4, P. 185–191
354 *Herrington R. M*. Jff 10 (Oct. 1979) P. 308–325
355 *Dokler T* J. of Hazardous Materials10 (1985) P. 73–87
356 *Vanspeybroek Et. Al*. Cellular Polymers 12 (1993) 5 P. 374
357 *Gross D* Interflam 85 März 85
358 *Stone H*. J. of Cell. Plsastics 23 (July 1987) p. 367
359 *Hillenbrand L. J. Wray J. A*. Battelle Columb. Laborat. July 1973
360 *Morikawa T. Yanai E*. J. Fire Of Sci. 4 (Sept/Oct 1086) P. 299
361 *Morikawa T*. J. Of Fire Sci. 6 (March/Avril 1988) P. 86
362 *Bankston C. P. Cassanova R. A. Powell E. Zinn B. T*. JFF 7 (April 1976) p. 165
363 *Steinert C*. VFDB 4/2000 p. 148
364 *Damantm G. H*. JFF / Consum. Prod. Flamm. 2 (june1975) p. 140
365 *Jin T*. Rep. Of Fire Res. Inst. Japan 33 (1971) P. 31
366 *Prager F. H*. VFDB 1/82
367 *Prager F. H. Wittbecker F. Egresi M. Sasse H. R. Zorgman H*. J. Of Fire Sci. 10 (March/Apr. 1992) P. 118
368 *Grayson St. Hume J. Smith D. A*. Cell. Polymers 3 (1984) p. 433–457
369 *Chien W. P. Seader J. D*. Fire Technology 8/74 P. 187
370 *Stone H*. J. of Cell. Plastics 23 (July 1987) P. 367
371 *Prager F. H*. Plastiques Modernes et Elastomeres Avril 1975
372 *Brochhagen F. K. Prager F. H*. Belgian Plastics 28 IV 1972
373 *Ames S. A. Fardell P. J*. FRS-BRE-CP 3180
374 *Woolley W. D. Ames S. A*. FRS Note 881
375 *Smith A. G*. Fire 71 (1979) Nr. 886 p. 553–555
376 *Bußmann B. Einbrodt H. J. Sasse R. H. Prager F. H*. J. of fire Sci. 10 (sept. /Oct. 1992) Nr. 5 p. 411–431
377 *ISO TC 92 SC3* TR 9122 part 4
378 *Morikawa T. Yanai E. Nishina S*. J. of Fire Sci. 58July/Aug. 1987) p. 248
379 *Prager F. H. Darr W. C. Wood J. F*. Cellular Polymers 3 (1984) p. 161–194
380 *Hurd r. King D. Powell D*. Cellular Polymers 10 (1991) Nr. 1
381 *Cornish H. Hahn K. Barth M*. Environm. Health Perspectives 11 (1975) p. 191–196
382 *Cornish H. Boettner E. Hartung R*. The Univers. of Michigan PRC 1974
383 *Rumberg E. Reploh H*. MPA NW Oct. 1978
384 *Prager F. H*. J. of Fire Sci. 6 (Jan. /Febr. 1988) Nr. 1 p. 3–24
385 *Lynch G*. Frs-Note 1035 May 1975
386 *IMO* FB/353 19. 1. 1984
387 *Einhorn I* Polym. Series Confer. Salt Lake City 1976
388 *Birky M. M*. Intern. Confer. Edinburgh Oct. 1975
389 *Alerie Y*. Tox. and Appl. Pharm. 51 (1979) p. 341–362
390 *Rasbash D. J*. Fire Safety J. 17 (1991) P. 85–93
391 *Rashbash D. D. Drysdale D. D*. Fire Safety J. 5 (1982) p. 77–86
392 *Baker R. R* Comb. and Flame 30 (1977) P. 21–32
393 *Babrauskas V*. Fire Safety J. 14 (1989) P. 135–142
394 *Babrauskas V. Levin B. C. Gann R. G*. ASTM Standardisation News 14 (1986) Nr. 9 p. 28–33
395 *Morikawa T*. Rep. Of Fire Res. Inst. Of Japan41 (1976)
396 *Oestmann B. Nußbaum H*. Fire and Materials 17 (1993) p0. 191–200
397 *Leonard J. E*. Fire and Materials 24 (2000) p. 143–150
398 *Abe K*. Zivilverteidigung II/1978
399 *N. N*. Fire Internat. 69 P. 44
400 *Berl W. G. ,Halpin B. M*. Fire J. Sept. 1979 Nr. 5 P. 105
401 *Morikawa T*. CIB Ssympos. 1979 Tokyo
402 *Christian S. D*. Thesis South Bank University July 1993
403 *Herpol C*. Fire And Mater. (1976) 1,P. 29–35
404 *Einhorn I. N*. Environm. Health Perspect. 11 (1975) P. 163
405 *Einbrodt H. J*. Forschungsbericht Nr. 101/77 Mai 1983
406 *Oettel Et. Al*. Z. Vfdb 3/1968 P. 79
407 *Fflury F. Zernik F*. Springer Verlag Berlin 1931
408 *Packham S. C. Hartzell G*. J. Of Test And Evaluation 9 (1981) P. 341
409 *Hilado C. Huttlinger P. A*. Fire Techn. 16. Nr. 3 8/1980
410 *Staak M. Kraemer R*. Med. Welt Bd. 30 (1979) Nr. 40 P. 1457 Baader E. W. Urban&Schwarzenberg Berlin,1961 Handbuch der gesamten Arbeitsmedizin
411 *Petry. H*. Urban&Schwarzenberg Berlin 1961 Handbuch der gesamten Arbeitsmedizin

412 *Einbrodt H. J. et al* Wissenschaft und Umwelt 4/1990 p. 191
413 *Müller W.* Dissertation Uni Münster CB 2015 1965
414 *ISO TC 92 SC3 TR 9122 part 2*
415 *Forbes Et. Al.* Am. J. Physiol. 143,594 (1945)
416 *Levin B. C. Paabo M. Gurman J. Harris S. Braun E.* Toxikology 1987
417 *Forbes W. H.* Amer. J. physiol. 14 (1945) p. 549
417 *Purser D.* STPE Handbook
418 *Einhorn I. Grunnet. M. L.* Fire Research 1 (1977/78) P. 143–169
419 *Petajan J. H. Et. Al.* Frc-Uu-16 10. Dec. 1973
420 *Wharton Et. Al.* JAMA Febr. 24 (1989) Vol. 221 Nr. 8 P. 1177
421 *Daunderer M.* Fortsch. Med. 97 (1978) Nr. 23
422 *Zirkria,Budd,Flach,Ferrar* Salt Lake City 1976
423 *Zirkria B.* 166 ACS Tagung Chicago 1973
424 *Morikawa T.* J. of Fire Sci. 11 (May 1993) P. 195
425 *Harland W. A. , Woolley W. D.* Fire Intern. 66-P. 37–42
426 *Harland W. A. Anderson R. A.* Qmc-1982
427 *Bonsall J. L.* Human Toxicol. (1984) Nr. 3 P. 57–60
428 *Sakurai T.* J. of Fire Sci. 7 (Jann. 1989)
429 *Hartzell G. E. Grand A. F. Switzer W. G.* J. of Fire Sci. 6 (Nov/Dec 1988) .
430 *Hartzell G. E.* Chapter 1 Advanc. In Combust. Toxicol.
431 *Einbrodt H. J.* Wissenschaft Und Umwelt 4 (1990) S. 191
432 *Higgins. ,Fiorca,Thomasw,Davies* Fire Techn. 18 (1972) P. 120
433 *Gray J.* A. M. A. Archiv. Of Ind,Health 19 (1959) May
435 *Oda H. Kusumoto S. Nogami H.* Chapter 5
436 *Plesser G.* Uni Würzburg
437 *Birky M. M.* Fire and Materials 10 (1986) P. 125–132
438 *Casida J. E Eto M. Moscioni A. Engel J. Milbrath D . Verkade J,G.* Toxic. and Appl. Pharmac. 36 (1976) p. 261–279
439 *Martin K. G. Dowling V. P.* Fire and Mat. 3 (1979) nr. 4 p. 202
440 *Morikawa T* Bul. of Jap. Ass. of Fire Sci. 41 (1992) p. 9
441 *Bryan J. L.* Fire Safety J. 11 (1986) P. 15–31
442 *Waritz R. S.* Environm. Health Persp. 11 (1973) P. 197–202
443 *Hilado C. ,Olcomendy E. Schneider J.* J. of Comb. Prod. Flamm. 6 (Sept. 1979) P. 189
444 *Kimmerle G.* Ann. Occup. Hyg. 19 (1976) P. 269–273
445 *Munn A* New Scientist 3. 4. 1975
446 *Munn A.* Ann. of Occup. Hygiene 8 (1965) May p. 143–160
447 *Einhorn I* Science 187 (Febr. 1975) p. 743
448 *Norpoth, Mangold,Brücher,Amann* Zbl. Bakt. Hyg. Orig. B. 156 (1972) P. 341–352
449 *Schildknecht H.* Sicherh. in Chem. und Umwelt 2 (1982) P. 121
450 *Effenberger E* Expert Report, Bayer AG 30. Jan. 1973
451 *Effenberger E* Expert Report, Bayer AG 10. Feb. 1978
452 *Sumi K. Tsuchiya Y.* Salt Lake City 1976
453 *Einhorn,Birky,Hileman,Ryan,Voorhees.* Frc/Uu 18a 1974
454 *Hileman,Voorhees,Wojcik,Birky,Tyan,Einhorn* J. Of Polym. Sci. Polym. Chem. 13 (1975) P. 571
455 *Kallonen R.* J. Of Fire Sci. 8 (Sept/Oct. 1990) P. 343
456 *Kallonen,Von Wright,Tikkanen,Kaustia* J. Of Fire Sci. 3 (May/June 1985) P. 145
457 *Merz,Neu,Kuck,Winkler,Gorbach,Muffler* Z. Anal. Chemie (1986) 325,449–460
458 *Sistovaris Et. Al.* Fresenius Z. Anal. Chem. (1089) 334 221–225
459 *Kübler R. et. al* VFDB 1/1991 p. 20
460 *Pohl K. D.* VFDB 2/1987 p. 73
461 *Mead J. A.* Qmc 1982
462 *N. N.* FP. /121 23. 06. 75
463 *Oettel H.* Moderne Unfallverhütung10 (1966) P95–119
464 *Prager F. H.* J. Of Fire Sci. 6 (Febr. 1988) P. 3
465 *ISO TC 92 SC3 TR 9122 part x*
466 *Klimisch H. Hollander W. Thyssen J.* J. Of Comb. Tox. 7 (Nov. 1980) P208
467 *Boudene C. Jouany S. Truhaut R* J. Makrom. Sci. Chem. An. (8)
468 *Behle Ch.* Thesis Gießen 1988
469 *Effenberger E.* Städtehygiene 12 (1972) p. 275
470 *Galster W. A. Hileman F. D.* FRC-Report Salt Lake City (Report to III)
471 *Einbrodt,Hupfeld,Prager,Sand.* J. of Fire Sci. 2 (1984) P.
472 *Einbrodt H. J. Sasse H. R. Prager F. H.* Materialprüf. 27 (1985) 4 P94
473 *Yusa S* Paper at ISO TC 92 SC 3
474 *Yusa S* Montedipe Confer. Milan June 1986
475 *Yamamoto K. Kato Y.* 15. Conf. of Jap. Leg. Med. Ass. Kinki 1968
476 *Hilado C. J.* J. of Con. Prod. Flamm. 4 (March 1977) p. 40
477 *Kaplan. H. Grand A. Switzer W. Gad S.* J. of Fire Sci. 2 (1984) Apr/May p. 153
478 *Effenberger E.* Städtehygiene 12 (1972) P. 275
479 *Sand H. Hofmann H* Österreich. Kunstst. Ztschr. 8 (1977) P. 37–45

480 *Sand H. Hofmann H.* Salt Lake C. 1976
481 *Fleischmann* Thesis RWTH Aachen 1993
482 *Prager F. H.* Fire and Materials 18 (1994) p. 131–149
483 *Galster Et. Al.* FRC-Report to III
484 *Alarie Y.* Ann. Rev. Pharm. Toxic. 25 (1985) P. 325–347
485 *Hilado C. J.* J. of Prod. Flamm. 3 (Dec. 1976) p. 288
486 *Babrauskas V. Harris R. H. Braun E. Levin B. C. Paabo M. Gann R. G.* J. of Fire Sci. 9 (March/Apr. 1991) P. 125
487 *Babrauskas V* Fire andMaterials 21 (1997) p. 53–65
488 *Purser D. A. Fardell P. J. Rowley J. Vollam S. Bridgman B. Ness E. M.* Flame retardants Conference 94
489 *Alexeeff G. V. Lee Y. Ch. White J. A. Putas N. D.* J. of Fire Sci. 4 (March 11986) p. 100
490 *Skydd 69* Plastics-Fire and Corrosion Stockholm 1969
491 *Sandmann Et. Al.* Fire and Materials 10 (1986) P. 11–19
492 *Nowlen S. P.* Fire Safety J. 15 (1989) p. 403–413
493 *Babrauskas Et. Al.* Fire and Flamm. Bulletin Dec. 1990 P. 7
494 *Humphreys H. Christopher A.* Fire and Mat. 9 (1985) Nr. 3 P. 120
Hupfeld J. Schadensprisma 14 (Nov. 1985) Nr. 4 P. 61
495 *Lawson Et. Al.* NBS IR 2787 1984
496 *Waterman Th. E.* IITRI Project J 8116 Nov. 1969
497 *Mcguire J. H. Cambell H. J.* Fire Techn. 16 (1980) 2 P. 133–141
Mcguire J. H. D'Souza M. V. Fire Techn. 15 (1979) 2 P. 102–106
498 *Martin K. G. Dowling V. P.* Fire and Mat. 3 (1979) Nr. 4 P. 202
499 *Martin K. G. Nicholl P. R.* CSIRO Building Research 1977
500 *Ramsayg. C. Dowling V. P.* CSIRO 36104/1984
501 *MPA's* (Braunschweig,Stuttgart, Dortmund, Leipzig) Report to IVPU
502 *Prager F. H.* Vogelbusch Katastrophenschutz in Arbeitsstätten 71 Erg. -49, 3/99
503 *ISO TC 61 TR 10353 – ISO 10093*
504 *Summerfield M. Ohlemiller T. Sandrusky H.* Combust. And Flame 33 (1978) P. 263–279
505 *Ramsay Et. Al.* Normvorschlag Australien
506 *Babrauskas,Harris,Braun, Levin,Paabo,Gaum* J. of Fire Sci. 9 (March/Apr. 1991) P. 125
507 *Damant G. H. Langford N:J.* Jff/Cons. Prod. Flamm. 2 (1975) Sept. P. 204
508 *Damant G. Nurbakhsh S* AFNA –Conference 19xx
509 *Prager F. H.* Vortrag Brüssel
510 *Day M. Suprunchuk T, Wilson D. M.* J. of Cons. Prod. Flamm. 6 (1979) P. 233
511 *Gippons J. A. Stevens G. C.* Fire Safety J. 15 (1989) P. 183–191
512 *Ettling B. V.* Fire Technol. 18 (Nov. 1982) Nr. 4,P. 344
512 *Hupfeld J.* Schadensprisma 14 (Nov. 1985) Nr. 4 P. 61
513 *Kahnau Et. Al* Elektrotechn. Ztg. B28 (1976) P. 2–9
513 *Nowlen S. P.* Fire Safety J. 15 (1989) P. 403–413
515 *Frey J. F. ,Lustig R. E.* FRS Techn. Paper Nr. 3,1962
516 *Pohl K. D.* Kriminalist H. 5 (1989)
517 *Ausobsky* Brandverh. Brandbekämpf. 4/66 P. 61
518 *Seekamp H. Roeske G.* Berichte aus der Bauforschung Heft 34
519 *Troitzsch J* Der Maschinenschaden 52 (1979) Nr. 6 P. 216
520 *Paul K. T.* J. of Fire Sci. 5 (May 1987)
521 *Mehkeri K. Dhawan K.* Fire Technology May 1982 P. 152–161
522 *Williamson R. B.* Polym. Ser. Conf. Salt Lake C. 1973
523 *Babrauskas V. Krasny J.* NBS Monograph 173 Nov. 1985
524 *Paul K. T.* BRE-Note 12/82
525 *Paul K. T.* Fire Safety J. 16 (1990) P. 389–410
526 *Paul K. T.* Fire Safety J. 14 (1989) p. 269–286
527 *Marchant R. P.* FIRA UK, May 1977
528 *Anderson R.* Fire Technol. 20 (2/1984) Nr. 1 P. 64
529 *Prager F. H.* PUR World Congress Nice 1991
530 *Paul K. Christian S.* J. of Fire Sci. 5 (May 1987) P. 178
531 *Bank* University Wuppertal 19.
532 *Paul K.* RAPRA-Conf. Report 8906
533 *N. N.* Boston Fire Departement Jann. 1975
534 *Abbot J. C,* Fire J. July 1971 p. 88–92
535 *Powers W. R.* SFPE Techn. Rep. 77–3 (1977)
536 *Kahnau H. ,Kieninger W.* Elektrotechn. Ztg. B28 (1976) P. 2–9
537 *Hölemann* Gfs-Sympos. 3. -5. 6. 1985 Ffm
538 *powell* J. of Fire Sci. 7 (May/June 1989) p. 145
539 *N. N.* StZ 27 (1975)
540 *Egeln O.* Schaden-Prisma 4 (1980) Nov. P. 62
541 *Cooksey P. N.* The Internat. Fire Chief 46 (1980) Nr. 9 P50

542 *Babrauskas V. Parker W. J.* Fire Safety J. 8 (1984/85) P. 199–200
543 *Ohlemiller T. J. Villa K.* NIST IR – 4348
544 *Ohlemiller T. J.* Fire Safety J. 18 (1992) P. 325
545 *Sundström B.* EUR 16477 EN
546 *Ueberall Th.* Re3port to DIBT/GKV/IVPU
547 *Nadeau H. G. ,Waszeciak P. H.* J. of Cell. Plast. Jul/Aug. 1976 P. 208
548 *Herzog W. ,Bauer H.* Melliand Textilber. 4 (1980) P. 362
549 *Williamson R. B. ,Baron F. M.* Jff 4 (Apr. 1973) P. 99
550 *Belles D, Fisher F. Williamso R. B.* Fire J. Jan/. Febr. 1988 P. 25
551 *Hirschler M. M.* Fire Safety J. 18 (1992) P. 305
552 *Briggs P.* The Plast. and Rubb. Inst. 92
553 *Ueberall H. Klingelhöfer H. G.* Forsch. Ber. MPA NW. T. 1891 1987
554 *N. N.* Fire Prevention 140 Febr. 1981
555 *N. N.* Fire Prevention Nr. 140 P. 32
556 *Hicks B:L.* The J. of Chem. Physics 22 (1954) Nr. 3 P414–429
557 *Hallman Et. Al.* Spe J. Sept. 28 (1972) P. 43–47
558 *Hicks B. L.* The J. of Chem. Physics 22 (1954) Nr. 3 P414
559 *Kashiwagi T.* Fire Safety J. 3 (1981) P. 185–200
560 *Berlin G. N.* Fire Technology 16 (1980) Nr. 4 P. 287
561 *Hansen Asibulkin M.* Combust. Sci. and Techn. 9 (1974) P. 173–176
562 *Sundström B, Kaiser I.* Sp-Rapp 1986 01 Issn 0280 2503
562 *Wickström U. Goranson U.* Cib-Vortrag-1986,Budapest
563 *Simmes D. L. Law M.* Combust. and Flame 11 (1967) P. 377–388
564 *Babrauskas* NBS IR 81–2271
565 *Babrauskas V. Parker W.* Fire and Mater. 11 (1987) P. 31–43
566 *Heffels P.* Vfdb 1/1986 P. 20
567 *Malhotra* Interflam 93
568 *Drysdale Et. Al.* Fire Safety J. 13 (1988) P. 185–196
569 *Robertsa. F.* Fire Technology Aug. 1971
570 *Papa A. J. ,Proops W. R.* J. of Appl. Polym. Sci. 16 (1972) P. 2361–2373
571 *O'neill T.* Fire Safety J. 15 (1989) P. 45–56
572 *NN.* Fire International 69 P. 60–66
573 *Oestmann B. Nussbaum R.* . Fire and Materials 10 (1986) P. 151–160
574 *Kashiwagi T.* Comb. Sci. and Techn. 8 (1974) P. 225–236
575 *Babrauskas V.* Fire Safety J. 14 (1989) P. 135–142
576 *Babrauskas Et. Al.* NBS Special Publication 749
577 *Bondel A.* Furniture Manufacturer Dec. 1988
578 *Finegan* Fire Prevention 327 Dec. 1999
579 *F. M.* FMRC Update
580 *Babrauskas V.* Fire and Flamm. Bull. Feb. 1992
581 *N. N.* Fire J. Jann. 1988 P. 53–63
582 *Friedman R.* Internat. Symposium Edinburg 1975
583 *Fernandez-Pello A. ,Williams F. A. The Combust. Inst. Symp. Pittsburgh 19??*
584 *Waterman D. E.* Fire Technology 2–1968
585 *Morikawa Et. Al.* Bul. of Jap. Ass. of Fire Sci. 41 (1992) P. 9
586 *N. N.* Eppma June 1976
587 *Lewis A.* Fire Prevention 216 Jann. 1989
588 *Quintiere J.* ASTM-STP 614
589 *Quintiere J.* Fire Technology Nov. 1988 P. 333
590 *Prusaczyk J. E. ,Boardway R. M. Jff 11 (Oct. 1980) P. 314*
591 *Budnik* NBS IR -79–172 March 1979
592 *Oedeen K.* 6. IBS Karlsruhe 1982 S. 93
593 *Takita T.* Schienen Der Welt Okt. 1977,S. 512
594 *Fang J. B.* Fire Technol. Feb. 1981 P. 5–16
595 *Tran H. C.* Interflam 90
596 *Malhotra H. L.* Int. Sympos. Edinburgh 1975
597 *Benjamin I. A.* Int. Sympos. Edinburgh 1975
598 *Paul K. T. King D. A.* Cellular Polymers
599 *Williamson R. B.* Salt Lake City 1973
600 *Beard A.* Fire Safety J. 18 (1992) P. 375
601 *Kawagoe K. , Hasemi Y.* Fire Safety J. 3 (1980/81) P. 149–162
602 *Thomas P. H.* FRS Note N. 176/83 Dec. 1983
603 *Bukowski R. W.* Fire Materials 9 (1985) P. 159–166
604 *Levin R. S.* Fire Technology May 1988 P. 163
605 *Lee B. T.* Jff 3 (1972) P. 164
606 *Jianmin Q* SP Report 1990:38
607 *Friedmann R.* Fire and Mater. 2 (1978) p. 27–33
608 *Woolley W. D. Fardell P. J.* 3. FR . Polym. Confer. Turin Sept. 1989
609 *Jolly S. Saito K.* Fire Safety J. 18 (1992) P139–182
610 *Wakamatsu T.* ASTM-STP 614 1976 P. 168
611 *Smith R. L.* Fire Technology 23 (1987) Nr. 1 P. 5
612 *Kunze E.* Schadensprisma Nr. 2 18 (1989) Mai P. 25
613 *Van De Leur, Kleijn, Hoogendoorn* Fire Safety J. 14 (1989) P. 287–302
614 *Waterman T. E. Domijanaitis R. A.* 30th Ann. Pur Techn. Mark. Conf. Oct. 1986

615 *Wickström U.* Fire Safety J. 16 (1990) p. 53–63
616 *Ku A. Doria M. Lloyd J.* 16. Symp. Comb. Inst. 1977
617 *Quintiere J. Denbraven K.* Nbsir-78–1512
618 *Quintiere J. Mccaffrey B. Rinkinen W.* Fire and Mat. 2/78 Nr,1 P. 18.
619 *Christian W. J.* Fire J. Jan. 1974
620 *Hinkley P. L.* Frs Note 807
621 *Zukowski E.* Salt Lake City FRC Conf. 1976
622 *Klote J. H.* Fire Safety J. 7 (1984) P. 93–98
623 *Fennel S.* Fire and Flammability Bulletin 10/1990 5
624 *NFPA* NFPA-Official Report Operation School Burning 1959
625 *Fackler J. B.* ASTM-STP-422 1966,P. 205–218
626 *N. N.* Fire Prevention 142 p. 12
627 *N. N.* Fire J. 5 (1980)
628 *Bayer A. G.* Sicherheitsmerkblatt PU 51003
629 *N. N.* Alt Nürnberg 14. 12. 1969
630 *N. N.* Stern 19. 7. 1979
631 *N. N.* Kölner Stadtanzeiger nr. 293 10. 12. 1983
632 *N. N.* Journal Officiel 6 Avril 1976 Nr. 17
633 *N. N.* Östereich. Kunststoff Ztschr. 11/80 1/2
634 *Seidel K. W.* Brandschutz,Deutsche Feuerwehr6/73-P. 167
635 *Seymor R. B.* Ind. and Engng. Chem. 46 (1954) Nr. 10 P. 2135
636 *Easton W. H.* J. of Ind. Hygien. and Toxic. 27 (1945) p. 211
637 *Carmack B. J:* Jff 7 (Oct. 1976) P. 559
638 *Carmack B. J.* JFF 7 (Oct. 1976) p. 559
639 *Powell Et. Al.* J. of Fire Sci. 7 (May/June 1989) P. 145
640 *Lathrop Et. Al.* Fire J. May 1975
641 *Teague P. E.* Fire J. Nov. 1978 P. 21–30
642 *Berlin G. Dutt A. Gruipta S.* Fire Technology Febr. 1982
643 *Stahl F. J.* Fire Technology Feb. 1982
644 *Roberts* Insulation Nov/Dec. 1974 P. 10
645 *Mepperley B. Sewell P.* Europ. Polym. J. 9 (1973) P. 1255–1264
646 *Harbst J. Madsen F.* The Danish Investment Foundation july 1, 1976
647 *Shorter* Canad. Build. Digist CBD 45
648 *Maguire* Fire Engineering 133 (1980) Nr. 9 P. 22–26
649 *Levin Et. Al.* Fire and Mat. 9 (1985) 3 P. 125
650 *Carter Et. Al.* Salt Lake 1976
651 *Pagni Et. Al.* Combust. And Flame 68 (1987) P. 131–142
652 *Melinek S. J.* Fire Prevention 140 Febr. 1981 P. 26
653 *Drysdale D. D.* Fire Prevention Sci. and Techn. 23 P. 18
654 *N. N.* Insulation März 1980 P. 22
655 *John R.* Agf 13 Karlsruhe Aug. 1969
656 *Ohlemiller T. J.* JFF 9 (Oct. 1978) p. 489
657 *Ohlemiller T. J. Et. Al.* Fire Safety J. 18 (1992) P. 325
658 *Ohlemiller T. Bellen; Rogers F.* Combust. and Flame 36 (1979) P. 197–215
659 *Ohlemiller T. J. Villa K.* Nist Ir 4348 June 1990
660 *Babrauskas V. Parker W. J.* NIST- 1991
661 *Ashida K. Oktani M. Ohkubo S.* J. Of Cell. Plastics 14 (1978) Nr. 6 P311–324
662 *Crossland B.* Fire Safety J. 18 (1992) P. 3–11
663 *Murrel J.* Fire 90 (1998) 1113 p. 19
664 *Simony Et. Al.* Railway Gazette Int. Jan. 1989 P. 23
665 *Roberts A. F, Clough G.* Combust. and Flam. 11 (1967) Oct. P. 365–376
666 *Roßmann G.* VFDB 4/1996
667 *Herzog W. Bauer H.* Melliand Textilbericht 4 (1980) p. 362
668 *Thomas P. H.* The Combust. Inst,14. Symp. Pittsburgh 1973
669 *Thomas P. H.* Fire Safety J. 3 (1980/81) P. 67–76
670 *Harmathy T. Z.* Combust. and Flame 31 (1978) P. 259–264
671 *Friedman R* Fire and Mater. 2 (1978) 1 P. 27–33
672 *Quintiere J. G,* Fire Safety J. 3 (1981) P. 201–214
673 *N. N.* Amtlicher Anzeiger Hamburg nr. 170
674 *N. N.* Ministerialbl. NW 17819649 nr. 40
675 *N. N.* Staatsanzeiger Nr. 27 p. 1204
676 *N. N.* NW Richtlinien Fassung Mai 1978
677 *Benjamin I.* Iso Tc 92 Sc1 N31 June 1983
678 *Smith A. G.* Foam, Fire Retardancy and Furnishing Textiles 1992
679 *Cornelissen A. A.* Interflam 90
680 *Birky M. M.* J. Comb. Toxicol. 3 (1976) Febr. P. 5
681 *Levin B. Paabo M. Fultz M. Bailey C.* Fire and Mat. 9 (1985) 3 p. 125
682 *Anderson. R. A. Harland W* Proc. of Europ. Meet. of Inst Ass. of Forens. Toxocol. 1979
683 *Grand A. F.* J. of Fire Sci. 2 (1984)
684 *Alarie Y Anderson D.* Fire and Materials 8 (1984) Nr. 1 P. 54
685 *Alarie Y. Kennash H. E. Stock M. F.* J. of Fire Sci. 5 (Jan. /Febr. 1987) p. 3–16
686 *Caldwell D. J. Alarie Y.* J. of Fire Sci. 9 (Nov/Dec1991) P. 470
687 *Alarie Y.* Ann. Rev. Pharmacol. Toxicol. 25 (1985) p. 325–347

688 Sand H. Hofmann. Th. Poymer Confer. Salt Lake City 1976
689 Norris J. C. Astm-Stp 1082 P. 57–71
690 Hilado C. J. Cumming H. J. Williams J. B. J. of Combust. Toxicol. 4 (Aug. 1977) p. 368
691 Edgerly P. G. Pettett K. BRE – N 39/1981
692 Purser D. STPE Handbook 1988 p. 200–245
693 Prager F. H. Sasse H. R, Cellular Polymers Vol. 20,Nr. 3 2001
694 Whiteley R. H. Flame Retardants 93 nr. 25–1
695 Hilado C. Smouse K. J. Of Comb. Toxic. 3 (Aug. 1976) P. 305
696 Packham S. C. Hartzell G J. of Test and Evaluation 9 (1981) p. 341
697 Packham St. Smith F. PaaboM. Stolte A. Levin B. Malek D. Birky M. M. Report to SPI 1981
698 Alexeeff. G. Lee Y. Thorning D. Howard M. ,Hudson L . J. of Fire Sci. 1986 p. 427
699 IMO FP/185 2. 5. 1978
700 Drysdale D ASTM-STP 882 – 1985 p. 285
701 Gaskill I. R. SPE J. 28 (Oct. 1972)
702 Hilado Fire + Flammability 10/1976
703 Teichgräber R. Topf P. VFDB 2 (1974) p. 52–57
704 Flisi U. Polymer Degradation and Stab. 30/90p. 153
705 Babrauskas V. Interflam 90
706 Caldwell D. J. Alarie. Y J. of Fire Sci. 8 (Jann. 1990) P. 23
707 Prager F. H. Sasse H. R. Cellular Polymers 20 (2001) p. 211–229
708 Szpilman J. Dan-Test Vdc 615. 9 614841. 41 Aug. 1983
709 Nisted T. UDC 615. 90 614. 841. 41 Nr5592–1987
710 Christian W. J. Fire J. Jan. 1974 p. 22
711 McNeil R. R. J. F. F. 9 (July 1978) P. 275
712 Prager F. H. Einbrodt H. J. Sasse H. R. VFDB 4/1994
712 Wills IIRS 11. Nov. 1981
713 Babrauskas V. Et. Al. NBS-Special Publication 749 1988
714 Mizuno T. Kawagoe K. Fire Sci. Techn. 6 (1986) Nr. 1,2 P. 29–37
715 Tran H. C. Fire and Mat. 12 (1988) p. 143–151
716 Sundström B. EUR 16477 EN
717 Shern J. H. STP-422–1966 P. 93–105
718 Derry L Fire J. Nov. 1977
718 Mickelson R. ,Traicoff R. Fire Techn. 8 (1972) 4 P. 301
719 Paul K. T. Fire Safety J. 16 (1990) p. 389–410
720 Paul K. T. Fire Safety J. 14 (1989) P. 269–286
721 Paul K. T. 2nd. Conf. Flam. Retard. 85
722 Paul K. T. Fire and Mater. 10 (1986) P. 29–39
723 Nelson G. Fire Technology 34 (1998) p. 222
724 Zirner F. Brandverhüt/Brandbekämpf. 20 (1970) 1 P. 9
725 Stoll A. M. Chianta M. A. Aerospace Med. 39 (1968) P. 1097–1100
726 Stoll A. M. Green L. C. J. of Appl. Physiol. 14 (3) 1959,P. 373–382
727 Halpin M. ,Radford P. Fisher R. Caplan Y. Fire J. 5/1975 P. 11
728 Berl W. Halpin B. M. ASTM-STP 614 1976 P. 26
729 Noguchi Th. Salt Lake City 1976
730 Gerson L. Am. J. Drug. Alcohol Abuse 8 (1979) 1 p. 125–133
731 Teige. B,Lundevahl. J,Fleischer. E Z. Rechtsmedizin 8 (1977) P. 17–21
732 Kishitani K J. of The Fac. Of Engng. Uni Tokyo Vol. XXXT Nr. 1,1971
733 Clarke F. B. ,Ottoson J. Fire J. Vol. 70 (1976) May P. 20
734 Kulgemeyer J. Diss. Aachen Nr. 161/84
735 V. Lüpke. H,Schmidt K. Z. Rechtsmedizin 79 (1977) P. 69–71
736 Tsuda Y. Japan. J. of Traumat. and Occup. Medicine 33 (1985) Nr. 3 p. 139
737 Tsuda Y Yokohama Igaku. 29 (1978) p. 163–168.
738 Miller A. Environm. Health Persp. 11 (1975) P. 243–246
739 Zirkria B. A. Polymer Confer. Salt Lake City 1976
740 Buettner K. J. Amer. Med. Ass. 144 (1950) 9 P. 732–738
741 Pauluhn J. J. of Fire Sci. 11 (1993) P. 109
743 Figge J Pieler J. Et. Al. Wissenschaft und Umwelt 3/1992
744 Pieler J Figge J Et. Al. Wissenschaft und Umwelt 3/1992
745 Herpol C. Report to VKE 1979
746 Oettel H. Hofmann T. H. Z. Vfdb. 3 (1968) P. 79–88
747 Anderson R. Willetts. P. Cheng K. Harland W Fire and Mat. 7 (1983) Nr. 2, p. 67
748 Anderson R. A. Harland W. A. Fire and Mat. 3 (1979) Nr. 2 p. 91
749 Radford,Pitt,Halpin,Caplan, Fisher,Schweda. Conf. Salt Lake City 1974
750 Harboe M. Scand. J. of Clin. Lab. Invest. 9/57 p. 317
751 Lily R. Cole P, Hawkins L. Brit. J. Industr. Med. 29 (1972) P. 454–457
752 Babrauskas V. Levin B. Gann R. Astm Standardis. News 14 (1986) Nr. 9 P. 28
752 Kishitani K. Nakamura K. Jff/Comb. Toxicol. 1 (1974) May p. 104

753 Levin B. C. Fire and Mat. 9 (1985) Nr. 3,p. 125
754 Levin B. Paabo M. Fultz M. Bailey C. Fire and Mat. 9 (1985) 3 p125
754 N. N. Fire Prevent. 176 Jan. Febr. 1985 p. 10
755 Alarie Y Ann. Rev. Pharm. Toxic. 25 (1985) p. 325–347
756 Hartzell,Packham,Grand,Switzer. J. of Fire Sci. 3 (May/June 1985) p195
757 Purser D. A. Qmc 1982
758 Sumi K. Fire and Materials 8 (1984) Nr. 1 p. 1
759 Sumi K. Taylor W Jff 12 (Oct. 1981) P. 258
760 Sakai T. Okukubo A. Europ. Confer. on Flamm. A. Fire Retard. 1977
761 Herpol C. Vandevelde P. Fire And Mater. 2 (1978) Nr. 1,P. 7
762 Alarie Y. Anderson. A. J. of Fire Sci. 9 (Nov/Dec1991) P. 470
763 Bayern Bayr. Gesetz u. Verordn. Bl. 20/7. 82 Nr. 18
764 NW Versammlungsstätten VO NW MBl 1976 § 110
765 NW Gesetz u. Verordnungsbl. NW 58 (2002) 8. Oct. Nr. 26
766 Prager F. H. Sasse H. R. Cellular Polymers Vol. 20 Nr. 3 2001
767 N. N. FEJ&FP Dec. 2002 p. 6
768 Steinhoff D. VFDB 3/1982
769 N. N. New Scientist 22 July 1982
770 N. N. Fire International 77 P. 23
771 N. N. Report to DOT (Departm. of Transport) 1974
772 N. N. Bau O NW §1
773 N. N. Section 5–45 und 5–44 von [1]
774 UK Upholstered Furniture (Safety) Regulations 1980
775 UK Amentments – Regulations 1983
776 UK 1988 Nr. 1324 The Furniture and Furnishings (Fire) (Safety) Regulations
777 Woolley W. D. CP 30 Febr. 1978
778 Woolley W. D. Fire Safety J. 2 (1979/80) p. 39–59
779 Berman A. M- New Scientist 22. 5. 1975
780 Lee B. T. Jff 3 (1972) p. 164
781 Woolley W. D. FRS Note 1038
782 Schmieder H. Unser Brandschutz 1 (1971) p. 11–19
783 Babrauskas V. Fire Safety J 8 (1984/85) p. 149–200
784 Fang J. B. JFF/ (1976) p. 368
785 Ramsay G. C. Int. Symposium Edinburgh 1975
786 Damant G. J. of Prod. Flamm. 5 (March 1978) p. 20
787 Seidel D. IBS Eindhoven 1967
788 N. N. Report University Sussex to DTI 1996
789 Wittbecker F. W Walter Et. Al. Interflam 93
790 Magnusson Et. Al. NBS STP 882 1985 Fire Safety Philadelph.
791 Klöker W. Niesel H. Prager F. H. Schiffer H. W. Bökenkamp, Klingelhöfer H. G. Kunststoffe 67 (1977) 8 p. 7–9
792 Jeffs G. M. Klingelhöfer H. G. Prager F. H. Rosteck H. Fire and Materials 10 (19869 p. 79–89
793 N. N. Fire J. Jan. 1982 p. 53–57
794 Prager F. H. Karst E. Rosteck H. Steuer W. VFDB Mai 1983 p. 58–63
795 Prager F. Cope B. H. Hildebrand Ch. Levio E. Report to ISOPA 1993
796 Hildebrand Ch. Bau-Zeitung 49 (1995) 12
797 Jagfeld P. VFDB1/1988 S. 10
798 Prager,Barthel,Decker,Rosteck,Steuer. Kunststoffe im Bau 15/1980,2,P. 61
799 Beitel J. ASTM Standardization News Jan. 1990
800 Thomas Ph. Combust. Inst. Pittsburgh 14 (1973)
801 Cope B. Hildebrand Ch. Levio E, Prager F. H. ISOPA Publication 1995
802 Hildebrand Ch. Bau-Zeitung 49 (1995) 1/2
803 N. N. Fire Prevention 307 march 1998
804 Babrauskas V. Interflam 90
805 Babrauskas V. J. of Fire Sci. 9 (March/Apr. 1991) p. 125
806 N. N. Fire prevention 2000
807 Purser D. A. ISO-TC 92-SC3 N14/WG5
808 Babrauskas Nbsir 80–2186
809 Hartzell G SWRI Interim Report 03–5921–001 1980
810 N. N. Fire Prevention 339 Dec. 2000
811 Todd C. Fire Prevention 337 Aug. 2001
812 Kirk P. G. ,Stark G. W. FRS,HM Factory Inspectorate UK
813 Karlsch D. VFDB 4/76
814 Karst H. F. Polyureth. World Congress Aachen 1987
815 Witte H. Brandsch/Deutsche Feuerwehrz. 12/1987
816 Wilde D. G. JFF 11 (Oct. 1980) p. 263
817 Prager F. H. Sasse H. R. Cellular Polymers Vol. 20 Nr. 3 2001
818 Prager F. H. Gutachten Dr. F. H. Prager 51379 Leverkusen
819 Fischer Zeuge Düsseldorf –Gutachten Prager Ref. 818
820 Sauerbrunn I. IVPU-Brandseminar 28/30. Okt. 1975
821 N. N. Ministerial Bl. NW Nr. 57 6. 6. 1978
822 N:N: Zeugin N. N. Düsseldorf – Gutachten Prager Ref. 818

823 *Klingelhöfer H. G.* VdS Symposium Düsseldorf 5 Jahre danach
824 *Dahlberg* Zeuge Düsseldorf – Gutachten Prager Ref. 818
825 *Babrauskas V.* Fire and materials 24 (2000) p. 113–119
826 *Schneider U* VdS Symposium Düsseldorf 5 Jahre danach
827 *Prager F. H.* Handelsblatt Mittwoch 3. 6. 1981 Nr. 105
828 *Babrauskas V.* NBS Monograph 173 Nov. 1985
829 *FRS* Report Maysfield Centre Belfast Her Majesty Stationary Office 1984
830 *Woolley W. D. Ames S. A. Pitt A. T. Buckland* Fire Safety J. 2 (1979/80) p. 31–59
831 *Piechota H.* J. of Cellular Plyastics Jan. 1965
832 *Szabat J. F.* J. of Fire Sci. 8 (March 1990) p. 109
833 *Gallagher J. S. Zich W. E.* J. of Fire Sci. 8 (Sept. 1990) p. 361
834 *Pohl K. D.* VFDB 1 (1986) p. 28
835 *Prager F. H.* Kunststoff J. 1–2 (1976)
836 *Prager F. H.* Thesis D82 RWTH Aachen 1985
837 *IfBt* Mitteilung IfBt 9 (1978)
838 *IfBt* Mitteilung IfBt 2 und 3 (1984)
839 *MPA's* Report to IVPU
840 *N. N.* Leverkusener Stadtanzeiger 20. 6. 1975
841 *Karlsch D.* VFDB 4/1976
842 *Brown J. R.* Fire and Mater. 19 (1995) 3 p. 109
843 *Robinson P.* Fire Engng. J. 55 (1995) 177 p. 29
844 *Rowling P.* Fire Engng. J. 55 (1995) 165
845 *Murrell J.* Fire 90 (1988) 1113 p. 19
846 *Jagfeld P.* Report to IVPU 19.
847 *USA* DOC FF 4–72
848 *Lawson J. R.* NBS IR 2787 1984
849 *Ihrig A. M. Rhyne A. C. Norman V. Spears A. M* J. of Fire Sci. 4 (July/Aug. 1986)
850 *Paul K. T.* J. of Fire Sci. 18 (2000) 1 p. 28–73
851 *Prager F. H.* J. of Consum. Prod. Flamm. 5 (March 1978) p. 3. 19
852 *Mc. Ree D. D.* J. of Fire Sci. 18 (2000) 3 p. 215–241
853 *Krasny J. T.* NBS IR 87–3509
854 *Blumenhagen H.* J. Schadensprisma 781978) Nr. 4 p. 62
855 *Sheldon M.* Fire Prevention 1978 Nr. 123
856 *N. N.* Brandverhütung 1980 Nr. 140 p. 25
857 *Butler J.* Fire Prevention 264 Nov. 1993 p. 26
858 *N. N.* Fire Prevention 332 May 2000
859 *Prager F. H.* Der Dachdeckermeister 8/1989
860 *Becker W.* Fire Safety J. 1081986) p. 139–147; 149–154
861 *Heany E. W.* Fire Command 45 (1978) Nr. 2 p. 30
862 *Klingelhöfer H. G.* Report to B ayer A. G.
863 *N. N.* Hamburger Anzeiger 01. 09. 1978
864 *2N. N.* Staatsanzeiger für das Land Hessen Nr. 6 06. 02. 1978
865 *3. N.* Ministerialbl. NW 31 (1978) Nr. 57 06. 07. 1978
866 *Scott A. Paul K. T.* RAPRA Members 1 (Oct. 1973) Nr. 10 p. 253
867 *Moulen A. W.* CEBS TR 52/75/386 1968
868 *Briggs P. Morgan A.* Fire Safety Engng. 6 (1999) 4 p. 18
869 *Benjamin J. A.* ISO TC 92 SC1 June 1983
870 *Prager F. H.* Vogelbusch Erg. Lfg 3/99
871 *Prager F. H.* VKE Vortragsveranst. Würzburg 1989
872 *Prager F. H.* Bauphysik 12 (1990) Nr. 6 p. 165
873 *Ries R.* VdS Sympos. 5 Jahre danach
874 *FAA* Notice Nr. 75–3, 40FR 6508 1975
875 *N. N* Journal Officiel 6 Avril 1976 Nr. 17
876 *N. N.* EG Doc. 1/246/70 E
877 *PAIII* IfBt 3 (1978) 5
878 *PAIII* IfBt 6 (1984)
879 *PAIII* IfBt 2 (April 1973)
880 *Prager F. H.* Bauphysik 12 (1990) Nr. 6 p. 165
881 *Szpilman J.* DAN Test VDC 615. 9. 614841. 41 Aug. 1983
882 *Prager F. H. Sand H.* FNM-Workshop RWTH Aachen 1986
883 *N. N.* California Assembly Bill 594 /CAL TB 133
884 *Jenkins Ch. E.* J. of Cons. Prod. Flamm. 9 (1982) p. 20
885 *Best R.* Fire J. May 1980 p. 51
886 *N. N.* Fire Prev. 239 May 1991
887 *Wallace D.* Fire and Mat. 16 (1992) p. 77–94
888 *Chien W. P.* J. of Fire Sci. 9 (May 1991) p. 173

Abbreviations list

Code	Explanation	Application	Area
AGt	Ausschuß für Gebrauchstauglichkeit	Furniture	
ALC	Acute Lethal Concentration	Toxicology	USA
APME	Association of Plastics Manufacturers in Europe	Association	Europe
AS	Australian Standard	Standardisation	Australia
ASTM	American Society for Testing Materials	Standardisation	USA
ATS	Airbus Transport Specification	Specification	Company
BDI	Union of German Industry	Association	Germany
BIS	Bayer-ICI-Shell	Cooperation	Europe
Bo Strab		Transport	Germany
BOAC	British Overseas Airtransport Company	Company	Worldwide
BOCA	Building Officials and Code Administration	Fire Protection	USA
BRI	Building Research Institute	Research	Japan. UK
BRMA	British Rubber and Manufacturing Association	Association	Great Britain
BRMA	British Rubber Manufacturing Association	Association	UK
BSI	British Standards Institute	Standardisation	UK
BVD	Brandverhütungsdienst	Fire Protection	Switzerland
CAL-TB	California – Technical Bulletin	Requirements	California
CAMI	Civil Aeromedical Institute	Aviation	USA
CAMI	Civil Aeromedical Institute	Institute	USA
CEN	Comite European de Normalisation	Standardisation	Europe
Cenelec	Comite European de Normalisation Electrot.	Standardisation	Europe
CIB	Conseil International du Batiment	Building	International
CNET	Centre National d'Etudes les Telecommun.	Electrotechnic	France
CO-HB	Haemoglobin – Carbon monoxide	Toxicity	Worldwide
CPSC	Consumer Product Safety Commission	Association	USA
CSE-RF	Centro Studie Esperience dei Vigili del Fueco	Standardisation	Italy
CSIRO	Commonwealth Scientific and Industrial Research Organisation	Association	Australia
DAN	Danish Standardisation	Standardisation	Denmark
DB	Deutsche Bundesbahn, Deutsche Bahn AG German Railway	Transport	Germany
DIN	Deutsches Institut für Normung	Standardisation	Germany
DIS	Draft International Standard	Standardisation	Worldwide
DOT	Department of Transport	Regulation	
DOT/	Department of Trade	Regulator	UK
DR AS	Draft Australian Standard	Standardisation	Australia
Ds	Specific Optical Density	Classification	Worldwide
DTLR	Department of Transport +++		

Polyurethane and Fire. Franz H. Prager and Helmut Rosteck
Copyright © 2006 WILEY-VCH Verlag GmbH & Co. KGaA, Weinheim
ISBN: 3-527-30805-9

Code	Explanation	Application	Area
EC	European Community	Cooperation	Europe
EC50	Effective Concentration of 59% incapacitation	Toxicity	Worldwide
EG	Europäische Gemeinschaft	Cooperation	Europe
EGOLF	European Group of Official Laboratories	Fire Protection	Europe
ETK	Standard Time–Temperature Curve	Fire Protection	Worldwide
EUFAC	European Upholstered Furniture Action Council	Furniture	Europe
FAR	Federal Aviation Regulation	Regulation	Worldwide
FED	Fractional Effective Dose	Toxicity	Worldwide
FM	Factory Mutual	Institute	USA
FMCR	Factory Mutual Research Corporation	Institute	USA
FMVSS	Transport	Regulation	USA
FPA	Fire Protection Association	Fire protection	UK
FRC	Fire Research Center (Salt Lake City)	Fire Reaction	USA
FRC	Fire Research Center (Warrington)	Fire Reaction	UK
FRS	Fire Research Station	Institute	UK
FSC	Flame Spread Classification	Fire Reaction	USA
FTC	Federal Trade Commission	Regulator	USA
FTIR	Fourier transform infrared spectroscopy	Analytic	Worldwide
GC	Gas chromatography		
Gost	Standardisation Comite Russia	Standardisation	Russia
HRR	Heat-release rate	Fire Reaction	Worldwide
IBCO	International Conference of Building Officials	Fire protection	USA
IBS	Internationales Brandschutz.Seminar	Fire reaction	Germany
ICUP	Inter Company Urethane Panel	Cooperation	Europe
IEC	International Electrotechnical	Standardisation	Worldwide
IfBt, DIBt	Institut für Bautechnik. Deutsches Inst. F.Bauen	Building	Germany
III	International Isocyanate Institure	Cooperation	
I-ILDA	Industriel ILDA	Cooperation	Europe
ILDA	International Laboratories Data Acceptance	Cooperation	Europe
IMO	International Maritime Organisation	Regulation	Worldwide
INERIS	Institut National de l'Environment Industriel et des Risques	Test Station	France
ISO	International Organisation for Standardisation	Standardisation	Worldwide
ISOPA	European Isocyanate Producers Association	Industry	Europe
IVPU	Industrieverband Polyurethan	Association	Germany
JIS	Japanese Institute of Standardisation	Standardisation	Japan
LBO	Regional Building Regulation (Germany)	Building	Germany
LBO	Landes BO (Regional Building Regulation)	Building	Germany
LC50	Effective Concentration of 50% Lethality	Toxicity	Worldwide
LNE	Laboratoire Nazionale d'Èssais		
LRL	Lawrence Research Laboratories	Testing	USA
MPA	Materialprüfungsanstalt (Official Testing Station of the Region)	Testing	Germany
MPA	Officiasi Testing Station	Testing	Germany
NASA	National Aeronautical and Space Administration	Regulator	USA
NBN	Normes Belgium	Standardisation	Belgium
NBS	National Bureau of Standards	Institute	USA
NEN	Nederlandse Nor	Standardisation	Netherlands
NES	Naval Engineering Standard	Standardisation	UK
NES	Naval Engineering Standard	Standardisation	UK
NF	Norm Francaise	Standardisation	France

Abbreviations list

Code	Explanation	Application	Area
NFPA	National Fire Protection Association	Fire Protection	USA
NHTSA	National Highway Transport Administration	Transport	USA
NIST	Nationalö Institute of Standards and Technology	Institute	USA
NW	North-Rhine Westfalia	Region	Germany
Ö-Norm	Ösrreichische Norm	Standardisation	Austria
OSU	Ohio State University	Research	USA
PP/LAT	Polypropylene Fabric *Latex Filling		
PRC	Product Research Committee	Fire Reaction	USA
PRC	Product Research Committee	Committee	USA
PSA	Property Service Agency	Fire Reaction	UK
PSA	Property Service Agency	Town Agency	UK
RAPRA	Rubber and Plastics Research Administration	Institute	UK
RATP			
SAFER	Special Aviation Fire and Explosion Reduction	Transport	USA
SBCCA	Southern Building Code Congress International	Fire Protection	USA
SBI	Single Burning Item	Building	Europe
SC	Subcommittee	Standardisation	Worldwide
SeeBG	Seeberufsgenossenschaft/Maritime Regulation	Transport	Germany
SNCF	Society Nazionale de Chemins de Fer	Transport	France
SNCF	Society Nazionale des Chemins de Fer	Transport	France
SNIP	Sovietskie Normy i Pravila	Standardisation	USSR
SOLAS	Safety of Life at Sea	Transport	Worldwide
SRI	Stanford Research Institute	Research	USA
SSVO	Schiffssicherheits VO		
SWRI	Southwest Research Institute	Testing	USA
TC	Technical Committee	Standardisation	Worldwide
TGL	Techn. Güte und Lieferbedingung	Standardisation	Germany/DDR
TNO	Official Testing Station Delft	Testing	Netherlands
TUF	Time of Useful Function	Toxicity	USA
UFAC	Upholstered Furniture Action Council	Furniture	USA
UIC	Union Internazionale des Chemins de Fer	Transport	International
UIC	International Union of Railways	Transportation	Europe
UL	Underwriters Laboratories	Classification	Worldwide
USB	Union Southern Building Code	Building	USA
USF	University of San Francisco	Toxicity	USA
UTE	Union Technique de l'Electricite	Standardisation	France
VdS	Assurers Association	Fire Reaction	Germany
VKE	Verband der Kunststofferzeugenden Industrie	Association	Germany
VKE	Association of the Plastics Producers	Association	Germany
VO	Verordnung/Regulation	Law	Germany
VO	Verordnung /Orders	Regulation	Germany
VTT	National Testing Station	Fire Reaction	Finland
WG	Working Group	Standardisation	Worldwide
WHO	World Health Organisation	Organisation	Worldwide
XP2	XP2 Smoke Density Chamber	Classification	USA

Norm list

Armtrak Specific. 352	Specification for flammability, smoke emission and toxicity
AS 1530.1-1994 1994-03-19	Methods for fire tests on building materials, components and structures – Combustibility test for materials
AS 1530.2-1993 1993-04-19	Methods for fire tests on building materials, components and structures – Test for flammability of materials
AS 1530.3-1989 1989-00-00	Methods for fire tests on building materials, components and structures – Simultaneous determination of ignitability, flame propagation, heat release and smoke release
AS 1530.4-1990 1990-00-00	Methods for fire tests on building materials, components and structures – Fire-resistance tests of elements of building construction
AS 1530.5-1989 1989-00-00	Methods for fire tests on building materials, components and structures – Test for piloted ignitability
AS DR 80 123	Method of test for the ignitability of upholstered seating
AS DR 80 124	Fire propagation of upholstered seating of the low fire hazard type
ASTM	Standard Test Method for Heat and Visible Smoke Release Rates for Materials and Products Using an Oxygen Consumption Calorimeter
ASTM D 568	Rate of burning in a vertical position
ASTM D 1929	Test method for the ignition properties of plastics
ASTM D 2843 1999-00-00	Standard Test Method for Density of Smoke from the Burning or Decomposition of Plastics
ASTM D 2863 2000-00-00	Standard Test Method for Measuring the Minimum Oxygen Concentration to Support Candle-Like Combustion of Plastics (Oxygen Index) (BS 2782 pt.1, NF T 51-071, ISO 4589
ASTM D 3675 2001-00-00	Standard Test Method for Surface Flammability of Flexible Cellular Materials Using A Radiant Heat Energy Source
ASTM D 3713	Standard method for measuring response of solid plastics to ignition by a small flame
ASTM D 635	Rate of burning and/or extent and time of burning of self-supporting plastics in a horizontal position
ASTM D4000	Gravimetric Smoke-Density Measurement
ASTM E 119a 2000-00-00	Standard Test Methods for Fire Tests of Building Construction and Materials / Note: 2. revision 2000
ASTM E 1354 2003-00-00	Standard Test Method for Heat and Visible Smoke Release Rates for Materials and Products Using an Oxygen Consumption Calorimeter
ASTM E 136 1999-00-00	Standard Test Method for Behavior of Materials in a Vertical Tube Furnace at 750 °C / Note: 1. editorial change

Polyurethane and Fire. Franz H. Prager and Helmut Rosteck
Copyright © 2006 WILEY-VCH Verlag GmbH & Co. KGaA, Weinheim
ISBN: 3-527-30805-9

ASTM E 162a 2002-00-00	Standard Test Method for Surface Flammability of Materials Using a Radiant Heat Energy Source / Note: 2. revision 2002
ASTM E 648 2003-00-00	Standard Test Method for Critical Radiant Flux of Floor-Covering Systems Using a Radiant Heat Energy Source
ASTM E 662 2003-00-00	Standard Test Method for Specific Optical Density of Smoke Generated by Solid Materials
ASTM E 800	Methods for the analysis of gases and vapors in fire effluents
ASTM E 84b 2003-00-00	Standard Test Method for Surface Burning Characteristics of Building Materials / Note: 3. revision 2003
ATS 1000.001	Fire-smoke-toxicity test specification – Airbus Industry
BS 2782	Methods of testing plastics rate of burning
BS 476 part 22	Fire resistance of nonload-bearing elements of construction
BS 476 part 3	External fire exposure roof tests
BS 476 part 5	Method of test for ignitability
BS 476-12:1991 1991-08-30	Method using a choice of seven flaming ignition sources for a variety of flame-application times
BS 476-4:1970 1970-01-26	Fire tests on building materials and structures. Noncombustibility test for materials
BS 476-6:1989 1989-03-31	Fire tests on building materials and structures. Method of test for fire propagation for products
BS 476-7:1997 1997-01-15	Fire tests on building materials and structures. Method of test to determine the classification of the surface spread of flame of products
BS 5852:1990 1990-12-31	Methods of test for assessment of the ignitability of upholstered seating by smouldering and flaming ignition sources
BS 5872	Code of Practice 1986
BS 5888	Fire precaution in the design, construction and use of buildings Part 1: for residential buildings Part 2: for shops and Part 3: for offices
BS 5940	Method of test for determination of the punking behavior of phenol-formaldehyde foam
BS 6853	British standard code of practice for rolling stock
BS 738	Flash-ignition temperature
CAL TB 116	Requirements, test procedure and apparatus for testing
CAL TB 117	Requirements, test procedure and apparatus for testing the flame retardance in upholstered furniture
CAL TB 121	Flammability test procedure for mattresses in high-risk occupancies
CAL TB 133	Flammability test procedure for seating furniture for use in high-risk and public occupancies
CSE RF 1/75/A	Small-burner test
CSE RF 2/75/A	Small-burner test classification
CSE RF 3/77	Radiant panel test (ISO DP 5658)
DIN 18230-1 1998-05-00	Structural fire protection in industrial buildings – Part 1: Analytically required fire-resistance time
DIN 18230-2 1999-01-00	Structural fire protection in industrial buildings – Part 2: Determination of combustion behavior of materials in storage arrangement – Combustion factor m
DIN 18230-3 2002-08-00	The document specifies calculation values for supporting DIN 18230-1
DIN 18232-1 2002-02-00	Smoke and heat control systems – Part 1: Terms, safety objectives

DIN 18232-4 2003-04-00	Smoke and heat control systems – Part 4: Heat exhaust systems (WA); Test methods
DIN 4102	Fire performance of building materials and components
DIN 4102-1:1998-05	Fire behavior of building materials and building components – Part 1: Building materials; concepts, requirements and tests
DIN 4102-14	Determination of the burning behavior of floor-covering systems using a radiant heat source
DIN 4102-15 1990-05-00	Fire behavior of building materials and elements "Brandschacht"
DIN 4102-16 1998-05-00	Fire behavior of building materials and building components – Part 16: "Brandschacht" tests
DIN 4102-2	Fire Behavior of Building Materials and Building Components; Building Components; Definitions, Requirements and Tests
DIN 4102-7	Part 7: Roofing; definitions, requirements and testing / Note: To be replaced by DIN EN 13501-5 (2002-03, t).
DIN 50 050	Prüfung von Werkstoffen-Brennverhalten von Werkstoffen-Kleiner Brennkasten
DIN 50 051 1977-02-	Testing of Materials; Burning Behavior of Materials; Burner
DIN 50 055	Lichtmeßstrecke für Rauchentwicklungsprüfungen
DIN 50 905-1 1987-01	Corrosion of metals; corrosion testing; principles
DIN 50 905-2 1987-01	Corrosion of metals; corrosion testing; corrosion characteristics under uniform corrosion attack
DIN 51 900	Bestimmung des Heizwertes mit dem Bombenkalorimeter
DIN 51 960	Prüfung von organischen Bodenbelägen
DIN 53 436	Toxicity tube furnace test
DIN 53 438-1 1984-06	Testing of combustible materials; response to ignition by a small flame; general data
DIN 53 438-2 1984-06	Testing of combustible materials; response to ignition by a small flame; edge ignition
DIN 53 438-3 1984-06	Testing of combustible materials; response to ignition by a small flame; surface ignition
DIN 54 332	Bestimmung des Brennverhaltens von textilen Fußbodenbelägen
DIN 54 341 1988-01-00	Testing of seats in railways for public traffic; determination of burning behavior with a paper pillow ignition source
DIN 54 837 2003-09-00	Testing of materials, small components and component sections for rail vehicles – Determination of burning behavior using a gas burner
DIN 5510-1 1988-10-00	Preventive fire protection in railway vehicles; levels of protection, fire-preventive measures and certification
DIN 5510-2 2003-09-00	Preventive fire protection in railway vehicles – Part 2: Fire behavior and fire side effects of materials and parts; Classification, requirements and test methods
DIN 5510-4 1988-10-00	Preventive fire protection in railway vehicles; vehicle design; safety requirements
DIN 5510-5 1988-10-00	Preventive fire protection in railway vehicles; electrical equipment; safety requirements
DIN 66 084	Klassifizierung des Brennverhaltens von Polsterverbunden
DIN EN 1363-1 1999-10-00	Fire-resistance tests – Part 1: General requirements; German version EN 1363-1:1999
DIN EN 1364-1 1999-10-00	Fire-resistance tests on nonload-bearing elements – Part 1: Walls; German version EN 1364-1:1999
DIN EN 1364-2 1999-10-00	Fire-resistance tests on nonload-bearing elements – Part 2: Ceilings; German version EN 1364-2:1999

DIN EN 1365-2	Fire-resistance tests for load-bearing elements – Part 2: Floors and roofs; German version
DIN EN 50267-2-1 VDE 0482 pt.267-2-1	Common test methods for cables under fire conditions – Tests on gases evolved during combustion of material from cables – Part 2-1: Procedures; determination of the amount of halogenacid gas; German version EN 50267-2-1:1998
DIN EN 50267-2-2 VDE 0482 pt.267-2-2	Common test methods for cables under fire conditions – Tests on gases evolved during combustion of material from cables – Part 2-2: Procedures; determination of degree of acidity of gases for materials by measuring pH and conductivity; German version EN 50267-2-2:1998
DIN EN 50267-2-3 VDE 0482 pt. 267-2-3	Common test methods for cables under fire conditions – Tests on gases evolved during combustion of material from cables – Part 2-3: Procedures; determination of degree of acidity of gases for cables by determination of the weighted average of pH and conductivity; German version EN 50267-2-3:1998
DIN IEC 695 pt 2-1	Glow-wire test
DIN IEC 695 pt 2-2	Test with the needle flame
DIN IEC 695 pt 2-3	Bad-connection test
DIN VDE 0266*VDE 0266 2000-03-00	Power cables with improved characteristics in the case of fire – Nominal voltages U<(Index)0>/U 0.6/1 kV
EN 1365-2:1999	DIN EN 50267-1*VDE 0482 Teil 267-1 2000-02-00 Common test methods for cables under fire conditions – Tests on gases evolved during combustion of materials from cables – Part 1: Apparatus; German version EN 50267-1:1998
FAR 25.853	Vertical bunsen-burner test
FAR part 25.853	OSU calorimeter test
GOST 12.1.012 1990-00-00	Occupational safety standards system. Vibrational safety. General requirements
ISO 1182	Fire tests – Building materials – Noncombustibility test
ISO 5660	Fire tests-reaction to fire – rate of heat release from building products
ISO 8191	Furniture-Burning behavior – part 1:Ignition source cigarette
ISO DP 5659	Fire tests – Smoke generated by solid materials
ISO DP 5924	Fire tests – Smoke generated by building products – dual chamber
JIS A 1321	Smoke-production test
JIS A 1321 1994-12-15	Testing method for incombustibility of internal finish material and procedure of buildings
JIS D 1201	Smoke-production test
NBN S 21-302	Cigarette- and match-type flame tests
NEN 3881	Non-combustibility of building materials
NEN 3882	Determination of fire hazard for roofing affected by spreading fire
NEN 3883	Combustibility of building materials
NEN 3884	Determination of fire resistance of building components.
NEN 6065:1991 1991-11-01	Determination of the contribution to fire propagation of building products
NEN 6066:1991 1991-11-01	Determination of the smoke production during fire of building products
NF C20-453 1985-01-01	Basic environmental testing procedures. Test methods. Conventional determination of corrosiveness of smoke
NF C20-902-1 1990-06-01	Fire-hazard testing. Test methods. Determination of smoke opacity without air-change. Part one: methodology and test devices

NF C32-070 2001-01-01	Insulated cables and flexible cords for installations – Classification tests on cables and cords with respect to their behavior to fire
NF F 16-001-3,NF EN 45545-3	Railway applications – Fire protection on railway vehicles – Part 3: fire-resistance requirements for fire barriers
NF F 16-101	Railway rolling stock, fire behavior, choice of materials
NF F 16-101 1988-10-01	Railway rolling stock. Fire behavior. Materials choice
NF F 16-102	Railway rolling stock. Application to electrical equipment
NF F 16-201 1990-03-01	Railway rolling stock. Fire resistance test for seats.
NF P 92 501	Epiradiateur Flammability test
NF P 92-503 1995-12-01	Safety against fire. Building materials. Reaction to fire tests. Electrical burner test used for flexible materials
NF P 92-504 1995-12-01	Safety against fire. Building materials. Reaction to fire tests. Flame persistence test and speed of the spread of flame.
NF P 92-505 1995-12-01	Safety against fire. Building materials. Reaction to fire tests. Test used for thermal melting materials. Dripping test.
NF P 92-506 1985-12-01	Safety against fire. Building materials. Reaction to fire tests. Radiant panel test for flooring
NF P 92-507 2004-02-01	Fire safety – Building – Interior fitting materials – Classification according to their reaction to fire
NF P 92-510 1996-09-01	Safety against fire. Building materials. Reaction to fire tests. Determination of upper calorific value
NF T 51 071	Oxygen index test
NF T 51 073	Fire tests – smoke-density measurement
NF X70-100-1 2001-09-01	Fire tests – Analysis of gaseous effluents – Part 1: methods for analysing gases stemming from thermal degradation.
NF X70-100-2 2001-09-01	Fire tests – Analysis of gaseous effluents – Part 2: tubular furnace thermal degradation method
NF X70-101 1987-06-01	Safety against fire. Fire-behavior tests. Analysis of gases resulting of combustion or pyrolysis. Test-chamber method
NT 033 (ISO 5657)	Ignitability of building products
NT Fire 014	Furniture upholstered seats ignitability part 1 Smokers material part 2 Flaming sources
NT Fire 025	Room fire tests in full scale
Ö Norm B 3800-1 2004-01-01	Burning behavior of materials excluding building products – Part 1: Requirements, tests and evaluations
Ö Norm B 3800-4 2000-05-01	Behavior of building materials and components in fire – Components: Classification of fire resistance
Ö Norm B 3800-5 2003-07-01	Fire behavior of building materials and components – Part 5: Fire behavior of facades – Requirements, tests
Ö Norm B 3810	Fire behavior of floor coverings
Ö Norm B 3825	Brandverhalten von Ausstattungsstücken. Prüfung von Möbelbezügen
PFF 6-76 part 1633	Proposed standard for the flammability (cigarette ignition)
TGL 10 685/11	Determination of the combustibility of building materials
TGL 10 687/12	Testing of the fire propagation of building components
UIC 564-2	Regulations relating to fire protection – paper-cushion test
UL 72 2001-05-00	Test for fire resistance of record protection equipment
UL 723 2003-08-00	Test for surface burning characteristics of building materials
UL 733 1993-08-00	Oil-fired air heaters and direct-fired heaters
UL 746 C	Flammability test for electrical appliances
UL 94	Horizontal test similar to ASTM D 1692
UL 94	Horizontal test similar to ASTM D 635

UL 94	Vertical test procedure
UL 94 1996-10-00	Test for flammability of plastic materials for parts in devices and appliances
ULC-S 102	Test for determination of the combustibility
ULC-S 102 M 83	Modified Steiner tunnel test for floor coverings
UTE C 20 452	Smoke-density determination of effluents
UTE C 20 453	Corrosivity determination of effluents
UTE C 20 454	Toxicity evaluation of effluents
VDE 0304 part 3.7	Glow-bar test (UNE 53-035, NF-T-51-015)
VDE 0345	Ignitability test rig, decomposition temperature
VDE 0470	Hot mandrel test specification
VDE 0470/1	Regeln für Prüfgeräte und Prüfverfahren
VDE 0471 part 2-1	Glow-wire test (DIN IEC 695–2 – 1
VDE 0471 part 2-2	Test with the needle flame
VDE 0471 part 2-3	Bad-connection test (DIN IEC 695 -2-3)
VDE 0471 part 5	Feuersicherheitliche Prüfung von elektrotechnischen Erzeugnissen -Prüfung mit Flammen
VDE 0471 part 6/7	Prüfung mit Kriechstrom als Zündquelle
VDE 0472 part 813	Korrosivität von Brandgasen
VDE 0472 part 814	Prüfung von Kabeln und isolierten Leitungen. Isolationserhalt bei Flammeneinwirkung
VdS 2052	Latex-Schäume, Sicherheitsvorschriften für die Herstellung und Verarbeitung
VdS 2053	Polyurethan-Weichschäume, Sicherheitsvorschriften für die Herstellung und Verarbeitung

Index

a

ABM-Test frame 79
acute lethal health, toxic-potency 416
acute toxicity, effluents 20, 139, 144
– fluoropolymers 221
additive-effect mechanism, fire-gas components 415
additives 187
– fire-retardants 300
– flame-retardant 204, 452
afterburning, fire protection 78
aldehyde concentration, effluents 398
ALH *see* acute lethal health
Al-lamination, requirement profiles 447
animal-experimental investigations, fire-gas components 234
arson, statistics 41
ASET *see* available safe egress time
asphyxiants, toxicological investigations 228
aspired fire reaction classification 116
autopsy findings, assessment criteria 391–394
available safe egress time 50
aviation sector, fire causes 134–137

b

B2-test, composite elements 83
bad connection test, electrical engineering 148
bioassay test results 68
– evaluation criteria 399
– field trials 439
– fire side effects 353
– fire testing 110
– polyurethane effluents 247, 263
– tetrafluorine ethylene 231
– toxicity index 415
BIS-decomposition apparatus, polyurethanes 233
blood analysis, fire victim 451
blood cyanide concentration, fire victims 395
blow torch test, polyurethanes 213, 216
Box procedure, fire testing 110

Brandschacht test, fire testing 73
BRI-decomposition model, fire testing 111
Brown'sche box, fire testing 101
BS 5852, fire protection regulations 126
building codes, fire protection 63
building products
– classification 109
– fire reaction testing 90, 191
– fire-reaction classification 73
building regulation 61
building section, fire protection 60
Bunsen burner, polyurethanes 234
burning behavior
– flame spread 14
– lamp covering 325
– polyurethanes 193
– upholstery combination 312
burning dripping
– electrical engineering 154
– evaluation criteria 371
– fire side effects 322
– fire testing 115
– ignition risk
– polyurethanes 193
– test configuration 93
burning duration, combustion systems 287
burning intensity, safety distance 299
burning rain, polyurethanes 194
burning rate, temperature influence 217
bus fires, ignition risks 55

c

cabin test, fire protection 92
cable test pieces, railway sector 142
cables, fire-performance testing 153
calibration body, fire protection 76
calorific value formula, complete combustion 218
candle flames, ignition risk 279
car fires, fire causes 53
CBUF, fire tests 328
ceiling construction, requirement profiles 449

Polyurethane and Fire. Franz H. Prager and Helmut Rosteck
Copyright © 2006 WILEY-VCH Verlag GmbH & Co. KGaA, Weinheim
ISBN: 3-527-30805-9

chair unit, fire side effects 313
chamber test 136
charcoal fire, smoke formation 18
chimney burner, irradiance profile 290
chimney height, test influence 291
chimney test 74, 141
– different configurations 192
– European harmonisation 112
– surface spread 291
– surface temperatures 290
chipboard, polyurethanes 198
chlorofluorohydrocarbons 157
christmas tree, ignition sources 295
cigarette
– smoldering behavior 272
– thermal potential 273
cigarette test 25
– fire testing 126
– polyurethanes 181
– upholstered furniture 133
classification criterion, fires 55–56
classification demands, requirement profiles 421
classification systems
– different national 59
– fire protection 57–156
climate, fire causes 43
CO concentration, requirement profiles 448
CO liberation, fire side effects 365
CO poisoning, polyurethanes 228–229
CO uptake, evaluation criteria 404
coating, flame spread 12
CO-Hb values
– effluents 403
– evaluation criteria 400–402
– polyurethanes 226, 229–230, 249, 394, 404, 415
CO_2/CO ratio
– field trials 436
– fire side effects 343, 357
– flashover situation 29
– flue gases 233
– ISO smoke box 358
– large scale tests 365
– PUR 170–171
– smoke toxicity 206
– wooden cribs 361
combustible building materials, fire protection 63
combustible materials, classification 86
combustion
– complete 23
– spatial 307
combustion behavior
– filling materials 317
– room fires 378

combustion gases, temperature 16
combustion products, isocyanate based foams 248
combustion systems
– DIN equipment 341
– fire side effects 363
– relevance 267–365
composite elements, B2-test 83
composite samples, test configuration 80
composites, testing 79–80
cone calorimeter
– combustion systems 297
– effluent generation 222
– fire protection 70, 104
– fire side effects 357
– flashover situation 31
– PUR 163
– risk relatedness 361
corrosimeter test method 155
– PUR 265
corrosive damage, gases 264
corrosive effects, effluents 21
cotton covering, fire side effects 330
course of fire 7
Crib test, upholstered furniture 133
critical flux, combustion systems 298
critical radiant flux test, fire protection 65

d

danger potential, reduction 418
debris, burning 78
decomposition
– oxidative 22
– thermal 148
– *see also* thermal decomposition
decomposition apparatus
– fire protection 77
– SRI 72
decomposition devices, fire protection 71, 76
decomposition model
– fire protection 70
– railway sector 146
– Skornik 72
– test parameters 237–238
– U-Pitt 69
decomposition products
– CO_2/CO ratios 359
– fire side effects 338
decomposition temperature
– determination 149
– Setchkin method 179
decomposition test equipment, fire protection 71
decompostion model furnace, polyurethanes 241

diffusion flames
- flame formation 6
- scheme 7
DIN equipment, combustion systems 341
diphenylmethane – diisocyanate see MDI
discoloration, smoke-density measurement 206
double-conveyor-belt process, PUR 159–160
dripping behaviour, classification 87
drum fire studies, PUR 162
drum tests, PUR 165–176
dual chamber box, fire testing 99
Düsseldorf Airport Fire
- fire disaster 364
- fire load 445
- interpretation 450
dynamic test procedures, effluent generation 219
dynamic toxicity index, evaluation criteria 414

e

early fire hazard test 106
EC classification scheme, European harmonisation 113
EC furniture directive, fire protection 59
effluents, acute toxicity 139, 144
- aldehyde concentration 398
- assessment criteria 391
- characterization 21
- chimney test 77
- CO_2/CO ratio 28, 170
- composition 223
- concentration 19
- corrosive effects 21
- head-nose exposure mode 348
- O_2 content 345
- polyurethanes 177
- PUR 159
- self-sustained smoldering fire 26
- test animals behaviour 242–243
- toxic potency 105, 217
- toxicity 20
- ventilation influence 403
effluents dilution, polyurethanes 246
effluents generation, polyurethanes 220
electrical blankets, fire causes 53
electrical engineering 146–156
electroburner test
- classification 95
- fire protection 93
enamel effects, fire side effects 356
enduse conditions
- fire side effects 364
- PUR 179
- PUR test results 267–472
epiradiateur test
- classification 94

- european harmonisation 112
- fire protection 92
- ignition risk 457
essential fire scenarios 21–34
Eurofic program
- European harmonisation 112
- fire reaction classification 116
European fire reaction testing, harmonisation 121
European harmonisation, fire protection regulations 112–121
European requirement and classification profiles, fire protection 73–104
evaluation criteria
- fire testing 115
- heat release 371–417
- relevance 366–417
external radiation
- ignition time 298
- optical density 383
extinction coefficient, smoke-density measurement 207
extinguishment, fire development 7

f

facade test 117
- field trials 425–430
- fire protection 96
- fire testing 104
- national test arrangements 118
face test, testing 109
failure criteria, B2 test procedure 188
fatalities, statistics 38
feedback effect, fire-gas temperatures
field trials, relevance 422
fire
- behavior 9–12
- definition 5
- fully developed 28–34
- well ventilated 205
fire causes 35–56
- relevance 441–452
- statistics 48
fire damage, PUR spray insulation 435
fire deaths, Scottish 40
fire detection units 46–55
fire development, combustion systems 307
fire disaster, Düsseldorf airport 364
fire effluents, toxicity 68
fire gas
- danger flooring material 466
- danger upholstery furniture 467
- danger wall lining 465
- maximum temperatures 326
fire gas analysis, PUR 169

fire gas characterisation, effluents   219
fire gas components
– classification   139
– main   136
– PUR   165
– suffocating   228
fire gas concentration   396
– nominal   385
– temporal course   349
fire gas danger, risk-relevant evaluation   387
fire gas formation, polyurethanes   205
fire gas liberation
– ignition risk   458
– PUR   336
fire gas spreading, uncontrolled   309
fire gas stream, field trials   438
fire gas temperature
– evaluation criteria   372
– room burn out   377
fire gas temperature evolution, combustion systems   304
fire gas toxicity
– evaluation   232
– national requirement   461
fire initiation   55–56
fire load surfaces   34
fire models, assessment   235
fire performance characteristics, PUR   157–266
fire performance criteria   5–4
fire performance parameters, polyurethanes   216
fire performance standards, assessment   270–271
fire point, flame formation   6
fire processes, analysis   443
fire propagation test   84
fire protection
– preventive   57–156
– rules   417
fire protection classification, material-specific   140
fire protection problems   3–34
fire reaction
– classification   88
– testing   1, 75
fire reaction classification, building products   73
fire reaction testing, fire protection   60
fire resistance   60
– classification   85
– overheated wiring   148–151
fire retardants
– PUR   202–203
– see also additives
fire risk   11, 37–46
– situations   119
– statistics   39
fire safety characteristics, christmas tree   296

fire scenarios
– definitions   268
– essential   21–34
– polyurethanes   205
– simulations   362, 468
– smoke formation   333
fire side effects   14, 154–156, 306–365
– bioassay test results   353
– burning dripping   322
– chair unit   313
– CO liberation   365
– CO_2/CO ratio   343, 357
– combustion systems   363
– cone calorimeter   357
– cotton covering   330
– decomposition products   338
– enamel effects   356
– enduse conditions   364
– flames-spread behavior   320
– flashover   332
– french combustion system   348
– furniture calorimeter   331–332
– glowing rate   311
– HCN   365
– head-nose exposure mode   348
– heat release   325
– ignition sources   321
– mass burning rate   327
– PMMA   327
– polyethylene   339
– primary air   337
– pure combustion   336
– radiant panel test   320
– room-corner procedure   320
– settee   314
– single point data procedures   336
– smoke formation   333
– smoke layer formation   308
– smoldering behavior   311
– spread of flame   362
– surface spread   311
– survival time   367
– test temperatures   340
– testing methods   326
– thermal decomposition   352
– thermal decomposition products   338
– toxic potential   340
– trench effect   318
– upholstery   312
– ventilation effects   329–331
fire spread   46–55
– combustion systems   302
– side effects   315
fire statistics   35
– experience values   391
– Finland   45

– Norway   45
– relevance   441
– UK   36, 442
fire test house
– field trials   424
– PUR   376
fire testing, evaluation criteria   115
fire trends   52
fire triangle, fire criteria   5
fire victims
– blood cyanide concentration   395
– causes   36
– fire statistics   36
– locations   48
– scenarios   40
– soot deposits   396
– statistics   36, 38
first-ignited items, fire causes   37
flame spread   12–31
– fire side effects   320
– roof constructions   118
– surface   84, 91
flames
– effect of size   7
– formation   6–9
– height   297
– radiation section   321
– resistance   286
– retardant additives   204
– temperature profile   280
– thermal stress   152–154
flammability test, fire protection   67
flash ignition, combustion systems   302
flash ignition temperature, electrical engineering   148
flashover   28–34
– combustion systems   303
– field trials   431
– fire development   8
– fire side effects   332
– passenger coaches   304
– transportation field   33
flexible PUR foam, heat-release   199
floor coverings
– combustible   78
– fire protection   66
flooring
– classification   87
– radiant panel test   95
flooring test, fire testing   103
flue-gas stream
– CO_2/CO ratios   171
– flashover situation   30
flue-gas toxicity, fire testing   110
fluoropolymers, acute toxic potential   221
foaming reaction, PUR   157–158

forest fire, wind effect   11
french combustion system, fire side effects   348
french nitrogen-chlorine law, requirement profiles   463–464
furnace protocol   77
furnace test, PUR-test results   268
furniture, fire testing   121–133
furniture calorimeter, fire side effects   331–332
furniture draft directive   131
furniture fires, statistics   39, 44
fuse effect   13

g

gangway, smoke density   462
gas concentration, fire protection   4
glow bar test, electrical engineering   150
glow resistance test, electrical engineering   149
glow wire test
– electrical engineering   148
– railway sector   143
glowing particles, ignition risk   184, 278
glowing rate, fire side effects   311
glowing zone, cigarette   273
GOST, fire testing norms   105
gravimetric method, smoke-density measurement   206

h

Haber's Rule, polyurethanes   226, 228
HCN
– asphyxiant gas   229
– concentrations   448
– fire side effects   365
head-nose exposure mode, fire side effects   348
heat, ignition risk   458
heat release
– aviation sector   135
– evaluation criteria   371–417
– fire side effects   325
– paper pillows   284
heat release determination, calibration setup   285
heat release per unit area, evaluation criteria   375
heat release potential, different insultants   196
heat release rate
– different materials   194
– fire testing   107
– flame spread   14
– ignition risk   458
– PUR   163, 200
– test method influence   332
heating rate, material surface   12
Herpol test procedure, fire testing   100
high intensity sources, igniting risk   282–283

horizontal test, fire protection 93
hospital fires, statistics 49
hot dripping, flame spread 13–14
hot gas stream, temperature profile 303
hot mandrel test, electrical engineering 148–150
hot wire ignition test, electrical engineering 148
human behavior, fire protection 58
humidity, effects 198
hydrocarbons, self-ignition 10

i

ignitability 9
– fire testing 107
– material 5
ignition
– fire development 7
– PUR 173
– statistics 39
ignition potential, secondary ignition sources 287
ignition risk
– aviation sector 135
– burning dripping 324
– electrical engineering 151
– evaluation criteria 368–371
– glowing particles 184
– high intensity sources 282–283
– low intensity sources 272, 282–283
– polyurethanes 177
– procedures 268–305
– PUR 161–176, 179
– requirement profiles 452–458
– transportation field 54, 300
ignition sources
– arson 42
– characterisation 284
– christmas tree 295
– combustion systems 296
– comparison 289
– EG characterization 130
– electrical engineering 146
– electrotechnical products 278
– field trials 436
– fire side effects 321
– low intensity 272, 454
– raised intensity 455
– resistance 124
– standardized 294
– statistics 44
ignition time
– evaluation criteria 369
– external radiation influence 297
incombustibility, testing method 108
incombustible materials, requirement profiles 464

indoor drum fires, gas analysis 169
industry, large fires 176
Ineris fire test gallery, PUR 174
Ineris test tunnel, PUR 172
insulants
– field trials 437
– heat-release potential 196
interior lining, heat release 135
irradiance level, evaluation criteria 369
irradiance profile, chimney burner 290
ISO Smoke Box, CO_2/CO ratios 358–359
isocyanate based foams, combustion products 248

j

jute curtain, combustion 301

k

kerosene burner test 135
Kings Cross Fire
– fire load 445
– PUR 318

l

lamp covering, burning behavior 325
latex foam, smoke temperature 26
LC_{50} data
– evaluation criteria 407
– see also toxic potency
life prevention measures 3
liquid, combustion process 351
low intensity ignition sources, resistance 454
low intensity sources, igniting risk 282–283
Lüscher Burner, fire testing 89

m

mass burning rate
– combustion systems 285
– evaluation criteria 376
– fire side effects 327
– upholstery combination 314
mass optical density see MOD
match, ignition risk 279
match test
– fire testing 126
– polyurethanes 186
material, influence on flames 10
material ignitability, fire criteria 5
material surface, heating rate 12
mattress test 124
mattress-cigarette test 122
mattresses
– failing 275
– polyurethane 216
– PUR 444

MDI drums, thermally stressed 166–168
metal calibration samples, combustion systems 290
methenamin tablet test 183
Michigan roll up test, burning behavior 218
mineral fibers
– insulation 447
– smoke temperature 27
– toxic potency 261
– toxic potential 451
MOD
– evaluation criteria 385
– polyurethanes 210
MOD values, radiation conditions influence 355
model building codes, fire protection 63
modified Australian cone calorimeter, polyurethanes 223
Moreton tests, flashover situation 30
mortality
– evaluation criteria 400
– temperature influence 245
motor vehicles
– fire causes 133
– ignition risks 54
multiple death fires 52

n

naval field, ignition risks 55
NBS chamber, railway sector 144
NBS test, flashover situation 32
needle flame, electrical engineering 151
net calorific value
– combustion 197
– evaluation criteria 372
– materials 15
noncombustibility, proof 74
noncombustibility furnace, fire protection 64
norms, list 495–500

o

O_2 index numbers, polyurethanes 189
O_2 index test 145
open air test, measuring points 170
open-cell structure, polyurethanes 184
optical density
– evaluation criteria 381
– polyurethanes 212
– PUR sandwich elements 359
– specific 382
– test-piece dimensions 224
optical measuring system, fire protection 66
overheated wire, ignition risk 278
overheated wiring 148–151
oxidative decomposition 21

oxygen availability, sumi indices 413
oxygen consumption, fire testing 104

p

paint layers, toxicity 246
panic, peoples behavior 225
paper, ignition risk 10
paper cushion test 130
paper pillows, heat-release 284
parameters, mass-burning rate 335
passenger coaches, flashover 304
PC vertical glazing, test results 293
peeling effect
– fire testing 127
– PUR 421
people, behavior in smoke 310
pH determination, wash bottles 337
pH measurement, electrical engineering 154
phases, fire development 7
photoelectric measurements, smoke-density measurement 206
plastics
– heat potential 201
– toxic potency 400, 405
PMMA, fire side effects 327
poisonousness, fire-gas components 226
polyether foam, ignition risk 281
polyethylene, fire side effects 339
polyurethane effluents, toxic potency 257, 261
polyurethane foams, pyrolysis products 406
polyurethanes 24, 176
– BIS-decomposition apparatus 233
– blow torch test 213, 216
– Bunsen burner 234
– burning behavior 193
– burning dripping 193
– burning rain 194
– chipboard 198
– cigarette test 181
– CO poisoning 228–229
– CO-Hb values 226, 229–230, 249, 394, 404, 415
– decompostion model furnace 241
– effluents 177
– effluents dilution 246
– effluents generation 220
– fire gas formation 205
– fire performance parameters 216
– fire scenarios 205
– Haber's Rule 226, 228
– ignition risk 177
– match test 186
– MOD 210
– modified australian cone calorimeter 223
– O_2 index numbers 189

- open-cell structure 184
- optical density 212
- radiation intensity 212
- recipe variations 202
- risk assessment 177
- small ignition sources 186
- smoke density records 209
- smoke formation 205
- smoldering fires 180
- suffocating fire-gas components 228
- supertoxicants 230
- thermal decomposition 198
- TMPP 231
- toxic hazard in fire 205
- toxicological results 248
- welding tests 184–185
- *see also* PUR
pool-fire configuration, combustion systems 289
pool-fire experiments, PUR 172
positive behavior, people 224
Potts pot procedure
- evaluation criteria 408
- fire protection 69
preheating, flame spread 12
premixed flames, flame formation 6
preventive fire protection 57–156
- electrical engineering 146
preventive fire protection measures
- Australia 106
- Austria 86–87
- Belgium 98–100
- Eastern Europe 104–112
- France 92–97
- Germany 73–83
- Great Britain 84–86
- Italy 89–91
- Japan 107
- Netherlands 97–98
- Scandinavia 100
- Switzerland 87–89
preventive fire safety 62
primary air
- fire side effects 337
- optical density 384
PUR
- effluents 255
- enduse conditions 179
- heat-release 199
- ignition risks 179
- insulation 420
- material-specific fire-performance characteristics 157–266
- production 157
- raw materials 161–163
- relative toxicity 252
- rigid foam 188
- smoke temperature 24
- testing 82
- *see also* polyurethanes
PUR foam
- charring process 240
- fire testing 128
PUR mattresses, fire development 444
PUR sandwich elements, optical density 359
PUR spray, fire damage 435
PUR steel-sandwich elements, field trials 426
PUR test results, interpretation 267–472
pure combustion, fire side effects 336
pyrolysis 21
- energy levels 335
pyrolysis products 22
- toxicity 406

r

radiant panel test, fire protection 89, 94
- fire side effects 320
radiated heat, ignition risk 180
radiation, high intensities 369
radiation distribution, fire testing 99
radiation intensity, polyurethanes 212
railway sector, fire causes 140–146
raised intensity ignition sources, resistance 455
reaction behaviour, training influence 405
reaction injection molding *see* RIM
real fire, influences 308
recipe variations, polyurethanes 202
red oak, smoke potential 381
reproducibility, reproducibility 276–277
requirement profiles, relevance 417
resistance, flames 286
rigid PUR, toxic potency 360
rigid PUR foam, carbonisation 195
RIM 188
risk and hazard, fire protection 4
risk assessment, polyurethanes 177
risk of ignition *see* ignition risk
risk related classification, fire protection 88
risk relatedness, cone-calorimeter procedure 361
road traffic, fire causes 133–134
rockwool
- field trials 432
- fire-penetration 433
- furnace test 453
- smoke formation 434–435
Roland test configuration, fire testing 114
roof construction
- burning outflow 322
- field trials 432
- fire load 445

– flame-spread tests   118
– smoke formation   434–435
room fires
– combustion behavior   378
– fire-gas temperature   377
– simulated   361
– simulation   301
room geometry, evaluation criteria   376
room-corner procedure
– field trials   423
– fire side effects   320
– fire testing   103
RTB-Decomposition Model, fire testing   111

s

safety distance, burning intensity   299
safety requirements, fire protection   4
sand-bed burner, fire testing   103
sandwich element samples, testing   81–82
sandwich elements, ignition risk   178
SBI test   113
– classification   117
Schlyter-test   86
seat test, aviation sector   135
secondary ignition sources, intensity   286
self-ignition, hydrocarbons   10
self-ignition temperatures, electrical engineering   148
self-sustained smoldering fire   24–28
seminoncombustible materials, testing   108
Setchkin method, decomposition temperatures   179
settee, fire side effects   314
shipbuilding sector, fire causes   137–140
ships, possible fire-gas dangers   469
side effects, assessment   306–365
single box method, NBSm-modified   357
single burning item see SBI
single chamber box, smoke-density measurement   138
single point data procedures, fire side effects   336
single point test methods, combustion systems   292
skin reaction, evaluation criteria   392
Skornik, decomposition model   72
slabstock production, PUR   159
small ignition sources, polyurethanes   186
small-burner test, fire protection   83, 88
smoke, danger   20
smoke box, fire testing   99
smoke components   50
– limiting values   470
– measurements   112
– modeling   409
smoke concentration   1

– PUR test samples   346
– wood test samples   344
smoke corrosivity
– classification criteria
– testing philosophy   264
smoke density   17
– fire testing   108
– material-related potential   211
– national requirements   440
– potential values   145
– recipe influence   214
smoke density measurement   19
– evaluation criteria   379
– fire protection   66–67, 76, 89
– fire testing   105, 107
– philosophy   206
smoke density records, polyurethanes   209
smoke density test, fire protection   75
smoke density values, ventilation influence   355
smoke detectors   51
– effectiveness   448
smoke formation   18
– fire models   236
– fire side effects   333
– fire testing   115
– ignition risk   459
– polyurethanes   205
– roof construction   434–435
– testing   109
smoke indices, evaluation criteria   386
smoke layer, oxidative decomposition   23
smoke layer formation, fire side effects   308
smoke layer temperature   17
smoke mass   132
– fire initiation   56
smoke movement, flashover situation   32
smoke potential, evaluation criteria   203, 378
– test conditions influence   355
smoke release
– PUR   164
– simulation   269
– upholstery combination   390
smoke release criteria   138
smoke requirements, borderline values   440
smoke temperature, time dependence   24
smoke toxicity, fire protection   4
smoke values, calculated   217
smoke volume
– evaluation criteria   388
– stochiometric combustion   472
smokers   49
– statistics   39
– see also cigarettes
smoldering behavior, fire side effects   311
smoldering fires
– igniting risk   272

– polyurethanes   180
– self-sustained   24–28
– spread   27
smoldering process, self-sustained   474
smoldering propagation rate   1
smoldering temperature, PUR foam   239
smoldering tendency, combustion systems   277
smoldering velocity   27
SO$_2$, toxic potential   398
soot deposits, fire victims   396
specific optical density, evaluation criteria   382
spontaneous combustion   178
spread of flame
– fire side effects   362
– fire testing   115
– horizontal   370
spread test, classification scheme   90
sprinklers   46–55
SRI, decomposition apparatus   72
stadium seats, fire spread   319
standardisation   473
– fire protection regulations   112–121
static test procedure, effluent generation   219
steel sandwich elements   82
– toxic potency   456
Steiner tunnel test, fire protection   64
stoichiometric combustion, smoke volume   472
suffocating fire-gas components, polyurethanes   228
Sumi indices, evaluation criteria   413
supertoxicants, polyurethanes   230
supplementary test, classification   95
surface coating, flame spread   13
surface flammability, fire protection   68
surface spread   192
– combustion systems   291
– european harmonisation   112
– fire side effects   311
surface temperatures, chimney test   290
survival time, fire side effects   367
Swiss facade testing station, fire protection   88
SWRI toxic-potency, PUR   258
synthetic materials, fire testing   96

t

TDI drums, thermally stressed   166–168
temperature
– combustion gases   16
– smoke concentration   1
– smoke-layer   17
test animals behaviour, effluents   242–243
test arrangement, fire protection   101
test classification criteria   74
test conditions, evaluation criteria   389
test furnaces, fire protection   61–62

test house
– field trials   437
– fire-load distribution   120
test methods, reproducibility   474
test parameters   473
– evaluation criteria   410
test pieces, segmentation   342–343
test temperatures
– animals response   404
– fire side effects   340
– transmission values   210
testing and evaluation criteria
– American   121–125
– Australia   132
– European   125–133
– Far East   132
– Germany   129
– United Kingdom and Ireland   126
– Western Europe   129
testing criteria, PUR   274
testing methods, fire side effects   326
testing philosophy
– burning dripping   193
– electrical engineering   147
– smoke corrosivity   264
Tewarson apparatus   16
– heat-release rate   201
– PUR   163–165
textile construction, flame spread   12
thermal decomposition
– electrical engineering   148
– fire side effects   352
– polyurethanes   198
– *see also* decomposition
thermal decomposition products
– assessment   469
– fire side effects   338
thermal stress
– critical radiant flux test   65
– fire protection   61
– flame test   91
– flames   152–154
thermoskin composites, field trials   428
three-sister model, European harmonisation   112
time dependence, smoke temperature   24–25
time to reaction, reduction   399
tire fire, smoke formation   18
TMPP, polyurethanes   231
TNO, fire-penetration   432
tobacco, smoldering   181
toluylene-diisocyanate *see* TDI
total released energy, fire testing   115
toxic hazard in fire, polyurethanes   205
toxic potency
– acute lethal health   416

– assessment 217
– comparability 408
– concentration-time relationship 471
– criterion 251
– fire testing 105
– plastics 400
– rigid PUR 360
toxic potential
– additive influence 256
– assessment criteria 391
– fire side effects 340
– reference basis influence 353
– SO_2 398
– test parameters 406
toxicity
– effluents 20
– fire effluents 68
– fire testing 110
toxicity index
– bioassay test 415
– dynamic 414
– evaluation criteria 410
toxicological investigations, animals 225
toxicological results, polyurethanes 248
transmission
– evaluation criteria 379
– temporal course 380
– test temperature 210
transportation, fire causes 133–146
transportation field, flashover situation 33
trench effect, fire side effects 318
trimethylpropane phosphate-bicyclic phosphate ester see TMPP
tube furnace, railway sector 144

u

Underwriters Laboratories, flame test 153
upholstered furniture, cigarette test 122, 133
– Crib test 133
– test 124
upholstery
– cigarette test 182
– fire side effects 312
– fire-gas temperatures 374
upholstery combination, fire resistance 288
U-Pitt 2 method, CO_2/CO ratios 363
U-Pitt decomposition model, fire protection 69
U-Pitt protocol, thermal loading 339

v

vandalism conditions, fire testing 128
ventilation effects 27
– burning rate 243
– combustion systems 305
– fire development 9
– fire side effects 329–331
– flashover situation 30
victims
– fire statistics 36
– Norway 46
– statistics 48
visibility, smoke-density measurement 208
Vlam overslag test protocoll, classification 97

w

wash bottles, pH-determination 337
washing water, electrical engineering 154
welding tests, polyurethanes 184–185
wind effect, forest fire 11
wire temperature, electrical engineering 150
wiring, overheated 148–151
wood
– burning rate 198
– toxic potency 400
wood tests, temperature profile 347–348
wooden cribs
– CO_2/CO ratios 361
– fire-gas formation 305
– radiation intensity 319

x

XP2-chamber
– fire testing 77
– smoke-density measurement 208